Management and Ecology of Freshwater Plants

Developments in Hydrobiology 120

Series editor
H. J. Dumont

Management and Ecology of Freshwater Plants

Proceedings of the 9th International Symposium on Aquatic Weeds, European Weed Research Society

Edited by

J.M. Caffrey, P.R.F. Barrett, K.J. Murphy & P.M. Wade

Reprinted from Hydrobiologia, vol. 340 (1996)

SPRINGER SCIENCE+BUSINESS MEDIA, B.V.

Library of Congress Cataloging-in-Publication Data

A C.I.P. Catalogue record for this book is available from the Library of Congress

ISBN 978-94-010-6441-5 ISBN 978-94-011-5782-7 (eBook)
DOI 10.1007/978-94-011-5782-7

Printed on acid-free paper

Contents

Hydrobiologia **340:** ix–xiii, 1996.
J. M. Caffrey, P. R. F. Barrett, K. J. Murphy & P. M. Wade (eds), Management and Ecology of Freshwater Plants.

Preface

The European Weed Research Society and the Management and Ecology of Freshwater Plants

J. M. Caffrey[1] & P. M. Wade[2]
[1]*Symposium Convenor, Central Fisheries Board, Balnagowan, Mobhi Boreen, Glasnevin, Dublin 9, Ireland*
[2]*Chairman Working Group on Aquatic Weeds, European weed Research Society, c/o International Centre of Landscape Ecology, Loughborough University, Loughborough, LE11 3TU, England*

Abstract

Attendance at the 9th International Symposium on Aquatic Weeds, held in Dublin in 1994, by 270 delegates from 35 different countries demonstrated the continuing interest in the management and ecology of freshwater plants. The relative importance of the various topics covered in this meeting is compared with that of the previous symposia (1967–1990) for which published proceedings are available. A shift of interest away from aquatic weed control towards ecology, plant-environment interactions and distribution is noted and demonstrates a growing recognition of the need for aquatic plant management. The interest in physical control has remained constant (5–12% of papers) whilst the interest shown in biological control over the period 1971 to 1982 has not been sustained in recent symposia. The international nature of the symposia has increased over the years with papers published rising from eight countries in the 1967, 1971 and 1974 symposia to 23, 18 and 20 in the last three. Consistent numbers of contributions have been made by delegates from the Netherlands, the Czech Republic and the United Kingdom with a significant and sustained increase since 1967 from the United States of America.

Introduction

'The control of aquatic weeds presents many specialised problems. So immense is the range of growth habits that a method of control appropriate to one is often completely unsuitable for another species growing in the same place. Moreover, techniques are most frequently required to control not pure colonies but mixed communities comprising several weeds of different life form' (Sculthorpe, 1967). This challenge of managing aquatic plants has been recognised for a long time and, since the 1960's, the European Weed Research Society (EWRS) (formerly the European Weed Research Council) has, through its Working Group on Aquatic Weeds, worked to better understand this diverse assemblage of plants and to assist in the development and implementation of appropriate management (Pieterse, 1978). The first aquatic weeds symposium was held in Oxford in 1965 followed by symposia at Oldenburg (1967), Oxford (1971), Vienna (1974) Amsterdam (1978), Novi Sad (1982) (in association with the 2nd International Symposium on Herbivorous Fish), Loughborough (1986) (in association with the Association of Applied Biologists), Uppsala (1990) and most recently the 9th International Symposium on Aquatic Weeds in Dublin in 1994.

The main objectives of the Working Group on Aquatic Weeds are to:

1. promote the interchange of information on aquatic weed problems between members of the Society by organising symposia at regular intervals;
2. encourage the co-operation between research scientists working in similar fields on common problems by forming working groups to consider specific topics (e.g. the invasion of aquatic and riparian habitats (de Waal et al., 1994; Pysek et al., 1995);
3. develop and maintain contact with other international organisations with similar interests (e.g. Aquatic Plant Management Society (USA));
4. stimulate and encourage members to contribute items of information on aquatic weeds and news

of individuals working on them to the EWRS Newsletter.

An important milestone in the life of the Working Group on Aquatic Weeds was the publication of Pieterse & Murphy (1990), a text comprehensively reviewing aquatic plant management.

A review of the contents of EWRS International Symposia on Aquatic Weeds (1967–1994)

The 270 delegates from 35 countries world-wide attending the 9th International Symposium on Aquatic Weeds illustrated the sustained interest in the management and ecology of freshwater plants and the continuing need for up-to-date information on methods to control aquatic weed species. Whilst the majority of delegates were from Europe, there was a significant number from North America. The content of the papers published in this volume have been compared with those of the previous seven symposia (for which published proceedings are available) and some interesting trends are apparent. The call for papers for the symposium has always been very open and the topics identified very similar. The division of the papers into the different topics is clear cut for most papers. Some have been allocated on a split basis between two topic headings, e.g., Engelhardt (1974) consider the regulation of the use of herbicides for aquatic weed control in the German Federal Republic and the paper has been included under 'Control: chemical' and 'Control: strategic and regional studies'. A minority of papers have been unclassified. In all cases the data are related to the year of the symposium which, except for the 9th symposium, is also the date of publication of the proceedings.

The numbers of papers concerned with ecology, plant-environment interactions and distribution has increased whilst those concerned with control *per se* have declined (Table 1). The largest number of papers concerning the former was at the 1986 meeting which is at least partly explained by symposium being run by the EWRS and the Association of Applied Biologists. This shift in emphasis is in part due to the recognition of the complexity of aquatic weed control so clearly spelt out by Sculthorpe (1967) such that 'weed control' is being displaced by 'plant management'. It is significant that the Hyacinth Control Journal in the United Sates of America became the Journal of Aquatic Plant Management Society. Another indicator of this shift in emphasis is the increase in numbers of papers concerned with the utilisation of aquatic plants, proportionately more in this current volume than in any previous one.

Pieterse (1978) commented on the decrease in the relative interest in chemical control of aquatic weeds, 25% of the papers being on that topic at the 1978 Amsterdam meeting compared to 50% in the previous proceedings. This trend did not hold good for the 1982 meeting but the percentage of papers on herbicides has subsequently decreased substantially (Table 1). Chemical control in or near water has always been a sensitive issue and currently many agencies and even states are reluctant to use chemicals and some prohibit such management. The high percentage of papers on herbicides in the earlier symposia tended to focus on up-and-coming and recently released herbicides. The agrochemical industry currently sees little potential in herbicides for use in or near water and there are no compounds which have generated any serious interest in recent years.

Given the decline, not only in investigations into the use and effects of herbicides but also in their actual use, one would have expected an increase in interest and research into other forms of management, notably physical and biological control. This is not borne out in Table 1. Physical control has, apart from 1971, always accounted for only 5–12% of the papers. Biological control was a particular focus in 1971, 1974 and 1978, and in 1982 the EWRS 6th International Symposium on Aquatic Weeds (1982) was held in conjunction with the 2nd International Symposium on Herbivorous Fish. Since then, however, there has been relatively little interest in the topic. This might be explained by this topic being reported in another publication but none has been identified.

Contributors to the EWRS Aquatic Weeds Symposia

Since 1967 the proceedings of the International Symposia on Aquatic Weeds have published contributions from 40 different countries, over half of which were European (Table 2). The pattern of these contributions is worthy of note. Table 2 shows the number of papers published in the proceedings by authors on the basis of their country, a straightforward assessment for most papers. Where a paper is written by authors from more than one country, this is recognised in the scoring. In a few cases the paper is written by an author from one country but is about aquatic plant management in

Table 1. Review of the contents of proceedings of the EWRS International symposia of aquatic weeds (1967–1994) (percentage of papers published per year)

Topics	1967	1971	1974	1978	1982	1986	1990	1994
Ecology of freshwater plants	3	11	5	11	26	25	15	14
Plant-environment interactions	–	11	3	17	3	22	20	13
Distribution: freshwater plants & aquatic weed problems	7	–	3	–	9	14	13	14
Control: strategic & weed control problems	13	–	13	10	14	14	2	13
Control: chemical	67	48	50	25	40	9	21	11
Control: physical	7	–	5	12	9	8	6	9
Control: biological & biomanipulative	3	15	18	21	*	5	11	7
Control of algae	–	15	–	–	–	–	2	5
Utilisation	–	–	–	4	–	3	6	13
Unclassified	–	–	3	–	3	–	4	–
Total no. papers	30	27	39	52	35	63	50	56

* 6th Int. Symp. Aquatic Weeds (1982) was held in conjunction with the 2nd Int. Symp. on Herbivorous Fish.

another country. Such a paper is scored according to the nationality of the author.

As might be expected the proceedings have attracted contributions from an increasing number of countries. This applies to both European and non-European countries (Table 2). From 1967 to 1982 only six or seven European countries were represented in the proceedings. This more or less doubled at and after the 1986 meeting in Loughborough. The increase in contributions from other parts of the world was initially very small and has increased steadily. This growth in contributions is probably a result of growing interest in the subject as also evidenced by the increase in the number of delegates attending the symposia, but also the reputation of the symposium as a valuable meeting has grown, encouraging more contributions. It is hoped that attempts to improve the quality of contributions to the proceedings will make the meeting even more attractive.

Certain nationalities have maintained a surprising consistency in their contributions, for example, the United Kingdom has always contributed 10 to 13 papers (except for 1990), the Netherlands three to eight (except for the meeting in Amsterdam, 20); and Czechoslovakia (now the Czech Republic) one to three papers for most of the symposia (Table 2). These countries have a traditional interest in the aquatic plants and pioneered a number of aspects of aquatic plant management (e.g. the Weed Research Organisation, Oxford (UK); the Agricultural University, Wageningen (the Netherlands), and the Botanical Institute, Trebon (Czech Republic). The United States of America has gradually increased the proportion of its contribution to a maximum of 14 at the Uppsala meeting of 1990, whereas Germany has seen a decrease from 13 papers in 1967 to zero in 1986, 1990 and 1994.

Non-European Weed Research Society activities

Outside the EWRS there have been a series of *ad hoc* symposia organised between the EWRS meetings focusing on the biology and ecology of aquatic macrophytes. These began in Illmitz (Austria) in May 1981 (Hammer et al., 1990) being followed by symposia in Brussels in September 1981 (Symoens et al., 1982); Nijmegen (the Netherlands) in 1983 (Anon 1983); University of Aarhus (Denmark) in 1988 (Jensen & Madsen, 1991) and in Daytona (Florida, USA) in 1992 (Haller et al., 1993). The latter meeting also had a substantial number of papers concerned with aquatic plant management.

A measure of the interest being taken in aquatic plants is the range of publications which deal with their identification. Wade (1987) reviewed the various manuals and texts dealing with the recognition of this diverse group of taxa. An imbalance was noticed with much better coverage of Europe, Australia and North America than other parts of the world but since Wade (1987) the publishing of accounts of the aquatic flora

Table 2. Review of the contents of proceedings of the EWRS International symposia of aquatic weeds (1967–1994) (percentage of papers published per year)

Nationality	1967	1971	1974	1978	1982	1986	1990	1994
European								
Austria	–	–	10	2	–	–	1	1
Belgium	–	-	–	3	1	4	2	–
Czechoslovakia*	–	2	1	–	1	2	1	3
Denmark	–	–	–	1	1	–	1	1
Finland	1	–	–	–	–	1	–	1
France	1	1	–	–	–	6	–	3
Germany**	13	1	6	1	1	–	–	–
Greece	1	–	–	–	–	-	–	–
Hungary	–	–	–	–	–	1	–	–
Ireland	–	–	–	–	–	1	1	4
Israel	–	–	–	–	–	1	1	–
Italy	–	–	–	–	–	–	–	1
Jugoslavia	–	–	–	–	3	1	2	–
Netherlands	3	5	5	20	5	8	7	5
Norway	–	–	–	–	–	–	–	1
Poland	–	–	–	–	–	2	–	–
Portugal	–	–	–	1	–	–	2	3
Romania	–	1	–	–	–	–	–	–
Spain	–	–	–	–	–	–	–	1
Sweden	1	–	–	–	–	2	3	–
Switzerland	–	1	1	–	–	2	–	–
United Kingdom	10	15	10	12	13	13	6	12
Non-European								
United States of America	1	2	4	5	5	7	14	7
Other countries***	–	–	1	7	9	12	9	11
Total number of countries (not papers)	8	8	8	12	15	23	18	20

* Now Czech Republic and Slovakia
** Includes both East and West Germany before unification
*** Argentina, Australia, Bangladesh, Canada, China, Egypt, India, Indonesia, Japan, Malaysia, Mexico, New Zealand, Nigeria, South Africa, Sudan, Venezuela and Zambia

of other parts parts of the world and the Indian subcontinent in particular has helped to redress the imbalance. These include: Japan (Kadono, 1994), Papua and New Guinea (Leach & Osborne, 1985), Venezuela (Velasquez, 1994), Kerala State, India (Sivarajan et al. 1995), North-east India (Islam, 1989), the Lower Ganga delta (Naskar, 1990) and Bangladesh (Khan & Halim, 1987)

The acknowledged success of the Dublin symposium was due in no small measure to the financial and other services generously provided by the many sponsors (listed on page **xv**). The European Weed Research Society is most grateful to the Central Fisheries Board who provided staff, equipment and facilities essential to the smooth running of the meeting. Expert referees (listed on page **xvi**) have helped to ensure a high standard of papers in this volume of *Hydrobiologia*, an important aim of the symposium and the first time the output of the meeting has been published in this way. An extra special commendation is given to the symposium secretary, Sandra Doyle, whose unstinting commitment to every aspect of the meeting, professionalism, good humour and friendly disposition guaranteed the success of the event.

References

Anon, 1983. Proceedings of the International Symposium on Aquatic Macrophytes, Department of Aquatic Ecology, Nijmegen, The Netherlands, 326 pp.

de Waal, L. C., L. Child, P. M. Wade & J. H. Brock (eds), 1994. Ecology and Management of Invasive Riverside Plants. John Wiley & Sons, Chichester.

Engelhardt, D., 1974. Die Rechtliche Situation beim Einsatz von Herbiziden zur Krautbekömpfung in und an Gewässern in der BRD. Proc. EWRC 4th Int. Symp. Aquatic Weeds: 260–266.

Haller, W. T., D. N. Riemer, G. E. Bowes, A. M. Fox, J. C. Joyce, T. V. Madsen & M. Rattray (eds), 1993. International Symposium on the Biology and Management of Aquatic Plants. Journal of Aquatic Plant Management 31: 1–226.

Hammer, L., J. Kvet & P. M. Wade (eds), 1990. 1st European Conference on Aquatic Macrophytes, Illmitz, Austria, Folia Geobotanica & Phytotaxonomica 25: 227–335.

Islam, M., 1989. Aquatic weeds of North-east India. International Book Distributor, Dehra Dun, 155 pp.

Jensen, A. & T. V. Madsen (eds), 1991. Physiological Ecology of Aquatic Macrophytes. Aquatic Botany 39: 1–230.

Kadono, Y., 1994. Aquatic Plants of Japan. (in Japanese), Tokyo: 179 pp.

Khan, M. S. & M. Halim, 1987. Aquatic Plants of Bangladesh. Bangladesh National Herbarium, Dhaka: 120 pp.

Leach, G. J. & P. L. Osborne, 1985. Freshwater Plants of Papua New Guinea. The University of Papua New Guinea Press, National Cape District: 254 pp.

Naskar, K. R., 1990. Aquatic and semi-aquatic plants of the Lower Ganga Delta. Daya Publishing House, Delhi: 408 pp.

Pieterse, A. H., 1978. Preface. Proc. 7th Int. Symposium on Aquatic Weeds, 1–2.

Pieterse, A. H. & K. J. Murphy (eds), 1990. Aquatic Weeds. Oxford University Press, Oxford.

Pysek, P., K. Prach, M. Rejmanek & P. M. Wade (eds), 1995. Plant Invasions: General Aspects and Special Problems. SPB Academic Publishing, Amsterdam: 263 pp.

Symoens, J. J., S. S. Hooper & P. Compere (eds), 1982. Studies on Aquatic Vascular Plants. Societe Royale de Botanique de Belgique, Brussels: 424 pp.

Sculthorpe, C. D., 1967. The biology of aquatic vascular plants. Edward Arnold, London: 610 pp.

Sivarajan, V. V., K. T. Joseph, A. Rajani, 1995. Fresh water aquatic and plants of Kerala State (India): 500 pp.

Velasquez, J., 1994. Plantas Acuaticas Vasculares de Venezuela. Universidad Central de Venezuela, Caracas: 992 pp.

Wade, P. M., 1987. A review of the provision made for the identification of aquatic macrophytes as an aid to the study and management of wetlands. Archive für Hydrobiologie Ergebnisse der Limnologie 27: 105–113.

Hydrobiologia **340:** xv–xvi, 1996.
J. M. Caffrey et al. (eds), Management and Ecology of Freshwater Plants.

Symposium Sponsors

A.I.B. Bank

Aer Lingus

Ashling Hotel

Bord Failte Eireann

Bord na Mona

Central Fisheries Board

Coillte Teoranta

Commercial Hygiene Services

Duphar Ireland Ltd.

Electricity Supply Board

Guinness Son & Co. (Dublin) Ltd.

Harcourt Printing & Office Supplies Co. Ltd.

Irish Dairy Board

Irish Distillers Group Ltd.

Jameson Whiskey

Monsanto Ireland Ltd.

Office of Public Works

R & A Baileys & Co.

The Regency Hotel

Tedcastles Oil

Tipperary Natural Mineral Water Co.

Viking Marine Ltd.

Zeneca Ireland Ltd.

Special Edition Referees

Agami, M.
Akinyemiju, O.
Ali, M.
Anderson, L.
Armstrong, W.
Arsenovic, M.
Balciunas, J.
Barko, J.
Barrett, P.
Beer, S.
Beltman, B.
Best, E.
Bowes, G.
Boylen, C.
Brandrud, T.D.
Caffrey, J.
Cave, G.
Chambers, P.
Child, L.
Coops, H.
Couch, R.
Davidson, D.
Dawson, H.
de Waal, L.
Denny, P.
Dominy, P.
Dunderdale, J.
Eaton, J.
Fernandez, O.
Ferreira, T.
Fox, A.
Getsinger, K.
Gopal, B.
Greaves, M.
Grosse, W.
Haller, W.
Hanbury, R.
Harley, K.
Harpley, J.
Haslam, S.
Haury, J.
Hellsten, S.
Holmes, N.
Hootsmans, M.
Howard-Williams, C.
Hoyer, M.
Janaeur, G.
Jones, I.
Joska, M.
Kautsky, L.
Krosch
Kvet, J.
Langeland, K.
Madsen, J.
Mantai, K.
Marrs, S.
Mitchell, D.
Moreira, I.
Murphy, K.
Nat, E.
Newman, J.
Nichols, S.
Otte, R.
Ozimek, T.
Pieterse, A.
Pietsch, W.
Pot, R.
Prach, K.
Prenderville, G.
Rattray, M.
Richardson, C.
Ridge, I.
Rorslett, B.
Roshon, R.
Schwarz, A.
Smith, C.
Sorrell, B.
Spencer-Jones, D.
Spink, A.
Stanley, R.
Steward, K.
Teixeira, G.
Thayer, D.
Tiley, G.
Torstensson, L.
van Dijk, G.M.
van Vierssen, W.
Wade, M.
Waisel, Y.
Wallsten, M.
Whitton, B.
Willby, N.
Wood, B.
Woodcock, D.
Zákrávský, P.
Zimba, P.

Hydrobiologia **340**: 1–5, 1996.
J. M. Caffrey, P. R. F. Barrett, K. J. Murphy & P. M. Wade (eds), Management and Ecology of Freshwater Plants.

Photosynthetic plasticity in *Potamogeton pectinatus* L. from Argentina: strategies to survive adverse light conditions

M. J. M. Hootsmans, A. A. Drovandi[1], N. Soto Perez[2] & F. Wiegman
International Institute for Infrastructural, Hydraulic and Environmental Engineering, P.O. Box 3015, 2601 DA Delft, The Netherlands
[1]*National Institute of Hydric Science & Technology, 5500 Mendoza, Argentina*
[2]*SEMAPA, P.O. Box 1647, Cochabamba, Bolivia*

Key words: light climate manipulation, *Potamogeton pectinatus*, photosynthesis, turbidity, weed management

Abstract

Argentine *Potamogeton pectinatus* L. was grown in The Netherlands under laboratory conditions at four light intensities (50, 100, 150 and 200 μE m^{-2} s^{-1}), and photosynthetic performance was evaluated after about 1, 2 and 3 months of growth. At these moments, chlorophyll-*a* and -*b* and tissue N and P content were also determined. During the growing period, plant lengths and number of secondary shoots were measured. In the field in Argentina, photosynthetic performance of *P. pectinatus* was also measured at different light intensities created by artificial shading at various times during the growing season. Field and laboratory photosynthetic results were in good agreement. *P. pectinatus* showed a significant plasticity in its photosynthesis, rather than in morphology. A fairly constant maximum photosynthetic rate with reduced light enabled the plants to maintain net production rates rather unaffected at low light intensities. Still, it can be predicted that increasing turbidity from 1–2 m^{-1} at present to 3 m^{-1} could lead to a strongly light-limited growth which should reduce the present weed problem considerably. Such a turbidity increase might be achieved by the introduction of a fairly dense bottom-feeding fish population like Common carp (*Cyprinus carpio* L.).

Introduction

In many irrigated areas around the world, aquatic macrophytes develop so profusively that a serious weed problem emerges (Pieterse & Murphy, 1990).

In 1992, an international scientific research project started to address this issue by combining more or less detailed knowledge of the life cycle characteristics of aquatic weeds with management techniques to deteriorate in an environmentally friendly way the conditions for growth and re-growth. In two irrigation schemes south of Bahia Blanca, Argentina, *Potamogeton pectinatus* L. was selected to be studied in detail. The species dominates the vegetation in the principal drainage channels of both schemes, and also is a serious problem in the irrigation channels in one of the schemes (Irigoyen, 1981; Fernández et al., 1987).

The main question addressed here is: does *P. pectinatus* show photosynthetic plasticity which might enable it to survive adverse light conditions? In this context, the term photosynthetic plasticity is used to indicate the capacity to acclimate several plant characteristics which can improve net production rates under low light conditions. Thus, not only physiological but also morphological acclimation is studied. This information is then used to evaluate the potential of increasing turbidity, e.g. by using benthivorous fish like Common carp (*Cyprinus carpio* L.), as a management strategy for *P. pectinatus*.

Materials and methods

Collection, culture and processing of plant material for laboratory experiments

In December 1992, tubers of *P. pectinatus* were collected in Argentina and stored at 4°C in the dark. Experimental cultures in The Netherlands were set up in June 1993 in four 100 liter aquaria filled with tapwater (low in nutrients, neutral pH) which were kept at 20°C, under a long-day (16–8 h light-dark) regime. Four different light intensities (200, 150, 100 and 50 μE m^{-2} s^{-1}) were created using Philips fluorescent light tubes (color 84) and neutral density netting material. Per aquarium, 39 *P. pectinatus* tubers were used from a standard size class of 0.1–0.2 g fresh weight.

During the experiment, weekly measurements were taken of the length of the first shoot and the total number of developing shoots per plant was counted. After 42, 70 and 105 days, plants were harvested and light response curves were determined using a slighty modified version of the oxygen method described by Hootsmans & Vermaat (1994).

Plants were separated in stems, leaves and below-ground biomass. A small leaf sample was taken and frozen until later analysis for chlorophyll-*a*, -*b* and pheopigment content following the 96% ethanol method described by Wintermans & De Mots (1965). A number of plants was used to obtain total N and P content of tissue following Novozamsky et al. (1983). From the remaining plant material, dry weight (DW) and ash-free dry weight (AFDW) were determined.

Field measurements of photosynthesis

During November–January of 1992–1993 and 1993–1994, field measurements of photosynthesis were done in a natural drainage channel (referred to as Gius). In 1993–1994 a main irrigation channel, Ramal Sur, was added (referred to as RSur). Both channels are situated at about 40 ° S, 62 ° W, 150 km south of the city of Bahia Blanca (Argentina).

Shoots were collected by hand from the upper layers of the vegetation immediately before incubation. Adhering periphyton could be removed by shaking and plants were put in glass bottles of known volume (about 300 ml). These bottles were filled with channel water and suspended at a depth of about 30 cm in the channel. In 1992, besides respiration, photosynthesis was measured at 0% and 70% shading. In 1993, measurements were made of respiration and of photosynthesis

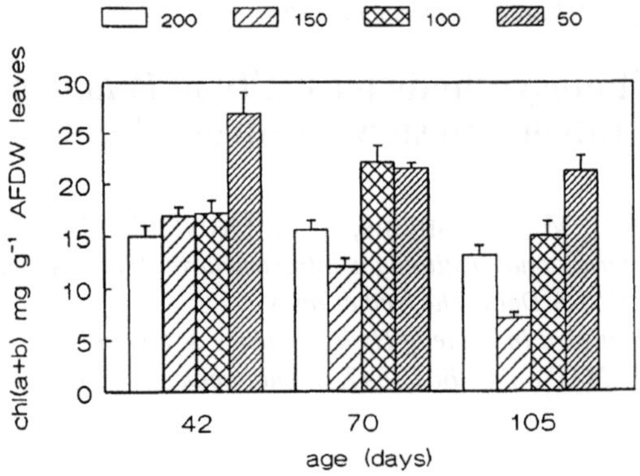

Figure 1. Chlorophyll ($a+b$) concentrations (mg g^{-1} AFDW) in leaves of *Potamogeton pectinatus* plants. In this figure and in the following figures, the four culture light levels (200, 150, 100 and 50 μE m^{-2} s^{-1}) are indicated by an increased hatching of the bars.

under 0%, 50, 75 and 88% shading. At the start and at the end of the experiment, the oxygen concentration in each bottle was measured using a clark type oxygen probe. At all light levels, 5 replicate bottles were used.

Continuous measurements were taken in the field with a Campbell 21 X datalogger connected with two LiCor 192SB underwater quantum light sensors at two different depths, to determine the light intensity during the various photosynthesis experiments.

Plant material was transported to the laboratory and stored at 4 °C in the dark until further analysis within a few days. Biomass of stems and leaves (DW, AFDW) and chlorophyll-*a* and -*b* concentrations in leaves were determined as described above.

Calculations and statistical analysis

All non-linear curves were fitted with an iterative non-linear regression procedure provided by the NLIN option in the SAS statistical package (SAS Institute Inc, 1985). The parameters from the light response curves (Pm, maximum gross rate of photosynthesis, Km, the half saturation light intensity and R, the respiratory rate) were used for production estimates. For the laboratory cultures, some other parameters were derived from these curve parameters, like α (the maximum quantum yield), the light compensation point (LCP) and net production rate at 200 μE m^{-2} s^{-1} (NP200) and at the actual light level during growth (NP). All analyses of variance (ANOVA) were performed using the GLM procedure in SAS.

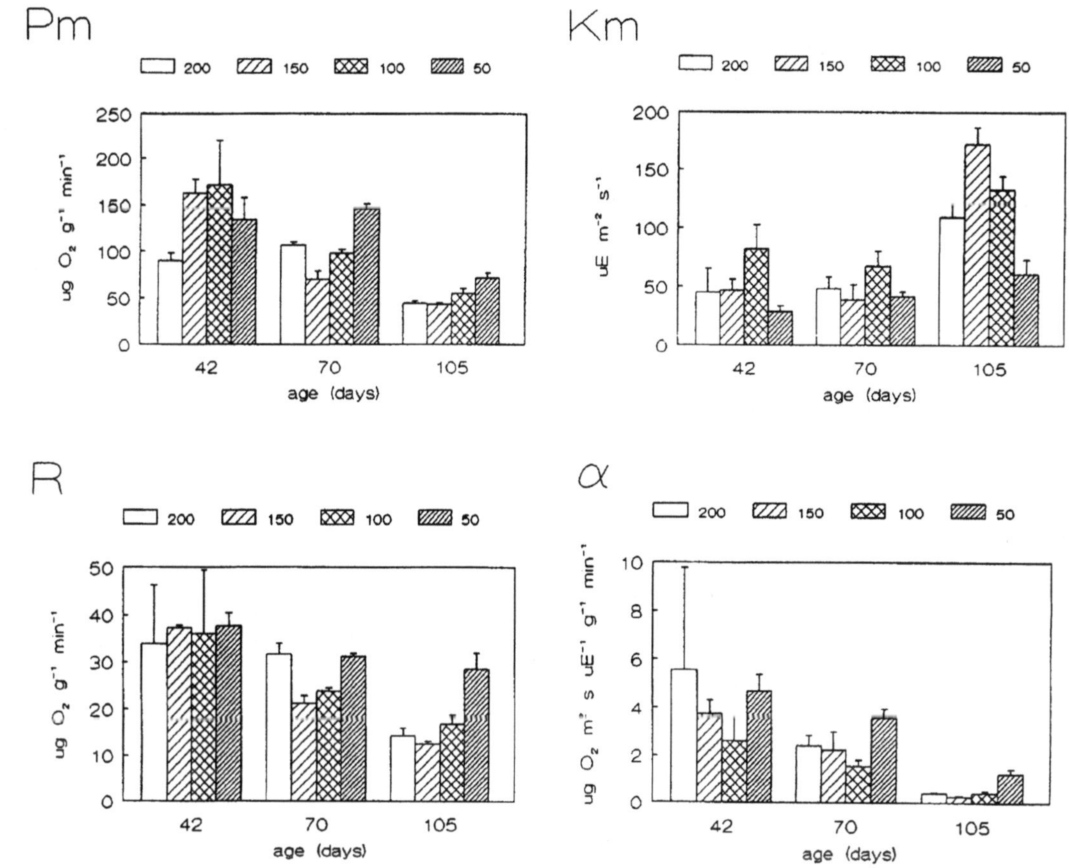

Figure 2. Parameters used in the P–I curves of *Potamogeton pectinatus* L. Pm = maximum gross rate of photosynthesis on aboveground biomass basis, Km = half-saturation constant, R = respiratory rate, α = maximum quantum yield.

Results

Plant morphology, nutrient and chlorophyll content

In the laboratory experiments, total biomass per plant increased significantly with age up to 600–900 mg AFDW after 105 days. A significant interaction was present between age and the culture light level: development at the lowest and highest light level appeared to be retarded as compared to the intermediate light levels. The ratio of aboveground to belowground biomass (2–3 g g^{-1}) was only slightly but significantly reduced at greater age, with an additional reduction at decreased light level especially at the highest age. The ratio of leaf biomass to total aboveground biomass (0.6–0.8 g g^{-1}) also showed a significant decrease with increased age, with a significant but in effect rather minor additional interaction between light and age.

The curves describing the development over time of both plant length and number of shoots all differed significantly from each other. Elongation was stronger with lower light levels during growth, and the number of shoots lagged behind for the lowest light level. However, although significant, these differences were usually relatively small (10–20%).

Chlorophyll ($a + b$) content of leaves decreased significantly with age and increased with lower light level (Figure 1). Pheopigment content remained low (3% of total chlorophyll; not shown) and only had a higher value (10–15%) for the lowest two light levels after 42 days. The fraction of chlorophyll-b in total chlorophyll showed a slight significant decrease with increased age but roughly remained constant around 30%. In the field, chlorophyll content of plant leaves only showed a slight significant fluctuation over time, more or less remaining around 10 mg g^{-1} AFDW; the fraction of chlorophyll-*b* in total chlorophyll hardly changed over time or between stations; like in the laboratory cultures, it remained around 30%.

N and P content of aboveground and belowground tissue showed a significant decrease with increased age, which was slightly more pronounced at the highest light level. On a total biomass basis, N decreased from 30 to 15 mg g^{-1} DW, P decreased from 5 to 2 mg g^{-1} DW.

Effects of light level and age on photosynthesis

In all cases, significant fits of the rectangular hyperbola model to the experimental light response curve data were obtained. Figure 2 shows the resulting mean values from the laboratory cultures for Pm, R, Km and α. With increased age, Pm, R and α significantly decreased while Km increased; light level interactively modified this effect for Pm and Km. For Pm, a more or less clear increase can be noted with decreased light level.

Predicted net production of laboratory cultures at 200 μE m^{-2} s^{-1} and at the light level of culture were influenced in the same way as Pm: a decrease with increased age, while they increased or remained constant with decreased light level. A remarkably high value can be observed for the lowest light level in all cases. LCP significantly increased from around 20 to 70 μE m^{-2} s^{-1} between 70 and 105 days of age.

Results from field measurements in the second season for Pm, Km and R are presented in Figure 3. Only Pm was significantly changing with time, decreasing in the irrigation channel RSur, and with a slight dip in the drainage channel Gius. Field results are in good agreement with the laboratory data.

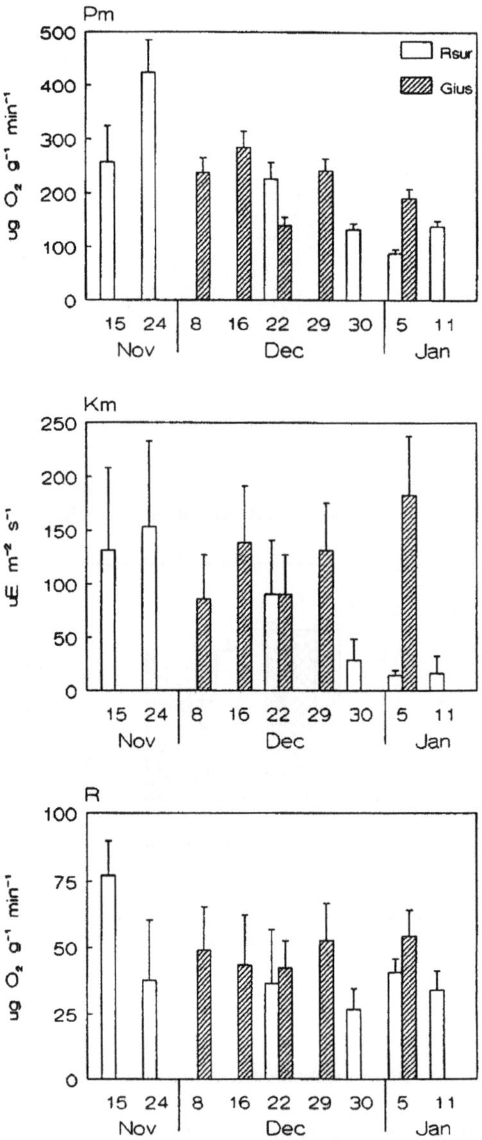

Figure 3. Photosynthetic parameters of light response curves fitted to field data expressed on an aboveground biomass basis. Rsur = irrigation channel, Gius = natural drainage channel.

Discussion

Morphological and photosynthetic acclimation

It can be concluded that Argentine *P. pectinatus* does not show a strong morphological acclimation to reduced light conditions in terms of elongation or changing allocation of biomass to leaf development. Its biomass appears much less affected by reduced light levels as compared with material from The Netherlands (Vermaat & Hootsmans, 1994).

Chlorophyll ($a+b$) in our material was at least two times higher under all conditions compared with Dutch plants grown under almost identical conditions (Hootsmans & Vermaat, 1994). Fraction of chlorophyll-*b* remained constant, as in Dutch plants, but was higher (0.30 against 0.20 on average). Nutrient content remained above the so-called critical values of Gerloff

& Krombholz (1966) and was comparable to values for Dutch plants. This indicates that serious nutrient-limited growth during the experiment is unlikely.

The findings for the limited morphological acclimation can be illucidated when the photosynthetic performance is taken into account. Laboratory material proved to be able to maintain its net photosynthetic rates at a relatively high level, even increasing with reduced culture light levels. This is mainly due to the rather constant or increasing values for Pm at different light levels, only reducing with age. The same holds for α, which remains more than twice as high as compared to Dutch material during the first two months of growth (Hootsmans & Vermaat, 1994). Consequently, LCP values observed in our present study are fairly low as compared to Dutch *P. pectinatus*, but still within the range of other findings, e.g. Van der Bijl et al. (1989).

It can be concluded that Argentine *P. pectinatus* shows a significant photosynthetic plasticity, mainly due to acclimation of its photosynthetic characteristics. This result contrasts clearly with findings for material of *P. pectinatus* from other localities, thus stressing the limited possibility for extrapolating results from one population of this cosmopilitan species to another.

Implications for channel management

With the data obtained on photosynthetic production rates, net daily production rates can be estimated. A detailed account of this will be published elesewhere. An important outcome is that daily net production of this population of *P. pectinatus* is expected to become zero or even negative if the extinction coefficient were to be raised to 3–4 m^{-1}. In spring (November) this does occur at depths of 75–100 cm, i.e. near the sediment surface. These results imply that regrowth under such conditions will be completely dependent on reallocation of biomass from the tubers, and thus will at least be hampered, if not successfully controlled. On the basis of data from Breukelaar et al. (1994), such turbidity levels can be achieved by a relatively high density of Common carp (*Cyprinus carpio* L.)

We conclude that although Argentine *P. pectinatus* is surprisingly able to acclimate its photosynthesis to reduced light levels, fish are probably able to keep turbidity high enough to limit plant growth considerably.

Acknowledgments

Professor Dr P. Denny, Prof. Dr W. van Vierssen and Dr J. E. Vermaat critically read the manuscript. This project was made possible through the European Union STD 3 programme contract no. TS3*-CT92-0125 and The Netherlands Government fellowship programme.

References

Breukelaar, A. W., E. H. R. R. Lammens, J. G. P. Klein Breteler & I. Tátra, 1994. Effects of benthivorous bream (*Abramis brama*) and carp (*Cyprinus carpio*) on sediment resuspension and concentrations of nutrients and chlorophyll *a*. Freshwat. Biol. 32: 113–121.

Fernández, O. A., J. H. Irigoyen, M. R. Sabbatini, & R. E. Brevedan, 1987. Aquatic plant management in drainage channels of southern Argentina. J. aquat. Plant Mgmt 25: 65–67.

Gerloff, G. C. & P. H. Krombholz, 1966. Tissue analysis as a measure of nutrient availability for the growth of angiosperm aquatic plants. Limnol. Oceanogr. 11: 529–537.

Hootsmans, M. J. M. & J. E. Vermaat, 1994. Light-response curves of *Potamogeton pectinatus* L. as a function of plant age and irradiance level during growth. Kluwer, Geobotany 21: 62–117.

Irigoyen, J. H., 1981. Crecimiento y desarollo de *Potamogeton pectinatus* en los canales de desagüe del Valle Bonaerense del Río Colorado. (Growth and development of *Potamogeton pectinatus* in drainage channels in the Bonaerense valley of the Colorado river; in Spanish, with English summary). II Reunión sobre malezas subacuáticas en los canales de desagüe de CORFO. Publ. CIC, La Plata: 47–69.

Novozamsky, I., V. J. G. Houba, R. Van Eck & W. Van Vark, 1983. A novel digestion technique for multi-element plant analysis. Comm. Soil Sci. Plant Anal. 14: 239–248.

Pieterse, A. H. & K. J. Murphy, 1990. Aquatic weeds. Oxford University Press, Oxford, UK, 593 pp.

SAS Institute Inc., 1985. SAS/STAT Guide for Personal Computers, version 6 edition. Cary, N.C., 378 pp.

Vermaat, J. E. & M. J. M. Hootsmans, 1994. Growth of *Potamogeton pectinatus* L. in a temperature-light gradient. Kluwer, Geobotany 21: 40–61.

Van der Bijl, L., K. Sand-Jensen & A. L. Hjermind, 1989. Photosynthesis and canopy structure of a submerged plant, *Potamogeton pectinatus*, in a Danish lowland stream. J. Ecol. 77: 947–962.

Wintermans, J. F. G. M & A. De Mots, 1965. Spectrophotometric characteristics of chlorophylls *a* and *b* and their pheophytins in ethanol. Biochim. Biophys. Acta 109.

& Krombholz (1966) and was comparable to values for Dutch plants. This indicates that serious nutrient-limited growth during the experiment is unlikely.

Our findings for the limited [illegible] of [illegible] cultures can be [illegible] when the [illegible] is taken into account. [illegible] proved to be able to maintain its net photosynthetic rates at a relatively high level, even increasing with reduced culture light levels. This is mainly due to the rather constant or increasing values for Pm at different light levels, only reducing with age. The same holds for α, which remains more than twice as high as compared to Dutch material during the first two months of growth (Hootsmans & Vermaat, 1994). Consequently, LCP values observed in our present study are fairly low as compared to Dutch *P. pectinatus*, but still within the range of other findings, e.g. Van der Bijl et al. (1989).

It can be concluded that Argentine *P. pectinatus* shows a significant photosynthetic plasticity, mainly due to acclimation of its photosynthetic characteristics. This result contrasts clearly with findings for material of *P. pectinatus* from other localities, thus stressing the limited possibility for extrapolating results from one population of this cosmopolitan species to another.

Implications for channel management

With the data obtained on photosynthetic production rates, net daily production rates can be estimated. A detailed account of this will be published elsewhere. An important outcome is that daily net production of this population of *P. pectinatus* is expected to become [illegible] or even negative if the extinction coefficient were to be raised to 3–4 m^{-1} in spring (November) [illegible] [illegible] ment surface. These results imply that regrowth under such conditions will be completely dependent on reallocation of biomass from the tubers, and thus will at least be hampered, if not successfully controlled. On the basis of data from Breukelaar et al. (1994), such turbidity levels can be achieved by a relatively high density of Common carp (*Cyprinus carpio* L.).

We conclude that although Argentine *P. pectinatus* is surprisingly able to acclimate its photosynthesis to reduced light levels, fish are probably able to keep turbidity high enough to limit plant growth considerably.

Acknowledgements

[illegible] Dr. [illegible], Prof. Dr. W. van Vierssen and Dr. J. E. Vermaat critically read the manuscript. This project was made possible through the European Union STD 3 programme (contract no. TS3*-CT92-0123) and The Netherlands Government fellowship programme.

References

Breukelaar, A. W., E. H. R. R. Lammens, J. G. P. Klein Breteler & I. Tátrai, 1994. Effects of benthivorous bream (*Abramis brama*) and carp (*Cyprinus carpio*) on sediment resuspension and concentrations of nutrients and chlorophyll a. Freshwat. Biol. 32: 113–121.

Fernández, O. A., I. H. Irigoyen, M. R. Sabbatini & R. E. Brevedan, 1987. Aquatic plant management in drainage channels of southern Argentina. J. aquat. Plant Mgmt 25: 63–67.

Gerloff, G. C. & P. H. Krombholz, 1966. Tissue analysis as a measure of nutrient availability for the growth of angiosperm aquatic plants. Limnol. Oceanogr. 11: 529–537.

Hootsmans, M. J. M. & J. E. Vermaat, 1994. Light response curves of *Potamogeton pectinatus* L. as a function of plant age and irradiance level during growth. Kluwer, Geobotany 21: 57–117.

Irigoyen, J. H., 1983. Crecimiento y desarrollo de *Potamogeton pectinatus* en los canales de desagüe del Valle Bonaerense del Río Colorado. (Growth and development of *Potamogeton pectinatus* in drainage channels in the Bonaerense valley of the Colorado river; in Spanish, with English summary.) II Reunión sobre malezas subacuáticas en los canales de desagüe de CORFO. Publ. CIC, La Plata: 17–46.

Novozamsky, I., V. J. G. Houba, R. Van Eck & W. Van Vark, 1983. A novel digestion technique for multi-element plant analysis. Commun. Soil Sci. Plant Anal. 14: 239–249.

[illegible] University Press, Oxford, UK, 290 pp.

SAS Institute Inc., 1985. SAS/STAT Guide for Personal Computers, Version 6 edition. Cary, N.C., 378 pp.

Vermaat, J. E. & M. J. M. Hootsmans, 1994. Growth of *Potamogeton pectinatus* L. in a temperature-light gradient. Kluwer, Geobotany 21: 40–61.

Van der Bijl, L., K. Sand-Jensen & A. L. Hjermind, 1989. Photosynthesis and canopy structure of a submerged plant, *Potamogeton pectinatus*, in a Danish lowland stream. J. Ecol. 77: 947–962.

Wintermans, J. F. G. M. & A. De Mots, 1965. Spectrophotometric characteristics of chlorophylls a and b and their pheophytins in ethanol. Biochim. Biophys. Acta 109: [illegible]

Hydrobiologia **340:** 7–10, 1996.
J. M. Caffrey, P. R. F. Barrett, K. J. Murphy & P. M. Wade (eds), Management and Ecology of Freshwater Plants.

Studies on vegetative production of *Potamogeton illinoensis* Morong in southern Argentina

Carlos Bezic, Armando Dall'Armellina & Omar Gajardo
Universidad Nacional del Comahue, Centro Regional Zona Atlántica, Ayacucho y Esandi (8500) Viedma, Rio Negro, Argentina

Key words: aquatic weeds, *Potamogeton illinoensis*, growth characteristics, survival strategies

Abstract

Potamogeton illinoensis Morong is a major submerged weed invading irrigation channels in the Lower Valley of the Río Negro, near Viedma, Argentina. Studies on morphology and growth characteristics of this species were conducted in an outdoor tank from August 1993 to May 1994 with the objective of increasing the knowledge of its ecology order to adjust control measures. The maximum aboveground biomass was reached in April, with a subsequent decrease to May when the water supply was cut off. Belowground biomass comprised two kinds of rhizomes. The first group (Rhizomes I) was produced from the beginning of the annual cycle causing both lateral shoots and new rhizomes I production. The second group (Rhizomes II) was distinguished as an enlargement of the extremes of rhizomes I from mid-November, producing only short overwintering sprouts. Plant parts production (DW in g/plant) in the first cycle was: 27.2 g leaves; 11.9 g stems; 17.4 g rhizomes I and 8.1 g rhizomes II. Vegetative propagation appeared to be an important survival strategy in this species. During the 3–4 month period without water only rhizomes with underground overwintering sprouts survive in the dry sediment.

Introduction

Potamogeton illinoensis Morong is an aquatic macrophyte that has developed profuse growths in irrigation channels of the Lower Valley of the Río Negro in southern Argentina. Impeding water movement, its development considerably reduces the water supply for agriculture and increases the irrigation network management costs.

Little is known about this weed in Argentina (Tur, 1982). Although listed by Pieterse & Murphy (1990) as an aquatic weed, Steward (1990) categorised the species as one which does not cause major weed problems in North America. It is a common lake species in areas such as Florida (e.g. Ewel & Fontaine, 1983). The species is named in the Applied Biochemists Inc. (1976) listing of aquatic weeds. Locally, it is known by the common name of 'lama' but until this study it had not even been properly identified in the area.

Plants of *Potamogeton illinoensis* begin growing from the first days of August, when the channels are refilled (Dall'Armellina et al., in press). Then a main shoot develops and, at the same time, two new rhizomes from the third and fourth nodes of the main shoot appear. These rhizomes have buds on alternate nodes over which they produce both a shoot and a new rhizome (branch). Each new rhizome develops like the previous one (Bezic, 1994). Seasonal sampling of natural populations of *Potamogeton illinoensis* shows different kinds of rhizomes (Dall' Armellina et al., in press). Bezic (1994) recognized two different types of rhizomes in this species: (a) Rhizomes I, produced from initial stages in the annual cycle and dying in the next winter, that appear to be structures adapted for colonization, and (b) Rhizomes II, that appear at the beginning of the summer as an enlargement of the Rhizome I extremes, probably by accumulation of carbohydrate reserves. They are whiter and bigger than the previous type developing one short belowground overwintering sprout at alternate nodes. Only Rhizomes II survive the dry season and initiate the next sprouting (Bezic, 1994). Aboveground parts are 92–93% represented by stems and submerged leaves (Bezic, 1994). Floating leaves have never been observed on *P. illinoensis* plants in the study area. Inflorescences are terminal spikes as in other *Potamogeton species* (Sculthorpe, 1967).

The objective of the study was to understand the seasonal growth and development of this macrophyte, in order to improve the efficacy of weed control measures directed against it.

Materials and methods

The study was performed in a concrete-lined 2 m deep circular tank, 50 m in diameter, between August 1993 and May 1994 at the IDEVI-INTA Experimental Station (40°48′ S; 63°05′ W; 4 m above sea level; annual mean temperature 14 °C) near Viedma, Rio Negro province, Argentina. The tank was topped up weekly with water, piped from the irrigation supply (derived from the Rio Negro). Dissolved oxygen concentration ranged between 8.6–14.6 mg. l^{-1}, mean electrical conductivity was 0.18 μS cm^{-1} and midday water temperature ranged between 10.3–25.4 °C during the study period.

Plants were grown in glass aquaria from pieces of rhizome (mean ± standard error: 0.8 ± 0.01 g fresh weight) with an overwintering sprout, collected from the bottom of the main irrigation channel in July 1993 and maintained in a refrigerator until 6 August. After 17 days growth in the laboratory (water mean temperature 18 °C; light intensity 100 μE m^{-2} s^{-1}; 16 h illumination per day) the plants were transferred to 54 wooden boxes (0.37 × 0.51 × 0.19 m), internally lined with a black plastic sheet and filled with soil free from propagules of other aquatic macrophytes. Three plants were placed in each box. Plant development was followed for a further 265 days. On each sampling occasion, 3 boxes were selected at random and the 9 plants which they contained were harvested. In the laboratory the plants were washed with clean water and separated into their main constituent parts (leaves, stems, rhizomes I and II), stem and rhizome length was measured, and then dried for 24 hours at 105 °C. Dry weight (DW) was taken with a 0.0001 g precision balance at the first stages and with a 0.01 g precision balance during the middle and at the end of the growing season. Biomass measurement and analysis were carried out following the recommendations of Madsen (1993). Relative growth rate (RGR) was calculated for total plant biomass over the main period of growth (measured at t1, 105 days after start and t2, 282 days after start). Although data were collected for other organs, we present here the results for leaves, stems, and rhizomes: these accounted for >90% of total plant weight in all specimens.

Results

Visible plant development from an overwintering sprout began when this was submerged in water. A main shoot then developed which had alternate submerged leaves with ligules. Within the belowground system the only components present were roots and rhizomes.

Rhizome production began at the first developmental stages with the production of two initial rhizomes at the third and fourth main shoot nodes. New rhizomes were produced at alternate nodes of the initial rhizomes together with the formation of lateral shoots (ramets) on the same node. From January (150 days) new rhizomes increased their size giving rise to the second identified type (Rhizomes II). No new rhizomes were produced by these, which only showed overwintering sprouts.

Plant production

Although the growing season began during the last days of winter, the main period of growth commenced 105 days after the start (mid-November), (Figures 1 and 2). Thereafter, growth was linear until 239 days, with the exception of rhizomes II that continued their growth until the end of the season, in May 1994. The calculated value of RGR for mean total plant biomass over the main growth period was 0.33 g g^{-1} d^{-1}.

The average growth rates for leaves and stems are presented in Table 1. The main growth period for Rhizomes I was the same as for leaves and stems, commencing in mid-November, and reaching maximum biomass at the end of April (Table 1). Rhizomes II production did not begin until the first week of January (150 days from start), with the maximum reached in May after 282 days.

Lateral shoots reached a length of almost 40 m per plant in March, with a very rapid increase during February. The same pattern was followed by rhizome I production, reaching a total length of 10 m per plant. Finally, rhizome II length only reached a maximum value of 1.8 m per plant in March (Table 1).

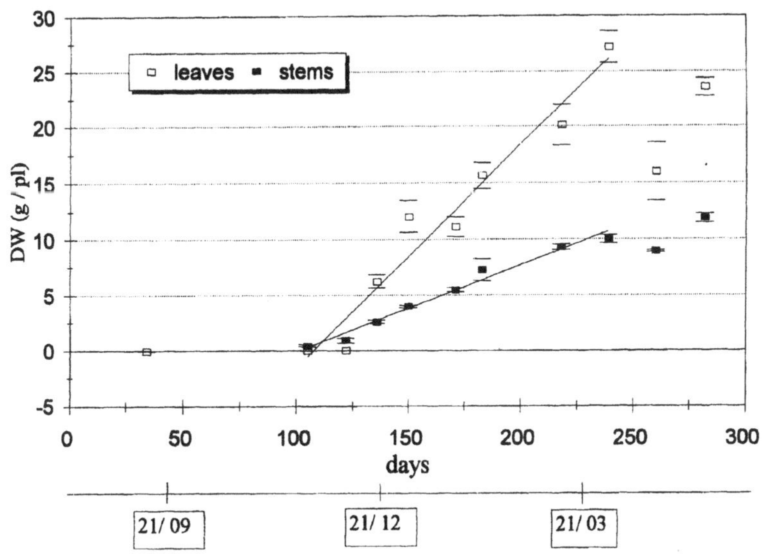

Figure 1. Development of main above-ground structures of *Potamogeton illinoensis* from overwintering sprouts in an outdoor tank during the 1993/94 season. Time axis is days after 6 August 1993. Data are mean dry weight per plant ± standard error ($n=9$). Linear regression comprises the active growing period between t1 = 105 and t2 = 239 days. —= fitted linear regression for rhizomes I ($y=0.087x-9.05$; $r=0.99$; $P<0.001$). - - - - =fitted linear regression for rhizomes II ($y=0.046x-5.70$; $r=0.96$, $P=0.003$).

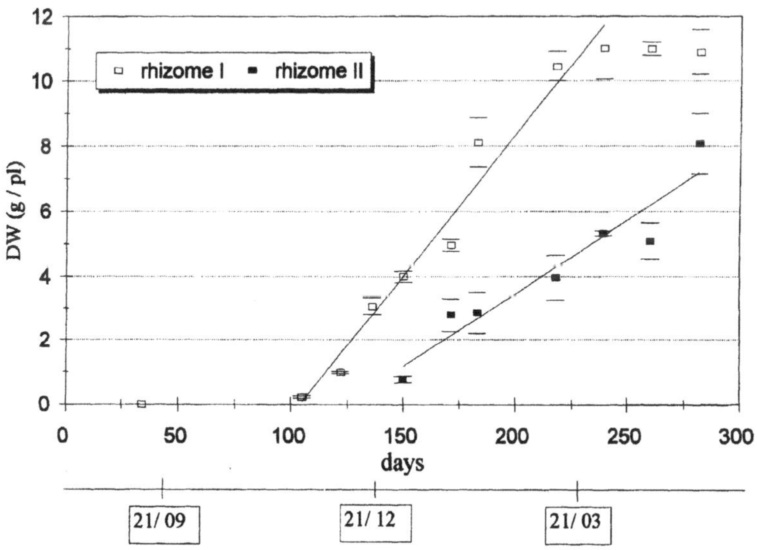

Figure 2. Development of main below-ground structures of *Potamogeton illinoensis* from overwintering sprouts in an outdoor tank during the 1993/94 season. Time axis is days after 6 August 1993. Data are mean dry weight per plant ± standard error($n=9$). Linear regression comprises the active growing period between t1 = 105 and t2 = 239 days for rhizomes I and between t1=150 to t2=282 days for rhizomes II. —= fitted linear regression for leaves ($y=0.199x-21.40$; $r=0.98$, $P<0.001$). - - - - = fitted linear regression for stems ($y=0.078x-7.83$; $r=0.99$, $P<0.001$).

Conclusions

Potamogeton illinoensis begins its development early in the season (August) when water covers the overwintering structures represented by an specific kind of rhizome which posesses a short belowground sprout. New shoots are produced by long rhizomes (Rhizomes I)

Table 1. Growth parameters of *Potamogeton illinoensis* Morong growing in experimental conditions in an outdoor tank in Viedma (Argentina) during the 1993/94 growing period.

	Leaves	Stems	Rhizomes I	Rhizomes II
AGR (g.day^{-1})	0.13	0.060	0.085	0.060
Max.DW (g.plant^{-1})	27.21±1.33	11.87±.38	17.4±3.22	8.09±0.68
Max.Length (m)		40	10	1.8

formed in the same season. From January, Rhizomes II production continues until the end of the growing period.

These organs showed varying growth rates, with the main period of growth being between November and April. Leaf biomass showed the fastest rate of increase, and rhizomes II the slowest, but these latter organs appear to play a key role in the regenerative strategy of *P. illinoensis*, permitting survival of annual periods of drought. Dependence on vegetative propagules is a common regenerative strategy in *Potamogeton* and many other submerged macrophytes (e.g. Kautsky 1988; Yeo 1965), but in *P. illinoensis* it appears to be of particular importance.

As rhizomes II did not develop until the mid-late phase of the growing season, control measures which prevented or reduced their formation would probably be a highly effective approach to managing this species. For example, the use of mechanical control, such as dredging, to destroy a high proportion of rhizomes I, before rhizomes II had a chance to form, could be very effective in reducing population survival into the next season. Alternatively, the introduction of control measures which severely depleted photosynthate supply for rhizome formation (by destroying foliage) during the period before production of rhizomes II might have the same effect. Possibilities here might include appropriate herbicide treatments (e.g. Fernández et al. 1987), or the use of grass carp. It is, however, clear that the current management methods, based on repeated use of a weed-clearing chain (Dall'Armellina et al., in press), are ineffective in reducing formation of rhizomes II, probably because the method has no direct effect on rhizomes and regrowth of foliage after treatment is too fast to permit much depletion of rhizome carbohydrate resources.

Acknowledgments

Funding for the study was provided by Secretaria de Investigacion de la Universidad Nacional del Comahue; Departamento Provincial de Aguas de la Provincia de Rio Negro; Instituto de Desarrollo del Valle Inferior del Rio Negro; and the STD3 Programme of the Commission of the European Communities. We thank Dr R. R. Haynes (University of Alabama, USA) for confirming the identity of *Potamogeton illinoensis*. Dr Kevin Murphy (Glasgow University, UK) assisted in revising this paper.

References

Applied Biochemists, Inc., 1976. How to identify and control water weeds and algae. Applied Biochemists, Inc. Mequon, Wisconsin, USA: 64 pp.

Bezic, C., 1994. Bioecolog¡a de *Potamogeton illinoensis*, principal maleza de los canales de riego del Valle Inferior del Rio Negro. Inf.Beca Invest., Univ. Nac. del Comahue, 115 pp (in press).

Dall'armellina, A., C. Bezic, A. Gajardo, E. Luna, A. Britto, J. Lizama & V. Dall'armellina (in press). Bioecolog¡a aplicada a la optimizacion del control de malezas acu ticas sumergidas en canales de riego. Univ. Nac. del Comahue, Inf. Av. Invest., 109 pp (in press).

Ewel, K. C. & T. D. Fontaine, 1983. Structure and function of a warm monomictic lake. Ecological Modelling 19: 139–161.

Fernandez, O. A., J. H. Irigoyen, M. Sabbatini & R. E. Brevedan, 1987. Aquatic plant management in drainage canals of southern Argentina. J. Aquat. Plant Mgmt 25: 65–67.

Kautsky, L., 1988. Life strategies of aquatic soft bottom macrophytes. Oikos 53: 126–135.

Madsen, J. D., 1993. Biomass techniques for monitoring and assessing control of aquatic vegetation. Lake and Reserv. Mgmt 7: 141–154.

Pieterse, A. H. & K. J. Murphy (eds), 1990. Aquatic Weeds. Oxford University Press, Oxford, UK. 593 pp.

Sculthorpe, C. D., 1967. The Biology of Aquatic Vascular Plants. Edward Arnold, London, 560 pp.

Steward, K. K., 1990. Aquatic weed problems and management in North America. Aquatic weed problems and management in the eastern United States. In (b) Aquatic Weeds (eds), A. H. Pieterse & K. J. Murphy, Oxford University Press, Oxford, UK: 391–405.

Tur, N. M., 1982. Revision del g,nero *Potamogeton* en la Argentina. Darwiniana 24: 217–265.

Yeo, R. R., 1965. Life history of sago pondweed.Weeds 13: 314–321.

Hydrobiologia **340:** 11–16, 1996.
J. M. Caffrey, P. R. F. Barrett, K. J. Murphy & P. M. Wade (eds), Management and Ecology of Freshwater Plants.

Diurnal carbon restrictions on the photosynthesis of dense stands of *Elodea nuttallii* (Planch.) St. John

J. Iwan Jones, Keith Hardwick & John W. Eaton
Department of Environmental and Evolutionary Biology, University of Liverpool, P.O. Box 147, Liverpool L69 3BX, United Kingdom

Key words: weedbed, growth, physiology, pH, oxygen, temperature

Abstract

In slow-moving and static eutrophic waters, submerged macrophytes growing in dense stands produce a highly structured environment, with reduced internal water flow. An afternoon lull in the net photosynthesis of such stands has been reported from a number of previous studies. This has been attributed to increased photorespiration caused by an accumulation of photosynthetically-derived, dissolved oxygen in the surrounding water. Results here demonstrate that even in a water quite rich in dissolved inorganic carbon (2.5 mmol l^{-1}), limitations on the supply of inorganic carbon will normally be more important in curtailing photosynthesis, with photorespiration playing only a minor role.

Introduction

Photosynthesis of submerged plants is affected by water flow around their leaves, but, conversely, the presence of the plants affects the flow of the water around them, especially when growing in dense stands. Then the flow within the vegetation is much reduced, with the majority of water diverted around the stand. The extent of resistance to flow and hence the degree of partitioning of water velocity, is determined by plant architecture and shoot density (Losee & Wetzel, 1993). Often different parts of a bed act in different ways, with the leafless parts allowing far more flow around them than the leafy parts (Dale & Gillespie, 1978; Losee & Wetzel, 1993).

A consequence of reduced flow within stands is decreased mixing and, since the plants have enormous potential for biological activity, water quality can be greatly altered as compared with that outside. Also warming by incident infra-red radiation reduces surface water density and curtails downward mixing (Dale & Gilespie, 1978). Further, as the plants strongly absorb the light incident on them, adding to the absorption of the water itself, there exists a strong vertical gradient of light through the bed, dependant on where the biomass is concentrated (Westlake, 1964; Van der Bijl et al., 1989; Frodge et al., 1990; Madsen & Maberly, 1991; Wychera et al., 1993). As a result of the structuring of the water and metabolic activity of the plant stand, vertical, diel and seasonal changes of pH, O_2 and temperature are generated (Buscemi, 1958; Goulder, 1970; Unni, 1972; Brown et al., 1974; Van et al., 1976; O'Niel-Morin & Kimball, 1983; Pokorny et al., 1984; Frodge et al., 1990).

The aim of this work was to analyse diel changes in water quality within stands of *E. nuttallii* (Planch.) St. John growing in a slowly-flowing, eutrophic channel, to determine the extent to which exchange of dissolved oxygen and inorganic carbon between the plants and their surrounding medium was controlling plant photosynthesis at this time.

Methods

The investigation was undertaken in the Leeds and Liverpool Canal at Aintree (SJ371990) on 17th/18th August 1993. Water samples were taken from within a stand of *E. nuttallii* with very occasional plants of *Myriophyllum spicatum* L., growing to the surface of 0.8 m of water. Profiles were taken from near the middle of the bed at 10:00 h, 13:00 h, 15:00 h, 20:00 h, 24:00 h and 07:30 h G.M.T. A profile was also measured at 13:30 h in open water 15 m upsteam from the stand and more than 5 m from any submerged plants.

Water samples were collected from known depths with a remote sampler, consisting of a 60 cm^3 hypodermic syringe attached to a graduated pole, and used for pH determination (CD66 field meter – WPA, Saffron Walden, Essex). An OXI 196 oxygen meter probe (WTW, Wielheim, Germany), was attached to a second graduated pole and used to measure the temperature and oxygen concentration in situ. Each profile took about 10 minutes to complete, with the approximate midpoint used to indicate when it was taken. Conductivity was measured at approximately 5 cm depth with a dip cell (Simac Instrumentation Ltd., Walton-on-Thames, Surrey). A water sample was also removed from the surface at 07:30 h on 18th August and the total DIC measured by Gran Titration (Mackereth et al., 1978).

Microscope investigation of plants showed very few periphytic microalgae (less than 50 cells per leaf), indicating that any changes were the result of the *E. nuttallii* plants alone. The plants had no marl deposits on them.

Results

The water within the stand of *E. nuttallii* developed a clear structure, with steep vertical gradients of pH, O_2 and temperature. The temperature profiles (Figure 1a) showed progressive surface warming during the day, particularly in the upper layers, reaching a maximum of 22.6 °C at 15:00 h, followed by cooling, again most rapidly in the upper layers, to a minimum at 07:30 h, after the night when the surface water was over 1 °C cooler than the deeper layers. Since the density of water above 4 °C is proportional to temperature, the upper water would be expected to sink by convection. The resistance to flow caused by the plants evidently slowed this movement, illustrating the greatly reduced mixing within vegetation stands. The diel variation in temperature at four selected depths through the stand (Figure 2a), shows clearly that the surface layers were prone to much larger fluctuations than water at depth. The temperature profile in open water showed a less marked vertical gradient than that in the stand at 13:00 h.

The pH profiles also followed a pattern of increase during the day and decrease at night, with maximum change again in the surface layers (Figure 1b). The maximum was at the surface at 13:00 h. In the deeper parts, 20 cm and below, the pH continued to rise until 15:00 h, later than the layers above (Figure 2b). As light declined respiration became the predominant process with minimum pH values at midnight, except near the highly anaerobic mud at the canal bottom. The slightly increased pH at 07:30 h, especially at the surface indicated that the cycle was re-starting the following day. When compared with the open water site, the pH differences were enormous. Apart from a slight increase near the canal bottom, perhaps due to epipelic algal activity, the open water was pH 7.8–7.9 throughout its depth, compared to pH 9.15 at the surface and pH 7.9 at 70 cm depth within the weedbed (Figure 1b).

Oxygen concentration (Figure 1c) followed much the same pattern as pH, increasing to a maximum at the surface at 13:00 h, then declining. The diel increase was again greatest in the upper layers (Figures 2c), with a maximum value of 18.3 mg l^{-1}. The high respiratory activity of the mud on the canal bottom in the present study can be deduced from the very low oxygen (Figure 1c) and decreased pH (Figure 1b) found at depth. At night respiration reduced the oxygen concentration throughout the water column (Figure 1c). The oxygen profile through the open water site showed little sign of photosynthetic activity, being below 10 mg l^{-1} throughout (Figure 1c).

CO_2^* (free CO_2, ie. dissolved CO_2 and H_2CO_3) concentrations were calculated by substituting values for total alkalinity (2.449 meq l^{-1}) and conductivity (527 $\mu S\ cm^{-1}$) measured in surface water at 07:30 h, and temperature and pH at the various depths and times, into the equations of Mackereth et al. (1978). The total DIC concentration, measured as 2.549 mmol l^{-1}, was not used since it is prone to variation due to the uptake or release of CO_2^*. Alkalinity on the other hand, is much more stable in the presence of photosynthesis and respiration (Stumm & Morgan, 1980). The uptake of ions by plants can alter the alkalinity of the water, but the extent is generally negligible (Stumm & Morgan, 1980), unless marl ($CaCO_3$) precipitation occurs as a result of HCO_3^- uptake. The almost constant conductivity throughout the diurnal cycle (Figure 2e), measured at the water surface where HCO_3^- utilisation would be expected to be greatest, indicated that no such precipitation occurred. Careful inspection of plants from the surface of the stand confirmed the

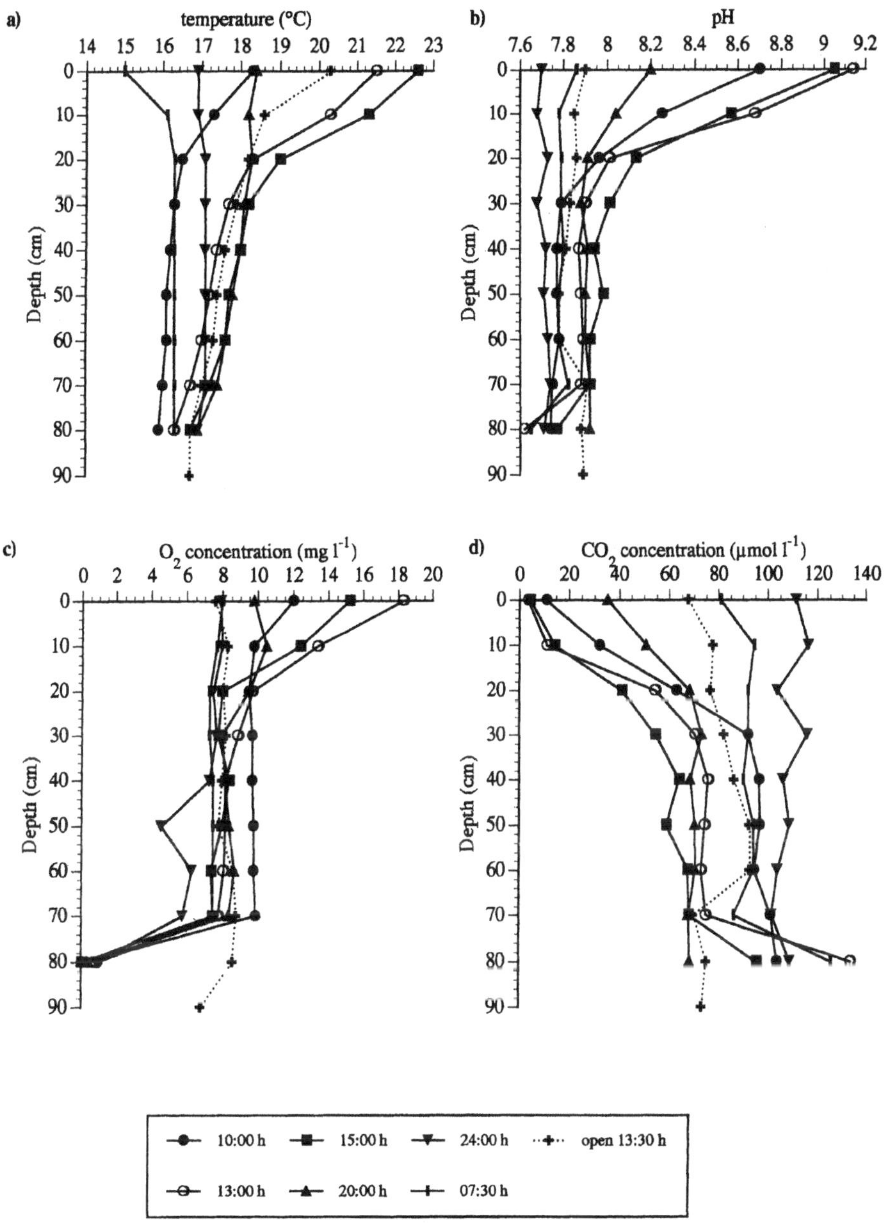

Figure 1. Vertical physico-chemical gradients measured in a stand of *Elodea nuttallii* (max. depth 80 cm) at six times over 24 hours. Also shown are vertical gradients measured in open water at 13:30 h. (a) temperature, (b) pH, (c) oxygen concentration, (d) calculated carbon dioxide concentration.

absence of marl on the leaves, so it was unlikely that the plants were utilising HCO_3^- (Jones et al., 1993).

CO_2^* showed greatest depletion in the upper layers (Figure 1d), with a minimum of 3.6 μmol l^{-1} at the surface at 13:00 h. In deeper layers CO_2^* continued to decrease until 15:00 h (Figure 2d). The CO_2^* concentration at the surface of the open water at 13:30 h was over twenty times that in the weedbed at 13:00 h and was higher at almost every other depth. There was a

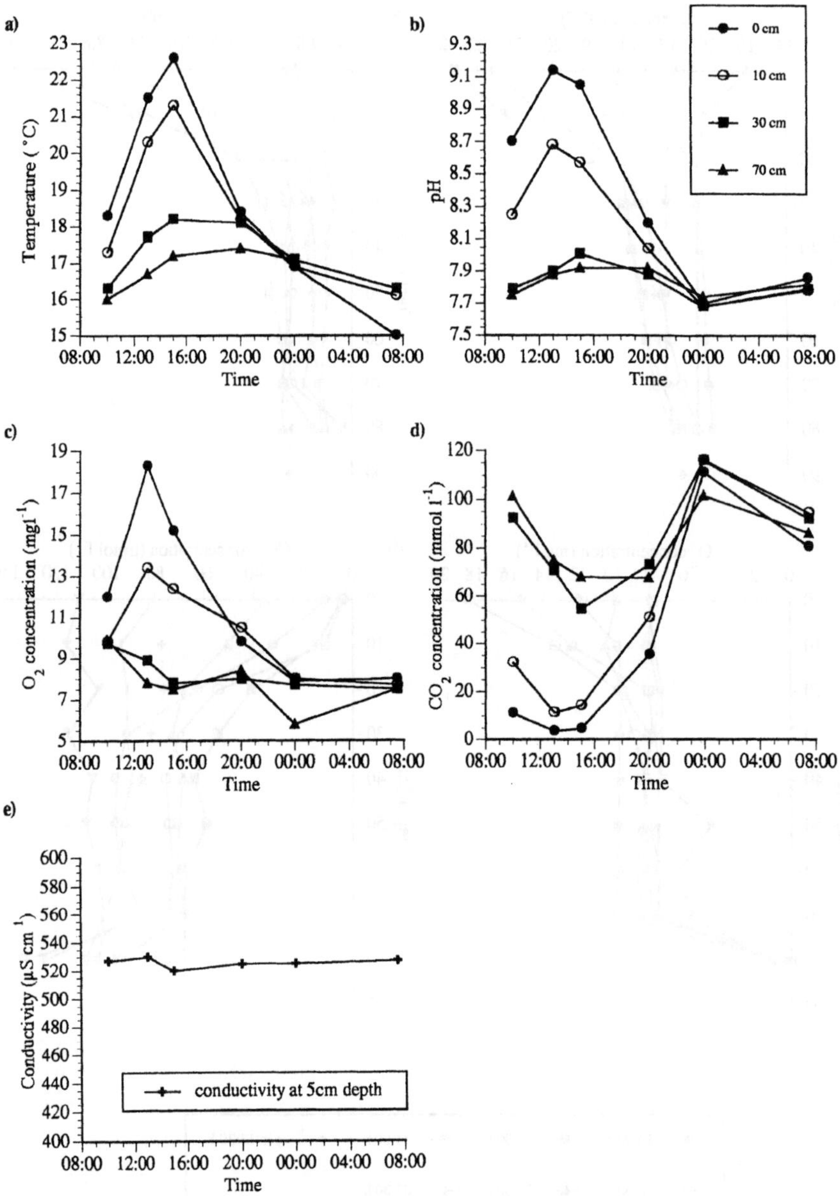

Figure 2. Diurnal changes in physico-chemical conditions measured in a stand of *Elodea nuttallii* (max. depth 80 cm) at four different depths. (a) temperature, (b) pH, (c) oxygen concentration, (d) calculated carbon dioxide concentration, (e) conductivity at 5 cm depth.

small indication of depletion by epipelic algae on the canal bed at the open site, however.

Discussion

The stand of *E. nuttallii* greatly restricted water movement which, together with the plants' photosynthetic and respiratory activity, resulted in a highly structured water column with very steep dissolved gas and pH gradients through the stand. Further evidence of reduced flow was provided by the changing temperature profile, particularly the inverted temperature gradient found at 07:30 h when the surface water was over 1 °C less than that of the deeper layers and might be expected to sink by convection. The resistance to flow caused by the plants evidently retarded such movement of water considerably.

The observed changes in pH were primarily caused by the uptake or release of the acidic gas CO_2^* during photosynthesis and respiration, and as such can be used as an indicator of biological activity. Hence most of the photosynthesis of the weedbed occurred in the morning, particularly in the upper parts (Figure 1b) and although the surface was receiving light in the afternoon, photosynthesis in the top 20 cm had essentially stopped. In the deeper parts, the pH continued to rise until 15:00 h, as a result of continued photosynthesis. With the onset of darkness, respiratory CO_2^* release showed as decreasing pH.

The low diel variation in oxygen in the deeper parts of the stand (Figures 2c) confirms that they were contributing very little to overall productivity. The oxygen concentrations in the deeper parts of this stand were not as low as has been described previously (Frodge et al., 1990), perhaps due to greater penetration of light. Frodge et al. (1990) also concluded that oxygen conditions within a stand are largely determined by plant architecture and density.

The CO_2^* concentrations showed greatest depletion in the upper layers. The 3.6 to 4.5 μmol l^{-1} range at the surface in early afternoon was of the same order as the compensation point determined for *E. nuttallii* in laboratory studies (results to be presented elsewhere) and at the lower end of the 0.7 to 25 μmol l^{-1} range reported for various submerged macrophytes (Bowes & Salvucci, 1989). Clearly plants in the upper parts of the stand were experiencing extreme carbon limitation over most of the daylight hours. The afternoon decrease in photosynthesis found in previous field studies has been attributed to increased photorespiration due to elevated oxygen concentrations (Hough, 1974), but the extremely low values of CO_2^* found in the surface layers of this carbon-rich water indicate that, although exacerbated by high oxygen, it was depletion of CO_2^* which restricted photosynthesis in the actively growing upper parts of the stand. Photosynthesis continued into the afternoon in deeper parts, where CO_2^* depletion was less as a result of slower, light-restricted photosynthesis. Greater CO_2^* supply upwards from the benthos seems unlikely to be important, in view of the great restriction on water movement already noted. It is therefore likely that light was more limiting in the deeper parts of the stand and CO_2^* in the upper parts, as suggested for *Callitriche cophocarpa* Sendt. and *Ranunculus peltatus* Schrank (Madsen & Maberly, 1991), but it is also apparent here that the upper parts of the bed are more active and contribute more to productivity. The observed daily cycle of CO_2^* depletion in the morning and replenishment at night was seen, especially in the upper layers (Figure 2d), and is consistent with previous findings that the majority of photosynthesis occurs in the morning (Goulder, 1970; Hough, 1974; Van et al., 1976).

As well as the plants themselves, other organisms are affected by the physico-chemical changes outlined above. The extremely reduced flow and mixing must affect non-motile phytoplankton by increased sinking (Schiemer & Prosser, 1976), which partly explains the reduced levels of plankton found in weedbeds which are often attributed to allelopathy (Hasler & Jones, 1949; Fitzgerald, 1969) although no allelopathic mechanism has been found. The locally high pH climate can adversely affect zooplankton such has *Daphnia longispina*, which actively avoid beds of *Chara, Potamogeton lucens* and *E. nuttallii* in the light but not in the dark (Dorgelo & Heyroop, 1985). Fish, on the other hand apparently benefit from the increased oxygen levels, which counteract the adverse effects of high pH, allowing them to remain and feed within weedbeds (Serafay & Harrell, 1993). It seems that the refuge theory of Timms & Moss (1984), which regarded all plant stands as functionally similar, requires refinement, since there are several differently functioning groups. Floating leaved plants like water-lilies reduce light, oxygen and pH beneath them (Unni, 1972; Frodge et al., 1990) and are avoided by fish (Caffrey, 1993). Elodeids reduce mixing to a greater extent and light to a lesser extent. In their upper parts they increase oxygen and pH and reduce CO_2^*. The reduced oxygen concentrations and low light under water-lilies may reduce predation of planktonic cladocera by fish (Timms & Moss, 1984), but differences between stands will occur (Venugopal & Winfield, 1983) dependant on the exact conditions which develop. Further, low resistance to flow of water-lily petioles may allow greater exchange with the open water, and enhance the effect of filtration by attached crustacea, such as *Sida* and *Simocephallus*. Perhaps it is the reduced mixing, as much as the increased grazing, which leads to reduced phytoplankton numbers in lakes dominated by submerged plants.

In conclusion, it is clear that at this time the major constraint on the photosynthesis of *E. nuttallii* growing in the Leeds & Liverpool Canal was reduced availability of CO_2^*. The stand greatly reduced water movement around its constituent plants, thereby restricting both vertical and horizontal replenishment. Although the deeper parts of the stand were less affected by CO_2^* depletion, they were probably limited by light and contributed little to overall production. This situation is likely to be typical of elodeid plants growing in moderately hard, slow-flowing or standing waters.

Acknowledgments

This work was funded by a Natural Environmental Research Council postgraduate studentship to J. I. Jones, for which we are grateful.

References

Bowes, G. & M. A. Salvucci, 1989. Plasticity in the photosynthetic carbon metabolism of submersed aquatic macrophytes. Aquat. Bot. 34: 233–266.

Brown, J. M. A., F. I. Dromgoole, M. W. Towsey & J. Browse, 1974. Photosynthesis and respiration in aquatic macrophytes. In R. L. Bieleski, A. R. Ferguson & M. M. Cresswell (eds), Mechanisms of Regulation of Plant Growth. The Royal Society of N.Z., Wellington, N.Z.: 243–249.

Buscemi, P. A., 1958. Littoral oxygen depletion produced by a cover of *Elodea canadensis*. Oikos 9: 239–245.

Caffrey, J. M., 1993. Aquatic plant management in relation to Irish recreational fisheries development. J. Aquat. Plant Mgmt 31: 162–168.

Dale, H. M. & T. J. Gillespie, 1978. Diurnal temperature gradients in shallow water produced by populations of artificial aquatic macrophytes. Can. J. Bot. 56: 1099–1106.

Dorgelo, J. & M. Heykoop, 1985. Avoidance of macrophytes by *Daphnia longispina*. Verh. int. Ver. Limnol. 22: 3366–3372.

Frodge, J. D., G. L. Thomas & G. B. Pauley, 1990. Effects of canopy formation by floating and submergent aquatic macrophytes on the water quality of two Pacific Northwest lakes. Aquat. Bot. 38: 231–248.

Goulder, R., 1969. Day-time variations in the rates of production by two natural communities of submerged freshwater macrophytes. J. Ecol. 58: 521–528.

Hasler, A. D. & E. Jones, 1949. Demonstration of the antagonistic action of large aquatic plants on algae and rotifers. Ecology 30: 359–364.

Hough, R. A., 1974. Photosynthesis and productivity in submersed aquatic vascular plants. Limnol. Oceanogr. 19: 9912–9927.

Jones, J. I., J. W. Eaton & K. Hardwick, 1993. Physiological plasticity in *Elodea nuttallii* (Planch.) St. John. J. Aquat. Plant Mgmt 31: 88–94.

Losee, R. F. & R. G. Wetzel, 1993. Littoral flow rates within and around submersed macrophye communities. Freshwat. Biol. 29: 7–17.

Mackereth, F. J. H., J. Heron & J. F. Talling, 1978. Water analysis: some revised methods for limnologists. F.B.A., Scientific Publication No. 36, Ambleside, 120 pp.

Madsen, T. V. & S. C. Maberley, 1991. Diurnal variation in light and carbon limitation of photosynthesis by two species submerged freshwater macrophyte with a differential ability to use bicarbonate. Freshwat. Biol. 26: 175–187.

O'Niel-Morin, J. O. & K. D. Kimbal, 1983. Relationship of macrophyte-mediated changes in the water column to periphyton composition and abundance. Freshwat. Biol. 13: 403–414.

Pokorny, J., J. Kvet, J. P. Ondok, Z. Toul & I. Ostry, 1984. Production-ecological analysis of a plant community dominated by *Elodea canadensis* Michx. Aquat. Bot. 19: 263–292.

Schiemer, F. & M. Prosser, 1976. Distribution and biomass of submerged macrophytes in Neusiedlersee. Aquat. Bot. 2: 289–307.

Serafy, J. E. & R. M. Harrel, 1993. Behavioural response of fishes to increasing pH and dissolved oxygen: field and laboratory observations. Freshwat. Biol. 30: 53–61.

Stumm, W. & J. J. Morgan, 1980. Aquatic chemistry an introduction emphasising chemical equilibria in natural waters. J. Wiley & Sons, New York, 780 pp.

Timms, R. M. & B. Moss, 1984. Prevention of growth of potentially dense phytoplankton populations by zooplankton grazing, in the presence of zooplanktivorous fish in a shallow wetland ecosystem. Limnol. Oceanogr. 29: 472–486.

Unni, K. S., 1972. An ecological study of the macrophyte vegetation of Doodhardari Lake, Raipur, M.P. – chemical factors. Hydrobiologia 40: 25–36.

Van, T. K., W. T. Haller & G. Bowes, 1976. Comparison of the photosynthetic characteristics of three submersed aquatic plants. Plant Physiol. 58: 761–768.

Van Der Bijl, L., K. Sand-Jensen & A. L. Hjermind, 1989. Photosynthesis and canopy structure of a submerged plant, *Potamogeton pectinatus* in a lowland stream. J. Ecol. 77: 947–962.

Venugopal, M. N. & I. J. Winfield, 1993. The distribution of juvenile fishes in a hypereutrophic pond: can macrophytes potentially offer a refuge for zooplankton? J. Freshwat. Ecol. 8: 389–396.

Westlake, D. F., 1964. Light extinction, standing crop and photosynthesis within weedbeds. Verh. int. Ver. Limnol. 15: 415–425.

Wychera, U., R. Zoufal, P. Christof-Dirry & G. A. Janauer, 1993. Structure and environmental factors in macrophyte stands. J. Aquat. Plant Mgmt 31: 118–122.

Hydrobiologia **340:** 17–22, 1996.
J. M. Caffrey, P. R. F. Barrett, K. J. Murphy & P. M. Wade (eds), Management and Ecology of Freshwater Plants.

Comparison of five media for the axenic culture of *Myriophyllum sibiricum* Komarov

R. D. Roshon[1], G. R. Stephenson[1] & R. F. Horton[2]
[1] *Department of Environmental Biology, University of Guelph, Guelph, Ontario, Canada*
[2] *Botany Department, University of Guelph, Guelph, Ontario, Canada*

Key words: aquatic plant media, *Myriophyllum sibiricum*, axenic culture, pesticide bioassay

Abstract

Myriophyllum sibiricum Komarov, an aquatic dicotyledonous macrophyte with a north temperate distribution, was assessed for use in a new bioassay to determine the effect of pesticides, agricultural runoff and municipal waste upon non-target aquatic macrophytes. An axenic culturing system was developed for which an optimal growth medium is required before a bioassay will be reliable. Five media (Murashige and Skoog, Hoagland's, Gaudet's, modified Andrew's, and Hard Water media), commonly used for aquatic plant culturing, were compared to determine the effect on *M. sibiricum* growth and development. Morphological endpoints for the assay included shoot length, total root length and number, fresh weight and plant area. Membrane integrity, chlorophyll *a*, chlorophyll *b* and carotenoid content were the physiological endpoints examined. Based upon these criteria, the modified Andrew's medium at a pH of 5.8, without the addition of a buffer was chosen as a medium which supported rapid and consistent development of *M. sibiricum* during the two week assay period.

Introduction

In order to preserve future aquatic resources, the aquatic plant community must be protected from the detrimental effects of environmental contaminants. When phytotoxic chemicals, including pesticides, are sprayed to control aquatic weeds and algal blooms or when they enter the waterway through runoff, atmospheric fallout, soil erosion, industrial effluent, sewage discharge, spills (McEwen & Stephenson, 1979) or drift from aerial or ground applications (Akesson & Yates, 1964), non-target plants can be adversely affected.

Presently, there are standardized laboratory tests for algae (*Selenastrum capricornutum*) (Environment Canada, 1991; U.S. EPA, 1971) and floating aquatic macrophytes (*Lemna gibba* G3) (ASTM, 1991; Federal Register, 1985) but no structured bioassay exists for non-target rooted aquatic macrophytes. The development of a test for a rooted aquatic macrophyte, like *Myriophyllum sibiricum* Komarov, could be a beneficial and essential addition to the existing bioassays since no single test method is likely to provide a comprehensive approach to environmental protection. Any new test should be quick, simple, inexpensive and reproducible (Swanson & Peterson, 1988). Since it is difficult for a bioassay with whole, rooted aquatic macrophytes to meet these criteria, an axenic test-tube culturing technique was developed. Before the results of pesticide testing with a bioassay are reliable, the laboratory culturing system and organism growth conditions must be established.

There are numerous media used for the laboratory culture of aquatic macrophytes. Five commonly used liquid culture media (Murashige and Skoog (M & S), Hoagland's, Gaudet's, modified Andrew's, and Hard Water (HW) media) were compared. M & S medium (Murashige & Skoog, 1962) was originally designed for tobacco tissue culturing but Kane & Gilman (1991) have used it successfully for the solution culture of other *Myriophyllum* species. Hoagland's medium (Hoagland & Arnon, 1938) has been used extensively for both terrestrial and aquatic plant culturing in its original (Hinman & Klaine, 1992) and modified forms (ASTM, 1991). Gaudet's medium was developed to examine the mechanism which forced an aquatic macrophyte to switch between the aquatic and land forms (Gaudet, 1963) but has subsequently been utilized to study aquatic plant nutrition (Bristow & Whitcombe, 1971). For the culture of *M. spicatum*, Gerloff's (Gerloff & Krombholz, 1966) medium was modified by Andrew (Selim et al., 1989). A bicarbonate-containing medium was developed specifically for aquatic plants normally found growing in hard water lakes (Smith, 1993) and has been used to study plant/pathogen interactions. The majority of studies on aquatic plant culturing have compared few media and have examined a limited number of endpoints. This study was undertaken to find a liquid medium which produced rapid and consistent development in axenically cultured *M. sibiricum*. Five different growth media and nine different morphological and physiological endpoints were evaluated.

Materials and methods

Myriophyllum sibiricum Komarov is an ecologically important submerged aquatic dicotyledon with a north temperate distribution which is readily cultured in test tubes in the laboratory. Lower autumn temperatures initiate the formation of turions on lateral branches which develop into new plants when environmental conditions become favourable (Gleason & Cronquist, 1991). For these studies, turions were collected in the fall from lakes in southern Ontario. Using aseptic technique under a Laminar Airflow Cabinet, the plants were disinfected in a 3% (w/v) sodium hypochlorite solution containing 0.01% Tween–20 for 20 minutes. Once an axenic culture of *M. sibiricum* was established, stems were cut into segments, transferred into sterile culture tubes (150 × 25 mm) containing 45 mls of half strength Murashige and Skoog Basal Salts with minimal organics (Sigma Chemical Company, St. Louis, MI, U.S.A.) and 30% sucrose (Kane & Gilman, 1991). Medium from randomly selected tubes was plated onto Typticase Soy Agar, allowed to incubate for a minimum of ten days and monitored for culture sterility. After ten days, one 3 cm long axillary bud, from each clonal stock plant was transferred into culture tubes containing 40 ml of sterile test medium and 3 g of sterile Turface®. A 15 cm length of Westergren Blood Sedimentation Tube, marked in mm, was placed in the centre of the test tube and held in place by a 3 cm piece of Tygon tubing glued into the cap. Each treatment was replicated five times. The results from two experiments are reported here. The first was a comparison of the five full strength media, for which the chemical composition can be found in Table 1. All stock solutions and final media were made with nanopure water (Barnstead Nanopure II) and each medium was supplemented with 30% sucrose. Preliminary experiments demonstrated that this species was unable to grow substantially in culture without the addition of an external carbon source, especially sucrose. The second experiment was a repeat of two types of media (modified Andrew's and Hoagland's). Racks of randomized plants were maintained in a growth cabinet set at 25 ± 2°C, a 16 hour day/8 hour night and a light intensity of 100 ± 5 μE m^{-2} s^{-1}.

Table 1. Elemental breakdown of the media tested in the *Myriophyllum* bioassay.

	Final concentrations of elements in media (mmol/L)				
Element	M+S with organics	Hoagland's	Gaudet's	Modified Andrew's	Hard water
C	1061	1057	1057	1057	5457
N	60.2	16.0	3.4	2.6	2.4
K	16.0	6.0	1.7	1.0	1.0
P	2.5	1.0	0.8	0.2	0.2
Ca	3.4	1.7	1.3	0.8	1.2
S	25.0	20.0	5.0	5.0	0.4
Mg	9.1	6.0	3.0	3.0	1.0
Cl	0.30	0.01	0.03	0.01	1.2
B	0.10	0.025	0.0243	0.0025	0.0025
Cu	1e–04	5e–04	1.6e–04	5e–05	5e–05
Co	9e–05	0	0	0	0
Fe	0.10	0.02	0.01	0.01	0.01
Mn	0.0999	0.002	0.0201	0.001	0.001
Mo	0.001	0.00055	3.3e–05	2e–05	2e–05
Zn	0.0299	0.002	0.0104	0.0004	0.0004

Growth curves were established by visually measuring the plant length every second day. The area under the growth curve was determined by Eq. (1), where IH is the increase in height from the start of the experiment and T is the time at each subsequent measurement point, in hours from time zero (Boutin *et al.*, 1993).

$$\text{Area} = \sum_{i=2}^{n} \frac{IH_{i-1} + IH_i}{2} \cdot (T_i + T_{i-1}); n = 8 \quad (1)$$

At the end of two weeks, shoot and total root length, plant area, chlorophyll and carotenoid content,

Table 2. Myriophyllum sibiricum growth in five media used in aquatic plant culture (mean ± s.d.). Any two means in the same column with the same superscript are not significantly different at $p = 0.05$ ($n = 5$, except for HW where $n = 4$).

	Area under the growth curve	Increase in shoot length (mm)	Root number	Total root length (mm)	Total fresh weight (mg)	Chl *a* content (mg/g fresh weight)	Chl *b* content (mg/g fresh weight)	Carotenoid content (mg/g fresh weight)	Membrane permeability (%)	Plant area (cm^2)
M & S	10290.4^{ab}	81.6^{a}	3.8^{a}	130.2^{a}	439.2^{a}	0.60^{a}	0.22^{a}	0.20^{a}	35.6^{a}	10.6^{a}
	±1568.9	±9.8	±1.6	±54.7	±116.4	±0.04	±0.02	±0.017	±9.5	±4.5
Hoagland's	12091.9^{a}	89.1^{a}	9.2^{b}	402.0^{b}	741.6^{c}	0.59^{ab}	0.21^{ab}	0.19^{ab}	12.4^{b}	18.3^{b}
	±724.5	±3.6	±0.8	±58.7	±72.6	±0.04	±0.02	±0.011	±0.5	±1.8
Gaudet's	8394.6^{c}	61.4^{c}	7.8^{b}	482.9^{b}	466.8^{ab}	0.52^{bc}	0.20^{ab}	0.18^{b}	9.4^{bc}	10.4^{a}
	±639.2	±6.6	±1.3	±138.5	±63.9	±0.03	±0.02	±0.008	±0.8	±2.0
Modified Andrew's	9781.8^{bc}	68.1^{bc}	9.6^{b}	420.5^{b}	560.2^{bc}	0.51^{c}	0.18^{b}	0.17^{b}	9.4^{bc}	13.1^{ab}
	±782.2	±3.9	±0.9	±37.1	±30.3	±0.01	±0.01	±0.003	±1.6	±1.2
Hard Water	9162.9^{bc}	78.5^{ab}	9.0^{b}	566.7^{b}	485.1^{ab}	0.50^{c}	0.17^{b}	0.17^{b}	7.5^{c}	10.3^{a}
	±631.4	±5.1	±0.8	±45.1	±49.7	±0.03	±0.02	±0.008	±0.5	±0.3

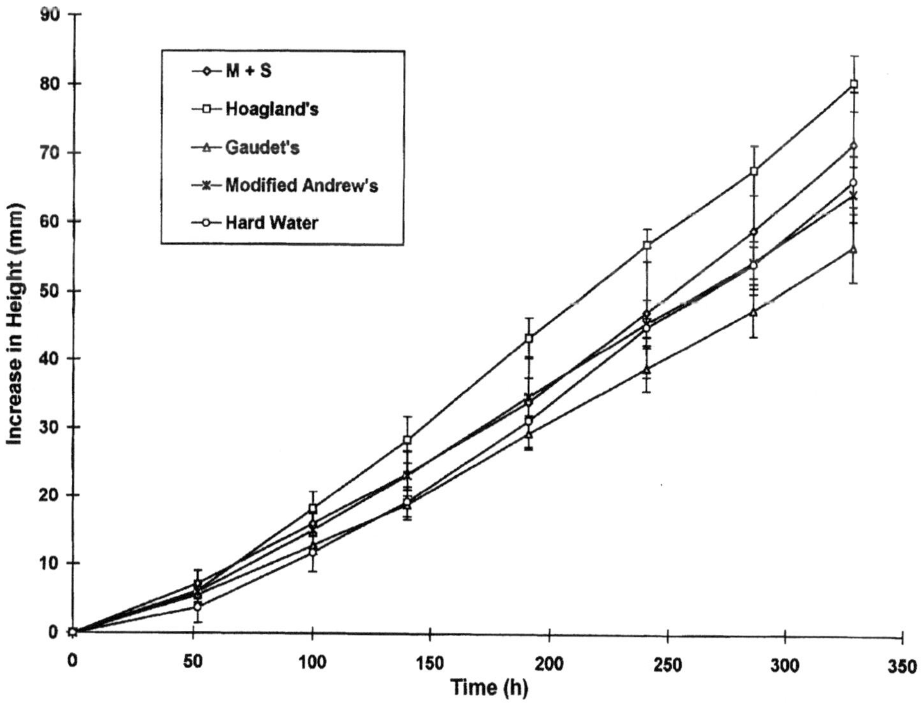

Figure 1. M. sibiricum growth in Murashige & Skoog, Hoagland's, Gaudet's, modified Andrew's and Hard Water media (mean ± s.d.) (n = 5, except for HW medium where n = 4).

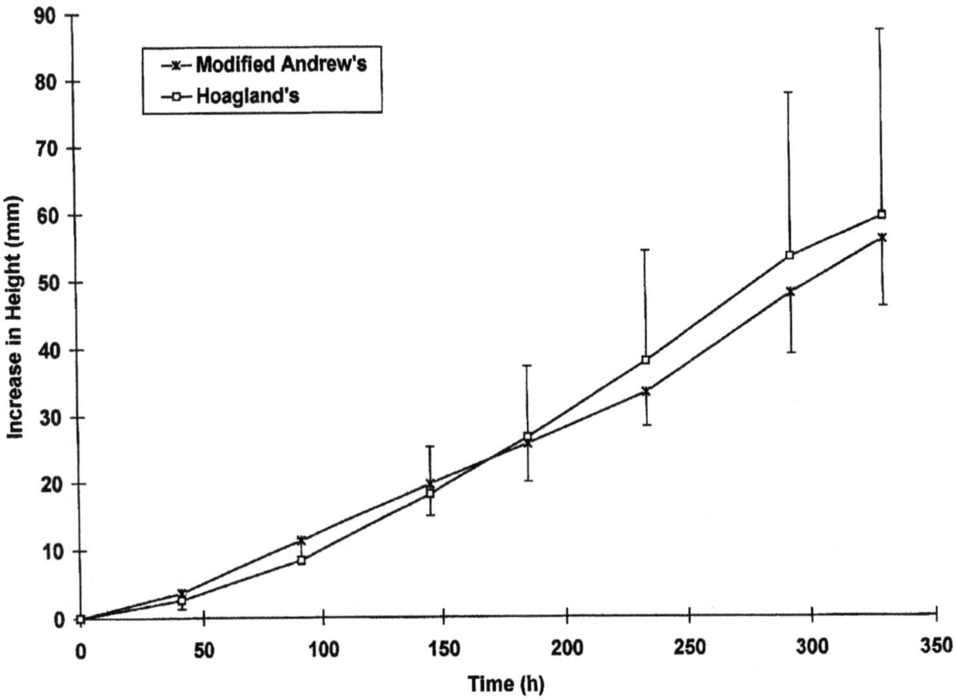

Figure 2. Growth curve for the repeated experiment of the comparison between Hoagland's and modified Andrew's media (mean ± s.d.).

fresh weight and membrane integrity were determined. These morphological and physiological parameters were selected because they should cover most of the possible effects caused by toxicants during pesticide testing. Dry weight on the remaining stem segments was also determined but is not presented here. Pigments (chlorophyll *a*, chlorophyll *b* and carotenoids) were extracted from each 50 mg apical segment into 10 ml of 80% ethanol (Lichtenthaler & Wellburn, 1983) and measured on a spectrophotometer (Beckman Du® –65) at 470, 647 and 663 nm. Modifying the method of Beckerson & Hofstra (1980), membrane integrity was determined by placing a 100 mg stem segment into 20 mL of nanopure water for 24 hours. The conductance of the solution was measured using a portable conductivity meter (Corning® Checkmate 90 Field System), the tubes were placed into boiling water for 20 min and allowed to cool. The conductance of the solution was measured again to determine the conductance after complete membrane disruption. Membrane integrity was determined as percentage of total electrolyte leakage:

$$\text{\% Membrane Leakage} = \frac{\text{Conductance before boiling}}{\text{Conductance after boiling}} \cdot 100 \qquad (2)$$

For the experiment comparing the five media, the data were tested for normality, log transformed (except for root number which was square root transformed) and analyzed using a single classification ANOVA. The Tukey-Kramer test was used to determine which means differed. The results from the second experiment were compared using the non-parametric Mann-Whitney U-test (Sokal & Rohlf, 1981).

Results and discussion

The medium selection process was an intensive study of the five media but only selected results are shown here.

Comparison of all five media

The increase in shoot length over the two week experiment for *M. sibiricum* grown in the five media is shown in Figure 1. The media separated into three groups; the Gaudet's producing the smallest increase in height, plants grown in Hoagland's and M&S produced the greatest increase in shoot height, and the remaining two media were intermediate in shoot height (Table 2). The area under the growth curve compares not only the final height but also the rate at which the plants grew (Boutin et al., 1993). From Figure 1 and Table 2, it is evident that the media divided into the same three categories. It is noteworthy that even though plants grown in HW media had a final height slightly greater than in modified Andrew's, they had a smaller area under the growth curve (not statistically significant at $p = 0.05$). This may be due to their slower initial growth rate.

Roots are important to aquatic plant growth because they absorb ions from the sediment (Bristow & Whitcombe, 1971; Mantai & Newton, 1982). The M & S medium (Table 2) produced the lowest number of roots and the shortest total root length while there was no statistically significant difference between the roots of plants grown in the other media ($p = 0.05$).

The pigment content of apices grown in M & S was significantly higher than for apicies grown in the other media. Chlorophyll *b* and carotenoid content of apices grown in the other four media were not significantly different.

Changes in membrane permeability are indicated by cellular leakage. The high membrane permeability value obtained with the M & S medium possibly indicates that this medium disrupts cell membrane function. The lower percentage leakage from plants grown in the other four media indicate normal membrane function (Beckerson & Hofstra, 1980; Dijak & Ormrod, 1982).

Hoagland's and modified Andrew's media produced plants with the greatest area and weight. The modified Andrew's medium was the only medium which induced the formation of branches.

Based upon this experiment, Gaudet's medium was eliminated from further consideration due to its small increase in shoot length. Due to the small plant area and the low fresh weight, the M & S medium was dropped from consideration. HW medium was not considered an acceptable medium because of its slow initial growth rate (Figure 1). Another problem observed with the HW medium, designed to examine plant-microbe interactions (Smith, 1993), was accidental colonization by bacteria and fungi. In this experiment one sample was lost due to bacterial contamination.

There was no significant difference between the Hoagland's and modified Andrew's medium for root number and length, total fresh weight, chlorophyll *b* and carotenoid content, membrane integrity and plant area.

Comparison of two media

Based on the above study it was decided to compare the Hoagland's medium with the modified Andrew's medium. Figure 2 displays the growth curve for this experiment. For plants grown in these two media, the two parameters which differed statistically ($p = 0.05$) were total root length and area under the growth curve. Because a larger standard deviation was obtained with the Hoagland's medium, the modified Andrew's medium was adopted for subsequent pesticide experiments.

Conclusions

From the existing media cited in the literature, there does not appear to be a 'perfect' medium for the axenic culture of *M. sibiricum*. However, modified Andrew's medium at a pH of 5.8 and without the addition of a buffer was selected for further use in pesticide experiments. Plants with large total root length, root number, total fresh weight and plant area could be produced consistently and variations in growth parameters between replicates were acceptably low.

Acknowledgements

This research was partially supported by a National Science and Engineering Research Council Postgraduate Scholarship and funding from the Canadian Network of Toxicology Centres. The laboratory assistance of K. Anderka, I. Gardner, L. King, F. Shafi and J. Glaser is gratefully acknowledged.

References

Akesson, N. B. & W. E. Yates, 1964. Problems relating to application of agricultural chemicals and resulting drift residues. In R. F. Smith & T. E. Mittler (eds), Ann. Rev. Ent. Ann. Reviews, Inc., California: 9: 285–318.

ASTM, 1991. Standard guide for conducting static toxicity tests with *Lemna gibba* G3. Annual Book of ASTM Standards. American Society for Testing and Materials. 11.04. E 1415–91.

Beckerson, D. W. & G. Hofstra, 1980. Effects of sulphur dioxide and ozone, singly or in combination, on membrane permeability. Can. J. Bot. 58: 451–457.

Boutin, C., K. E. Freemark & C. J. Keddy, 1993. Proposed Guidelines for Registration of Chemical Pesticides: Non target Plant Testing and Evaluation. Technical Report Series No. 145. Environment Canada, Ottawa, Ontario: 92 pp.

Bristow, J. M. & M. Whitcombe, 1971. The role of roots in the nutrition of aquatic vascular plants. Am. J. Bot. 58: 8–13.

Dijak, M. & D. P. Ormrod, 1982. Some physiological and anatomical characteristics associated with differential ozone sensitivity among pea cultivars. Envir. Exp. Bot. 22: 395- 402.

Environment Canada, 1991. Biological Test Method: Growth Inhibition Test Using the Freshwater Alga *Selenastrum capricornutum*. Environmental Protection. Conservation and Protection. Environment Canada. Report EPS. 42 pp.

Federal Register, 1985. *Lemna* acute toxicity test. Federal Register Rules and Regulations. 50: 39331–39334.

Gaudet, J. J., 1963. *Marsilea vestita*: Conversion of the water form to the land form by darkness and by far-red light. Science. 140: 975–976.

Gerloff, G. C. & P. H. Krombholz, 1966. Tissue analysis as a measure of nutrient availability for the growth of angiosperm aquatic plants. Limnol. Oceanogr. 11: 529–537.

Gleason, H. A. & A. Cronquist, 1991. Manual of Vascular Plants of Northeastern United States and Adjacent Canada. Second Ed. New York Botanical Garden, New York, 910 pp.

Hinman, M. L. & S. J. Klaine, 1992. Uptake and translocation of selected organic pesticides by the rooted aquatic plant *Hydrilla verticillata* Royle. Envir. Sci. Technol. 26: 609–613.

Hoagland, D. R. & D. I. Arnon, 1938. The water-culture method for growing plants without soil. Agricultural Experiment Station Circular 347. Berkeley, California: 1–35.

Kane, M. E. & E. F. Gilman, 1991. *In vitro* propagation and bioassay systems for evaluating growth regulator effects on *Myriophyllum* species. J. Aquat. Plant Mnmt. 29: 29–32.

Lichtenthaler, H. K. & A. R. Wellburn, 1983. Determinations of total carotenoids and chlorophylls a and b of leaf extracts in different solvents. Biochem. Soc. Trans. 11: 591–592.

Mantai, K. E. & M. E. Newton, 1982. Root growth in *Myriophyllum*: A specific plant response to nutrient availability? Aquat. Bot. 13: 45–55.

McEwen, F. L. & G. R. Stephenson, 1979. The Use and Significance of Pesticides in the Environment. John Wiley & Sons, New York, 538 pp.

Murashige, T. & F. Skoog, 1962. A revised medium for rapid growth and bio assays with tobacco tissue cultures. Physiol. Pl. 15: 473–497.

Selim, S. A., S. W. O'Neal, M. A. Ross & C. A. Lembi, 1989. Bioassay of photosynthetic inhibitors in water and aqueous soil extracts with Eurasian watermilfoil (*Myriophyllum spicatum*). Weed Sci. 37: 810–814.

Smith, C. S., 1993. A bicarbonate-containing medium for the solution culture of submersed plants. Can. J. Bot. 71: 1584–1588.

Sokal, R. R. & F. J. Rohlf, 1981. Biometry: The Principles and Practices of Statistics in Biological Research. W.H. Freeman & Company, New York, 859 pp.

Swanson, S. & H. Peterson, 1988. Development of Guidelines for Testing Pesticide Toxicity to Non-Target Plants. SRC Publication No. E–901–20-E–88. Environment Canada. 148 pp.

U.S. EPA, 1971. Algal Assay Procedure, Bottle Test. National Eutrophication Program. Environmental Protection Agency, Corvallis, Oregon. 82 pp.

Hydrobiologia **340**: 23–26, 1996.
J. M. Caffrey, P. R. F. Barrett, K. J. Murphy & P. M. Wade (eds), Management and Ecology of Freshwater Plants.

The effects of floating mats of *Azolla filiculoides* Lam. and *Lemna minuta* Kunth on the growth of submerged macrophytes

Rachel A. Janes, John W. Eaton & Keith Hardwick
Department of Environmental and Evolutionary Biology, University of Liverpool, P.O. Box 147, Liverpool L69 3BX, United Kingdom

Key words: Elodea nuttallii, Potamogeton crispus, interference, shading

Abstract

In laboratory experiments, the growth characteristics of the submerged species *Elodea nuttallii* (Planch.) St. John and *Potamogeton crispus* L. were assessed in the presence and absence of floating mats of *Azolla filiculoides* Lam. and *Lemna minuta* Kunth. Light penetration and the development of pH and dissolved oxygen differences were monitored. The growth of *P. crispus* was suppressed much more than that of *E. nuttallii* and the effects of *A. filiculoides* were more severe than those of *L. minuta*. Findings are related to possible field responses of submerged plants under floating mats, especially their abilities to compensate for the potential suppressive effects of floating mats under natural conditions.

Introduction

Azolla filiculoides Lam., Water Fern, and *Lemna minuta* Kunth, Least Duckweed, are both alien species in Britain. The former was introduced in the late 19th century (Sculthorpe, 1985) and the latter by 1977 (Leslie & Walters, 1983). Both have spread rapidly, causing weed problems in recent years. Associated fish kills and decreased diversity of invertebrate and submerged plant species are indicated from a preliminary analysis of water industry records. If important conservation habitats were to be invaded by these two species, serious losses of biodiversity could ensue.

We report here an experimental study of effects of *A. filiculoides* and *L. minuta* mats on the growth of two common submerged plants, namely *Elodea nuttallii* (Planch.) St. John, an introduced species which our field studies showed to often persist under floating mats, and *Potamogeton crispus* L., a native species which did so less often. The aims were to determine whether *P. crispus* was suppressed more than *E. nuttallii* and whether *A. filiculoides* and *L. minuta* differed in their suppressive influences.

Methods

Three clean, healthy, 10 cm apical shoots (without side shoots) of *E. nuttallii*, collected from the Leeds and Liverpool Canal (Map reference OS SJ 371 990) on 4/4/93 were planted into a small pot filled with canal sediment. Four such pots were placed into each of nine 10 l plastic containers. These contained filtered canal water to a depth of approximately 300 mm.

A surface covering of 1 kg m^{-2} fresh biomass of *L. minuta* (cultured from ponds near Chorley SD 568 218) was added to three of the containers and 1 kg m^{-2} fresh biomass of *A. filiculoides* (cultured from ponds at Worcester SO 838595) was added to another three containers. The remaining three containers were controls without floating mats. The containers were placed in a latin square design in a growth room at 16 ± 1 °C and illuminated at 140 μmol m^{-2} s^{-1} surface photosynthetically active radiation (PAR) under a 12h:12h; Light:Dark cycle for 14 days.

Starting dry biomass was estimated by weighing further 10 cm lengths of *E. nuttallii* ($n = 24$) after oven drying at 60 °C for 48 hours.

At the end of the experiment and before the floating mats were disturbed, light and pH were measured using

Table 1. Responses of *E. nuttallii* and *P. crispus* to the presence of floating mats of *A. filiculoides* and *L. minuta*. Results of ANOVA and Tukey's analysis are shown (treatment means linked by a line are not significantly different). Values shown are means (n=3) with standard errors in parentheses.

	Elodea nuttallii					*Potamogeton crispus*				
	+Azolla	*+Lemna*	Control	ANOVA	Tukeys	*+Azolla*	*+Lemna*	Control	ANOVA	Tukeys
Total length (mm)	231	270	262	p=0.6277	Not sig.	136	145	189	p=0.0168	*Az Lm* C
	(28.00)	(15.54)	(30.21)			(11.49)	(8.24)	(9.05)		
Height (mm)	167	208	153	p=0.0709	Not sig.	132	142	169	p=0.0479	*Az Lm* C
	(13.29)	(18.39)	(7.66)			(10.83)	(8.33)	(5.45)		
Internode length (mm)	5.92	4.95	3.45	p=0.0230	*Az Lm* C	11.53	10.56	10.22	p=0.4117	Not sig.
	(0.52)	(0.25)	(0.54)			(1.00)	(0.57)	(0.54)		
Number of side shoots	1.00	1.12	2.81	p=0.0031	*Az Lm* C	0.29	0.14	1.10	p=0.0116	*Lm Az* C
	(0.03)	(0.16)	(0.21)			(0.08)	(0.08)	(0.25)		
Dry biomass (g)	0.015	0.024	0.041	p=0.0006	*Az Lm* C	0.020	0.025	0.080	p<0.0001	*Az Lm* C
	(0.003)	(0.003)	(0.002)			(0.002)	(0.002)	(0.005)		
	(Estimated starting dry biomass = 0.019 g)					(Estimated starting dry biomass = 0.046 g)				
Chlorophyll (μg per whorl or leaf)	0.009	0.008	0.006	$p = 0.1581$	Not sig.	0.034	0.038	0.026	$p = 0.4176$	Not sig.
	(0.001)	(0.001)	(0.0003)			(0.007)	(0.004)	(0.006)		

Table 2. Physico-chemical conditions at 150mm depth at the end of the experiments. Results of ANOVA and Tukey's analysis are shown (treatment means linked by a line are not significantly different). Values shown are means (n = 3). The pH and equilibrium pH of the canal water used were 8.2 at the start of the experiment.

	Elodea nuttallii					*Potamogeton crispus*				
	+Azolla	*+Lemna*	Control	ANOVA	Tukeys	*+Azolla*	*+Lemna*	Control	ANOVA	Tukeys
pH	7.51	8.17	8.93	$p = 0.0389$	C *Lm Az*	7.25	8.57	8.12	$p = 0.0001$	*Lm* C *Az*
Equilibrium. pH	8.18	7.99	7.87	$p = 0.0001$	*Az Lm* C	8.16	8.05	7.91	$p = 0.0002$	*Az Lm* C
Oxygen (g m^{-3})	5.9	7.6	12.6	$p = 0.0006$	*Az Lm* C	6.8	10.1	11.3	$p = 0.0338$	*Az Lm* C
Light (% Surface PAR)	7.3	9.3	68.7	$p = 0.0001$	*Az Lm* C	7.0	14.0	66.7	$p = 0.0001$	*Az Lm* C

submersible probes and water samples were extracted using a syringe for estimation of equilibrium pH (i.e. after vigorously sparging with air for 30 minutes) and dissolved oxygen concentration (Winkler method, Mackereth et al., 1978), all from 150 mm depth.

One *E. nuttallii* plant from each container was used for chlorophyll determination. Total chlorophyll in the fifth discernable whorl of leaves was measured, by the method of Arnon (1949). The following were measured for each of the remaining plants; total length (including side shoots), longest shoot length (i.e. height), number of side shoots, the lengths of the first ten internodes, counted down the shoot from the uppermost level at which they were distinguishable and dry biomass.

The experiment was repeated using *P. crispus*, collected from the Trent and Mersey Canal SJ 707 656 on 12/5/93. Only two 10 cm apical shoots per pot were used, as the plants were larger. At harvest, the same data as for *E. nuttallii* were collected, but total chlorophyll was measured in the fifth leaf from the apex in one plant per container and 5 internodes were measured since most plants did not achieve 10.

Data were analysed by Analysis of Variance (ANOVA). If significant effects were detected, Tukey's multiple comparison test was used, following Zar (1984). Data described as percentage were arcsine transformed before ANOVA was carried out and pH values were analysed as absolute H^+ ion concentrations.

Results

Neither floating plant species increased significantly in biomass during the course of the experiments.

At the end of the experiments, *E. nuttallii* plants under mats differed from those in controls (Table 1), typically being taller, due to longer internodes and with fewer branches. However, some plants were very short because of poor growth or death and inclusion of these in the mean lengths and heights makes interpretation of these data difficult. *P. crispus* plants under the floating mats were no taller and much less branched than the controls. Biomass gain per plant for both submerged species was much less under floating mats than in the controls. However, biomass reduction was much more severe under mats for *P. crispus* than for *E. nuttallii*. Total chlorophyll was not significantly different between treatments for either submerged species.

Water quality results in Table 2 showed marked rises in end pH in the controls and decreases in the *+Azolla* treatment with *E. nuttallii*. Marl deposition occurred on the surfaces of *E. nuttallii* leaves in the controls only. In the *P. crispus* experiment, the *+Lemna* treatment gave the highest end pH; again the *+Azolla* treatment was lowest. The equilibrium pH fell in the controls and to a lesser extent in the *+Lemna* treatment during the course of both experiments. Final dissolved oxygen concentrations were significantly lower under mats than in the controls. Similarly, light levels were substantially less under mats as compared to controls.

Discussion

Mats of free-floating plants are well known to cause physico-chemical changes in the water beneath them similar to those reported here (e.g. Pokorny & Rejmankova, 1983). Due to their position at the water/air interface, they interfere with light penetration and gaseous exchange, causing the predominance of respiratory activity beneath the mats and the observed reduction in dissolved oxygen, increase in carbon dioxide and reduction in pH. Equilibrium pH was probably reduced in the controls by submerged plants taking up bicarbonate, precipitating calcium carbonate (seen as marl) and hence changing the ionic concentration in the water.

From the same starting mat densities, *A. filiculoides* was associated with greater change in submerged plant growth than *L. minuta*, perhaps because of the former's greater effects upon underwater conditions.

Both *E. nuttallii* and *P. crispus* showed reduced branching and biomass production beneath mats, the reduction being greatest in *P. crispus*. *E. nuttallii* elongated markedly, presumably gaining increased light, but *P. crispus* showed no such response. Various workers have reported similar responses to shading in submerged species (Barko & Smart, 1981; Maberley, 1993; Ozimek et al., 1991; Tobiessen & Snow, 1984).

The physico-chemical changes under mats are not all deleterious to submerged plant growth. Increased carbon dioxide may benefit photosynthesis and decreased oxygen concentrations reduce photorespiration (Simpson et al., 1980). Different species may have different capacities and strategies to survive conditions below floating mats. Fast-growing, phenotypically plastic elodeid species may respond to shading by elongation, so that they colonise the subsurface layer where the greatest amount of the remaining light is available. Species with large food reserves may endure periods of cover by floating mats with little morphological response. Thus, short-term survival may be possible but, since the number of perennating propagules formed by submerged plants is determined by their growth performance (Maberley, 1993), long-term survival may be jeopardised. Thus thick floating mats of vegetation could reduce submerged plant diversity by selecting for a few tolerant species.

Acknowledgments

This work was funded by a collaborative award from the Science and Engineering Research Council and British Waterways to R. A. Janes, for which we are grateful.

References

Arnon, D. L., 1949. Copper enzymes in isolated chlorolasts, Polyphenol oxidase in *Beta vulgaris*. Plant Physiol. 24: 1–15.

Barko, J. W. & R. M. Smart, 1981. Comparative influences of light and temperature on the growth and metabolism of selected submersed freshwater macrophytes. Ecol. Monogr. 51: 219–235.

Leslie, A. C. & S. M. Walters, 1983. The occurrence of *Lemna minuscula* Herter in the British Isles. Watsonia 14: 243–248.

Maberley, S. C., 1993. Morphological and photosynthetic characteristics of *Potamogeton obtusifolius* from different depths. J. Aquat. Plant Mgmt 31: 34–39.

Mackereth, F. J. H., J. Heron & J. F. Talling, 1978. Water analysis: some revised methods for limnologists. Freshwater Biological Association, Ambleside, 120 pp.

Ozimek, T., E. Pieczynska & A. Hankiewicz, 1991. Effects of filamentous algae on submerged macrophyte growth: a laboratory experiment. Aquat. Bot. 41: 309–315.

Pokorny, J. & E. Rejmankova, 1983. Oxygen regime in a fishpond with duckweeds (Lemnaceae) and *Ceratophyllum*. Aquat. Bot. 17: 125–137.

Sculthorpe, C. D., 1985. The biology of aquatic vascular plants. Second Edition. Koeltz Scientific Books. Kšnigstein, Germany, 610 pp.

Simpson, P. S., J. W. Eaton & K. Hardwick, 1980. The influence of environmental factors on apparent photosynthesis and respiration of the submersed macrophyte *Elodea canadensis*. Plant Cell Envir. 3: 415–423.

Tobiessen, P. & P. D. Snow, 1984. Temperature and light effects on the growth of *Potamogeton crispus* in Collins Lake NYS. Can. J. Bot. 62: 2822–2826.

Zar, J. H., 1984. Biostatistical analyses. Second edition. Prentice Hall International, Inc. Englewood Cliffs, New Jersey, 718 pp.

Hydrobiologia **340:** 27–30, 1996.
J. M. Caffrey, P. R. F. Barrett, K. J. Murphy & P. M. Wade (eds), Management and Ecology of Freshwater Plants.

The biology of *Butomus umbellatus* in shallow waters with fluctuating water level

Zdenka Hroudová, Anna Krahulcová, Petr Zákravský & Vlasta Jarolímová
Institute of Botany, Czech Academy of Sciences, Průhonice CZ-252 43, Czech Republic

Key words: wetland plants, spreading ability, polyploidy

Abstract

Butomus umbellatus L. is a plant species typical of littoral communities of river and stream shores. It can form continuous stands in shallow reservoirs with fluctuating water level. Their expansion is promoted by: (a) intensive vegetative reproduction of plants, (b) crowded sprouting from rhizome fragments on emerged pond bottom, (c) shallow water layer in the year following summer drainage. Expansion of *B. umbellatus* depends on ploidy level: two cytotypes were found in the Czech and Slovak Republics, differing in their reproductive ability. Seed production of triploids is strongly limited (they are self-incompatible within clones), while diploids can be fully fertile. Nevertheless, even in diploids, the efficiency of seed reproduction under natural conditions is low. Triploids spread by intensive vegetative reproduction, which is decisive for clonal growth of populations and their regeneration after scraping of bottom surface. During seasonal development, maximum of aboveground biomass is produced in early summer, while underground biomass increases till autumn. Growth of the plants is limited by cutting before maximum underground biomass is attained, or by duck grazing.

Introduction

Butomus umbellatus occurs in shallow standing waters (fishponds or other reservoirs, temporarily flooded field depressions), in small streams or on river banks, mostly on sites with fluctuating water level. These fluctuations promote the development of *B. umbellatus* populations (Hejný, 1960, Hroudová, 1980). In small fishponds, where fluctuations of water level are caused by fishpond management, *B. umbellatus* can form continuous stands and become an undesirable weed.

Our work was aimed at finding what biological properties of the plant and what habitat conditions lead to expansion of *B. umbellatus* populations.

By studying its ecology and biology, we found the spreading ability of *B. umbellatus* to be dependent on ploidy level. Within *B. umbellatus*, two cytotypes were found in the Czech and Slovak Republics: diploids ($2n = 26$) and triploids ($2n = 39$). Triploids occur more frequently (82 localities) than diploids (17 localities) (Krahulcová & Jarolímová, 1993, Hroudová & Zákravský, 1993b). Thus, we tried to find different properties of diploids and triploids conditioning their expansibility and causing the difference in their frequency of occurrences. Besides, we studied habitat conditions affecting both cytotypes of *B. umbellatus* as regards soil chemistry and water level dynamics. The project was supported by Grant Agency of CZAS (grant no. 60543) and Grant Agency of the Czech Republic (grant no. 204/93/1178).

Methods

Plants of *B. umbellatus* were sampled from the localities over the whole Czech and Slovak Republics together with soil samples of their habitats. Chemical soil analyses were performed (see Hroudová & Zákravský, 1993b).

In natural habitats, seed production and biomass production in selected populations were studied; at the same time, seasonal changes in biomass production were studied in a cultivation experiment (Hroudová, 1980).

To test the dependence of growth of *B. umbellatus* plants on fluctuating water level, the plants were cultivated under stable water level for six years (Hroudová, 1989).

Chromosome numbers were counted from 99 localities from the whole Czech and Slovak Republics. Karyotype morphology was studied in both cytotypes. To explain a cause of different fertility of either cytotype, pollen fertility was studied and crossing experiments performed (Krahulcová & Jarolímová, 1993).

Seed germination, seedling viability and response of both cytotypes to increased trophic impact were studied experimentally (Hroudová & Zákravský, 1993a).

Results

Life cycle

Successful development of *B. umbellatus* populations need not be conditioned by fluctuations of water level. When cultivated at a stable water level, the plants grew well for several years (Hroudová, 1989). In natural habitats, however, *B. umbellatus* can be suppressed by reed-bed species (Koreljakova, 1977). Thus, *B. umbellatus* is able to use temporarily available habitats created by water level fluctuations: it often forms a belt along the inner edge of reed stands when water level has fallen.

When the bottom is drained, soil surface is exposed to sun rays and warms up. Increased temperature promotes sprouting of *B. umbellatus* which leads to multiplication of shoots and establishment of new plants from rhizome fragments (Figure 1, point A).

In a subsequent growing season in which water level remains low (Figure 1, B), the populations of *B. umbellatus* may expand. Released from competitive pressure, the stands may thus persist for several years.

Further rise of the water level suppresses fertility and growth of the plants, leading to the formation of sterile stands of *B. umbellatus*, or to their survival at the stage of dormant rhizomes (Figure 1, C–D). But the stands can establish themselves again during the next fall of water level; repeated emergence of the bottom makes possible the renewal of *B. umbellatus* populations by vegetative propagation.

Ways of reproduction

The most efficient way of propagation is vegetative spreading by rhizome fragments. It has been described by Woodhead & Kirchner (1908), Glück (1911), Weber (1950), Hejný (1960) and Lieu (1979). *B. umbellatus* produces several times as much underground biomass (mainly rhizomes) as aboveground shoots (Figure 2). Its dense rhizome network with numerous lateral buds represents a considerable reproductive potential. Rhizome fragments or broken lateral buds are spread easily, namely by water streams. This explains the frequent occurrence of *B. umbellatus* along rivers.

Most of the rhizome biomass is produced late in the growing season (Figure 2). Decrease in photosynthetic activity caused by cutting or grazing of the leaves limits rhizome biomass production.

A complementary way of vegetative reproduction provide bulbils formed in the inflorescences of *B. umbellatus* (Luther, 1951; Lohammar, 1954). Their formation is irregular and infrequent (Hroudová, 1980, Krahulcová, unpublished data); this way of reproduction is thus probably not very important.

Seedling establishment (Figure 1, E) does not occur frequently in natural habitats. Germinating seeds were observed on an emerged fishpond bottom (Hroudová & Zákravský, 1993a); seedlings are probably weak competitors owing to their small size and slow growth. High mortality was observed in cultivated seedlings and probably also occurs in natural habitats of *B. umbellatus*.

Great differences exist in seed production within the species depending on ploidy level (see next paragraph). As a result of a high seedling mortality (even in fully fertile plants), sexual reproduction of *B. umbellatus* is probably inefficient in natural habitats. Nevertheless, sexual reproduction enables the spreading of *B. umbellatus* to geographically distant localities (Hroudová & Zákravský, 1993b).

Influence of ploidy level

On the basis of previous studies on breeding behaviour (Krahulcová & Jarolímová, 1993), comparison betveen sexual and vegetative reproduction, biomass production under different nutrient levels (Hroudová & Zákravský, 1993a) and habitat conditions (Hroudová & Zákravský, 1993b) we can outline the following differences between the two cytotypes of *B. umbellatus*:

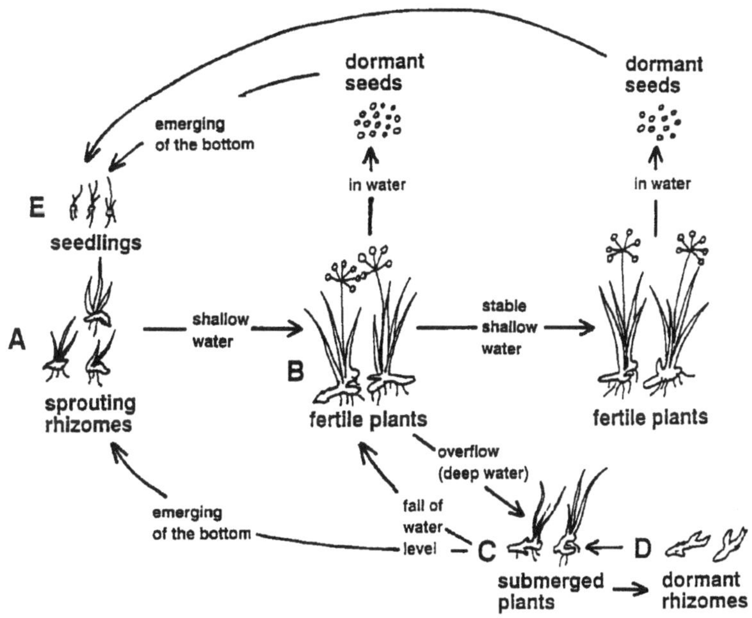

Figure 1. Cyclic development of *Butomus umbellatus* populations in natural habitats in dependence on changes of water level.

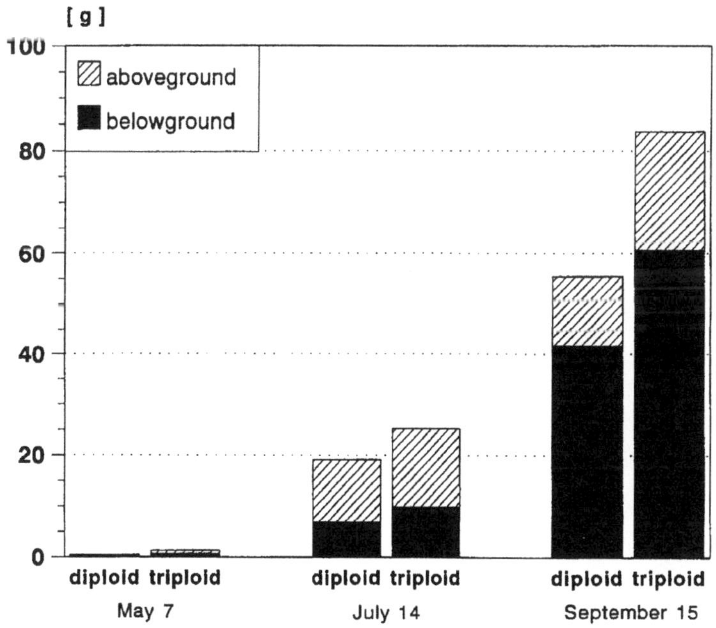

Figure 2. Biomass production in both cytotypes of *B. umbellatus* during the growing season. Plants cultivated in optimal concentration of nutrient solution (according to Hroudová & Zákravský, 1993a).

Conclusions

Spreading by rhizome fragments represents the most efficient way of propagation of *B. umbellatus*. A fall of water level promotes the multiplication of *B. umbellatus* individuals by sprouting of rhizomes. For a control of *B. umbellatus* populations, we recommend to keep

	Diploids $2n = 26$	Triploids $2n = 39$
(a)	self-compatible within clones;	self-incompatible within clones;
(b)	in natural habitats fully fertile;	seed production in natural habitats strongly limited;
(c)	less numerous lateral rhizome buds;	high production of lateral rhizome buds;
(d)	do not produce bulbils in inflorescences;	can produce bulbils in inflorescences;
(e)	lower biomass production;	higher biomass production;
(f)	habitats with more acidic and base-poor soils;	habitats with less acidic and more base-rich soils;
(g)	less tolerant of high nutrient inputs;	more tolerant of high nutrient inputs.

a high water level (more than +0.8 m). When fluctuations of water level are associated with reservoir management, cutting of shoots in early summer limits population development of *B. umbellatus*; repeated cutting gradually weakens the plants. The most efficient way of controlling of this species is grazing by ducks.

Triploid plants of *B. umbellatus* are more robust, and spread easily by vegetative propagation. This may facilitate their spreading by water flow and strengthen their competitive ability, which explains the frequent occurrence of the triploids in river-bank communities.

Higher tolerance of an increased nutrient level enables the triploids to survive better in eutrophicated habitats and may lead to their prevalence in man-influenced habitats and waters.

References

Glück, H., 1911. Biologische und morphologische Untersuchungen über Wasser- und Sumpfgewächse. III. Die Uferflora. Jena.

Hejný, S., 1960. Ökologische Charakteristik der Wasser- und Sumpfpflanzen in der Slowakischen Tiefebenen. Vydav. SAV, Bratislava, 492 pp.

Hroudová, Z., 1980. Ekologická studie druhû *Sagittaria sagittifolia* L., *Butomus umbellatus* L., *Bolboschoenus maritimus* (L.) Palla, *Oenanthe aquatica* (L.) Poir. Ms., [Theses; Institute of Botany CZAS, Prûhonice, 256 pp.].

Hroudová, Z., 1989. Growth of *Butomus umbellatus* at a stable water level. Folia Geobot. Phytotax. 24: 371–385.

Hroudová, Z. & P. Zákravský, 1993a. Ecology of two cytotypes of *Butomus umbellatus* II. Reproduction, growth and biomass production. Folia Geobot. Phytotax. 28: 413–424.

Hroudová, Z. & P. Zákravský, 1993b. Ecology of two cytotypes of *Butomus umbellatus*. III. Distribution and habitat differentiation in the Czech and Slovak Republics. Folia Geobot. Phytotax. 28: 425–435.

Koreljakova, I. L., 1977. Rastitelnost Krementschugskogo vodochranilischtscha. Naukova Dumka, Kijev.

Krahulcová, A. & V. Jarolímová, 1993. Ecology of two cytotypes of *Butomus umbellatus* I. Karyology and breeding behaviour. Folia Geobot. Phytotax. 28: 385–411.

Lieu, S. M., 1979. Growth forms in the Alismatales. II. Two rhizomatous species: *Sagitttaria lancifolia* und *Butomus umbellatus*. Can. J. Bot. 57: 2353–23.

Lohammar, G., 1954. Bulbils in the inflorescences of *Butomus umbellatus*. Svensk Bot. Tidsskr. 48: 485–488.

Luther, H., 1951. Wasserpflanzen im Brackwasser der Ekenas-Gegend II. Acta Bot. Fenn. 50: 1–370.

Weber, H., 1950. Über das Wachstum des Rhizoms von *Butomus umbellatus* L. Planta 38: 196–204.

Woodhead, T. W. & O. Kirchner, 1908. *Butomaceae*. In Kirchner, O., E. Loew & C. Schröter (eds): Lebensgeschichte der Blütenpflanzen Mitteleuropas, Stuttgart, Vol. 1/1: 648–664.

Hydrobiologia **340**: 31–35, 1996.
J. M. Caffrey, P. R. F. Barrett, K. J. Murphy & P. M. Wade (eds), Management and Ecology of Freshwater Plants.

Growth response of *Bolboschoenus maritimus* ssp. *maritimus* and *B. maritimus* ssp. *compactus* to different trophic conditions

Petr Zákravský & Zdenka Hroudová
Institute of Botany, Academy of Sciences of the Czech Republic, CZ-252 43 Průhonice, Czech Republic

Key words: wetland plants, eutrophication, clonal plants, biomass production

Abstract

Bolboschoenus maritimus (L.) Palla (=*Scirpus maritimus* L.) forms extensive stands in the littoral zone of small fishponds and as a weed in rice and maize fields. Within the species, two subspecies are distinguished: *Bolboschoenus maritimus* subsp. *maritimus, B. maritimus* subsp. *compactus*. They differ in ecology, especially in their relationships with trophic conditions and salinity of habitats. To determine growth response of these two types to different nutrient levels, we compared their seasonal development under experimental cultivation at four controlled nutrient levels. Some differences between the subspecies were found to be stable, regardless of nutrient level, namely greater amount of smaller underground tubers and more extensive rhizome system in subsp. *compactus* compared to less numerous larger tubers and simpler rhizome system in subsp. *maritimus*. In response to trophic conditions, the plants of subsp. *compactus* were more resistant to the conditions of the highest trophic level than those of subsp. *maritimus*, which were stressed. This demonstrates better adaptability and spreading ability of *B. maritimus* subsp. *compactus* at high trophic levels.

Introduction

Within *Bolboschoenus maritimus*, two subspecies occur in Central Europe: *B. maritimus* subsp. *maritimus* and *B. maritimus*. subsp. *compactus* (G. F. Hoffmann) Hejný in Dostál (Casper & Krausch, 1980). They differ in ecology and in occurrence in plant communities (Hejný, 1960; Hejný in Moravec et al., 1983; Hejný & Husák, 1978; Zahlheimer, 1979), *B. maritimus* subsp. *maritimus* inhabiting fishponds or other impoundments with acidic, base-poor substrates, while *B. m.*. subsp. *compactus* is found in habitats with base-rich, often saline soils, such as oxbows, reservoirs or temporarily flooded field depressions and often also as a weed in rice and maize fields (Hejný, 1960).

Different tolerance to salinity is evident in both subspecies. Our goal was to find whether a similar physiological distinction exists regarding the total trophic status of habitats, which might cause different reactions of the two subspecies to an increased nutrient input to their habitats. Reaction of the two subspecies to increased nutrient inputs was studied in a cultivation experiment with controlled concentrations of nutrient solution.

Materials and methods

For cultivation, plants of *B. maritimus* subsp. *maritimus* were brought from the Kačležský fishpond (alt. 533 m, Jindřichův Hradec district, South Bohemia), and *B. maritimus* subsp. *compactus* from a flooded field depression near the Nesyt fishpond (alt. 175 m, Břeclav district, South Moravia). Plants were cultivated in an experimental garden for one year to provide a satisfactory amount of vegetative offshoots for planting. Before planting, a sample of 20 plants of each subspecies was measured and individual dry weights were recorded, to define starting plant size. In the spring (May 30, 1984), vegetatively reproducing plants (one or two tubers with shoots) were planted. Four treatments were applied:

1. in spring water without nutrient solution;

2. basic nutrient solution according to Šetlík (Šetlík, 1968; Véber & Zahradník, 1986) in basic concentration modified by using $Ca(NO_3)_2$ instead of NH_4NO_3; (nutrient contents: $CO(NH_2)_2$ 300 mg l^{-1}, $Ca(NO_3)_2$ 576 mg l^{-1}, $MgSO_4 . 7H_2O$ 986 mg l^{-1}, KH_2PO_4 340 mg l^{-1}, Chelaton III 25.5 mg l^{-1}, H_3BO_3 3.1 mg l^{-1}, $MnSO_4 . 4H_2O$ 1.1 mg l^{-1}, $CuSO_4 . 4H_2O$ 1.25 mg l^{-1}, $ZnSO_4 . 7H_2O$ 1.45 mg l^{-1}, $(NH_4)_2MoO_4$ 1.0 mg l^{-1}, $CoSO_4 . 7H_2O$ 1.4 mg l^{-1}, Fe-chelatonic 22.02 mg l^{-1}).

3. double-strength basic nutrient solution (only macronutrients);

4. four times concentrated basic nutrient solution (only macronutrients);

Each treatment consisted of 40 replicates of each subspecies. Each replicate was represented by one plant planted in a perforated plastic bag containing sand which was saturated with the nutrient solution. Water level was maintained approximately at the sand surface. Nutrient solution was supplied after two weeks of planting and renewed at approximately 3–4 week intervals depending on decrease in ion concentration (checked by measurement of conductivity change). The experiment lasted two years. The plants were sampled and measured twice (in July and September) in each year, using ten plants in each sampling.

The following data were recorded for each plant: shoot height, number of offshoots, number of inflorescences, total length of all parts of the rhizome system, number of tubers, size of tubers [g of dry weight per tuber], biomass of underground and aboveground organs and the ratio between rhizome system and shoot biomass (R:S ratio). Plant biomass was dried at 90 °C. In addition, the proportions of living and dead plants were recorded.

Differences in plant size and biomass production were tested by two-way analysis of variance (Snedecor & Cochran, 1971).

Table 1. Dry mass production and R:S ratio of *Bolboschoenus maritimus*: effect of subspecies and nutrient level (two-way ANOVA). n = 80 in each sampling. T = sequence of samplings, n.s.= not significant, * = $P < 0.05$, ** = $P < 0.01$, *** = $P < 0.001$, **** = $P < 0.0001$.

	T	Subspecies	Nutrient level	Interaction
Aboveground biomass	1	n.s.	****	n.s.
	2	****	****	*
	3	***	****	**
	4	***	***	n.s.
Belowground biomass	1	****	n.s.	n.s.
	2	n.s.	***	n.s.
	3	n.s.	****	*
	4	n.s.	***	n.s.
Total biomass	1	****	n.s.	n.s.
	2	*	****	n.s.
	3	*	****	*
	4	*	***	n.s.
R:S ratio	1	***	**	n.s.
	2	****	***	***
	3	****	****	***
	4	****	***	***

Results

Variables dependent on nutrient level (see Tables 1, 2)

1. Biomass production (Figure 1). Treatments 2 and 3 appeared to favour growth; in treatments 1 and 4 plant growth was reduced. During the experiment, the ratio between the two subspecies changed; plants of subsp. *compactus* became more productive, especially in treatment 4.

2. Plant mortality. While in treatments 1, 2 and 3, 100% of plants survived during the whole experiment, in treatment 4 only 50% of them remained alive till the end in both subspecies.

3. Maximum height of plants. The tallest shoots were found in treatment 2. At the lower nutrient levels, the shoots of subsp. *maritimus* were more than 0.1 m taller than those of subsp. *compactus*.

4. Number of inflorescences. Treatment 2 was associated with the most prolific flowering in both subspecies. In treatment 4, subsp. *maritimus* remained sterile, while subsp. *compactus* formed about 20% of fertile shoots.

Variables dependent on subspecies:

1. Tuber size. Plants of subsp. *maritimus* formed bigger tubers than those of subsp. *compactus* independently of nutrient level.

2. Number of tubers. Plants of subsp. *compactus* produced more tubers than subsp. *maritimus* in all treatments.

3. Number of offshoots. Number of aboveground shoots was higher in subsp. *compactus*.

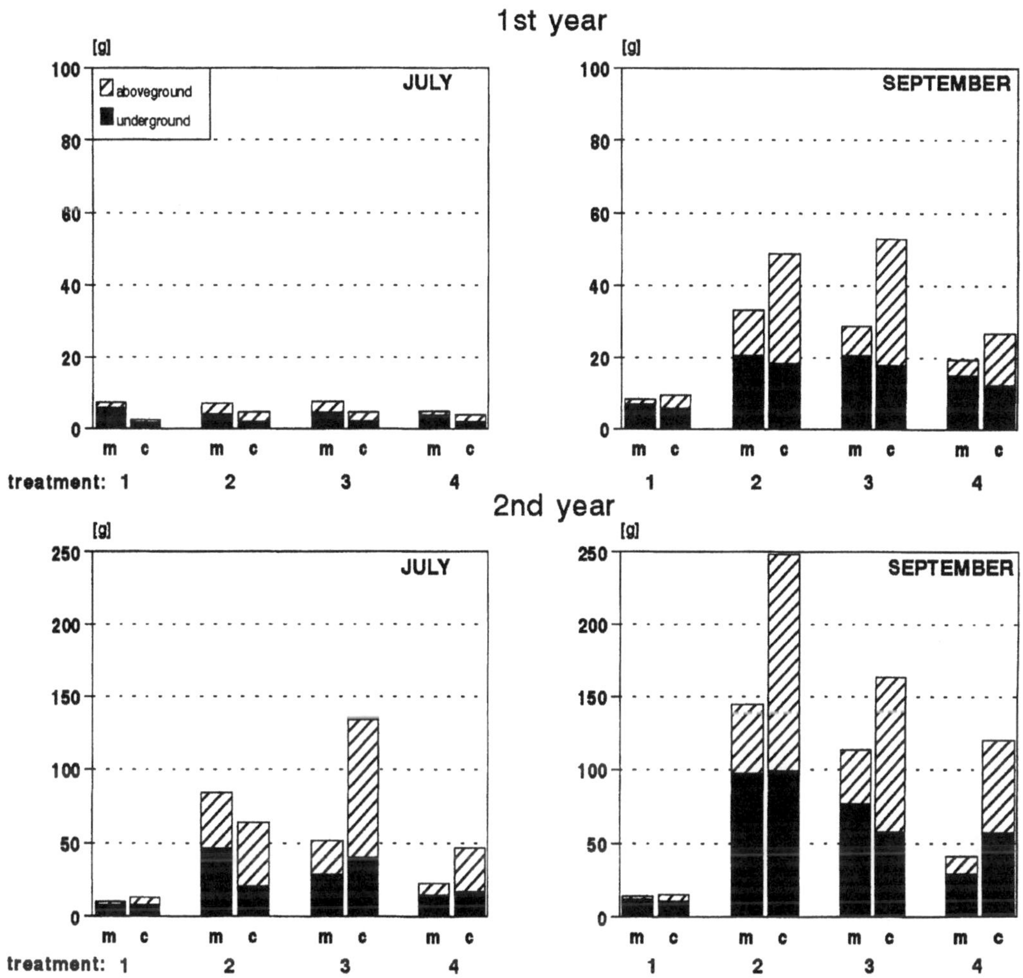

Figure 1. Dependence of biomass production of the two subspecies of *B. maritimus* on nutrient input during a two-year experiment (dry mass weight per plant). m = *B. maritimus* subsp. *maritimus*, c = *B. maritimus* subsp. *compactus*.

4. Total length of rhizomes in rhizome systems. Plants of subsp. *compactus* formed more complicated and more extensive rhizome systems.

Variable dependent on both sources of variation – subspecies and nutrient level

R:S ratio. This was >1 in subsp. *maritimus* in all treatments, this being higher than in subsp. *compactus* during the whole experiment. Under stress (treatments 1 and 4) the R:S ratio was higher than at favourable nutrient levels. The interaction between both sources of variation resulted in lower R:S values in subsp. *compactus* (R:S <1) in treatments 2 and 3 compared to subsp. *maritimus*.

Discussion

Some variables measured appeared to be stable at different nutrient levels conditions and may be considered as biological features of the respective subspecies.

B. maritimus subsp. *maritimus* is characterized by a lower number of bigger tubers, simpler rhizome system and a sparser aboveground stand formed by taller stems. This habit is well adapted to littoral habitats

Table 2. Total number and size of tubers, length of rhizomes, number of offshoots, maximum shoot height and number of inflorescences per plant of *Bolboschoenus maritimus*: effect of subspecies and nutrient level (two-way ANOVA). n = 80 in each sampling. T = sequence of samplings, n.s.= not significant, * = $P < 0.05$, ** = $P < 0.01$, *** = $P < 0.001$, **** = $P < 0.0001$.

	T	Subspecies	Nutrient level	Interaction
Total number of tubers	1	**	**	n.s.
	2	****	***	n.s.
	3	****	***	*
	4	****	*	n.s.
Size of tubers	1	****	***	**
	2	****	*	n.s.
	3	****	***	****
	4	****	**	n.s.
Length of rhizomes	1	****	****	n.s.
	2	****	***	*
	3	****	***	*
	4	***	*	n.s
Number of offshoots	1	****	***	n.s.
	2	****	****	*
	3	****	***	**
	4	***	**	n.s.
Maximum height of plant	1	****	****	*
	2	n.s.	****	n.s.
	3	n.s.	****	n.s.
	4	n.s.	****	n.s.
Number of inflorescences	1	n.s.	n.s.	n.s.
	2	**	****	**
	3	****	****	**
	4	***	****	n.s.

where a higher biomass production is needed for the plants to emerge above the water surface.

Plants of *B. maritimus.* subsp. *compactus* produce numerous tubers of smaller size connected by a complicated network of rhizomes, and dense aboveground stands are formed by shorter stems. They seem to be well adapted to growth in terrestrial habitats temporarily flooded (Hejný, 1960); field cultivation usually promotes vegetative propagation and spreading by cutting the rhizome systems and dispersing of tubers.

The two subspecies exhibit different ecological amplitudes in relation to increased nutrient inputs, which is in agreement with the preliminary results given by Dykyjová (1986). *B. maritimus.* subsp. *compactus* was able to survive better at high nutrient levels, while under nutrient limitation the two subspecies did not differ distinctly. This supports the hypothesis that eutrophication may limit the occurrence of *B. maritimus* subsp. *maritimus* to nutrient-poor habitats. Disappearance of *B. maritimus.* subsp. *maritimus* stands along the polluted and highly nutrient-loaded shores of the Rožmberk fishpond in S. Bohemia (Hroudová et al., 1988) may be just one example of its lower tolerance of eutrophy.

Conclusions

B. maritimus subsp. *compactus* is more tolerant of high nutrient inputs. This may explain the observed decrease in abundance or disappearance of *B. maritimus.* subsp. *maritimus* in highly eutrophic natural habitats.

In some plant characters (size and number of tubers, length of rhizomes, shoot length and density), differences were found between the two subspecies, which were independent of nutrient level applied. These differences thus may be considered as biological taxonomic features distinguishing the two subspecies from each other.

Acknowledgments

Our sincere thanks are due to Dr D. Dykyjová CSc. for initiating this study and valuable comments and to Mrs E. Zamazalová and Mr I. Ostrý for technical assistance throughout the project. The work was supported by the Grant Agency of the Czech Republic (grant no. 204/93/1176).

References

Casper, S. J. & H.-D. Krausch, 1980. Süsswasserflora von Mitteleuropa, Bd. 23–24: Pteridophyta und Anthophyta. G. Fischer, Jena, 944 pp.

Dykyjová, D., 1986. Production ecology of *Bolboschoenus maritimus* (L.) Palla (*Scirpus maritimus* L. s.l.). Folia Geobot. Phytotax. 21: 27–64.

Hejný, S., 1960. Ökologische Charakteristik der Wasser- und Sumpfpflanzen in der Slowakischen Tiefebenen. Vyd. SAV, Bratislava, 492 pp.

Hejný, S. & Š. Husák, 1978. Higher plant communities. In Dykyjová, D. & J. Květ (eds), Pond littoral ecosystems. Springer-Verlag. Berlin, Ecol. Stud. Vol. 28: 23–64.

Hroudová, Z. & S. Hejný, P. Zákravský, 1988. Littoral vegetation of the Rožmberk fishpond. In Hroudová, Z. (ed.), Littoral vegetation of the Rožmberk fishpond and its mineral nutrient economy. Studie ČSAV, Academia, Praha 9/88: 23–60.

Moravec, J. & E. Balátová-Tuláčková, E. Hadač, S. Hejný, J. Jenik, J. Kolbek, K. Kopecký, F. Krahulec, Z. Kropáč, R. Neuhäusl, K. Rybniček, J. Vicherek, 1983. Rostlinná společenstva Česk, Socialistické, republiky a jejich ohroženi. Severočeskou přírodou, Příl. 1983/1, Litoměřice, 130 pp.

Snedecor, G. W. & W. G. Cochran, 1971. Statistical methods. Ames, Iowa.

Šetlík, I., 1968. Growth and photosynthetic characteristics of algae. In Nečas, J. & O. Lhotský (eds), Ann. Rep. Algolog. Lab. Třeboň for 1967: 71–140.

Véber, K. & J. Zahradník, 1986. Dočišťování vod autotrofními mikroorganismy a vyššími rostlinami. Studie ČSAV, Academia, Praha 24/86: 1–156.

Zahlheimer, W. A., 1979. Vegetationsstudien in den Donauauen zwischen Regensburg und Straubing als Grundlage für den Naturschutz. Hoppea 38: 3–398.

Hydrobiologia **340**: 37–41, 1996.
J. M. Caffrey, P. R. F. Barrett, K. J. Murphy & P. M. Wade (eds), Management and Ecology of Freshwater Plants.

Mineralogical and microscopic analyses of material deposited on submersed macrophytes in Florida lakes

Paul V. Zimba[1] & Stan R. Bates[2]
[1]*Department of Fisheries and Aquatic Sciences, 7922 N.W. 71st St., Gainesville, FL 32653, USA (Current address: U.S.D.A., P.O.B. 19678, New Orleans, LA 70179, USA)*
[2]*Materials Science and Engineering, P.O.Box 116400, University of Florida, Gainesville, FL 32611, USA*

Key words: marl, Florida, lakes, carbonates, x-ray diffraction, scanning electron microscopy

Abstract

Attached material on submersed vegetation from 18 lakes was analyzed by x-ray diffraction and scanning electron microscopy to identify constituent components. Lake trophic state ranged from oligo-to hyper-eutrophic. Minerals present on 11 submersed taxa included calcite, various salts (KCl and NaCl), silicon dioxide (both biogenic and sand) and hematite. Abundance of deposited material was not related to concentrations of precursor elements in the water column. Resuspended sediments and diatom frustules both contributed to the silica fraction of marl and should be compartmentalized.

Introduction

We report results of chemical and microscopic examination of material found on submerged plant species from 17 Florida lakes and one pond. Our goal was to determine whether calcite was the primary carbonate species present in these subtropical freshwater lakes, and to document the presence of other minerals in this material. We analyzed samples to partition abiotic silica compounds from biotic forms.

Lewis & Weibezahn (1981) and Lewis (1981) examined sediment carbonate species present in tropical lakes and found have only calcite present in superficial sediments, although saline lakes also may have aragonite. For temperate lakes carbonate speciation results have been reported only descriptively (literature cited below).

Most research on carbonate deposition has occurred in temperate North American and European lakes. These lakes are of glacial origin, and are typically deep water systems. Otsuki & Wetzel (1972) reported calcite as the only form of carbonate in Lawrence Lake, whereas Lewis & Weibezahn (1981) reported differential layering of calcite and aragonite as a function of salinity changes in a Venezuelan lake. Waisel et al. (1990) identified quartz, aragonite, calcite, and apatite in leaf encrustations from Sweden.

Material deposited on submersed macrophytes (i.e. marl *sensu* Wetzel (1983)) results from several biological and physical processes: photosynthesis, sinking of dead plankton, sediment resuspension, and growth of attached biota. Hassack (1888) identified carbonates as the major component of marl layers formed by photo synthesis. Jensen et al. (1985) documented macroalgal (*Halimeda* spp.) deposition of carbonates in Caribbean water, estimating that 20–40% of assimilated carbon dioxide was deposited onto the cell wall. Brammer (1978) observed greater calcium carbonate deposition in areas colonized by the freshwater emergent *Stratiotes aloides* relative to noncolonized areas. Intensive photosynthesis depletes available carbon dioxide and shifts pH to basic conditions, enhancing carbonate formation (Cole, 1975).

Aggregates consisting of microalgae, bacteria, protozoa, and decomposing macrophytes are common components of shallow-water systems. Dense vegetation reduces water movement, causing increased sedimentation of particles (Joyce et al., 1993; Kemp et al., 1984). These aggregates (termed 'marine snow' or nepheloid layers) are well-known centers of intense biological and chemical activity (Haines, 1977).

Attached biotic components (such as epiphytic algae and invertebrates) on macrophytes can be a significant component of lake communities (Phlips et al., 1993; Cattaneo & Kalff, 1979). In eutrophic Lake Okeechobee, epiphytes accounted for over 25% of the bound nutrients in primary producers (Zimba, 1995). Shallow lakes are subject to greater wind mixing than deeper systems. Large shallow-water lakes (e.g., Lake Okeechobee, Florida) have greater resuspended benthic microalgae in the water column during storm periods relative to calm periods (Phlips et al., 1993). It is likely that resuspension of inorganic sediment particles (especially silts and clays, but also diatom frustules) follows a similar pattern (Aalderink et al., 1985).

Calcite is the predominant form of carbonate formed in freshwater systems. This is due in part to the low dissolution rate of calcite relative to other carbonate species (Borowitzka, 1987). Coprecipitation of cations with calcite is common, with preferential chelation of magnesium, iron, manganese, zinc, and to a lesser extent copper (Borowitzka, 1987). Phosphate and vitamin B_{12} are also coprecipitated with calcite (Otzuki & Wetzel, 1972). The presence of humic acids reduces rates of iron-calcium carbonate formation (Wetzel & White, 1985), possibly due to competition by non-calcium cations for chelating ligands.

Lowenstam (1981) has proposed two mechanisms for carbonate mineralization: a biologically induced process and an organically matrix-mediated process (typically biochemically mediated). The latter has been observed in coralline macroalgae (Walker & Woelkerling, 1988) and coccolithophorids (Young et al., 1991). Biologically induced processes have been observed in *Chara* and *Halimeda* (Borowitzka, 1987). Precipitation of carbonates is enhanced by the presence of ions or other charged particles that may serve as nuclei for crystallization. Cell wall surfaces often serve as sites for carbonate precipitation, possibly by lowering nucleation energy requirements under suboptimal (low super-saturation) conditions. This is particularly true for members of the algal division Charophyta in which carbon incorporation and calcification are both passive and facilitated processes (McConnaughy, 1991). Cellular carbon dioxide and calcium are often calicified extracellularly in *Chara*, in part due to alternating zones of alkaline and acidic conditions.

Methods

Study areas

Samples of submersed vegetation were collected during near maximal biomass (June 1992) from 17 lakes and from the Department of Fisheries and Aquatic Sciences' Crayfish Pond located in Gainesville, FL. Monotypic plant samples were returned to the laboratory and maintained at 4 °C until processed, this interval never exceeded 48 hours. Lakes were located throughout subtropical Florida.

Sampling methodology

Water column samples were collected from one or more mid-lake locations within each lake. Water sample processing methodology is given within Lakewatch (1992) annual reports available through the University of Florida. Submersed plant biomass sampling along 10 equidistant transects per lake included collection of 0.25 m^2 samples. Monotypic samples of submersed plants were removed from these biomass samples and stored on ice until return to the laboratory. Laboratory processing occurred within 24 hours of sample collection.

Separation and analysis techniques

Water samples were transported on ice to the laboratory and stored at 4 °C until analyzed. Standard methodology for nutrient analyses were used for chemical measures (APHA, 1985). Elemental analyses were made with a Perkin-Elmer #703 atomic absorption spectrophotometer.

Mechanical shaking was used to dislodge attached material from submersed macrophytes. Typically 2–300 ml of deionized water was added to ca. 50 g of wet plant material, then shaken for 60 seconds @ ca. 2 revolutions per second. This technique typically removed 85–98% of attached material as determined by dry weight and chlorophyll analyses.

Attached material was dried at 60 °C for 3–4 days, then ground with a mortar and pestle. Subsamples were examined by x-ray diffraction analysis of powdered bulk samples. Samples were analyzed using a Philips Electronics #3720 diffractometer equipped with Cu Kalpha (1.54 angstroms radiation) detector. Field samples were compared with known reference compounds to confirm identification.

Table 1. Physical characteristics of study lakes.

Lake name	Area (hectares)	Secchi (m.)	Chl (μg/L)	Macrophyte species	Total				
					N (μg/L)	P (μg/L)	Ca (μg/L)	Fe (mg/L)	Na (mg/L)
Alligator	1378	1.95	3.0	*Mayaca fluviatilis*	610	11	8.2	0.2	11.5
Blue Heron	N.D. [1]	0.85	51.0	*Utricularia purpurea*	940	55	4.6	N.D.	2.8
Crayfish Pond	<1	0.30	44.0	*Potamogeton pusillus*	1755	87	76.2	0.8	21.2
Diane	26	2.53	3.0	*Bacopa caroliniana*	300	13	N.D.	N.D.	N.D.
Eaton	124	0.64	6.0	*Ceratophyllum demersum*	1130	28	38.3	0.7	5.3
Elbert	70	2.53	7.0	*Nitella prolonga*	530	18	33.0	N.D.	N.D.
Emporia	N.D.	0.64	8.5	*Nitella prolonga*	720	11	N.D.	N.D.	N.D.
Fannie	335	2.53	24.0	*Hydrilla verticillata*	1020	47	11.7	0.1	38.8
Frederica	29	3.20	5.0	*H. verticillata*	440	13	N.D.	N.D.	N.D.
Harris	5580	0.70	64.0	*Vallisneria americana*	1780	30	54.9	0.0	10.2
Hartridge	176	3.02	8.0	*Utricularia purpurea*	500	13	13.8	N.D.	N.D.
Iamonia	2330	1.19	9.0	*Eleocharis baldwinii*	530	17	24.9	N.D.	N.D.
Ivanhoe	N.D.	0.98	32.0	*V. americana*	760	31	N.D.	N.D.	N.D.
Mound	32	4.57	2.0	*M. fluviatilis*	350	6	N.D.	N.D.	N.D.
Okeechobee	18600	0.75	22.0	*Potamogeton illineonsis*	1450	78	25.4	0.4	7.2
Panasoffkee	1805	1.19	8.0	*Nitella prolonga*	690	21	100.7	0.0	5.7
Spring Garden	211	N.D.	13.0	*C. demersum*	770	43	N.D.	N.D.	N.D.
Tibet	485	2.83	4.0	*Cabomba pulcherrima*	450	9	N.D.	N.D.	N.D.

N.D.[1] - not determined

Table 2. Mineralogy and SEM observations of marl constituents.

Lake	Trophic classification	Macrophyte species	Minerals present	SEM observations
Alligator	Oligotrophic	*Mayaca fluviatilis*	silica	epiphytic diatoms
Blue Heron	Eutrophic	*Utricularia purpurea*	calcite, hematite	sand
Crayfish Pond	Eutrophic	*Potamogeton pusillus*	silica, calcite	sand
Diane	Oligotrophic	*Bacopa caroliniana*	silica, calcite	benthic diatoms
Eaton	Mesotrophic	*Ceratophyllum demersum*	calcite	N.D.
Elbert	Mesotrophic	*Nitella prolonga*	NaCl, KCl	N.D.
Emporia	Mesotrophic	*Nitella prolonga*	silica, calcite	benthic diatoms
Fannie	Eutrophic	*Hydrilla verticillata*	silica	epiphytic diatoms
Frederica	Mesotrophic	*Hydrilla verticillata*	silica	sand, benthic diatoms
Harris	Eutrophic	*Vallisneria americana*	silica, NaCl, KCl	sand
Hartridge	Mesotrophic	*Utricularia purpurea*	silica, calcite	sand
Iamonia	Mesotrophic	*Eleocharis baldwinii*	calcite	N.D.
Ivanhoe	Mesotrophic	*Vallisneria americana*	calcite	N.D.
Mound	Oligotrophic	*Myaca fluviatilis*	calcite	N.D.
Okeechobee	Eutrophic	*Potamogeton illineonsis*	calcite	N.D.
Panasoffkee	Mesotrophic	*Nitella prolonga*	calcite	N.D.
Spring Garden	Mesotrophic	*Ceratophyllum demersum*	silica, calcite	epiphytic and planktonic diatoms
Tibet	Oligotrophic	*Cabomba pulcherrima*	silica	sand, benthic diatoms

Subsamples of dried material were mounted on stubs and viewed with a Cambridge Leica stereoscan 240 SEM. Operating voltage was 25 kv. Samples were sputter coated with gold- palladium.

Results

Lakes analyzed in this work ranged from <1 to >18 000 hectares in surface area. Lake trophic state (using criteria in Rast et al., 1989) based upon chlorophyll *a*, nitrogen, and phosphorus concentrations ranged from oligotrophic to hyper-eutrophic (Table 1). Nitrogen:phosphorus ratios ranged from ca. 5 to 60, chlorophyll *a* values followed similar trends.

A total of 11 species of submersed aquatic vegetation were analyzed in this work. Species included *Nitella prolonga*, a new record for Florida. This species was common in six lakes sampled as part of the LAKEWATCH program (Lakewatch, 1992), however sufficient epiphytic biomass was available only from three of these lakes for further analyses. Five other submersed plant species were found in at least two of the lakes surveyed.

Calcite was the only carbonate species identified from these 18 water bodies, and was found in 12 samples. Silica was identified from many of the samples, both as biogenic forms (diatoms) and as crystals-sand (Table 2). Biogenic forms of silica included planktonic, epiphytic, as well as diatoms more typical of sediments. Less than half of the samples contained biotic forms of silica recognizable by SEM. Blue Heron Lake contained hematite whereas Lake Elbert marl contained both sodium chloride and sylvite (KCl).

Discussion

In contrast to previous reports of marl composition from sub-tropical lakes, we have found high diversity of chemicals present in submersed plant encrustations in Florida lakes. We have identified hematite and chloride salts as constituent components of submersed plant encrustations in Florida lakes in addition to calcium (as calcite) and both biogenic and abiogenic silica. Hopefully this work will refocus interest in this metabolically active microhabitat.

The only carbonate species identified from these 18 water bodies was calcite. These results differ from those reported by Waisel et al. (1990) for a high altitude temperate lake. No chemical information regarding water quality was provided in this report, making resolution of these differing results difficult. The only other report of aragonite formation is in saline waters.

Recently Conley & Schelske (1993) proposed use of digestion time courses to distinguish between diatoms and sponge spicules. This work analyzed 82 lake surface sediments and concluded sponge spicules were an important source of remineralized silica. We were unable to identify a single sponge spicule in the 18 lakes examined. If our results are typical, time course digestion procedures should be used in conjunction with microscopic examination of sediments to identify the form of silica present.

Acknowledgments

Water quality data was provided by the University of Florida's LAKEWATCH program directed by Dr Dan Canfield. Sample collections were made by Dan Willis. Financial support for x-ray diffraction analyses provided by South Florida Water Management District and the Department of Materials Science, University of Florida. Scanning electron microscopy was done at the Florida Marine Research Institute (Florida Department of Environmental Protection), St. Petersburg with the able assistance of Dr Earnest Truby. The thorough review of this manuscript by two unidentified reviewers contributed to a better product.

References

Aalderink, R. H., L. Lijklema, J. Breukelman, W. van Raaphorst & A. G. Winkelman, 1985. Quantification of wind induced resuspension in a shallow lake. Water Sci. Technol. 17: 903–914.

APHA., 1985. Standard methods for the examination of Water and Wastewater. 16th edn. American Public Health Association. Washington, D.C.

Borowitzka, M. A., 1987. Calcification in algae: mechanisms and the role of metabolism. In A. W. D. Larkum (ed.), CRC Critical Reviews in Plant Sciences. CRC Press, Boca Raton, (FL) 6: 1–45.

Brammer, E. S., 1978. Phytogenic precipitation of calcium carbonate as a source of sedimentation. Pol. Arch. Hydrobiol. 25: 49–59.

Cattaneo, A. & J. Kalff, 1979. Primary production of algae growing on natural and artificial plants: a study of interactions between epiphytes and their substrate. Limnol. Oceanogr. 24: 1031–1037.

Cole, G. A., 1975. Textbook of Limnology. C. V. Mosby Co. St. Louis, USA, 283 pp.

Conley, D. J. & C. L. Schelske., 1993. Potential role of sponge spicules in influencing the silicon biogeochemistry of Florida lakes. Can. J. Fish. aquat. Sci. 50: 396–402.

Davis, C. A., 1901. A contribution to the natural history of marl. J. Geol. 8: 485–497.

Haines, E. B., 1977. The origins of detritus in Georgia salt marsh estuaries. Oikos 29: 254–260.

Hassack, C., 1888. Über das Verhältnis von pflanzen zu bicarbonaten und ber kalkincrustation. *Unersuch. Bot. Inst. Tübingen* 2: 467–473.

Joyce, J. C., K. A. Langeland, T. K. Van & V. V. Vandiver, 1993. Organic sedimentation associated with hydrilla management. J. Aquat. Plant Mgmt 30: 20–23.

Kemp, W. M., W. R. Boynton, R. R. Twilley, J. C. Stevenson & L. G. Ward, 1984. Influences of submersed vascular plants on ecological processes in upper Chesapeake Bay. In V. S. Kennedy (ed.), The Estuary as a Filter. Academic. New York: 367–394.

LAKEWATCH, 1992. Annual Report. University of Florida, Gainesville, FL.

Lewis, W. M. & F. H. Weibezahn, 1981. Chemistry of a 7.5 m sediment core from Lake Valencia, Venezuela. Limnol. Oceanogr. 26. 907–924.

Lewis, W. M., 1981. Precipitation chemistry and nutrient loading by precipitation in a tropical watershed. Wat. Resour. Res. 17: 169–181.

McConnaughy, T., 1991. Calcification in Chara carolina: CO_2 hydroxylation generates protons for bicarbonate assimilation. Limnol. Oceanogr. 36: 619–628.

Otsuki, A. & R. G Wetzel, 1972. Coprecipitation of phosphate with carbonates in a marl lake. Limnol. Oceanogr. 17: 763–767.

Phlips, E. J., P. V. Zimba, M. S. Hopson & T. L. Crisman, 1993. Dynamics of the plankton community in submerged plant dominated regions of Lake Okeechobee, Florida, USA. Verh. int. Ver. Limnol. 25: 423–426.

Rast, W., V. H. Smith & J. A. Thornton, 1989. Characteristics of eutrophication. In S. Ryding & W. Rast (eds), The control of eutrophication of lakes and reservoirs. UNESCO, Paris: 37–64.

Waisel, Y., J. J. Oertli & A. Stahel. 1990. The role of macrophytes in phosphorus turnover: sources and sinks. EWRS Symposium on Aquatic Weeds 8: 243–248.

Walker, D. I. & W. J. Woelkerling, 1988. Quantitative study of sediment contribution by epiphytic coralline red algae in seagrass meadows in Shark Bay, Western Australia. Mar. Ecol. Prog. Ser. 4: 71–77.

Wetzel, R. G., 1983. Limnology. Saunders College Pub. Philadelphia, 860 pp.

Wetzel, R. G. & W. S. White, 1985. Alteration of iron-$CaCO_3$ precipitation by yellow organic acids of aquatic angiosperm origin. Arch. Hydrobiol. 104: 247–251.

White, W. S. & R. G. Wetzel, 1985. Association of vitamin B_{12} with calcium carbonate in hardwater lakes. Arch. Hydrobiol. 104: 305–309.

Young, J. R., J. M. Didymus & S. Mann, 1991. On the reported presence of vaterite and aragonite in coccoliths of *Emiliania huxleyi*. Bot. Mar. 34: 589–591.

Zimba, P. V., 1995. Epiphytic biomass in the littoral zone of Lake Okeechobee, Florida. Arch. Hydrobiol. 45: 233–240.

Hydrobiologia **340:** 43–49, 1996.
J. M. Caffrey, P. R. F. Barrett, K. J. Murphy & P. M. Wade (eds), Management and Ecology of Freshwater Plants.

Assessing functional typology involving water quality, physical features and macrophytes in a Normandy river

Jacques Haury
Institut National de la Recherche Agronomique, Laboratoire d'Ecologie aquatique & Ecole Nationale Supérieure Agronomique, Département Environnement et Exploitation des Ressources Naturelles, Laboratoire d'Ecologie et Sciences phytosanitaires, 65, rue de Saint Brieuc, 35052 Rennes Cedex, France

Key words: Multicompartment typology, stream, methodology, macrophyte, bioindication

Abstract

To carry out a functional typology of a Normandy watercourse, three compartments (water quality, physical features and macrophytes) were studied in 74 stations. The data were analysed with multidimensional methods: Principal Component Analysis (P.C.A.), Hierarchical Cluster Analysis (H.C.A.) and Multiple Factor Analysis (M.F.A.).

P.C.A. analysis of the 3 compartments ordered significant variables as following :
– stream order, water velocity and light,
– conductivity, pollution parameters as opposited to mineralisation ones,
– upstream/downstream, sciaphilous/heliophilous, rheophilous, ditch species.

All the compartments showed a longitudinal zonation pattern which characterised the first axis of M.F.A.. The second factor of M.F.A. was mainly due to water quality and macrophytes, and the third one to physical features and macrophytes: macrophytes appeared to be reliable ecological integrators of the spatial and functional heterogeneity.

Four clusters obtained in H.C.A. defined functional parts of the watercourse.

This example of typology, and the relationships identified between the 3 compartments, assessed the interest of macrophytes for characterising watercourses, and diagnosing their equilibrium.

Introduction

Typing the functioning of rivers with their ecological compartments is increasingly used in hydrobiological research either to get a diagnosis of river stations or to understand the relationships between these compartments. On such topics, it is necessary to assess the 'normal' situation, and to compare it to abnormal situations (Holmes, 1983; Haslam, 1987).

Thus, to carry out a functional typology of a Normandy watercourse, with only one survey and mainly field research, three compartments were involved: vegetation, physical features, water quality. The results of each compartment were discussed in regard to the whole system to assess ecological processes and internal stability.

Study area

The River Oir is an estuarine tributary of the River Sélune which joins the sea in the Mont Saint Michel Bay. Its catchment is about 87 km^2, and the total network is 90 km long Baglinière et al. (1993) (Figure 1). Only acidic substrata (schists and a few granites) are present, but the whole basin is mainly influenced by thick soils.

Methods

A stratified plan was chosen to draw up an inventory of macrophyte vegetation (Haury, 1985). The relevant parameters were the size of the watercourse, light, local hydrodynamics and known pollution inputs. Each of the 74 studied stations was 50 m-long. Field data

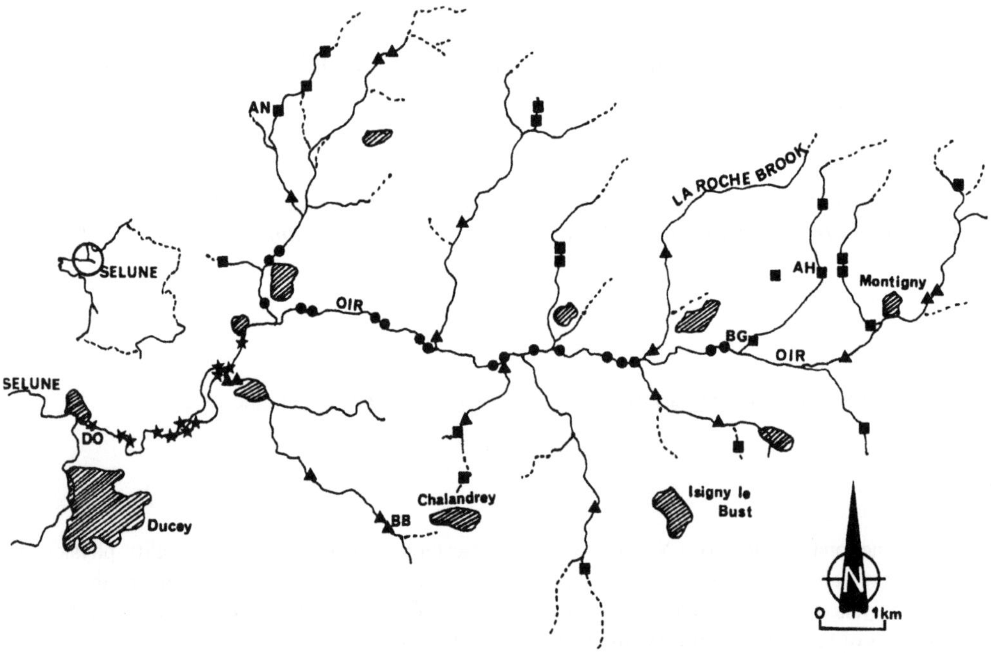

Figure 1. The River Oir and its 74 stations. The squares represent order 1 stations, the triangles order 2 stations, the circles order 3, and the stars order 4. AH, AN, BB, BG and DO were particular stations studied in the text.

were collected between June and July 1988 (low water-level and period of maximum macrophyte cover in Normandy conditions), by one survey for each compartment, and two other verifications for macrophyte distributions.

Size of watercourses was expressed by Strahler's stream order and Horton's cumulative order and by length of watercourses upstream from each station. All these orders and slope were recorded from 1/25 000 National Geographical Institute's maps. Width and bank height were measured on 5 transverse sections. The ratio between width and bank height was also calculated (embankment). Depth was measured on these sections with regular observations pinning down a gauge (point method: Haury, 1985). At each point, the particle size of apparent substratum was also noted using the Cailleux scale; then the relative cover of each class of size substratum was estimated along the station and corrected with the obtained percentages on the transerve sections. Current velocities were measured with a OTT current-meter. Light was estimated using five classes from 1 (very shaded: >80% of the stream shaded) to 5 (less than 20% shaded). Percentage of bank with trees also indicated the available light. This compartment of physical features included 29 variables.

All the measurements of water quality, and the samples were obtained on the same day. Conductivity (Metrohms conductimeter) and pH (OSI pHmeter) were measured in the field. Four parameters were analyzed by spectrocolorimetry (HACH): PO_4 (samples collected in glass bottles and stocked in a fridge), NO_3, NO_2 and NH_4 (samples fixed with sulfuric acid).

The bryophytes and phanerogams present either in the submerged-bed or on the subaquatic zone were all recorded by wading and using a bottom-glazed box. Their cover was estimated after calibration with the frequency obtained with the point method, Only the 32 most hydrophilous species were involved in data analyses.

Three multivariate methods were involved. Principal Component Analysis (P.C.A.-CISIA, 1991) was used to study the correlations between variables and the groups of them determining the similarities between individuals. Hierarchical Cluster Analysis (H.C.A.-CISIA, 1991) was used to establish clusters of individuals, ordinate and demonstrate the determining variables. Multiple Factorial Analysis (M.F.A.) enables a simultaneous analysis of the three compartments (Escofier & Pagès, 1988): the method calculates common factors as far as possible, equalizing the importance of each comparment; thus, functional units can be demonstrated (Pagès et al., 1991).

Results

Data matrix and general results

The data matrix corresponded to 74 individuals (stations) and 67 active quantitative variables. Mean results characterized the type of stream we studied.

The general features of the network assessed the River Oir is a medium-size system (90 km of total watercourse length, maximum stream order: 4, and maximum cumulative order: 40), narrow (mean width: 1.88 m), scarcely shaded (mean percentage of forested banks: 32,5%). Bed substratum was characterized by the importance of fine particles, with a large amount of silt (30.8%) and sands (22.4%), few gravels (13.7%), more stones and pebbles (26.2%) and very few boulders (3.7%). During the study, mean depth was only 15.6 cm; mean current velocity was 52 cm s^{-1}.

The mean water quality corresponded to acidic (mean pH: 6.6) and poorly mineralised waters (mean conductivity: 163 μS cm^{-1}). The mean values of nitrate, nitrite, ammonia and orthophosphate contents were respectively 54.3 mg l^{-1}, 0.05 mg l^{-1}, 0.12 mg l^{-1} and 0.23 mg l^{-1}.

Dominant macrophytes (cover >4%) were: *Callitriche platycarpa, C. hamulata, C. obtusangula, Phalaris arundinacea, Apium nodiflorum, Conocephalum conicum, Lunularia cruciata, Leptodictyum riparium, Chiloscyphus polyanthus.*

Relationships between variables and the structure of matrix

The variance of physical compartment data was mainly represented par the axes 1 to 4 of P.C.A. (59.7% of total variance). On the first axis, stream order, depth and width appeared as opposed to width variation and slope. The second axis was due to current velocity and coarse granulometry versus silt. The third corresponded to light versus bank trees and percentage of plant detritus. The fourth corresponded to stream embankment.

Table 1. The eigenvalues of the M F A, and the correlation coefficients between factors of the global configuration and projection of the three compartment configurations.

Factor N° of the whole system	1	2	3	4	5
Eigenvalues	2.26	1.07	0.83	0.73	0.63
% variance	18.5	8.7	6.8	6.0	5.2
Correlation coefficients between compartments and general factors					
Physical mesology	0.938	0.523	0.821	0.713	0.477
Chemistry	0.806	0.837	0.303	0.788	0.597
Macrophytes	0.907	0.769	0.905	0.854	0.825

The variance of water quality data was mainly shown by the two first axes (57.9% of the variance). Conductivity and pH chiefly characterized the first axis. Pollution parameters (orthophosphates and nitrites) contributed mainly to the second (21.0%), (while nitrates characterized the third axis).

Macrophyte data gave a four-dimensional structure (41.3% of total variance). On axis one, *Fontinalis antipyretica, Leptodictyum riparium, Callitriche hamulata, C. platycarpa* and *Phalaris arundinacea* opposed to *Glyceria fluitans,Mentha aquatica, Hygroamblystegium fluviatile* and *Apium nodiflorum.* The second axis was characterized by most bryophytes. *Ranunculus penicillatus* and *Nasturtium officinale* contributed mainly to the third axis. The fourth axis was due to sunny ditch or upstream species, such as *Myosotis scorpioïdes, Apium nodiflorum, Callitriche stagnalis.*

The 1 × 2 factorial plane of M.F.A. (Figure 2) illustrated the relationships between all the variables. It assessed the first axis expressed longitudinal zonation, and showed the particular position of pollution chemical variables, NO_2, NH_4, PO_4, as opposed to most bryophytes, specially *Scapania undulata.* The general structure of the stream appeared to be three (or five)-dimensional, but each of the compartments did not contribute equally to the whole structure, as assessed by Table 1. Mesology contributed mainly to the first and third axes, while chemistry was predominant in the second and first ones; macrophytes contributed to a large extent to each axis and appeared to show a more regular and strong structure than the other compartments. Thus, the structure given by the first axis (longitudinal zonation) was the only one common to the three compartments.

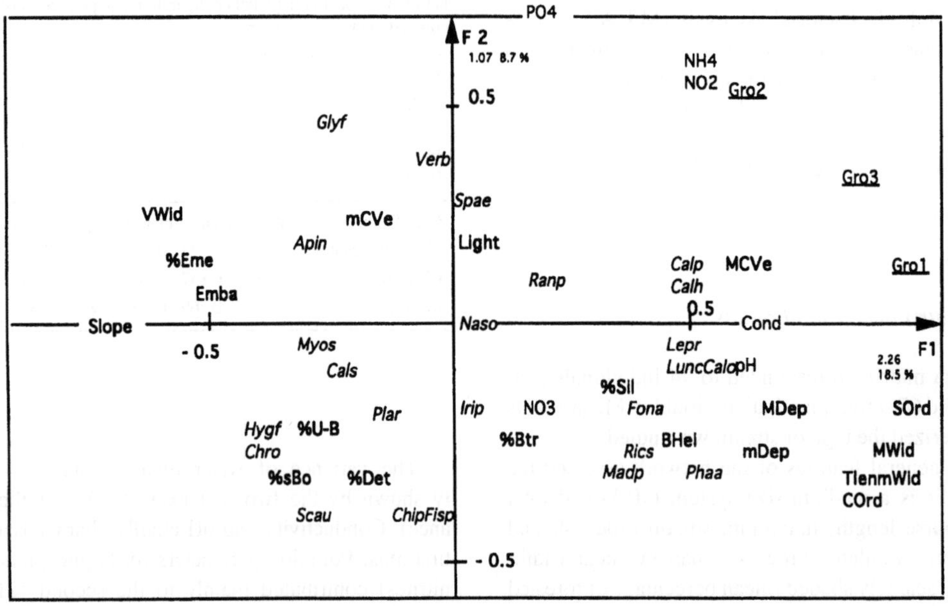

Figure 2. The F1xF2 M.F.A. plane with active variables and position of groups. *Gro1*: physical features; *Gro2*: water quality; *Gro3*: macrophytes. Only the variables most distant from the centre were presented. **Physical features: COrd**: cumulative order; **MWid**: maximum width; **mWid**: mean width; **VWid**: width variation; **mDep**: mean depth; **MDep**: maximum depth; **BHei**: bank height; **%Sil**: % silt; **Light**: light; **Slope**: slope; **%Eme**: % emerged parts; **MCVe**: maximum current velocity; **mCVe**: mean current velocity; **%SBo**: % small boulders; **%Det** : % detritus; **Emba**: embankment; **Tlen**: total lenght upstream; **%Btr**: % bank trees; **SOrd**: stream order; **Water quality:** Cond: Conductivity; NH_4: ammonia; NO_2: nitrites; NO_3: nitrates; pH: pH; PO_4: orthophosphates. **Macrophytes – Bryophytes:** *Chip: Chiloscyphus polyanthus* (L.) Corda; *Fisp: Fissidens pusillus* Wils.; *Fona: Fontinalis antipyretica* Hedw.;*Hygf: Hygroamblystegium fluviatile* (Sm.) Loeske; *Lepr: Leptodictyum riparium* (Hedw.) Warnst.; *Lunc: Lunularia cruciata* (L.) Dum.; *Madp: Madotheca porella* (Dicks.) Nees; *Plar: Platyhypnidium rusciforme* (Neck.) Fleisch.; *Rics: Riccardia sinuata* (Dicks.) Trev.; *Scau: Scapania undulata* (L.) Dum.. **Phanerogams:** *Apin: Apium nodiflorum* (L.) Lag.; *Calh: Callitriche hamulata* Kütz. ex Koch; *Calo: C. obtusangula* Le Gall; *Calp: C. platycarpa* Kütz.;*Cals: C. stagnalis* Scop.; *Chro: Chrysosplenium oppositifolium* L.; *Glyf : Glyceria fluitans* (L.) R. Br.; *Irip: Iris pseudacorus* L.; *Myos: Myosotis scorpioides* L.; *Naso: Nasturtium officinale* R. Br.; *Phaa: Phalaris arundinacea* L.;*Ranp: Ranunculus penicillatus* (Dum.) Bab.; *Spae: Sparganium erectum* L.; *Verb: Veronica beccabunga* L.

Functional clusters assessed with H.C.A.

The dendrogram of H.C.A. gave four clusters determined by active variables of each compartment (Table 2). The first cluster corresponded to upstream, narrow and sunny stations with *Glyceria fluitans* and *Myosotis scorpioïdes.* The second cluster gathered upstream oligotrophic and shaded stations with many bryophyta such as *Scapania undulata*, but also *Hygroamblystegium fluviatile*, and the spermaphyta *Chrysosplenium oppositifolium.* The third one gathered swift, medium-size stations, with *Callitriche hamulata, Ranunculus penicillatus* and mosses, with possible ammonia. The last one corresponded to downstream eutrophicated stations.

Study of individuals

Groups of individuals were demonstrated with the H.C.A. and the M.F.A. (Figure 3) and assessed longitudinal zonation. Particular stations had great internal inertia: their compartment representations showed discordances between each other which can assess internal disequilibrium (Figure 4) (Pagès et al., 1991). The forested station AH assessed good water quality with *Scapania undulata*: in the factorial plane, its botanical representation appeared more upstream than its physical features. In the downstream station DO, much *Phalaris arundinacea, Lunularia cruciata* and *Leptodictyum riparium* assessed eutrophicated water and width. Some stations had a greater concentration of nitrites or ammonia and showed either a local lack of

Table 2. Results of the Cluster Analysis.

Cluster N°	Physical features	Water quality	Macrophytes
1	Embanked Sunny Shallow	Low conductivity	*Glyceria fluitans* *Mysosotis scorpioides*
2	Coarse granulometry Shaded	Good water quality	*Chrysosplenium oppositifolium* *Hygroamblystegium fluviatile* *Scapania undulata*
3	High current velocity A lot of stones	High content of Ammonia	*Callitriche hamulata* *Ranunculus penicillatus* *Platyhypnidium rusciforme* *Nasturtium offricinale*
4	3 & 4 Stream order Deep Wide High bank	High pH High conductivity	*Lunularia cruciata* *Phalaris arundinacea* *Callitriche obtusangula* *Leptodictyum riparium* *Riccardia sinuata*

oxygen in ditches due to still water and/or agriculture pollution (AN) or the influence of sewage pollution (BB: Chalandrey). The station BG had a fair cover of *Ranunculus penicillatus* and *Nasturtium officinale*: it was a straightened and sunny station.

Discussion

Macrophyte bioindicators and phytocenosis organization

Ecological tendancies of most macrophytes were in accordance with previous results (Newbold & Holmes, 1987), unless some species appeared to have slightly differing distribution, such as *Hygroamblystegium fluviatile* which belonged to a good water quality group in our study. The assessed macrophyte groups were also in accordance with literature (Holmes, 1983; Wiegleb, 1983), unless they gathered few species. The hierarchy of macrophyte distribution appeared to be mainly due to mesology versus chemistry. These differences were due to the smallness of the network, to the homogeneity of the river basin, and to the absence of heavy pollution. It means that a water quality diagnosis with macrophytes cannot ignore the physical context (Haury & Muller, 1991).

Macrophyte typologies

Such typologies involving macrophyte cover can assess 'normal' vegetation with an ordered importance of species versus 'abnormal' situation characterized either by some species proliferation or by the lack of other ones. Nevertheless, few attempts gave 'normal cover', though Haslam (1987) involved such a parameter in her Damage Rating. The links of 'normal' phytocenoses with biodiversity are not as reliable as usually thought, for, in oligotrophic and acid systems, macrophyte cover and species richness increase with downstream enrichment. As river macrophytes can be easily mapped, their communities lead to river types and maps, as shown by multivariate analysis (Grasmuck et al., 1993) or by floristic surveys relating them either to geological subtrata and width (Haslam, 1987) or to water quality (Haury & Muller, 1991).

Multitable analyses can order and hierarchize ecological factors for a better understanding of the functioning of the system (Haury et al., 1995); they lead to a general quantification of relationships between biotope and phytocenoses. Thus, the four cluster types corresponded to different ecological processes within each zone: development of small paludous amphiphytes in sunny brooks where slope induced erosion and coarse granulometry, predominance of sciaphilous mosses growing on boulders and stones in forested upstream parts, rheophilous *Callitriche-Ranunculus* vegetation bearing ammonia with *Nas-*

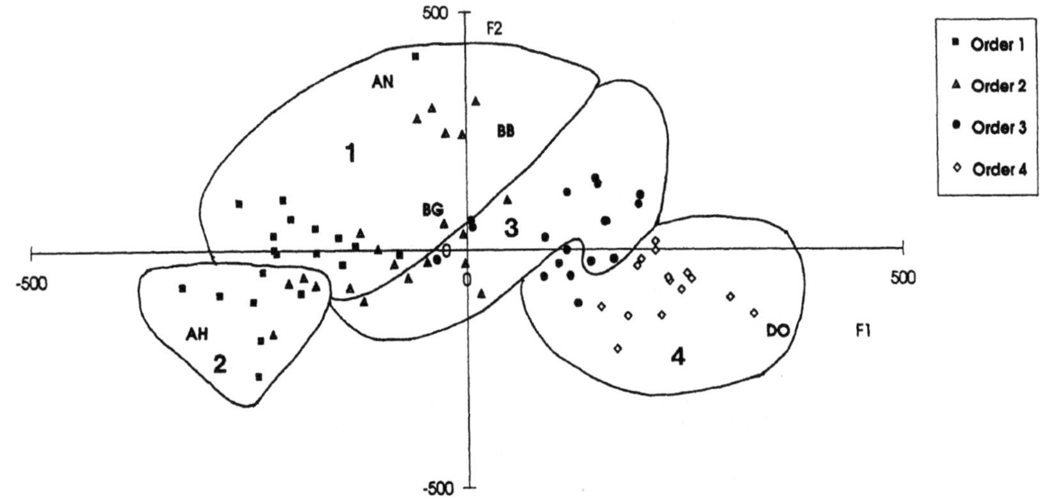

Figure 3. Distribution of the 74 stations (whole compartments) within the F1 × F2 M.F.A. plane and Clusters 1 to 4; place of the 5 particular stations AH, AN, BB, BG & DO.

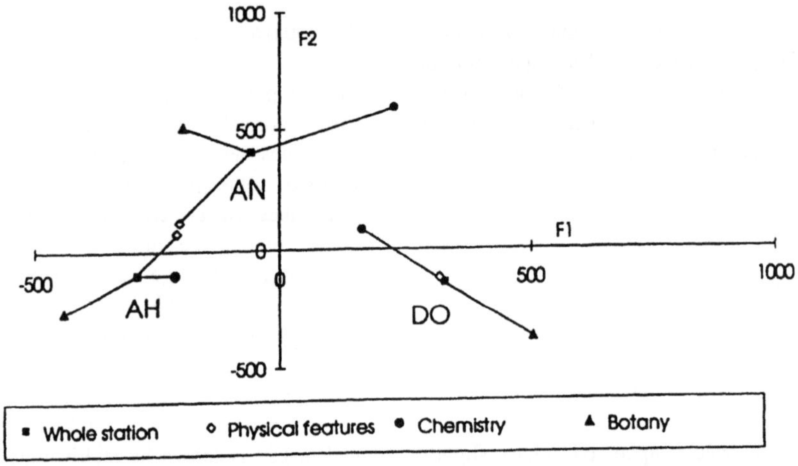

Figure 4. Distribution of three particular stations and of their compartment representations in the F1 × F2 M.F.A. plane.

turtium officinale in straightened and/or eutrophicated parts, deep low-flowing and wide downstream stretches with helophyte stands of *P. arundinacea* or bank moss communities with *L. cruciata*, these both species bearing silting.

For each station, the differences between compartment representations may point out an abnormal situation. They measure the disequilibrium between potential flora, depending on the position inside the network, and the observed one which expresses harnessing or pollution. Such a study should be undertaken from time to time to assess the evolution of differences between compartment representations and measure the quickness of evolution.

Conclusion

In the general topic of assessing the functioning of rivers, macrophyte vegetation is an useful tool, because it is easily and quickly explored, while chemical analyses only give instantaneous figures. Macrophyte vegetation appears as a reliable bio-integrator of mesological features and water quality; thus, all macrophyte indices (Haury & Peltre, 1993) should integrate both compartments in the diagnosis. Research on reference phytocenoses must be undertaken and carried on with in Western Europe, including at least bryophyta whose development and distribution is linked to aquatic Spermaphyta. In France, a field method involving macrophytes is already undertaken to assess general river quality; it will complete other biological indices using benthic invertebrates and diatoms.

Acknowledgments

I thank F. Marchand (INRA Ecologie) who performed mesological relevés and water analyses, and C. Beouan (Rennes University) for her help in English verification.

References

Baglinière, J.-L., G. Maisse & A. Nihouarn, 1993. Comparison of two methods of estimating Atlantic Salmon (*Salmo salar*) wild smolt production. In R. J. Gibson & R. E. Cutting (eds), The production of juvenile Atlantic salmon, *Salmo salar*, in natural waters. Can. J. Fish. aquat. Sci., spec. Bull. 118: 189–201.

CISIA (Centre International de Statistique et d' Informatique Appliquées), 1991. SPAD N.2.0 intégré. St Mandé (France).

Escofier, B. & J. Pagès, 1988. Analyses factorielles simples et multiples – Objectifs, méthodes et interprétation. Dunod Paris, 241 pp.

Grasmück, N., J. Haury, L. Léglize & S. Muller, 1993. Analyse de la végétation aquatique fixée des cours d'eau lorrains en relation avec les paramètres d'environnement. Ann. Limnol. 29: 223–237.

Haslam, S. M., 1987. River plants of Western Europe. Cambridge Univ. Press, Cambridge, 512 pp.

Haury, J., 1985. Etude écologique des macrophytes du Scorff (Bretagne-Sud). Thès. Dr.-Ing. Ecologie Univ. Rennes I, 243 pp.

Haury, J. & S. Muller, 1991. Variations écologiques et chorologiques de la végétation macrophytique des rivières acides du Massif Armoricain et des Vosges du Nord (France). Rev. Sci. Eau 4: 463–482.

Haury, J. & M.-C. Peltre, 1993. Intérêts et limites des 'indices macrophytes' pour qualifier la mésologie et la physico-chimie des cours d'eau: exemples armoricains, picards et lorrains. Annls Limnol. 29: 239–253.

Haury, J., J.-L. Baglinière, A.-I. Cassou & G. Maisse, 1995. Analysis of spatial and temporal organisation in a salmonid brook in relation to physical factors and macrophytic vegetation. Hydrobiologia 300-301: 269–277.

Holmes, N. T. H., 1983. Focus on Nature Conservancy. 4 – Typing British rivers according to their Flora. Shrewbury: Nature Conservancy Council.

Newbold, C. & N. T. H. Holmes, 1987. Nature conservation: water quality criteria and plants as water quality monitors. Water Pollution Control 86: 345–364.

Pagès, J., B. Escofier & J. Haury, 1991. Multiple factor analysis: a method to analyse several groups of variables measured on the same set of individuals. In J. Devillers & Karcher (eds), Applied Multivariate Analysis in SAR and ENVIRONMENTAL Studies, ECSC, EEC, EAEC, Brussels and Luxembourg: 33–83.

Wiegleb, G., 1983. A phytosociological study of the macrophytic vegetation of running waters in Western Lower Saxony (Fed. Repub. Ger.). Aquat. Bot. 17: 251–274.

Hydrobiologia **340:** 51–57, 1996.
J. M. Caffrey, P. R. F. Barrett, K. J. Murphy & P. M. Wade (eds), Management and Ecology of Freshwater Plants.

The effects of a record flood on the aquatic vegetation of the Upper Mississippi River System: some preliminary findings

Andrew Spink[1] & Sara Rogers[2]
[1] *Illinois Natural History Survey, 794 N Schrader Ave, Havana, IL 62644, USA*
Present address: Department of Plant Ecology & Evolutionary Biology, University of Utrecht, P.O. Box 800.84, 3508 TB Utrecht, The Netherlands
[2]*EMTC, 575 Lester Ave, Onalaska, WI 54650, USA*

Key words: Flood, Mississippi, Salix, Potamogeton, Vallisneria, Myriophyllum

Abstract

During 1993 the Upper Mississippi River System experienced floods of exceptional magnitude and duration, especially at its more downstream reaches. The flood had widespread effects on the vegetation. Submerged species such as Potamogeton pectinatus significantly decreased in abundance, especially at sites with more severe flooding. However, many species were able to regenerate in 1994 from seeds or storage organs. Emergent species such as *Scirpus fluviatilis* were similarly affected, but in the upstream reaches were able to regrow in the autumn following the flood and at many sites showed exceptionally high productivity in the following year, probably due to nutrient-rich sediment deposition by the flood. Many tree species were very severely impacted, although *Acer saccharinum* and *Populus deltoides* have shown some seedling regeneration on newly deposited sediment beneath stands of mature trees, which would have out-shaded the seedlings if they had not been killed by the flood.

Introduction

The flood of 1993 had widespread effects on the vegetation of the Mississippi River and its major tributaries. In this paper the results of measurements and observations throughout the Upper Mississippi are brought together and summarized.

The watershed of the Upper Mississippi River System (UMRS: Figure 1) covers 500 000 km^2. It has been altered to support commercial navigation by the construction of locks and dams, wing dikes, and through dredging. This has resulted in dramatic effects to the river ecosystem. The locks and dam, constructed in the 1930s, created relatively stable water levels immediately upstream of the dams and increased water surface areas due to inundation of the floodplain (Chen & Simons, 1986). However, the dams have increased the trapping efficiency of fine sediments in off-channel areas (Peck & Smart, 1986), which can lead to wide-ranging problems for aquatic macrophytes (Sparks et al., 1990). It is predicted that at present sedimentation rates, many backwater areas will become marshes within the next 50 to 100 years (Chen & Simons, 1986). The dams are not managed to store water: during the 1993 their gates were left open to increase water conveyance.

Flooding acts as another stress that can affect macrophytes in a variety of ways depending on the timing, duration and magnitude of the event. Although flood waters may provide additional nutrients via suspended materials to rooted macrophytes (Barko & Smart, 1983), negative consequences can also occur. These include burial or coverage by sediments reduced light availability (Van Dijk, 1992; Tanner et al., 1993), reduced in oxygen supply (when leaves of emergents are submerged: Coutts & Armstrong, 1976, Nielsen, 1993), increased herbicide supply (in agricultural catchments: Goolsby et al., 1994), and uprooting due to high velocities (Spink, 1992) or wind- generated waves because of the increased fetch can also occur.

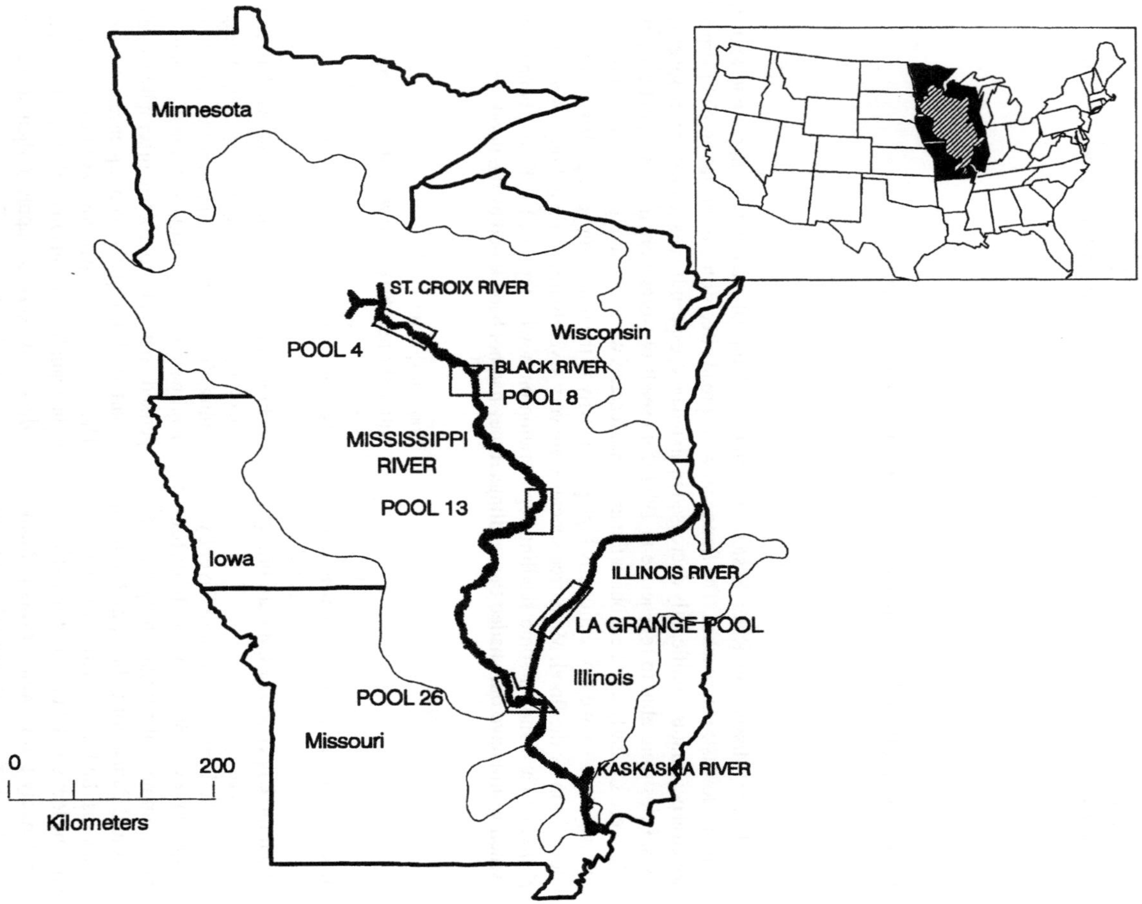

Figure 1. Map of the Upper Mississippi River System (UMRS). Pools are named after the number of the lock and dam at the downstream (southern) end. The UMRS consists of the Illinois River and upper 26 pools of the Mississippi River.

The flood

The flood was caused by a long period of exceptionally high rainfall over much of the upper Mississippi watershed. During June 1993, rainfall was generally twice average: several states had their highest July rainfalls since 1895. In addition to the huge amount of water carried by the 1993 flood, it is suspected that tremendous amounts of sediment were also carried as suspended load and as bed load. In addition to being moved downstream, massive quantities of sediment were deposited onto the floodplain and back channels. Newly created sandbars are evident in most pools, and deposition among the floodplain forest reached several centimeters thick in lower reaches of the river.

The flood of 1993 was notable because of the wide area it affected, large peak discharges, and exceptional duration (Figure 2A). It set new record peaks at nearly every station along the Mississippi. Peak water levels in late June through August were preceded by a period of high river levels that were frequently above flood stage for nearly four months (IFMRC, 1994). Extremely large amounts of agricultural chemicals and sediments were flushed into the Mississippi River and tributaries, including mean atrazine concentrations of 2.2 μgl^{-1} (Goolsby et al., 1994).

Not all places along the river were affected equally by the flood; the more downstream sites had higher floods lasting for longer (Figure 2B).

Methods

The data in this paper are taken from three sources;

(a) Vegetation transects of the Long Term Resource Monitoring Program (LTRMP). The LTRMP monitors the biology and chemistry of the UMRS in order to provide information on which to base management decisions. Thirty locations in five of the 26 reaches upstream of a navigation dam are surveyed twice every growing season. Within each location, transects are positioned perpendicular to shore at 50 m intervals. Samples are taken every 15–30 m with a long-handled rake to determine species composition, frequency of occurrence and to estimate relative abundance. In 1993, the timing of the transect sampling coincided with flood events. In all pools, spring/early sampling was completed after early spring high water and just before water levels rose to flood stage in late June. The second sampling began in the upper pools as water levels receded and in Pool 26 while water was still high.

(b) Field observations recorded by LTRMP and other biologists.

(c) Published reports (especially Dieterman, 1993).

Results

Most of the species in the UMRS (e.g. *Elodea canadensis, Vallisneria americana*) show a typical increase in density and biomass throughout the season, with a peak biomass towards mid/late summer (Madsen & Adams, 1988). However, the introduced species *Potamogeton crispus* starts turion production followed by senescence at water temperatures around 20 °C, showing an early decline, especially in the more northern pools (Nichols & Shaw, 1986).

In 1993, a different pattern was evident along transect sites between the early sampling period, which occurred between May 15 and June 15, and the second sampling period, which occurred between July 15 and August 30 (Figure 3). In the spring of 1993, before the record flood had started, the vegetation was growing more abundantly than in the previous year. However, as the flood waters rose many of the species were unable to cope with the resultant stresses, and by the summer their cover was significantly (*t*-tests) less than in the previous year. This was particularly marked at sites which had experienced more severe flooding (Figure 2A, 3B): generally more downstream reaches.

Dieterman (1993) found a similar pattern in reaches 3, 5, 5A and 6. He observed declining populations of plants in nearly all backwaters during June–August 1993. In Pool 26 there were no submerged plants observed at all by mid-August 1993, though *Ceratophyllum* demersum was observed in tree branches after the floodwaters began to recede (J. Nelson, in litt.).

Response to flooding (specific species)

Myriophyllum spicatum

M. spicatum apparently became established during the mid 1980s within many of the northern reaches (3 through 13) of the UMRS. Following a widespread drought in 1988, *M. spicatum* appeared in some reaches as large monotypic beds. In backwaters of reaches 7 and 8 many of the plants that were present before the 1993 flood survived when they responded to high water levels by producing stem lengths up to 3.3 m long, reaching the water's surface within two weeks of higher water levels (S. Rogers, pers. obsv.). Once

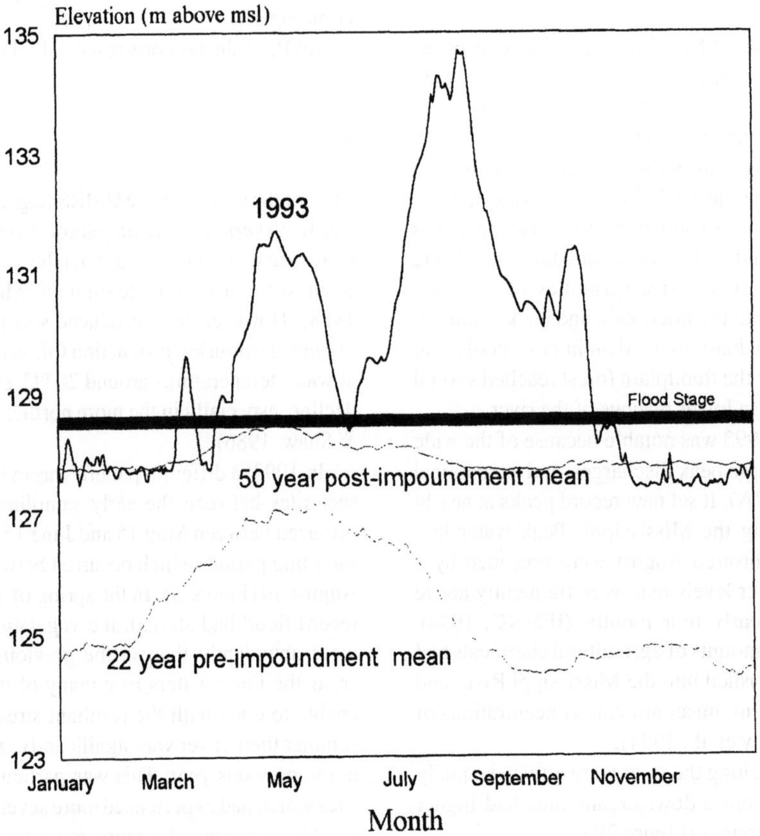

Figure 2a. Hydrograph of 1993 water levels of the Mississippi River at Alton, Illinois (Pool 26), with mean water levels for comparison. Flood stage (128.6 m above mean sea level) is also shown.

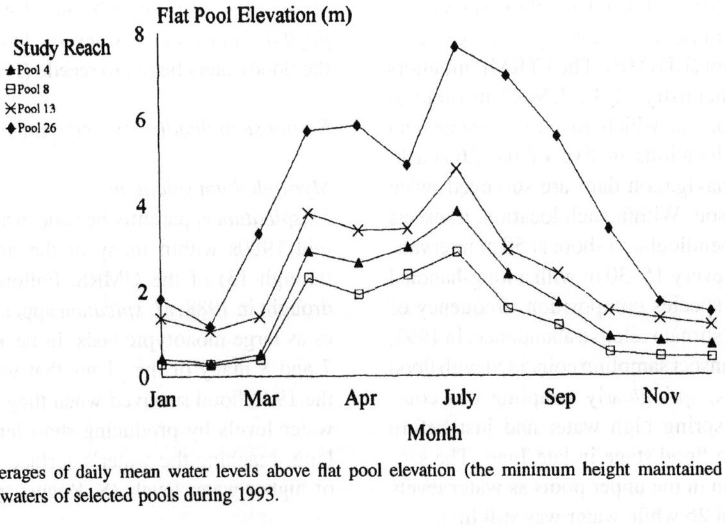

Figure 2b. Monthly averages of daily mean water levels above flat pool elevation (the minimum height maintained for navigation) in the Mississippi River at tailwaters of selected pools during 1993.

water levels fell, thick canopies of *M. spicatum* provided substrate for *Lemna* spp, which collected up to several centimetres thick in *Myriophyllum* beds in Lake Onalaska (Pool 7). In contrast, monotypic beds occurring in the impounded reaches of Pools 8 and 13 (where it was the dominant species in 1991 and 1992), and where velocities and turbidities were probably higher than in backwaters, disappeared by late July. By July 1994, beds of *M. spicatum* in Lake Onalaska persisted in some areas. However, in other portions of the lake, former stands of *M. spicatum* have been partially or completely replaced by *Vallisneria americana* and/or *Zosterella dubia*.

Nelumbo lutea
Initially during the flood, *N. lutea* grew up with the increasing water level, forming stems of at least 4 m in Pool 26 (J. Nelson, in litt.). However, as the water level increased, it was no longer able to keep its leaves above water and suffered severe die-back, especially in the lower pools. Very large beds of *Nelumbo lutea* completely disappeared from the Pools 19 and 26 during the flood event (R. Anderson pers. comm.) as well as other sites including backwaters adjacent to the Illinois River (A. Spink, pers. obsv.). In more upstream pools, where the flooding was less severe, it was able to make a recovery when water levels dropped in July (T. Blackburn, pers. comm.). During 1994 *N. lutea* has been re-establishing from seeds in the sediments deposited during the flood.

Potamogeton pectinatus
During the flood *P. pectinatus* showed a decrease in abundance throughout the UMRS (Figure 3A), and this was more pronounced towards the downstream end of the system, where the flooding was more intense. In the Illinois River it was almost completely eliminated by the 1993 flood. However, during 1994 it has grown again to biomass levels approaching previous years (A. Spink & T. Cook, pers. obsv.), presumably from turions or tubers. This is also the case for more upstream pools, where it has replaced *Myriophyllum spicatum* in some places.

Vallisneria americana
This species was abundant throughout the upper pools until a period of drought in 1988. The drought was associated with periods of low flow (and therefore low nutrient supply) rates as well as higher rates of epiphytic algal growth. During the 1993 flood the remaining *Vallisneria* beds grew well. During 1994 existing beds increased in size, and new beds have appeared in locations where it had been common before 1989. The flooding has also enabled this species to disperse to some areas where it has not been found previously (e.g. backwaters of Reach 26; J. Tucker pers. obsv.), and it is possible this has been the case for some other species as well.

Scirpus fluviatilis
The shoots of this emergent species were completely eliminated during the 1993 flood in many reaches of the lower UMRS. However, its dead stems act as efficient sediment traps, and during 1994 it has shown exceptionally high growth rates at many sites (e.g. a 3 m increase in stem length during a period of one month) (A. Spink, pers. obsv.). In the upper section of the river its growth was decreased during the flood (apparently due to sedimentation), but by mid-September 1993 new growth was appearing and in the summer 1994 the species had shown luxuriant re-growth.

Woody species

Floodplain forest is the most extensive plant cover type in the UMRS and the following summary is intended as a brief overview. Many flood tolerant tree species (e.g. *Salix nigra*, *Acer saccharinum*) have suffered very high mortality rates (especially among saplings) in the lower portion of the UMRS. In the Illinois River most *Salix* survived the flood, but in the spring of 1994 many leafed-out, only to loose their leaves and die within a few weeks. Less tolerant species showed higher mortality, especially during the flood itself. For example, 96% of *Celtis occidentalis* and 100% of *Carya laciniosa* were killed in Pool 26 (J. Nelson & Y. Yin, unpublished data). However, for at least some of these species (e.g. *A. saccharinum*, *Populus deltoides*), the death of shade-forming adult trees and deposition of new sediment has provided the opportunity for extensive seedling regeneration. Further upstream the effects were less severe, with most tree deaths occurring due to uprooting by shoreline erosion.

Discussion and conclusions

The majority of species in the river showed a clear north-south gradient in terms of response to the 1993 flood. The reduction in growth and increase in mortal-

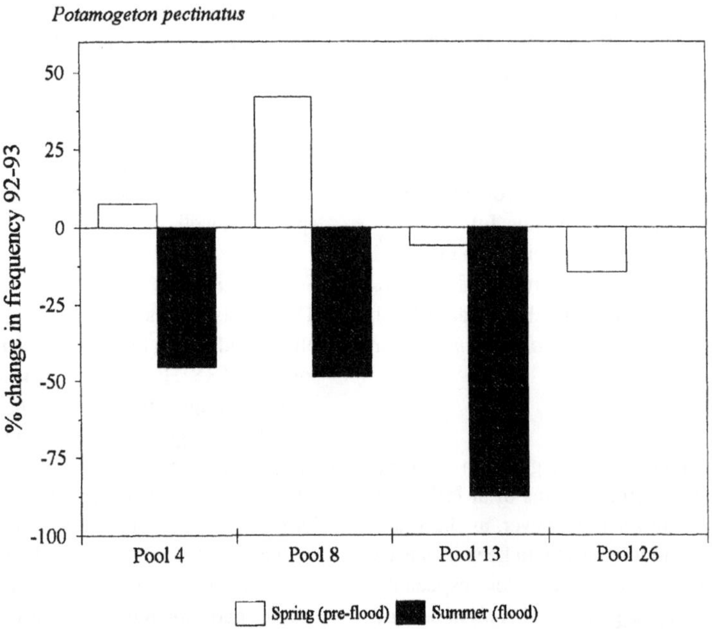

Figure 3a. Comparison of abundance of Potamogeton pectinatus in 1992 and 1993. Bars show % change between 1992 and 1993 in sites with *P. pectinatus* present. Open bars are comparisons between spring sampling (in 1992 before the flood) and summer sites (in 1993 during the flood). There was no *P. pectinatus* in Pool 26 in summer 93.

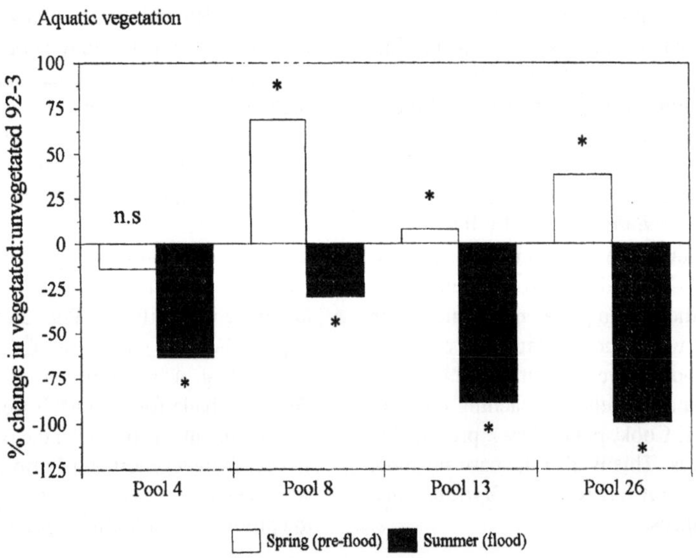

Figure 3b. Change in total submerged vegetation between 1992 and 1993. Bars show % change in the ratio of vegetated to unvegetated sites from 1992 to 1993. Open bars represent spring, closed summer. n.s. = no significant change from 1992-3, * significant change ($p<0.05$).

ity was much greater at the sites further downstream (Figure 3), where flooding was most severe (Figure 2). At the upstream sites many of the indigenous species displayed strategies for flood tolerance by stem etiolation, tolerance of low light levels and a capacity for rapid regrowth. However, at more southerly sites the duration and magnitude of the flood was such that many species suffered mortality. Trees, such as *Salix*, which tolerated limited inundation upstream were unable to cope with their roots being submerged for in excess a year at more severely affected sites. This was probably due to a combination of stresses including prolonged anaerobiosis (causing root death), sediment deposition and an accumulation of toxic agricultural chemicals. Those species which did survive did so by avoidance – *P. pectinatus* re-established from tubers, *N. lutea* from seeds and *S. fluviatilis* from rhizomes. At many of those locations the productivity during 1994 was exceptionally high, probably due to very high nutrient levels (A. Spink & M. Rijks, unpublished data) caused by fresh organic-rich sediment brought in during the flood. However, at sites where dikes broke, floodwaters moved across at high velocities, scouring out both vegetation and sediments, and what deposition did occur was course infertile sand sediment unsuitable either for tree seedling germination or plant growth. High water clarity in the spring of 1994 probably also played a role in the high productivity during the growing season.

The affects of the 1993 flood will influence the vegetation dynamics for many years to come, especially in the downstream reaches, and provides a clear example of how low frequency but high magnitude disturbance events can play a major role in structuring ecosystems.

Acknowledgments

The results presented in this paper are mostly derived from data obtained by the LTRMP vegetation crews. The authors would particularly like to thank Theresa Blackburn, Thad Cook, Heidi Langrehr, Robert Maher, John Nelson and Yao Yin. We also thank Rick Anderson and John Tucker for their observations, and Yao Yin and Jenny Sauer for their help in data analysis. K. D. Blodgett, J. Nelson, C. Smith, R. E. Sparks and Y. Yin provided helpful comments on an earlier draft of this paper.

References

Barko, J. W. & R. M. Smart, 1983. Effects of organic matter additions to sediments on the growth of aquatic plants. J. Ecol 71: 161–175.

Chen, Y. H. & D. B. Simons, 1986. Hydrology, hydraulics, and geomorphology of the Upper Mississippi River System. Hydrobiologia 136: 5–20.

Coutts, M. P. & W. Armstrong, 1976. Role of oxygen transport in the tolerance of trees to waterlogging. In Cannell, M. G. R. & F. T. Last (eds), Tree Physiology and Yield Improvement. Academic Press, London: 361–385.

Dieterman, D., 1993. Major river survey. Backwaters of the Mississippi River, Pools 3, 5, 5A & 6. Minnesota Department of Natural Resources, Division of Fish & Wildlife, 14.

Goolsby, D. A., W. A. Battaglin & E. M. Thruman, 1993. Occurrence and transport of agricultural chemicals in the Mississippi River basin, July through August 1993. United States Geological Survey Circular 1120-C. pp 22.

IFMRC (Interagency Floodplain Management Review Committee), 1994. Sharing the Challenge: Floodplain Management into the 21st Century. US Government Printing Office, Washington DC. 189 pp.

Madsen, J. D. & M. S. Adams, 1988. The seasonal biomass and productivity of the submerged macrophytes in a polluted Winsconsin stream. Freshwat. Biol. 20: 41–50.

Nichols, S. A. & B. H. Shaw, 1986. Ecological life history of three aquatic nuisance plants, *Myriophyllum spicatum*, *Potamogeton crispus* and *Elodea canadensis*. Hydrobiologia 131: 3–21.

Nielsen, S. L., 1993. A comparison of aerial and submerged photosynthesis in some Danish amphibious plants. Aquat. Bot. 45: 27–40.

Peck, J. H. & M. M. Smart, 1986. An assessment of the aquatic and wetland vegetation of the Upper Mississippi River. Hydrobiologia 136: 57–76.

Sparks, R. E., P. B. Bayley, S. L. Kohler & L. L. Osborne, 1990. Disturbance and recovery of large floodplain rivers. Envir. Mgmt 14: 699–709.

Spink, A. J., 1992 The ecological strategies of aquatic *Ranunculus* species. Ph.D. Thesis, University of Glasgow: 360.

Tanner, L. C., J. S. Clayton & R. D. S. Wells, 1993. Effects of suspended solids on the establishment and growth of *Egeria densa*. Aquat. Bot. 45: 299–310.

Van Dijk, G. M., 1992 Impact of light climate history on seasonal dynamics of *Potamogeton pectinatus* L. during a three year period (1986–1988). Aquat. Bot. 43: 17–42.

Hydrobiologia **340:** 59–65, 1996.
J. M. Caffrey, P. R. F. Barrett, K. J. Murphy & P. M. Wade (eds), Management and Ecology of Freshwater Plants.

Monitoring watercourse vegetation, a synecological approach to dynamic gradients

Roelf Pot
Advisory Group on Vegetation Management, National Reference Centre for Nature Management, Bornsesteeg 69, 6708 PD Wageningen, The Netherlands

Key words: syntaxonomy; community type assignment; cleaning frequency

Abstract

Changes in vegetation under reduced control measures over 3 to 5 years in watercourses in a rural environment in The Netherlands were evaluated. A method to deal with slow changes on a steep gradient is presented. The gradient with various vegetation types between the middle of the watercourse and the bank-top was split up into zones. Species composition of each zone was evaluated using literature on syntaxonomy. Cover of character species, multiplied by the width of the zones, was used to quantify the contribution of various syntaxa in the vegetation. Changes in these contribution data were used to evaluate changes over the years. The method was applied to two experiments in which cleaning frequency was reduced. Submerged vegetation of 'Callitriche-Ranunculetum penicillati' in one and of 'Potamogetonetalia pectinati' in the other case, hardly changed. Emergent vegetation of 'Nasturtio-Glycerietalia' or 'Sparganio-Glycerietum fluitans' tended to expand into the submerged zone. Bank vegetation began to show signs of development into ruderal vegetation, as a shift from 'Molinio-Arrhenatheretea' into 'Artemisietea' was detected. The method allowed the conclusion that conditions were too eutrophic in both experiments for a diverse brook vegetation development without additional habitat improvement.

Introduction

There is a growing demand for evaluation of vegetation response to control measures in watercourses in The Netherlands. Mechanical control has been developed to a very high level of efficiency. Vegetation removal frequency has increased through the years up to more than five times a year in many of the main watercourses to guarantee unimpeded discharge. The vegetation became so disturbed that it is now dominated by fast growing opportunist species, so that frequent intensive control measures remain necessary. The solution seems to be management rather than removal of vegetation. Experiments were made with timing, with different machines, wider and shallower bank profiles and control frequency to reduce the impact of the vegetation control on the ecosystem. In most such experiments the vegetation changes very slowly. Monitoring studies therefore demand very sensitive analyses techniques.

A special difficulty in research on vegetation in watercourses is the method of describing the plant communities. There is a strong cross gradient in watercourses from the deepest point up to the bank-top, with species that have a range of environmental demands. The Braun-Blanquet approach to vegetation ecology (Westhoff & Van der Maarel, 1973) is commonly used to describe vegetation types that can be related to environmental factors, but this method needs homogeneous sample plots. Describing the whole gradient in one sample makes evaluation of the vegetation very hard because of the heterogeneity. Slight changes in only part of the present plant communities or shifts in their mutual relations would not be noticed at all in time-series analyses of whole gradient samples.

A method to deal with the heterogeneity problem by splitting the gradient up into zones has been published elsewhere (Pot, 1993). In this paper the method has been worked out to detect slow changes in watercourse vegetation. Results of monitoring the vegetation of two watercourses after lowering cleaning frequency are presented.

Method

The vegetation on the cross gradient of all watercourses was split up into zones that were recognizable through their structure or dominant species. In all cases at least a dry bank-slope zone and a submerged zone were described. In most cases a more detailed sequence of zones could be distinguished, such as a zone dominated by emergent species or a zone just above the level of the water and strongly affected by the water. Within each zone, the vegetation is considered to be more or less homogeneous. Although zones mutually influence each other, mostly because of species that invade adjacent zones, boundaries of the zones are obvious in most cases. This is because the main factors determining the gradient are discontinuous, such as water availability and cleaning techniques which act differently on different parts of the gradient.

The length of the sampled area was determined by the minimum area to cover all the variance within each zone sample but not including transverse gradients. All macrophyte species for each of the zones were recorded. Presence of species was estimated using the combined cover-abundance scale of Braun-Blanquet, modified by Barkman et al. (1964) and Maarel (1979) into a scale with nine ordinal classes.

Sites were re-sampled every year or every second year. Location and length of the sampled area was kept the same, but boundaries and width of the zones were re-considered every time, in order to keep maximum homogeneity in the zones. Width of the zone is treated as a measure for the weight of the communities found in the zone, as described below.

Species composition of every zone sample was compared with vegetation types described by Westhoff and Den Held (1969), Preising (1990), Pott (1992) or Oberdorfer (1977) in the given order. Species were listed syntaxonomically, i.e. according to their quality as character species of syntaxa (hierarchically related community types). Only syntaxa which were represented by a reasonable number of species were considered. Species that are character species for more than one syntaxon were listed more than once if both syntaxa were also represented by other character species. For example *Glechoma hederacea* is a character species of some woodland communities, but also of 'Arrhenatheretalia'.

A measure of the contribution of the syntaxa in each vegetation zone was calculated by the cumulative cover-abundance of the species representing every syntaxon, multiplied by the width of the zone. These values could be calculated over every zone, and also allowed the contribution of the various syntaxa on the cross gradient as a whole to be expressed. Comparison of sets of zone samples from two or more succeeding years was then possible by comparing the calculated contribution of the various syntaxa. This method resulted in a relatively small number of data, which simplified interpretation. Changes in contribution could be interpreted as indications of changes in the vegetation, using the information on synecology from the literature cited above.

The Keersop experiment

An elaborated example of the application of the method is derived from experiments in the Keersop, a more or less winding relatively natural brook in a rural landscape on pleistocene sandy soil, near Valkenswaard in the south of The Netherlands. Until 1988 the brook was cleaned from bank-top to bank-top using a mowing bucket twice a year. Vegetation on the banks was cut very short, channel vegetation was removed completely. Since 1989 twice a year only half of the profile was cleaned, leaving bank vegetation on one side and a strip of vegetation in the water channel of at least one meter in width untouched. Sides were cut alternatively however, so effectively the profile was cleaned once a year. Five sites, equally distributed over the brook, were monitored from 1989 to 1993. Sampling was omitted in 1992. Results of two of the sites will be discussed.

Table 1 shows all samples taken at site 4. Species are listed in the order of the syntaxa to which they are character species. Syntaxa that are only represented by high hierarchical order character species (1, 6–9) are regarded as less important to the vegetation processes. Table 2 summarises the contribution of the syntaxa in the zones and the gradient.

'Callitriche-Ranunculetum penicillati', as listed by Pott (1992), was well developed in the submerged zone. The impact of changing the cutting regime seems to be negligible, although a slight shift from association- to alliance-level or even to order-level can be seen in the totalised columns, which could be interpreted as an impoverishment of the community.

Table 1. Site 4 of the Keersop experiment. Species in 16 samples listed in syntaxonomical order. First column: syntaxa to which the species are character species, see table 2 for full names and explanation of numbers. Abundance in the samples: scales 1–4: cover < 5%; 1 = 1–2 indiv. per 10 m^2, 2 = 3–20, 3 = 21–100, 4 = > 100. scales 5–9: number of indiv. irrelevant; 5 = 5–12 % cover, 6 = 13–25%, 7 = 26–50%, 8 = 51– 75%, 9 = 76–100%.

	Side	Left						Middle				Right					
	Zone	Bank				Em.		Subm.				Em.		Bank			
	Year	89	90	91	93	91	93	89	90	91	93	91	93	89	90	91	93
1	Lemna minor	·	·	·	·	·	·	3	5	·	·	·	·	·	·	·	·
1	Spirodela polyrhiza	·	·	·	·	·	·	·	4	·	·	·	·	·	·	·	·
21	Potamogeton pectinatus	·	·	·	·	·	·	·	·	6	·	·	·	·	·	·	·
21	Potamogeton natans	·	·	·	·	·	·	5	6	5	8	·	4	·	·	·	·
21	Potamogeton crispus	·	·	·	·	·	·	·	2	5	·	·	·	·	·	·	·
21	Potamogeton trichoides	·	·	·	·	·	·	·	2	·	5	·	·	·	·	·	·
211	Callitriche platycarpa	·	·	·	·	·	4	4	6	7	8	·	5	·	·	·	·
211	Sparganium emersum	·	·	·	·	·	·	·	3	3	5	·	·	·	·	·	·
211	Sagittaria sagittifolia	·	·	·	·	·	1	1	2	1	·	·	·	·	·	·	·
2111	Ranunculus peltatus v. heterophyllus	·	·	·	·	·	·	5	8	5	7	·	·	·	·	·	·
2111	Callitriche hamulata	·	·	·	·	·	·	·	6	·	·	·	·	·	·	·	·
3	Glyceria maxima	·	·	·	·	·	8	·	·	·	·	·	·	·	·	·	·
3	Myosotis laxa	·	·	2	·	·	·	·	·	·	·	·	·	·	·	2	·
3	Mentha aquatica	·	2	2	1	·	·	·	·	·	·	·	2	·	·	·	·
3	Rorippa amphibia	1	·	·	·	·	·	·	·	·	·	·	·	·	·	·	·
3	Galium palustre	1	·	3	·	·	·	·	·	·	·	·	·	·	·	·	·
3	Peucedanum palustre	·	·	1	·	·	·	·	·	·	·	·	·	·	·	·	·
31	Rorippa microphylla	·	4	·	·	2	2	·	3	2	·	1	3	·	1	·	·
31	Myosotis palustris	4	5	2	·	·	·	·	3	·	·	·	4	3	2	3	·
31	Berula erecta	·	·	·	·	·	·	·	·	·	·	1	·	·	·	·	·
3111	Glyceria fluitans	·	4	·	·	8	7	4	5	4	6	6	·	·	·	·	·
3112	Phalaris arundinacea	5	6	·	6	7	7	·	4	·	·	8	8	6	7	·	·
4	Lolium perenne	3	·	·	·	·	·	·	·	·	·	·	·	·	2	·	·
41	Agrostis stolonifera	5	7	3	2	7	·	·	·	·	·	6	·	5	5	·	·
41	Alopecurus geniculatus	3	2	·	2	·	·	·	·	·	·	·	·	5	·	·	·
5	Holcus lanatus	7	4	3	2	·	·	·	·	·	·	·	·	·	7	4	·
5	Vicia cracca	·	3	2	1	·	·	·	·	·	·	·	·	·	·	·	·
5	Poa trivialis	·	·	·	·	·	·	·	·	·	·	·	·	·	6	·	·
5	Rumex acetosa	1	2	·	·	·	·	·	·	·	·	·	·	·	·	1	·
5	Trifolium pratense	·	·	·	·	·	·	·	·	·	·	·	·	·	2	·	·
51	Achillea ptarmica	·	3	4	2	·	·	·	·	·	·	·	·	·	·	·	·
51	Angelica sylvestris	2	2	1	·	·	·	·	·	·	·	·	·	2	2	·	1
51	Lysimachia vulgaris	·	1	1	·	·	·	·	·	·	·	·	·	·	1	·	·
51	Lotus uliginosus	·	2	2	·	·	·	·	·	·	·	·	·	·	·	·	·
51	Equisetum palustre	·	·	·	·	·	·	·	·	·	·	·	·	2	1	·	1
511	Stachys palustris	·	·	1	·	·	·	·	·	·	·	·	·	4	2	3	3
511	Eupatoriun cannabinum	2	·	·	·	·	·	·	·	·	·	·	·	2	2	1	·
511	Filipendula ulmaria	·	1	·	·	·	·	·	·	·	·	·	·	·	·	·	·
52	Arrhenatherum elatius	8	7	8	9	·	·	·	·	·	·	·	·	8	6	8	8
52	Achillea millefolium	·	3	1	1	·	·	·	·	·	·	·	·	·	3	3	·
52	Dactylis glomerata	5	1	·	·	·	·	·	·	·	·	·	·	3	5	·	·
52	Glechoma hederacea (1)	2	·	·	·	·	·	·	·	·	·	·	·	2	2	2	4
52	Cerastium fontanum ssp. vulgare	·	·	·	·	·	·	·	·	·	·	·	·	·	2	·	·
52	Stellaria graminea	2	·	·	·	·	·	·	·	·	·	·	·	·	2	·	2
6	Polygonum mite	2	·	·	·	·	·	·	·	·	·	·	·	2	·	·	2
6	Bidens frondosa	·	·	·	·	·	·	·	·	·	·	·	·	·	·	·	2
7	Stellaria media	·	·	3	·	·	·	·	·	·	·	·	·	·	3	3	2
7	Sisymbrium officinale	·	·	·	·	·	·	·	·	·	·	·	·	·	2	1	·
7	Capsella bursa-pastoris	·	·	·	·	·	·	·	·	·	·	·	·	·	2	·	·
7	Polygonum aviculare	·	·	1	·	·	·	·	·	·	·	·	·	·	·	·	·
8	Urtica dioica	3	5	4	7	·	·	·	·	·	·	·	·	3	5	6	5
8	Galium aparine	·	4	·	·	·	·	·	·	·	·	·	·	·	·	·	·
8	Cirsium arvense	·	·	·	·	·	·	·	·	·	·	·	·	·	2	3	·
8	Rumex obtusifolius	·	2	·	·	·	·	·	·	·	·	·	·	·	1	·	·
9	Holcus mollis	5	6	·	·	·	·	·	·	·	·	·	·	8	7	7	8
9	Dactylis glomerata	5	1	·	·	·	·	·	·	·	·	·	·	3	5	·	·
9	Glechoma hederacea (2)	2	·	·	·	·	·	·	·	·	·	·	·	2	2	2	4
–	Agrostis capillaris	2	·	3	·	·	·	·	·	·	·	·	·	·	3	4	·
–	Cardamine pratensis	·	2	·	·	·	·	·	·	·	·	·	·	·	·	·	·
–	Elymus repens	·	4	·	·	·	·	·	·	·	·	·	·	·	·	5	·
–	Epilobium tetragonum	·	1	·	·	·	·	·	·	·	·	·	·	·	1	·	·
–	Festuca rubra	·	6	5	·	·	·	·	·	·	·	·	·	·	3	·	·
–	Galeopsis tetrahit	·	·	1	·	·	·	·	·	·	·	·	·	·	2	·	·
–	Juncus effusus	1	4	1	·	·	·	·	·	·	·	·	·	·	·	·	·
–	Matricaria maritima	·	·	·	·	·	·	·	·	·	·	·	·	·	1	·	·
–	Polygonum amphibium	·	·	·	·	·	·	·	·	·	·	·	2	·	·	2	·
–	Ranunculus repens	2	3	·	·	·	·	·	·	·	·	·	·	·	2	2	2
–	Rubus species	1	1	2	·	·	·	·	·	·	·	·	·	·	·	1	·
–	Vicia hirsuta	1	·	·	·	·	·	·	·	·	·	·	·	·	·	·	·

Table 2. Profile data and contribution of syntaxa to the vegetation at site 4 of the Keersop experiment, calculated as explained in the text. First column: indication of hierarchy, digit 1 = Class, digit 2 = Order, digit 3 = Alliance, digit 4 = (As-)sociation. Profile data: Level = position of zone on gradient between two marks relative to current water level (negative = below water); Filling = vertical cover at cross section.

	Side	Left						Middle				Right						Totalised and standardised (%)			
	Zone	Bank				Emerg.		Submerged				Emerg.		Bank							
	Year	89	90	91	93	91	93	89	90	91	93	91	93	89	90	91	93	89	90	91	93
	Width of the zone (cm)	100	130	150	150	50	200	900	900	800	650	40	50	100	120	140	110				
	(at places)			(95)		(100)															
	Level: maximm (cm)	110	120	120	120	0	0	0	−15	−20	−40	10	0	110	120	120	90				
	Level: minimm (cm)	0	0	0	0	−10	−40	−30	−40	−40	−60	−20	−40	0	0	10	0				
	Height of vegetation (cm)	80	85	20	50	30	90	30	40	40	50	50	80	50	60	30	50				
	Total cover (%)	100	95	100	100	95	100	20	·	75	90	90	70	100	·	100	95				
	Filling of the profile (%)							20	·	50	80	100	100								
1	Lemnetea minoris	·	·	·	·	·	·	27	45	·	·	·	·	·	·	·	·	8	6	·	·
21	Potamogetonetalia pectinati	·	·	·	·	·	·	45	90	144	78	·	2	·	·	·	·	13	12	27	20
211	Ranunculion fluitantis	·	·	·	2	·	8	45	99	99	78	·	3	·	·	·	·	13	13	18	22
2111	Callitricho-Ranunculetum penicillati	·	·	·	·	·	·	45	126	45	42	·	·	·	·	·	·	13	17	8	10
3	Phragmitetea, several levels	2	3	12	2	·	16	·	·	·	·	·	1	·	·	3	·	1	·	3	5
31	Nasturtio-Glycerietalia	4	12	3	·	1	4	·	54	18	·	1	4	3	4	4	·	2	9	5	2
3111	Sparganio-Glycerietum fluitantis	·	5	·	·	4	14	36	45	36	36	2	·	·	·	·	·	11	7	8	12
3112	Phalaris arundinacea-sociation	5	8	·	9	4	14	·	36	·	·	3	4	6	8	·	·	3	7	1	7
4	Plantaginetea majoris	3	·	·	·	·	·	·	·	·	·	·	·	·	2	·	·	1	·	·	·
411	Lolio-Potentillion	8	12	5	6	4	·	·	·	·	·	2	·	10	6	·	·	5	2	2	1
5	Molinio-Arrhenatheretea	8	12	8	5	·	·	·	·	·	·	·	·	·	18	7	·	2	4	3	1
51	Molinietalia	2	10	12	3	·	·	·	·	·	·	·	·	4	5	·	2	2	2	2	1
511	Filipendulion	2	1	2	·	·	·	·	·	·	·	·	·	6	5	6	3	2	1	1	1
52	Arrhenatheretalia	17	14	14	15	·	·	·	·	·	·	·	·	13	24	18	15	9	5	6	7
6	Bidentetea tripartiti	2	·	·	·	·	·	·	·	·	·	·	·	2	·	·	4	1	·	·	1
7	Chenopodietea	·	·	6	·	·	·	·	·	·	·	·	·	·	8	6	2	·	1	2	1
8	Artemisietea vulgaris	3	14	6	11	·	·	·	·	·	·	·	·	3	10	13	6	2	3	3	4
9	Woodland communities	12	9	·	·	·	·	·	·	·	·	·	·	13	17	13	13	7	3	2	3
–	non-character species	7	27	18	·	·	·	·	·	·	·	·	1	·	14	20	2	2	6	7	1
	Totals	75	127	84	51	12	56	198	495	342	234	9	14	60	121	88	48	100	100	99	99

'Nasturtio-Glycerietalia' species were also found in the submersed zone. During the research period this vegetation type, with two subordinate associations, was concentrated at the fringes of the water, forming emergent zones. 'Phragmitetea' species were also found in the lower part of the bank zones, but not abundantly enough to justify separation into another zone. Because of the reduced cutting regime, 'Nasturtio-Glycerietalia', but also *Glyceria maxima* increased in prominence. According to Westhoff and Den Held (1969) the sociation of *Phalaris arundinacea* is a dominance community, developing in strips along brooks and small rivers, and indicating strong human impact, both from disturbance and phosphate loading. 'Glycerietum maximae' could be distinguished for the same reason, but according to Pott (1992) this syntaxon develops on hypertrophic sapropelium which was not present at this site. More probably, *Glyceria maxima* simply behaved like an opportunist species that could invade successfully after unintentedly extensive cleaning measures in autumn 1991, to which it is highly tolerant.

On the left bank zone, and to a lesser extend also on the right bank, several species were out-competed. 'Molinietalia' were reduced heavily, three 'Phragmitetea' and all non-character species disappeared. *Arrhenatherum elatius* increased, but other species of the same syntaxon decreased. It must be concluded that the bank vegetation became impoverished but did not shift to another type during the experiment. Three tall species: *Arrhenatherum elatius*, *Urtica dioica* and *Phalaris arundinacea* compete for space. The one that wins will set the trend for further community development. In this case *Urtica dioica* seems to have a good chance of leading the vegetation to 'Artemisietea'.

Table 3 shows the developments at site 1, which was much further upstream than site 4. The contribution of the waterplant communities was less because the submerged zone was much narrower. 'Sparganio-Glycerietum fluitans' was favoured by the changed cut-

Table 3. Profile data and contribution of syntaxa to the vegetation at site 1 of the Keersop experiment. See also Table 2.

	Side	Left							Middle				Right						Totalised and standardised (%)			
	Zone	Bank				Wet	Emerg.		Submerged				Emerg.		Bank							
	Year	89	90	91	93	89	91	93	89	90	91	93	91	93	89	90	91	93	89	90	91	93
	Width of the zone (cm)	110	140	160	130	30	60	90	300	210	200	200	30	15	110	140	140	110				
	(at places)						(120)				(130)											
	Level: maximm (cm)	130	120	120	95	30	10	0	0	−10	−10	−10	30	15	130	130	130	90				
	Level: minimm (cm)	30	0	10	0	0	−10	−10	−30	−20	−20	−30	0	0	30	0	15	0				
	Height of vegetation (cm)	100	40	30	50	30	10	70	30	10	5	30	30	2	100	30	25	30				
	(at places)			(100)			(80)				(20)							(70)				
	Total cover (%)	100	90	95	100	95	100	80	20	10	5	80	95	10	100	100	100	100				
	Filling of the profile (%)								20	10	5	70										
1	Lemnetea minoris	·	·	·	·	·	·	2	·	·	·	4	·	·	·	·	·	·	·	·	·	3
211	Ranunculion fluitantis	·	·	·	·	·	5	6	12	8	8	14	·	·	·	·	·	·	7	3	6	9
2111	Callitricho-Ranunculetum penicillati	·	·	·	·	·	2	3	18	8	4	6	·	·	·	·	·	·	11	3	3	4
3	Phragmitetea, several levels	·	4	·	3	·	1	·	·	·	·	·	·	·	·	·	·	1	·	2	1	2
31	Nasturtio-Glycerietalia	2	7	·	·	1	2	2	·	8	·	·	·	·	·	6	·	2	2	8	1	2
3111	Sparganio-Glycerietum fluitantis	·	·	·	·	·	4	5	3	11	6	14	·	·	·	·	·	·	2	4	5	9
3112	Phalaris arundinacea-sociation	·	7	·	5	1	4	6	·	11	4	·	·	·	·	·	·	2	1	6	4	6
411	Lolio-Potentillion	·	13	·	5	2	3	8	·	·	·	8	2	·	·	3	·	·	3	6	2	9
5	Molinio-Arrhenatheretea	7	8	6	5	·	·	·	·	6	·	·	·	·	8	7	10	4	9	8	8	4
51	Molinietalia	1	14	13	4	·	·	·	·	·	·	·	·	·	1	·	1	4	1	5	7	4
511	Filipendulion	2	6	8	5	1	·	·	·	·	·	·	·	·	2	1	3	1	3	3	5	3
52	Arrhenatheretalia	12	14	18	10	3	·	·	·	·	·	·	2	·	12	15	13	13	17	11	15	10
6	Bidentetea tripartiti	·	·	·	1	·	·	·	·	·	·	·	·	·	·	·	·	1	·	·	·	1
7	Chenopodietea	7	7	5	·	·	·	·	·	·	·	·	1	·	7	4	6	3	9	4	5	1
8	Artemisietea vulgaris	8	13	5	5	2	·	·	·	·	·	·	3	1	17	15	15	7	17	10	10	5
9	Woodland communities	7	27	24	12	·	·	·	·	·	·	·	·	1	4	15	8	14	7	15	16	11
–	non-character species	10	20	18	22	2	1	·	·	·	·	·	1	·	6	13	8	12	11	12	13	15
	Totals	55	139	96	78	12	22	32	33	53	22	46	8	3	56	80	64	66	99	99	99	98

Table 4. Profile data and contribution of syntaxa to the vegetation at the discussed site of the Schoonebeekerdiep experiment. See also Table 2.

	Zone	Banktop			Bank			Amphibic			Submerged			Total (%)		
	Year	92	93	94	92	93	94	92	93	94	92	93	94	92	93	94
	Width of the zone (cm)	130	180	130	180	160	140	50	90	80	150	200	200			
	Level: maximum (cm)	130	130	110	130	130	110	30	50	40	−10	−10	−30			
	Level: minimum (cm)	130	130	110	30	50	40	−10	−10	−30	−80	−100	−100			
	Hight of vegetation (cm)	40	50	80	10	30	30	70	60	70	80	100	100			
	Total cover (%)	100	100	100	80	100	95	100	100	100	5	5	7			
11	Potamogetonetalia pectinati	·	·	·	·	·	·	·	·	·	3	4	10	1	1	3
2111	Sagittaria sparganietum	·	·	·	·	·	·	·	·	·	9	16	16	3	5	5
3111	Phalaris arundinacea-sociation	·	·	·	·	·	·	5	5	6	·	·	6	2	1	3
411	Lolio-Potentillion	20	22	14	13	11	17	3	9	10	·	·	·	13	13	12
5	Molinio-Arrhenatheretea	27	38	21	36	24	25	2	5	6	·	·	·	24	21	15
51	Molinietalia	7	9	7	4	3	7	4	9	10	·	·	·	5	7	7
52	Arrhenatheretalia	18	31	21	27	27	18	·	·	·	·	·	·	17	18	11
6	Chenopodietea	·	18	3	13	5	7	·	·	2	·	·	·	5	7	3
7	Artemisietea vulgaris	10	14	14	4	10	10	3	5	9	·	·	·	6	9	10
8	Woodland communities	10	20	29	18	24	27	·	·	10	·	·	·	10	13	19
–	non-character species	·	7	16	38	11	27	·	·	·	·	·	·	14	6	12
	Totals	92	158	124	151	115	137	16	32	52	12	20	32	100	100	100

ting regime because of the shallowness of the brook at this side. *Glyceria fluitans* might become a problem for weed control, but other species of the community grow nearby and probably also colonize thus preventing monospecific wild growth. *Phalaris arundinacea* also gained in prominence, but is well controlled, and is expected not to expand more.

'Lolio-Potentillion' tends to increase at the edges of the channel. Since this community occurs typically in inundating pastures and other wet grass-lands (Sýkora, 1982), it suggests that a narrow typical river vegetation zone is developing. Micro-habitat and silt content for such a zone are apparently favourable. Further development can lead to enrichment of the community at this site.

'Artemisietea' were strongly represented on the right bank by a high cover of *Urtica dioica*. On the adjacent land, maize was grown which usually is associated with high nitrogen losses. In 1993 the land was turned into pasture, resulting in a drastic reduction of *Urtica dioica* growth.

On the left bank, the number of species found in 1990 was higher than in other years. This is due to the changed cutting methods. In autumn 1989 the vegetation had grown longer and was cut shorter than before. Gaps in the vegetation were easily filled by seedlings of opportunist species.

The Schoonebeekerdiep experiment

A second example is derived from experiments in the straightened channel of the Schoonebeek brook. This brook forms the border of The Netherlands and Germany east of Coevorden in the Northern part of The Netherlands. It merely drains relatively recently dug up bog land. The vegetation on the banks used to be cut twice a year and the vegetation in the water itself was cut by a mowing boat two to four times a year. The emergent zone was left unmown for the whole summer to benefit aquatic macrofauna, as directed by German law. In the experiment, frequency and timing of cutting of the banks varied. Mowing by boat was minimised to once or twice a year, depending on the weather conditions.

Changes in the vegetation in one of the eight sites is presented in Table 4, analogous to Table 2 and 3 of the Keersop experiment. At this site the bank and emergent vegetation has not been cut since autumn 1991. Mowing boats cut the submersed zone in autumn only. Slight increase in water vegetation, especially of 'Sagittaria sparganietum' can be the effect of lower cutting frequency. The 'Potamogetonetalia pectinati' merely respond on water quality, which is poor but varying through the years.

The bank vegetation is slowly but clearly changing from 'Arrhenatheretalia' into 'Artemisietea'. Both syntaxa are represented by several species (e.g. *Holcus lanatus*, *Plantago lanceolata* resp. e.g. *Urtica dioica*, *Rumex obtusifolius*) making the indication relatively strong. Also cover of several species of woodland communities, among which *Holcus mollis*, increase or colonize on the bank. This indicates that formation of woodland, which is natural in non-managed sites, can be expected on short terms.

The effect of reduced cleaning frequency

Reduced cleaning frequency of watercourses that used to be cut twice a year has led to slow changes in the vegetation. Cover of submersed vegetation increased, but the community composition was hardly affected by the changed cutting regime in the first three to five years. Emergent communities took advantage of the lowered frequency and their contribution to the watercourse vegetation increased, mostly by widening of their zone. Depth of the channel limits the spread however: only in the shallow site of the Keersop could 'Sparganio-Glycerietum fluitans' increase all over the channel. At the edges the beginning of a river edge vegetation could be detected. Bank vegetation clearly started to shift from hay-land types of 'Molinio-Arrhenatheretea' into the ruderal 'Artemisietea'. In the Schoonebeekerdiep also the start of a shift into woodland types could be detected. These changes indicate that conditions were too eutrophic in both experiments for a diverse brook vegetation development without additional habitat improvement.

Conclusion

Bio-indication with plants mostly depends on preferences of species recorded elsewhere (e.g. Ellenberg, 1979; Ellenberg et al., 1992). On steep gradients such as watercourses, indicators of several contradictory states or processes are usually found together. Sample means of indicator values for species are very often meaningless if all species contribute equally to the results. Syntaxa are not clearly separated either, but with increasing number of character species, including weak ones, the likelihood that the detected syntaxa play a role increases. Syntaxa that are represented only by a small number of character species are therefore negligible on the criterion of low saturation. Up to half of the species in a sample can be disregarded if they are indicative for different syntaxa. Applied to dynamic gradients such as watercourses, this method can reduce the influence on interpretation of species that are present because of chance, historical reasons, or too easy dispersion from other parts of the gradient.

The way in which samples were assigned to syntaxa partly depends on professional judgement. Computer aided techniques are under development (Hill, 1989; Van Tongeren, pers. comm.; Kopecký et al., 1995) but are not satisfactory yet.

The method described allows changes in sample plot dimensions. In monitoring studies of vegetation the sample size and position usually does not change, even when this conflicts with the rule on homogeneity. The goal of monitoring is to detect changes in the vegetation. When the vegetation moves along a gradient it is more efficient to follow the vegetation and measure the way it moves then to stick to plot dimensions. When new vegetation types overgrow the former, they should be sampled themselves.

The presented results show that assignment to syntaxa can lead to successful interpretation of slow changes in the vegetation. Allocating zones facilitates interpretation of changes on the gradient, but also makes it possible to quantify extension or shrinking of (sub-)communities in the vegetation.

Acknowledgments

This paper was sponsored by the TAO foundation for Applied Aquatic Research in Amsterdam.

References

Barkman, J. J., J. Doing & S. Segal, 1964. Kritische Bemerkungen und Vorschläge zur quantitativen Vegetationsanalyse. Acta Bot. Neerl. 13: 394–419.

Ellenberg, H., 1979. Zeigerwerte des Gefäszpflanzen Mitteleuropas. Scripta Geobotanica 9, Erich Goltze, Goettingen, 122 pp.

Ellenberg, H., H. E. Weber, R. Duell, V. Wirth, W. Werner & D. Paulissen, 1992. Zeigerwerte von Pflanzen in Mitteleuropa. 2. Aufl. Scripta Geobotanica 18, Erich Goltze, Goettingen, 258 pp.

Hill, M. O., 1989. Computerized matching of relevés and association tables, with an application to the British National Vegetation Classification. Vegetatio 83: 187–194.

Kopecký, K., J. Dostálek & T. Frantík, 1995. The use of the deductive method of syntaxonomic classification in the system of vegetational units of the Braun-Blanquet approach. Vegetatio 117: 95–112.

Maarel E. van der, 1979. Transformation of cover-abundance values in phytosociology and its effects on community similarity. Vegetatio 39: 97–114.

Oberdorfer, E., 1977. Süddeutsche Pflanzengesellschaften. Teil I. Gustav Fischer, Stuttgart, 2. Aufl. 317 pp.

Pot, R., 1993. Vegetation zones along watercourses: interrelationships and implications for mechanical control. J. aquat. Plant Mgmt 31: 157–162.

Pott, R., 1992. Die Planzengemeinschaften Deutschlands. Eugen Ulmer, Stuttgart, 427 pp.

Preising, E., 1990. Die Pflanzengesellschaften Niedersachsens. Salzpflanzengesellschaften der Meeresküste und des Binnenlandes & Wasser- und Sumpfplanzengesellschaften des Süßwassers. Naturschutz und Landschaftspflege in Niedersachsen 20 (7–8). Niedersächsisches Landersverwaltungsamt, Hannover, 163 pp.

Sýkora, K. V., 1982. Syntaxonomy and synecology of the Lolio-Potentillion Tüxen 1947 in the Netherlands. Acta Bot. Neerl. 31: 65–95.

Westhoff, V. & A. J. Den Held, 1969. Plantengemeenschappen in Nederland. Thieme, Zutphen, 324 pp.

Westhoff, V. & E. van der Maarel, 1973. The Braun-Blanquet approach. In R. H. Whittaker (ed.), Ordination and classification of communities. Handbook of vegetation science 5: 617–726.

Hydrobiologia **340**: 67–76, 1996.
J. M. Caffrey, P. R. F. Barrett, K. J. Murphy & P. M. Wade (eds), Management and Ecology of Freshwater Plants.

A reference system for continental running waters: plant communities as bioindicators of increasing eutrophication in alkaline and acidic waters in north-east France

F. Robach[1], G. Thiébaut[2], M. Trémolières[1] & S. Muller[2]
[1] *Laboratoire de Botanique et Ecologie végétale, CEREG, URA 95 CNRS, IFARE, Institut de Botanique, 28 rue Goethe, F-67083 Strasbourg cedex, France*
[2] *Laboratoire de Phytoécologie, Faculté des Sciences, Ile du Saulcy F-57045 Metz, France*

Key words: acidic water, alkaline water, autoecology, bioindication, macrophyte community, eutrophication

Abstract

Two bioindication scales of the degree of eutrophication based on aquatic macrophyte communities were established in two types of running waters free of organic matter, the one in acidic "soft" waters (pH 5.5–7.0, conductivity 40–110 μS.cm^{-1}), the other in alkaline hard waters (pH 7–8, conductivity 500–900 μS.cm^{-1}). We show that the main determining factor of the macrophyte distribution is the nutrient level (trophy), especially the level of phosphate and ammonia. The acidic scale, with increasing pH, includes four stages ranging from oligotrophic to eutrophic level (traces to 300 μg.l^{-1} N-NH_4^+ and P-PO_4^{3-}), while the alkaline scale at constant pH comprises six stages of a trophic gradient. For the most part, the floristic composition found in the two sequences is different and depends on conductivity and alkalinity variation. However, some species occur in the two scales and may reflect differences in the trophic level, depending on whether the waters are alkaline or acidic. This change of trophic level for these species is discussed.

Introduction

The occurrence of aquatic vascular macrophyte is related unambiguously to the water chemistry as shown by e.g. Kohler (1975), Wiegleb (1984), Klosowski (1985), Lachavanne (1985), Konold et al. (1990). Developing the plant species or communities as indicator method (Iserentant & de Sloover, 1976) has been an objective for surveying water quality. Until now biological methods have been carried out especially using invertebrate macrofauna, but some authors such as Newbold & Holmes (1987), Haslam (1982) and Harding in Standing Committee of Analysts (HMSO 1985–1986) have constructed biological indices based on aquatic plants for assessing water quality. More recently bioindication scales of the nutrient level (phosphorus and nitrogen) based on aquatic macrophyte communities have been performed and tested in running water streams of the Alsace floodplain (Carbiener et al. 1990) and the Vosges mountains, in Eastern France (Muller, 1990).

The aim of this paper is to compare the two bioindication scales, one in acidic water, the other in alkaline waters, in order to define a hierarchy of variables which explains the distribution of plants.

We focused on the study of distribution of species which appear in these two types of water, to determine the best ecological conditions for their occurrence.

Study Site

In the east of France, two large hydrological networks of running waters were selected in two types of geochemistry. The first concerned streams in the Alsace Rhine floodplain. Most of them are fed by groundwater which flows in a gravelly calcareous aquifer. The groundwater-fed streams present a large homogeneity of characteristics, organic matter-free, calcareous hard water with a pH 7.5–8 and a conductivity higher than 500 μS.cm^{-1}. The remainder are surface waters connected to the Rhine or to the Ill, the main tributary of the Rhine in the Alsace floodplain. The second network flows in the Northern Vosges mountains on acidic substrate (sandstone), and has a pH of 5.5–7 and a conductivity between 40–110 μS.cm^{-1}. The variation in altitude from the spring to the bottom of valley is 100 to 150 m. Their downstream course where they reach the Alsace plain is characterized by a neutral pH and a conductivity of 150–350 μS.cm^{-1}. Both slightly and highly mineralized systems exhibit a large trophic range from oligotrophic to eutrophic. 129 sites were investigated, 91 in the alkaline network and 38 in the acidic network.

Materials and methods

Vegetation analysis

Vegetation surveys have been carried out more and less regularly since 1970, in the streams of the Alsace floodplain and since 1988 in those of the Vosges mountains. The vegetation was investigated using Flora Europaea (Tutin et al., 1964–1993). When no perturbation occurs, no change of vegetation was observed for many years. However, if vegetation changed, the site was considered as a new one. In each site a phytosociological relevé was recorded by the method defined by Braun-Blanquet (1964), using a coefficient of abundance-dominance. Aquatic vascular and non vascular (Bryophytes and algae) plants were listed over a minimal distance of 50 m and usually 100 m of a stream and assigned a coefficient of abundance-dominance.

Water analysis

In parallel with the vegetation survey, water was sampled in streams, monthly in the Alsace floodplain and every three months in the Vosges mountains. Several sites per stream were selected according to the change of vegetation. A vegetation relevé corresponded to each sample of water. pH, conductivity, dissolved oxygen and temperature were measured in situ. Phosphate, nitrate, ammonia, chloride and water hardness were analysed in the laboratory by using the procedures (APHA, 1985) previously described by Trémolières et al. (1993).

Data analysis

The mean of each variable was calculated over one year, integrating the yearly fluctuations. The reference year was determined by the existence of a maximum of data (>4) covering the whole year.

Statistical analyses such as c-PCA (centred Principal Component Analysis) and FDA (Factorial discriminant analysis) were applied, using the Statitcf program. A first PCA was processed on 129 sites and 7 physico-chemical variables, a second one on species affected of a coefficient of abundance. FDA was used to integrate the new variable, the plant community which we defined for each site. The objective was to compare the plant communities and to verify their classification realized *a priori*.

7 Classes for each variable were determined *a posteriori* on the whole data, in order to have an homogeneous distribution of the n sites among the classes (each class represents around n/7 sites). Thus, the distribution of a species was specified according to a given variable.

Results

Water chemistry

The running waters of acidic geochemistry in the Northern Vosges mountains have a pH which varies from 5.5 to 7 and a low conductivity from 40 to 110μS.cm^{-1}. The trophic level related to phosphate and ammonia changes from oligotrophic (annual mean 50 μg.l^{-1} N-NH_4^+, 25 μg.l^{-1} P-PO_4^{3-}) upstream to eutrophic (150 μg.l^{-1} N-NH_4^+, 150 μg.l^{-1} P-PO_4^{3-}) downstream. The nitrate concentrations remain almost constant around 0.5 mg.l^{-1}. In the Alsace floodplain, the streams consist of buffered bicarbonated water with alkaline pH (7.5–8.2) and high conductivity varying from 400 to 1000 μS.cm^{-1}. The nitrate nitrogen varies greatly between 0.5 and 7.5 mg.l^{-1}, the streams with the highest values being located in agricultural areas. Phosphate phosphorus and ammonia nitrogen range from 3 to 350 μg.l^{-1} (annual mean).

Physico-chemical variables hierarchy

According to the results of c-PCA (129 individuals) on 7 physico-chemical variables (pH, conductivity, hardness, chloride, phosphate, ammonia, nitrate) two zones can be distinguished on the PCA ordination diagram

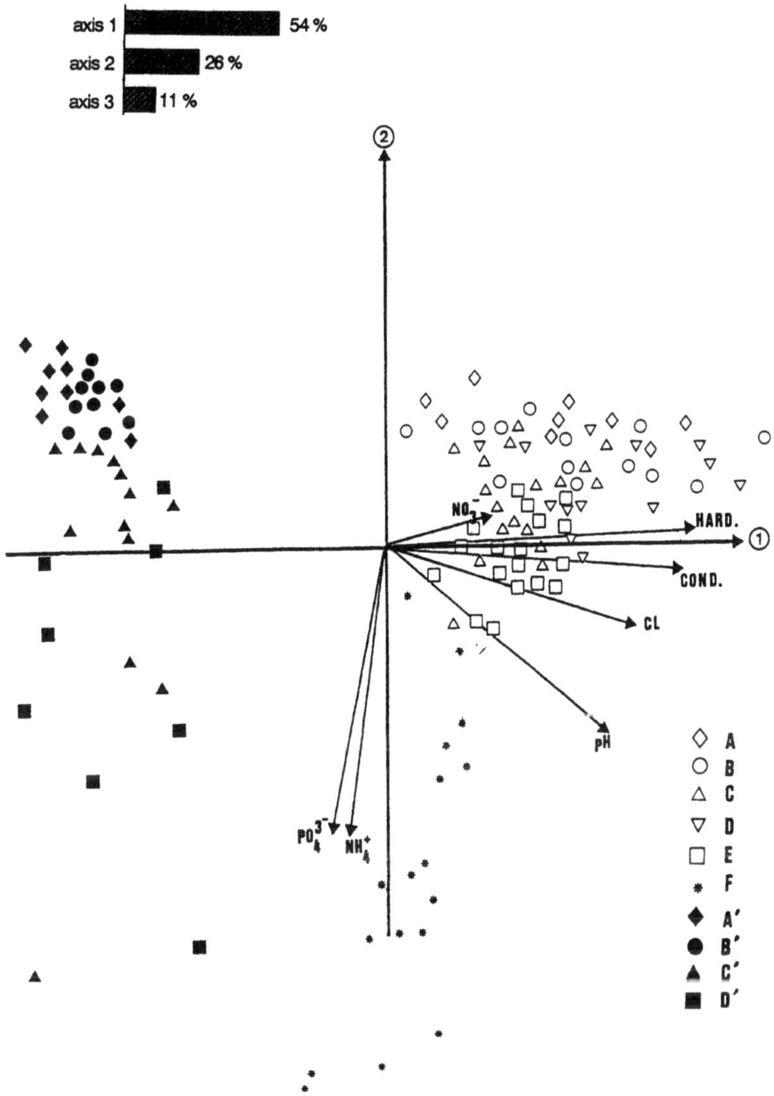

Figure 1. Ordination of samples plots using PCA of physico-chemical variables with reference to communities A to F in the Alsatian Rhine floodplain, and A′ to D′ in Northern Vosges.

(Figure 1). The first axis is highly correlated with conductivity (0.95), hardness (0.94) and less markedly with pH (0.6). The second axis reflects the trophic level, phosphate and ammonia nitrogen with a correlation of 0.83 and 0.78 respectively. The sampling sites are clearly distributed along these two axes, the Vosges streams with negative scores on the first axis and the Alsatian streams with positive scores. On the second axis the sites can be arranged along a trophic level gradient, on two parallel lines to the second axis.

These results demonstrate a first hierarchy of variables: conductivity, hardness and pH as the first significant parameters of water quality, and secondarily the trophic level.

Vegetation data

A PCA on all aquatic plant species, linking presence of species and abundance scores shows a distribution of species which is homologous to the distribution of

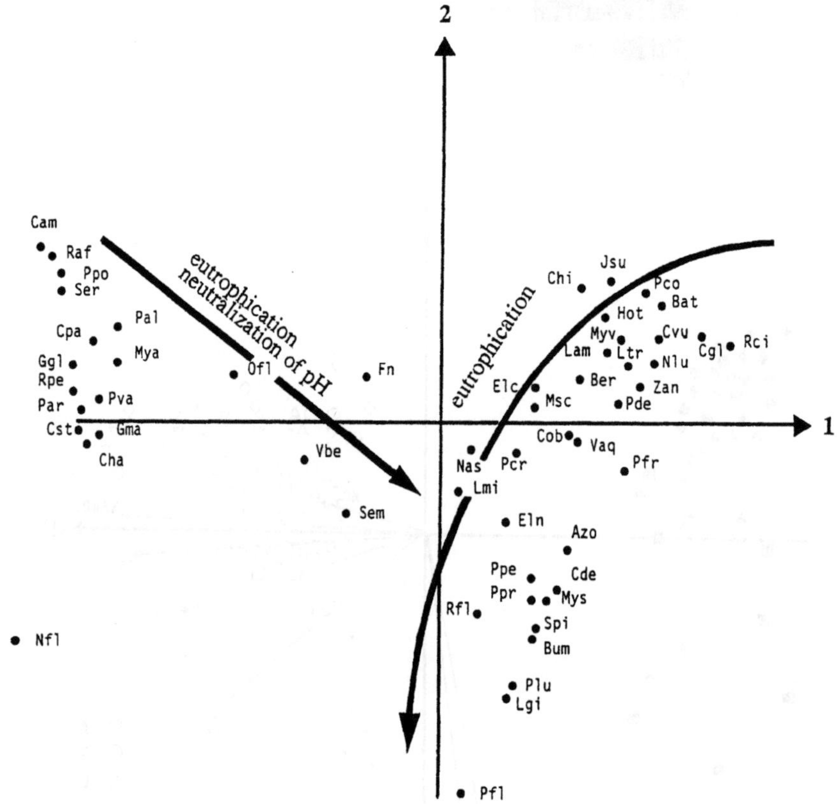

Figure 2. Ordination of samples plots using PCA of floristic data (for the abbreviations of the species names, see Table 1a and 1b), with reference to communities A to F in the alsatian Rhine floodplain, and A′ to D′ in Northern Vosges

sites according to physico-chemical variables (Figure 2). The species characteristic of calcareous waters have positive scores on the first axis, whereas species of acidic waters have negative scores. In contrast, the oligotrophic species such as *Potamogeton coloratus* in calcareous waters and *P. polygonifolius* in acidic waters present positive scores on F2 axis and eutrophic species such as *Potamogeton pectinatus* have negative scores. The species which are indifferent to pH and/or conductivity are distributed in the centre of the PCA diagram.

The oligotrophic species in acidic waters such as *P. polygonifolius* have high negative scores on the first axis and positive on the second axis. The other species, *Oenanthe fluviatilis, Fontinalis antipyretica* or *Sparganium emersum* are distributed on the first axis towards the centre of diagram and are correlated with an increase of pH, and on the second axis towards negative scores, without being comparable to the component scores of eutrophic species (see *Potamogeton lucens* or *Potamogeton nodosus*) characteristic of hard water. In acidic streams, plant species are distributed along a double gradient of pH and trophy, which we do not observe in alkaline buffered waters.

In calcareous streams a sequence of six plant communities was defined previously through a trophic gradient (Table 1b) (Carbiener et al., 1990; Robach et al., 1991). *Potamogeton coloratus*, with an algae, *Batrachospermum moniliforme* are characteristic of oligotrophic waters. *Berula erecta* and the disappearance of oligotrophic species are discriminant of the second step of the sequence. The appearance of *Callitriche obtusangula* defines the third community named C, with *Lemna trisulca, Fontinalis antipyretica* and *Elodea canadensis. Zannichellia palustris, Nasturtium officinale* and *Groenlandia densa* have negative component scores on the second axis (Figure 2). At higher negative scores on the F2 axis, are recorded

Table 1a. Phytosociological table based on frequence of occurence of species in community (V 75–100%, IV 50–75%, III 25–50% II 10–25%, I <10%) and physico-chemical variables correlated (Mean and Std caculated on one year). a: acidic streams, b: calcareous streams.

Abbr.	Associations (n)	A′(10)	B′(10)	C′(12)	D′(6)
Ppo	Potamogeton polygonifolius	V(2)	V(1)	–	–
Cam	Cardamine amara	I(1)	II(1)	–	–
Gfl	Glyceria fluitans	IV(2)	V(1)	V(1)	V(+)
Fn	Fontinalis antipyretica	I(1)	II(1)	II(1)	III(2)
Maq	Mentha aquatica	I(+)	I(1)	II(+)	II(+)
Sem	Sparganium emersum	I(+)	III(1)	IV(1)	III(1)
Ser	Sparganium erectum	IV(1)	III(1)	I(+)	I(+)
Vbe	Veronica beccabunga	I(1)	–	III(+)	I(+)
Cst	Callitriche stagnalis	–	IV(+)	IV(+)	III(+)
Cha	Callitriche hamulata	–	III(1)	V(1)	V(2)
Lmi	Lemna minor	–	II(+)	III(+)	IV(+)
Rpe	Ranunculus peltatus	–	II(1)	II(2)	II(3)
Cpa	Callitriche platycarpa	–	III(2)	III(2)	II(1)
Ber	Berula erecta	–	II(1)	I(1)	I(2)
Pcr	Potamogeton crispus	–	–	I(1)	–
Mya	Myriophyllum alterniflorum	–	–	II(+)	–
Pal	Potamogeton alpinus	–	–	I(+)	–
Pva	Potamogeton variifolius	–	–	I(+)	–
Par	Phalaris arundinacea	–	–	V(1)	V(+)
Gma	Glyceria maxima	–	–	III(+)	II(+)
Elc	Elodea canadensis	–	–	III(3)	I(1)
Eln	Elodea nuttallii	–	–	I(2)	III(3)
Nas	Nasturtium officinale	–	–	III(1)	III(+)
Ofl	Oenanthe fluviatilis	–	–	I(2)	I(+)
Pbe	Potamogeton berchtoldii	–	–	I(2)	I(+)
Msc	Myosotis scorpioides	–	–	–	II(+)
Cob	Callitriche obtusangula	–	–	–	V(3)
Nfl	Nitella flexilis	–	–	–	I(+)
		mean(std)	mean (std)	mean (std)	mean (std)
	pH	6(0.2)	6.5(0.2)	6.9(0.3)	6.8(0.3)
	Conductivity (μS/cm)	59(14)	49 (5)	74(18)	80(21)
	Hardness (meq/l)	0.3(0.1)	0.3(0.1)	0.5(0.2)	0.6(0.3)
	N-NH_4^+ (μg/l)	49(16)	47 (6)	111(99)	142(72)
	P-PO_4^{3-} (μg/l)	25(11)	26(13)	96(79)	150(66)
	N-NO_3^- (mg/l)	0.6(0.2)	0.3(0.2)	0.5(0.2)	0.7(0.3)

Potamogeton pectinatus, Elodea nuttallii, Myriophyllum spicatum and *Potamogeton perfoliatus.* The most eutrophic species, *Ranunculus fluitans, Potamogeton nodosus* and *Potamogeton lucens* have low positive scores on the F1 axis and the lowest score on the F2 axis. The communities at the extremes of the sequence are species-poor with around 4–5 species per site and community, the richest is a mesotrophic community with *Nasturtium, Zannichellia* and *Groenlandia*, corresponding to mean concentrations of 29 μg.l^{-1} P-PO_4^{3-} and 33 μg.l^{-1} N-NH_4^+.

The sequence of acidic waters comprises four communities (Table 1a). The oligotrophic and most acidic one is characterised by *Potamogeton polygonifolius* and the absence of *Callitriche.* In the following community of the scale, *Callitriche hamulata, C. stagnalis, C. platycarpa, Ranunculus peltatus* and *Berula erecta* appear with lower negative score on the F1 axis than the previous species (Figure 2). The third com-

Table 1b.

Abbr.	association (n)	A(9)	B(11)	C(15)	D(17)	E(20)	F(16)
Poo	Potamogeton coloratus	V(2)	–	–	–	–	–
Bat	Batrachospermum monoliforme	II(1)	–	–	–	–	–
Jsu	Juncus subnodulosus fo. subm.	II(+)	–	–	–	–	–
Cvu	Chara vulgaris	II(+)	–	–	–	–	–
Chi	Chara hispida	II(+)	–	I(2)	–	–	–
Lam	Lamprocystis roseo persicina	II(+)	–	I(3)	I(+)	–	–
Ber	Berula erecta (Sium erectum)	V(2)	V(4)	V(2)	III(1)	III(1)	I(+)
Cob	Callitriche obtusangula	I(+)	III(+)	V(2)	V(2)	IV(1)	III(+)
Ltr	Lemna trisulca	–	I(1)	III(1)	III(1)	II(1)	–
Fn	Fontinalis antipyretica	–	I(+)	I(1)	II(1)	II(+)	–
Elc	Elodea canadensis	–	–	IV(1)	III(1)	II(1)	I(+)
Sem	Sparganium emersum	–	–	II(+)	II(1)	II(+)	IV(1)
Lmi	Lemna minor	–	–	II(1)	III(1)	V(2)	V(1)
Pfr	Potamogeton friesii	–	–	II(+)	II(1)	II(1)	II(1)
Eln	Elodea nuttallii	–	–	I(2)	II(2)	IV(2)	IV(1)
Rci	Ranunculus circinatus	–	–	I(+)	–	–	–
Nas	Nasturtium officinale	–	–	–	V(1)	III(+)	II(+)
Spi	Spirodela polyrhiza	–	–	–	I(1)	II(1)	IV(l)
Azo	Azolla filliculoides	–	–	–	I(1)	I(1)	II(1)
Pde	Groenlandia densa	–	–	–	II(1)	II(1)	–
Pcr	Potamogeton crispus	–	–	–	III(1)	II(1)	I(2)
Myv	Myriophyllum verticillatum	–	–	–	I(1)	I(+)	–
Zan	Zannichellia palustris	–	–	–	II(1)	II(1)	–
Hot	Hottonia palustris	–	–	–	I(1)	–	–
Hip	Hippuris vulgaris	–	–	–	I(1)	–	–
Ppe	Potamogeton pectinatus	–	–	–	II(1)	IV(3)	IV(1)
Mys	Myriophyllum spicatum	–	–	–	II(1)	IV(1)	V(1)
Ppr	Potamogeton perfoliatus	–	–	–	I(+)	I(2)	III(1)
Cde	Ceratophyllum demersum	–	–	–	I(+)	V(2)	V(1)
Efl	Oenanthe fluviatilis	–	–	–	–	I(r)	–
Rtr	Ranunculus trichophyllus	–	–	–	–	I(+)	–
Ppu	Potamogeton pusillus	–	–	–	–	I(+)	–
Rfl	Ranunculus fluitans	–	–	–	–	II(2)	III(+)
Plu	Potamogeton lucens	–	–	–	–	–	III(1)
Pno	Potamogeton nodosus	–	–	–	–	–	III(2)
Maq	Mentha aquatica fo. subm.	I(r)	I(+)	I(I)	I(+)	I(1)	–
Vaq	Veronica anagallis aq.	I(+)	–	I(+)	I(+)	II(+)	–
Msc	Myosotis scorpioides	I(r)	–	I(1)	II(+)	II(+)	–
Vbe	Veronica beccabunga	–	–	–	I(+)	I(+)	I(+)
		mean (std)	mean (std)	mean (std)	mean (std)	mean (std)	mean (std)
	pH	7.4 (0.1)	7.5 (0.2)	7.5 (0.1)	7.6 (0.2)	7.9 (0.2)	7.9 (0.2)
	Conductivity	608 (115)	736 (112)	740 (99)	657 (66)	657 (63)	508 (52)
	Hardness (meq.l^{-1})	4.8 (1.4)	4.7 (0.9)	5 (0.7)	3.9 (0.4)	3.8 (0.5)	3.2 (0.3)
	N-NH_4^+ (μg/l)	13.7 (7.3)	22.2 (13.8)	45.3 (27.8)	33.8 (31.3)	61.2 (40)	255 (107)
	P-PO_4^{3-} (μg/l)	7.2 (1.7)	13 (5.5)	14.9 (6.8)	29.4 (23.6)	39.9 (33)	191.5 (116)
	N-NO_3^- (mg/l)	5.5 (1.4)	5.1 (1.8)	4.7 (2.1)	2.9 (2.5)	1.6 (1.1)	2.5 (0.9)

Table 2. Distribution of *Potamogeton coloratus, Potamogeton polygonifolius, Berula erecta* and *Callitriche obtusangula* according to the 7 classes of phosphate, ammonia, nitrate concentrations and conductivity values. The distribution is expressed in % of occurence frequence.

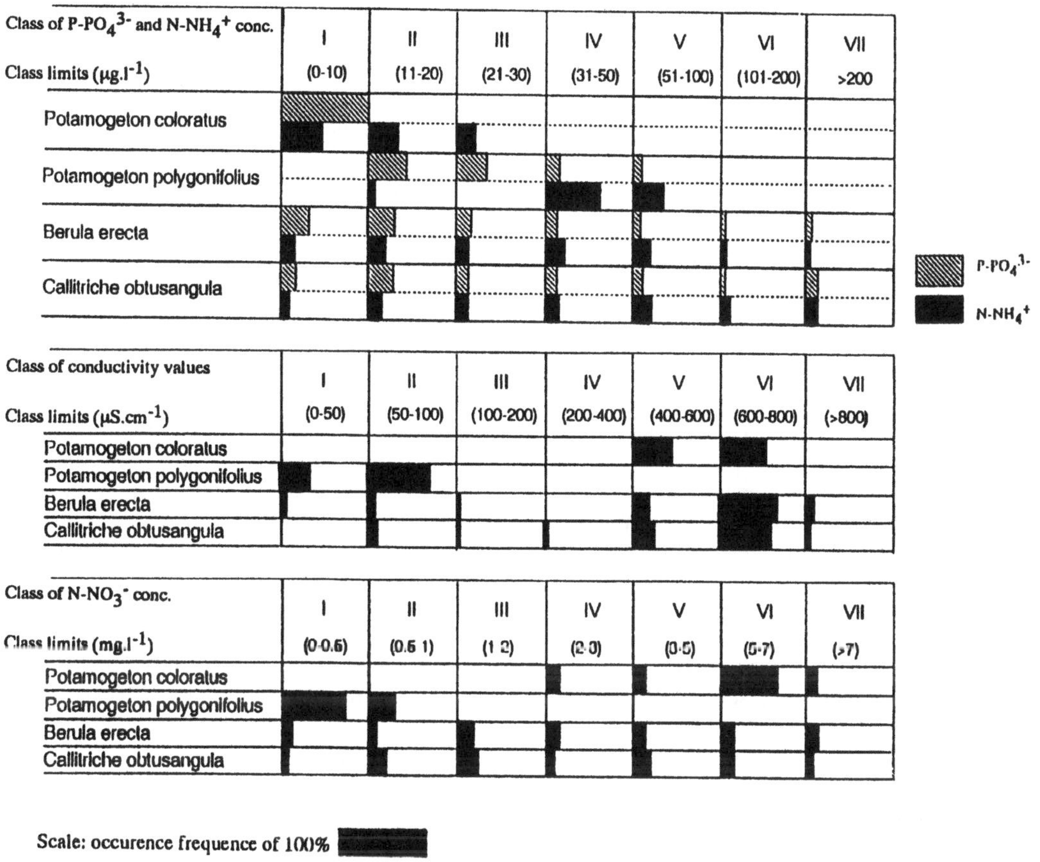

munity is defined by the disappearance of *Potamogeton polygonifolius* and the appearance of *Elodea canadensis, Elodea nuttallii, Nasturtium officinale* and *Myriophyllum alterniflorum. Callitriche obtusangula* and a Characean *Nitella flexilis* characterize the fourth community, the most eutrophic of the sequence. In these two latter communities some species occur in both acidic and alkaline sequences and seem indifferent to mineralization of water (*Elodea nuttallii, Elodea canadensis, Callitriche obtusangula, Nasturtium officinale, Berula erecta*). It should be noticed that they are not present in the oligotrophic step of both sequences. Most of them grow better in water with high levels of phosphate (from 50 μg.l^{-1} P-PO$_4^{3-}$).

In order to compare the trophic sequences according to the mineralization of water, two FDA were processed. The first one was processed on 88 sites of the alsatian floodplain and 9 variables (pH, conductivity, hardness, Cl^-, PO_4^{3-}, N-NH$_4^+$, N-NO$_3^-$, N-NO$_2^-$, plant communities). The first axis (67.4%) is highly correlated with N-NH$_4^+$ (0.97), P-PO$_4^{3-}$ (0.88) and N-NO$_2^-$ (0.93). The second axis (17.7%) is mainly correlated with N-NO$_3^-$ (0.67) and pH (–0.61). According to the vegetation composition, we previously assigned to each site one of our 6 plant communities. The FDA showed that 64% of the sites were well classified. This FDA also exhibits a sequence of 6 groups of plants communities arranged along the trophic gradient.

A second FDA was processed on 126 sites (acidic and calcareous streams) and 8 variables (pH, conductivity, hardness, Cl^-, P-PO$_4^{3-}$, N-NH$_4^+$, N-NO$_3^-$, plant communities). The first axis (85.4%) was highly correlated to the hardness (0.97), the conductivity

(0.99) and the chloride (0.99). The second axis is mainly correlated to $P\text{-}PO_4^{3-}$ and $N\text{-}NH_4^+$. 56.3% of the sites were well classified. This FDA also exhibits two homologous sequences of vegetation arranged along the trophic gradient, the first one corresponding to the acidic waters, the second one to the calcareous water.

Discussion

Comparison of trophic sequences according to the mineralisation of water

A relation between communities and trophic level was well established in the two types of waters. Only four steps named A′, B′, C′ and D′ were identified in acidic waters and six steps, A, B, C, D ,E and F in calcareous waters.

These two homologous sequences present different floristic compositions because of a highly different mineralisation. Thus the species, indicators of eutrophication steps D and E of the calcareous sequence (*Zannichellia palustris, Myriophyllum spicatum, Ranunculus fluitans, Ceratophyllum demersum*) are missing in the same steps of the slightly mineralized sequence. However, the species linked to the meso-/eutrophic step of both sequences (*Elodea canadensis, Elodea nuttallii*) show a similar behaviour towards trophy in the two sequences, but with some variations of optimum for *Oenanthe fluviatilis* and *Callitriche obtusangula*, for example. *Oenanthe fluviatilis* appears in the eutrophic stage (E) in the calcareous sequence and from C′ in the acidic sequence. *Callitriche obtusangula* characteristic species of community D' in the Eastern Vosges appears in the Alsace floodplain waters as from B with its optimum in C, corresponding to a trophic level lower (15 $\mu g.l^{-1}$ $P\text{-}PO_4^{3-}$) than the stage D′ (150 $\mu g.l^{-1}$ $P\text{-}PO_4^{3-}$) in the Vosges.

The two sequences present a different trophic gradient with levels of steps higher in acidic waters than in alkaline waters, i.e. for example the calcareous oligotrophic community A is related to concentrations of ammonia and phosphate equal to or lower than 10 $mg.l^{-1}$, whereas the acidic community A' corresponds to a mean concentration of 50 $\mu g.l^{-1}$ $N\text{-}NH_4^+$ and 23 $\mu g.l^{-1}$ $P\text{-}PO_4^{3-}$ (see also the trophic level of communities with *Oenanthe fluviatilis* in acidic waters, community C′: 96 $\mu g.l^{-1}$ $P\text{-}PO_4^{3-}$ and 111 $\mu g.l^{-1}$ $N\text{-}NH_4^+$ and in alkaline waters, community E: 40 $\mu g.l^{-1}$ $P\text{-}PO_4^{3-}$ and 61 $\mu g.l^{-1}$ $N\text{-}NH_4^+$).

The species of acidic streams seem less sensitive to eutrophication than the calcareous ones. This result could be explained differently as a function of the variable tested. Thus it is well known that ammonia is more toxic in alkaline water than in acidic water where it is present as the non toxic ionized form (Glänzer et al., 1977; Dendène et al., 1993). In the case of phosphorus, we suggest two hypotheses: the phosphorus nutrition of plants is provided mainly from water but it could be supplemented by release from sediments (root absorption from interstitial water or foliar absorption when there is exchange between sediment and water under certain conditions of pH and redox potentiel, e.g. Gächter et al., 1988; Furumai & Ohgaki, 1989; Manning 1989); this exchange process between sediment and water might be more efficient in calcareous alkaline waters, where most of the phosphate is retained on the colloids through the Ca^{2+} cation and/or stored in the form of tricalcium phosphate in the sediment which acts as a buffer. In acidic waters, phosphate is mainly in soluble form, most of the phosphorus nutrition is obtained directly from the water and plants need high concentrations of phosphorus in the water. The second hypothesis concerns a more efficient absorption of phosphorus by plants in alkaline waters; the absorption would be facilitated by the Ca^{2+} cation fixed on cellular membranes which ensures active transport of both phosphate and calcium.

Particular cases of common species and vicariant species

The identification of discriminant variables of plant communities distribution allows the response of some vascular species to the physico-chemical variables level to be specified. We chose to analyse the behaviour of two types of plants, vicariant species such as *Potamogeton coloratus* and *Potamogeton polygonifolius*, segregated according to the mineralization, and species with a large ecological width, appearing both in alkaline hard waters and in soft waters (common to the two bioindication scales).

Case of vicariant species

These species segregate mainly according to hardness and conductivity (Table 2). *Potamogeton coloratus* and *Potamogeton polygonifolius* are characteristic species of oligotrophic waters, the first one of calcareous waters and the other one of non calcareous waters. However, we observe that *P. polygonifolius* has a larger ecological width than *P. coloratus*. This species is able to exist in rather different microhabitats with a great number of morphological modifications without systematic value (Roweck, Risse & Kohler, 1986). *Potamogeton coloratus* is more strictly oligotrophic since it occurs in waters with very low level of phosphate (10 $\mu g.l^{-1}$ annual mean) and tolerates up to 30 $\mu g.l^{-1}$ N-NH_4^+ (Table 2) as shown also by Roweck et al. 1986). *P. polygonifolius* grows in water with a mean concentration of N-NH_4^+ of 50 $\mu g.l^{-1}$ and of P-PO_4^{3-} of 25 $\mu g.l^{-1}$ (Table 2).

Case of common species

Berula erecta is usually considered as an eurytopic species (Kohler et al., 1974; Haslam, 1978). However, it seems to be characteristic of oligotrophic and mesotrophic waters (community B), it appears in all communities of the alkaline sequence with a high frequency except in F (Table 1b). We find this species throughout the whole range of phosphate and ammonia, however its occurrence frequency decreases from 100 $\mu g.l^{-1}$ of P-phosphate and ammonia nitrogen whatever the degree of water mineralisation (conductivity, hardness, table 2). Its frequency maximum is related to high conductivity (class V between 600 and 800 $\mu S.cm^{-1}$). It is less frequent in the level of N-NO_3^- superior to 8 $\mu g.l^{-1}$ (Table 2).

Callitriche obtusangula appears as from community C of which it is a characteristic with *Berula erecta* in hard waters, and as from the community D′ in acidic soft waters (Haury & Muller, 1991). Like *Berula* it is more frequent in the conductivity class ranging from 600 to 800 $\mu S.cm^{-1}$ (Table 2). It seems to be fairly indifferent to the levels of phosphate and ammonia if we do not take into account the pH (Table 2). However, it seems that *Callitriche* grows better in highly eutrophic and calcareous waters (our results and Kohler 1975; Kahnt et al., 1989). According to Krause (1971) it prefers the warm waters of winter like *P. coloratus* or *Nasturtium officinale*, which explains why we find it downstream in the Vosges streams and also in the groundwater streams where the water never freezes.

Nasturtium has a behaviour rather comparable to *Callitriche*, i.e. more abundant in waters of high conductivity and low nitrate content, indifferent to the trophic level since it is distributed over the whole range of nutrient level. Casper & Krausch (1981) note that it grows in organic-free but mineral nutrient-rich waters. It must be specified that the distribution of species takes into account only presence or absence of the species and not abundance.

These three species taken in isolation seem not to be trophic bioindicators. However, they have a large ecological spectrum, and in association with other species they are characteristic of a community. For example the communities with *Callitriche obtusangula* and *Berula* (C in alkaline waters and D′ in acidic waters) or with *Nasturtium* (D) provide a well defined bioindication of water quality (mineralization and trophy).

Conclusion

The comparison of the two trophic sequences in acidic and alkaline streams of North Eastern France reveals significant differences of floristic composition between these two systems. Thus the first discriminant variable of the floristic composition is the mineralization related to conductivity and alkalinity. Within a water type, aquatic macrophytes are distributed according to a trophic gradient from oligotrophic to eutrophic water. However, species richness and the number of communities related to the trophic gradient are a little higher in alkaline waters than in acidic waters, just as observed in terrestrial ecosystems. The species appear more sensitive to trophic level in alkaline waters. In acidic waters, the trophic gradient upstream to downstream interacts with the gradient of pH neutralization. It is often difficult to separate the effect of the two gradients.

The ecological spectra of species compared for a range of concentrations of phosphate and ammonia reflect the differences between the trophic optima under different conditions of mineralization. They illustrate the notion of the limiting factor, as yet little studied in the case of aquatic macrophytes. The bioindicator value of a species can change according to the value assumed by other variables, as shown by species studied (*Callitriche obtusangula* or *Berula erecta*). This study will be continued by the comparative analysis of other taxa (e.g. *Elodea, Myriophyllum*).

References

APHA, 1985. Standard Methods for the Examination of Water and Wastewater. 16th edition. American Public Health Association, New York. 1268 pp.

Braun-Blanquet, J., 1964. Pflanzensoziologie. Springer Verlag Wien, New York. 865 pp.

Carbiener, R., M. Trémolières, J.L. Mercier & A. Ortscheit, 1990. Aquatic macrophyte communities as bioindicators of eutrophication in calcareous oligosaprobe stream waters (Upper Rhine plain, Alsace). Vegetatio 86: 71–88.

Casper, S. J., & H. D. Krausch, 1981. Süsswasserflora von Mitteleuropa, Tome 24. Gustav Fisher Verlag Stuttgart, New York. 943 pp.

Dendène, M. A., T. Rolland, M. Trémolières & R. Carbiener, 1993. Effect of ammonium ions on the net photosynthesis of three species of *Elodea*. Aquat. Bot. 46: 301–315.

Furumai, H. & S. Ohgaki, 1989. Adsorption-desorption of phosphorus lake sediments under anaerobic conditions. Wat. Res. 23: 677–684.

Gächter, R., J. S. Meyer, & A. Mares, 1988. Contribution of bacteria to release and fixation of phosphorus in lake sediments. Limnol. Oceanogr. 33: 1542–1558.

Glänzer, U., W. Haber & A. Kohler, 1977. Experimentelle Untersuchungen zur Belastbarkeit submerser Fliessgewässer-Makrophyten. Arch. Hydrobiol. 79: 193–232.

Haslam, S. M., 1978. River plants. Cambridge Univ. Press, Cambridge, 396 pp.

Haslam, S. M., 1982. A proposed method for monitoring river pollution using macrophytes. Envir. Techn. Letters 3: 19–43.

Haury, J. & S. Muller, 1991. Variations écologiques et chorologiques de la végétation macrophytique des rivières acides du Massif Armoricain et des Vosges du Nord (France). Rev. Sc. de l'Eau 4: 463–482.

Iserentant, R. & J. de Sloover, 1976. Le concept de bioindicateur. Mém. Soc. Roy. Belgique (Bruxelles) 7: 15–24.

Kahnt, U., W. Konold, G. H. Zeltner & A. Kohler, 1989. Wasserpflanzen in Fliessgewässern des Ostalb. Verbreitung und Ökologie. In Ökologie in Forschung und Anwendung, Hrsg. D. Knuth 2, 148 pp.

Klosowski, S., 1985. Habitat requirements and bioindicator value of the main communities of aquatic vegetation in north-Eastern Poland. Polskie Arch. Hydrobiol. 32: 7–29.

Kohler, A., R. Brinkmeier & H. Vollrath, 1974. Verbreitung und Indikatorwert der submersen Makrophyten in den Fliessgewässern der Frieberger Au. Ber. Bayer. Bot. Ges. 45: 5–36.

Kohler, A., 1975. Submerse Makrophyten und ihre Gesellschaften als Indikatoren der Gewässerbelastung. Beitr. Naturkd. Forsch. Südwestdtschl. 34: 149–159.

Konold, W., O. Schäfer & A. Kohler, 1990. Wasserpflanzen als Bioindikatoren, dargestellt am Beispiel kleinerer Stillgewässer Oberschwabens und der Franche-Comté. Ökologie und Naturschutz, 3: 167–180.

Krause, E., 1971. Die Makrophytische Wasservegetation der Südlichen Oberrheinebene. Die Äschenregion. Arch. Hydrobiol. Suppl. 37, 4: 387–465.

Lachavanne, J. B., 1985. The influence of accelerated eutrophication on the macrophytes of Swiss lakes. Verh. Int. Ver. Limnol. 22: 2950–2955.

Manning, P. G., 1989. Iron, phosphorus and lead relationships in suspended sediments from Lake St Clair and the Detroit river. Can. Mineralogist 27: 247–255.

Muller, S., 1990. Une séquence de groupements végétaux bioindicateurs d'eutrophisation croissante des cours d'eau faiblement minéralisés des Basses Vosges gréseuses du Nord. C.R.Acad. Sci. Paris. 310: 509–514.

Newbold, B. Sc. & N. T. H. Holmes, 1987. Nature conservation: water criteria and plants as water quality monitors. Wat. Poll. Control. 1987, 345–363.

Robach, F., I. Eglin & R. Carbiener, 1991. L'hydrosystème rhénan: évolution parallèle de la végétation aquatique et de la qualité de l'eau (Rhinau). Bull Ecol. 22: 227–241.

Roweck, H., S. Risse & A. Kohler, 1986. Zur Verbreitung, Standortsökologie und morphologischen Variabilität von *Potamogeton polygonifolius* in den Fliessgewässern des Südlichen Pfalzerwaldes. Mitt. Pollichia 73: 289–374.

Roweck, H., K. Weiss & A. Kohler, 1986. Zur Verbreitung und Biologie von *Potamogeton coloratus* and *P. polygonifolius* in Bayern und Baden-Württemberg. Ber. Bayer. Bot. Ges. 57: 17–52.

Standing Committee of Analysis (H.M.S.O.) 1987. Methods for the use of aquatic macrophytes for assessing water quality 1985/1986. Her Majesty's Stationery Office, London, 176 pp.

Trèmolières, M., I. Eglin, U. Roeck & R. Carbiener, 1993. The exchange process between river and groundwater on the Central Alsace flooplain (Eastern France). I The case of the canalised river Rhine. Hydrobiologia 254: 133–148.

Tutin, T. G., N. A. Burges, A. O. Chater, J. R. Erdmondson, V. H. Heywood, D. M. Moore, D. H. Valentine, S. M. Walters & D. A. Webb, 1964–1993. Flora Europaea. 5 volumes. Cambridge University Press.

Wiegleb, G., 1984. A study of habitat conditions of the macrophytic vegetation in selected river systems in Western lower Saxony (Fed. Rep. of Germany). Aquat. Bot. 18: 313–352.

Hydrobiologia **340:** 77–83, 1996.
J. M. Caffrey, P. R. F. Barrett, K. J. Murphy & P. M. Wade (eds), Management and Ecology of Freshwater Plants.

The impact of three industrial effluents on submerged aquatic plants in the River Nile, Egypt

M. M. Ali[1] & M. E. Soltan[2]
Department of Botany[1] and Chemistry[2], Faculty of Science at Aswan, Assiut University, Aswan 81528, Egypt

Key words: River Nile, industrial pollution, submerged macrophytes, canonical ordination

Abstract

The submerged vegetation growing in the drainage channels taking effluent from three factories (two processing sugar cane plus one producing chipboard or paper pulp; and one large fertilizer plant) into the River Nile in Upper Egypt, and in the river itself upstream and downstream of the discharge points, was studied during 1994. The main pollutants from the sugar cane factory effluents comprised organic matter, including carbohydrates; from the fertilizer plant ammonia was the principal pollutant. The study investigated the effect of these different pollutants on aquatic plant standing crop and distribution, in relation to physico-chemical characteristics of water and hydrosoil. In the effluent channels, dominated by large growths of sewage fungus, submerged vegetation was absent, although some emergent vegetation survived. In the most polluted river sites, up to 2 km downstream of discharge points, the flora was restricted to *Potamogeton pectinatus* L. Elsewhere in the river, a more diverse submerged flora was present, including *Ceratophyllum demersum* L. and *Potamogeton crispus* L.

Introduction

Pollution is generally associated with heavy industrialisation and dense population and is one of the principal ecological problems of the River Nile system. Aquatic environment pollution comes from both natural and anthropogenic sources and occurs in many different forms: sediments, sewage, disease-causing agents, inorganic plant nutrients, organic compounds, inorganic chemicals, radioactive substances and thermal pollution. The four major sources of human induced water pollution are industry, domestic activities, shipping and agriculture.

Factors affecting the distribution of submerged aquatic macrophyte communities in the Nile in Upper Egypt have been described by Ali (1992). This present study aims to detect the principal pollutants produced by three factories that discharge their effluents into the River Nile, and to describe the effect of these point-source pollutants on submerged aquatic macrophyte communities.

Description of sites studied

Effluents of three factories (two processing sugar cane and producing chipboard or paper pulp; and one large fertiliser plant) which discharge into the River Nile in Upper Egypt (Figure 1), were selected for study. The Edfu sugar cane (ESC) and paper pulp (EPP) factory, 100 km North of Aswan, discharges its effluents directly into the Nile. The Kom Ombo sugar cane (KSC) and chipboard (KCB) factory, 45 km North of Aswan, discharges its effluents by way of an uncovered discharge channel that also receives leaching from the surrounding agricultural lands. The Aswan Kima fertilizer (AKF) factory, discharges its effluents into the River Nile through an uncovered channel that passes through the city and receives untreated domestic wastes (human and sewage wastes). Ammonium nitrate 34.8% (N concentration 99.8%) is the main product of the factory.

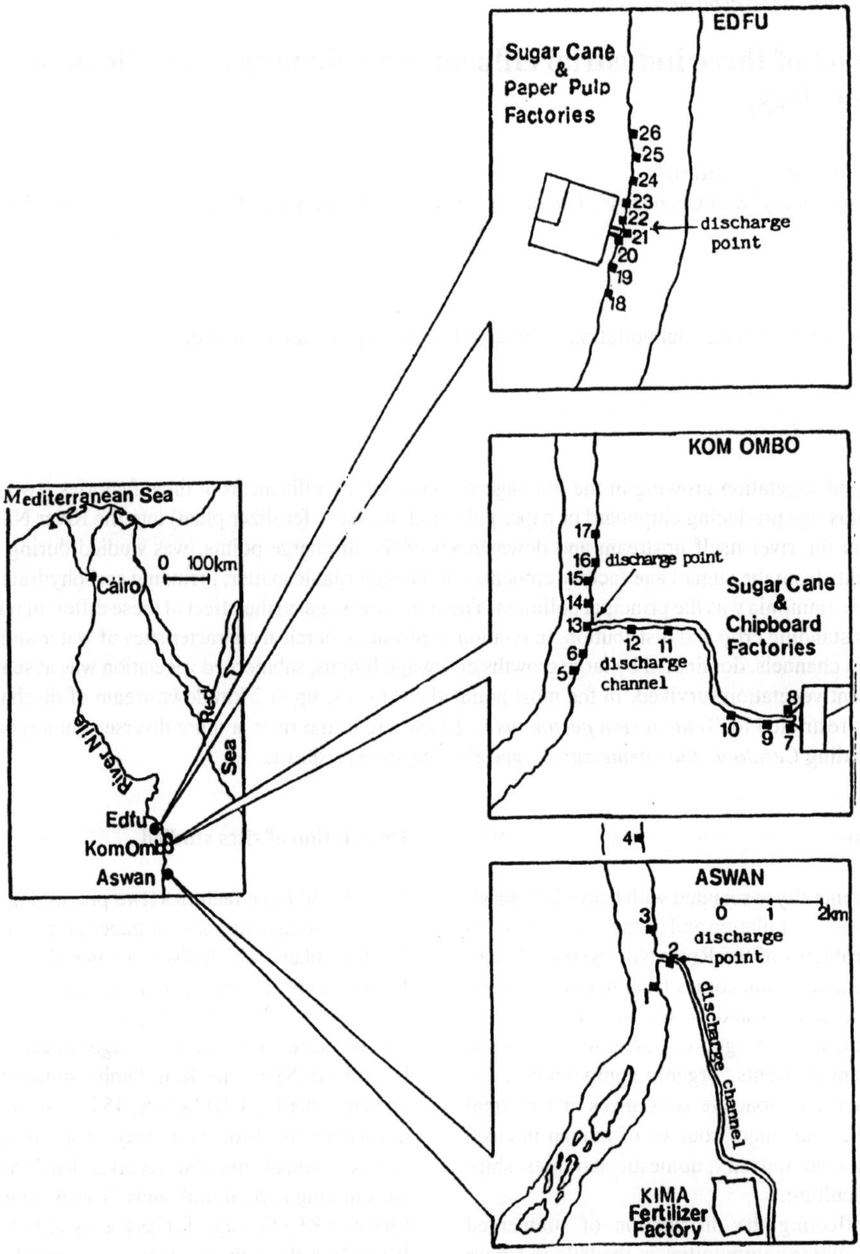

Figure 1. Location map of the River Nile showing the sites surveyed (between February and April 1994)

Materials and methods

Sampling regime

Samples were collected between February and April 1994 from sites upstream (two sites, each 0.5 km upstream, except for Kima factory one site only) and downstream (four sites, each 0.5 km downstream, except Kima factory one site only) of the discharge points, as well as, from the discharge channels.

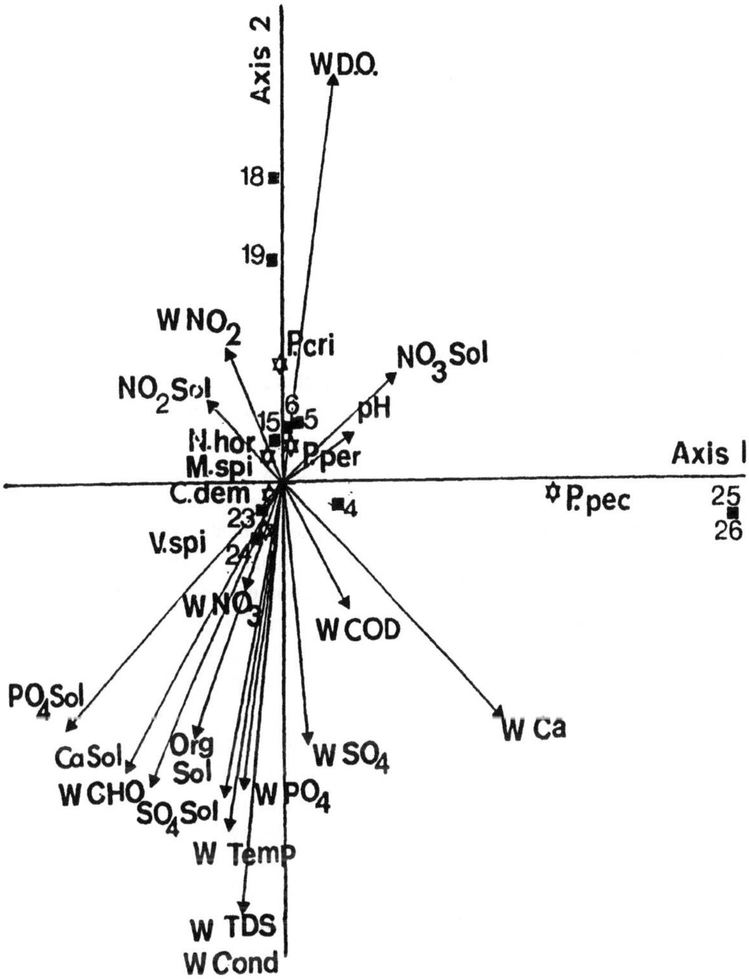

Figure 2. The Distribution of the submerged aquatic macrophytes, pollutant factors and sampling sites in the River Nile, at Aswan, Egypt. Canonical Correspondence Analysis (CCA) ordination diagram with 7 species (✡), 10 sampling sites (■), 17 quantitative environmental variables (arrows). The submerged species are: C. dem = *Ceratophyllum demersum*, N. hor = *Najas horrida*, M. spi = *Myriophyllum spicatum*, P. cri = *Potamogeton crispus*, P. per = *Potamogeton perfoliatus*, P. pec = *Potamogeton pectinatus* and V. spi = *Vallisneria spiralis*. The environmental factors are: W Temp = *water temperature*, pH = water pH, W D.O. = water dissolved oxygen, W TDS = water total dissolved solids, W Cond = water conductivity, W Ca = water calcium, W PO_4 = water phosphate, W SO_4 = water sulphate, W NO_3 = water nitrate, W NO_2 = water nitrite, W COD = water chemical oxygen demand, Ca Sol = hydrosoil calcium, PO_4 Sol = hydrosoil phosphate, SO_4 Sol = hydrosoil sulphate, NO_3 Sol = hydrosoil nitrate, NO_2 Sol = hydrosoil nitrite, Org Sol = hydrosoil organic matter. Site locations (1–26) as shown in Figure 1.

Site codes were numbered in ascending order from upstream to downstream in the river (Figure 1).

Vegetation

Standing crop samples of submerged aquatic vegetation were collected from the 0 to 3 m depth zone, using a grapnel to collect 5 grapnel hauls per sampling site. Plants were separated into different species. Dry weight standing crop (DWSC) of each species per grapnel haul was determined after air drying at ambient temperature.

Water analysis

In the field, samples (1 l each) were collected from surface waters and measurement made of water temperature and pH (using a Jenway portable combined thermistor-pH meter, model 3070) and of dissolved oxygen (using a Jenway Oxygen meter, model 9070). Conductivity and total dissolved solids (T.D.S., using ESD model 76) were carried out during the day time, in situ. In the laboratory, water samples were analysed for the following determinants: chemical oxygen demand COD; calcium and magnesium (titration with EDTA); potassium (by flame photometer); nitrate (chromotropic acid method); nitrite (Griess-Ilosvay method); phosphate (molybdenum blue method); sulphate (turbidimetric method). The methods of analysis were taken mainly from 'Standard Methods for the Examination of Water and Waste Water' (Americam Public Health Association (A.P.H.A.), 1980).

Hydrosoil analysis

A 1/10 w/v dry hydrosoil/1M sodium acetate extract was used for the determination of calcium, magnesium, potassium, nitrite, nitrate, phosphate and sulphate in hydrosoil samples (A.P.H.A., 1980). Hydrosoil organic matter was measured by the loss-on-ignition method of Allen et al. (1986).

Data analysis

The field data were drawn up in the form of plant species × samples, and sample × environmental variable matrices. Ordination analysis of the two matrices was carried out by Canonical Correspondence Analysis (CCA), release 3.12 of CANOCO (Ter Braak, 1987).

Results

Plant distribution

Submerged vegetation was absent from sites that directly receive the factories' effluents. *Ceratophyllum demersum* L. was the dominant species in sites upstream (DWSC of 8.4 and 9.0 g sample^{-1}) and downstream (DWSC 12.7–14.3 g sample^{-1}) at the discharge point into the Nile at Kom Ombo. where each of *Vallisneria spiralis* L. and *Najas horrida* A. Br. ex Magn. was recorded only once. At Edfu, *Potamogeton crispus* L. dominated sites upstream (DWSC of 4.44 and 0.65 g sample^{-1}) of the discharge point on the Nile and *Potamogeton pectinatus* L. was the only species recorded at sites 25 and 26 (DWSC 1.38 and 1.8 g sample^{-1}, respectively).

Water quality

Waters of high temperature flow out from the three factories, where cooling is required (ranging from 24.5 °C to 34.6 °C). The effluents from the KSC and KCB factory and from the ESC factory, had almost neutral pH values (6.9, 7.02 and 7.4, respectively). The EPP and AKF factories had alkaline effluents, (pH values 9.7 and 9.75, respectively). Dissolved oxygen was low (range from 0 to 6.1 mg l^{-1}) in water samples either from the main pollution sources, or from the discharge channels. Conductivity was very high at the factories discharge points and in the channels, particularly water from Kima drain (2250 μS cm^{-1}). The Kima drain contained 10 mg l^{-1} nitrite-N and 30 mg l^{-1} nitrate-N, which were diluted down to 0.21 mg l^{-1} and 1.8 mg l^{-1}, respectively, by the River Nile water. High phosphate concentration (3.2 mg l^{-1}) was recorded downstream of the effluent's channel of Kima, where sewage and domestic wastes discharge along its length. Effluent from ESC factory contained 1.76 mg l^{-1} phosphate, but lower concentrations of phosphate (0.17–0.30 mg l^{-1}) were recorded in water samples from Kom Ombo effluent's channel. Sulphate concentrations were high in Kima drain (49 mg l^{-1}), and both the EPP and ESC factories (52 and 30 mg l^{-1}, respectively). Relatively high calcium concentrations were recorded in samples from the two sugar cane factories (28.86 mg l^{-1}, at Kom Ombo and 30 mg l^{-1}, at Edfu). However, samples from the Kima discharge channel and sites at the Nile had lower concentrations of calcium (from 6.14 to 22.44 mg l^{-1}). Relatively small variations were recorded for both magnesium and potassium concentrations.

Carbohydrate (polyhydroxy aldehydes) concentrations were very high (100 mg l^{-1}) in the sample of the EPP factory, since large amounts of cellulose are released as a result of the processing of the raw materials. Effluent from the paper pulp factory had predictably high COD (1311 mg l^{-1}), while effluents from the KSC and KCB factories (where oxidizing reagents are used in large quantities) had low COD (43.52 and 30.72 mg l^{-1}, respectively).

Hydrosoil analysis

Organic matter percentages ranged from 0.63% (at site 17) to 23.42% (at site 12). Calcium concentrations ranged from 144 g g^{-1}, at site 19 to 1683.36 μg g^{-1}, at site 24, where the highest sulphate concentration (1070 μg g^{-1}) was also recorded. Hydrosoil samples from the discharge channel at Kom Ombo also had

Table 1. CCA ordination characteristics for the first four axis.

	1	2	3	4
Eigenvalues	1.00	1.00	1.00	1.00
Species-environmental				
variables correlations	0.60	0.33	0.13	0.04
Inter set correlation of:				
Ca Sol	-254	-425		
PO4 Sol	-349	-359		
SO4 Sol	-95	-459		
Org Sol	-141	-371		
W Temp	-89	-509		
W Cond	-68	-631		
W TDS	-77	-619		
W D.O.	84	596		
W Ca	356	-350		
W SO4	43	-386		
W CHO	-213	-445		

high sulphate concentrations (up to 720 μg g^{-1}, at the sites closest to the river). A relatively high sulphate concentration (495 g g^{-1}) was noticed at site 1.

A sample from site 3, at Kima, contained high phosphate concentration (4.95 μg g^{-1}). In samples from Kom Ombo, phosphate concentrations were high (2.03 μg g^{-1}) at site 13 and (1.35 μg g^{-1}) at site 16. At Edfu, high concentrations of phosphate (1 μg g^{-1}) were observed in hydrosoil samples from sites 23 and 24.

CCA ordination

The species-environmental biplot (Figure 2) shows the relations of the species and the environmental variables with the ordination axes. The length of the arrow indicates the relative importance of the environmental variable in determining the axes. The position of the species centres (points) along the ordination axes represent their respective optima along the environmental gradient.

Table 1 lists the ordination characteristics. The species-environmental correlation (1.0 for each axis) is high. This indicates that the environmental variables measured described well the impact of the industrial wastes on the growth and distribution of the submerged aquatic macrophytes. CCA yields two dominant ordination axes, with eigenvalues 0.60 and 0.33 (0.13 and 0.04 for axes 3 and 4). Axis 1 of the CCA is strongly aligned with water and hydrosoil calcium ion concentrations and hydrosoil phosphates. The second canonical axis is most associated with water temperature, conductivity, total dissolved salts, dissolved oxygen, phosphates, carbohydrates, sulphates and calcium, as well as, hydrosoil calcium, phosphates and sulphates (Table 1).

Figure 2 shows that *Ceratophyllum demersum*, *Vallisneria spiralis*, *Myriophyllum spicatum* L., *Najas horrida* and *Potamogeton perfoliatus* L., occurred near the origin of the ordination, which indicates that these species tend to occur in habitats with low levels of the pollutant factors. *Potamogeton crispus* is associated with highly oxygenated sites (e.g. site 18 and site 19). Although, these sites were characterised by clean and well-oxygenated water; their hydrosoil samples contained high concentrations of nitrite and nitrate anions. *Potamogeton pectinatus* is found towards the extremes of the environmental gradients measured and is the only species that relates strongly to axis 1 and tending to occur most often at highly polluted sites.

Discussion

The factories studied produce different products, but their effluents have common pollutant factors. These are: (i) Thermal pollution: occurs when water, produced during the industrial processes and heated by as much as 5–10 °C, is released into the river. A rise in temperature of a body of water has a number of chemical, physical, and biological effects. Chemical reactions, include fast decomposition of organic matter and depletion of oxygen. Moreover, as less oxygen dissolves in warm water than in cool water, such low dissolved oxygen concentrations may be one factor responsible for disappearance of the submerged aquatic vegetation, at the discharge points. (ii) Organic pollution: was a major form of pollution, brought about mainly by the factories discharging industrial wastes into the river. Pollution by organic matter stimulates bacterial and fungal growth which deoxygenates water. Its effects depend upon the amount of discharged and the speed and volume of clean water present to dilute it. Sugar cane factories release substantial amounts of organic plant material. Seager et al. (1992) concluded that in extreme cases where the pollution is very heavy, the dissolved oxygen present is completely utilised for decomposition of the organic matter and recovery of such a polluted system becomes impossible. Such waters contain only bacteria and sewage fungus. (iii) Sewage: effluents from the three factories also receive domestic wastewaters from the surrounding houses containing pollutants that increase organic matter and inorganic salts. (iv) Inorganic chemicals: effluents from the factories contain different concentrations and types of inorganic chemical pollutants. Phosphate, sulphates, nitrate and nitrite are the main pollutants in Kima effluent. Effluent from sugar cane factories contains high concentrations of sulphate and calcium, the later being due to the use of $Ca(OH)_2$ to neutralize the acidic sugar cane juice and to react with the other mineral salts producing calcium salts that precipitate easily and thereby facilitate disposal. Samples from sugar cane factories often contain relatively high phosphate concentrations according to the chemicals used in the treatment of the raw materials and as a result of excessive use of $Ca_3(PO_3)_2$ in the sugar cane juice treatment process.

These point sources of pollution have a pronounced influence on the River Nile system. However, the effects of non-point source pollution should not be ignored. High phosphate concentration (3.2 mg l^{-1}) was recorded downstream in the effluent channel of Kima as a result of receiving sewage and domestic discharges (including synthetic detergents) along its length.

Effluents from the factories produced a measurable effect on hydrosoil quality. For example, in the sugar cane factories of Kom Ombo and Edfu, hydrosoil samples, up- and downstream of the discharge point, were characterized by high concentrations of calcium. Also, high sulphate concentration in the hydrosoil samples resulted from the use of sulphur oxide (SO) in the sugar cane juice treatment process, NaOH (12%) and Na_2S in the treatment of the sugar cane plant material in the EPP factory, and H_2SO_4 in the AKF. High hydrosoil nitrate concentrations up- and down stream of Kom Ombo factories resulted from fertilizer runoff, from adjacent agricultural lands rather than from factory effluents (Raven et al., 1993). Also, the concentration of high organic matter in hydrosoil and very low oxygen levels may lead to phytotoxic compounds being produced during anaerobic decomposition (Drew & Lynch, 1980).

Rather low macrophyte dry weight standing crop (DWSC) was observed at pollution-influenced sites compared with cleaner sites upstream of the factories discharge points, e.g. *Ceratophyllum demersum* up to 21.8 g sample^{-1}, *Potamogeton crispus* up to 23.5 g sample^{-1}, *Myriophyllum spicatum* up to 42.6 g sample^{-1}, *Potamogaton perfoliatus* up to 52.6 g sample^{-1}, *Potamogeton pectinatus* 2.1 g sample^{-1}, (Ali, 1992). Some submerged macrophytes cannot tolerate high stress levels that may occur due to industrial pollution. Sites that directly receive the factories' effluents had no submerged vegetation either at discharge channels or at the point of discharge into the River Nile. Pollution stress and disturbance coinciding may be responsible for the disapearance of submerged vegetation from these sites which also suffer from high disturbance by river traffic.

From the CCA ordination diagram (Figure 2) a number of environmental variables were shown to be potentially important indicators in an industrially polluted river. The significant water pollutant indicators were temperature, dissolved oxygen, calcium, sulphate, phosphate and carbohydrates. The important hydrosoil pollutants were calcium, sulphate, phosphate and organic matter. These results are consistent with those found by previous researchers (e.g. Caffrey, 1985; Jones & Cullimore, 1973).

The ordination diagram (Figure 2) suggests that most of the species escape from the most heavily polluted environment. However, *Potamogeton pectinatus* tends to occur most often at highly polluted sites. Previous work has demonstrated that *P. pectinatus* is a very phenoplastic species with a wide ecological amplitude, found even in eutrophicated waters (van Wijk, 1986).

From the CCA diagram (Figure 2), water and hydrosoil phosphate are important factors affecting the submerged aquatic plant growth, distribution, and community structure. Jones & Cullimore (1973) stated that *P. pectinatus* was increasingly abundant as total phosphorus rose from 0.1 to 0.6 ppm, while *Potamogeton richardsonii* (Benn.) Rydb. showed a reverse trend dominating water with 0.1 ppm phosphorus, but declined as the level rose to 0.7 ppm. Also the diagram suggests that water and hydrosoil calcium are also important pollutant factors that affect the distribution of the submerged macrophytes. Kadono (1982) stated that high concentrations of water calcium ($<$ 50 mg l^{-1}) do not inhibit the growth of *Myriophyllum spicatum* L. and *C. demersum*. These results imply that hydrosoil calcium may be a more important pollutant determinant factor than water calcium.

Following Grime (1979), *P. pectinatus* is a stress-tolerant competitive plant (CS); while *C. demersum*, *M. spicatum* and *P. perfoliatus* L. are competitive (C) species (Kautsky, 1987; 1988). Murphy et al. (1990) classified *P. pectinatus, P. perfoliatus, P. crispus, M. spicatum* and *C. demersum* as having a stress-tolerant disturbance strategy (SD). Either CS or SD submerged macrophytes are likely to inhibit an industrial polluted aquatic system, such as described in this paper, which causes stress rather than disturbance pressure on submerged macrophytes.

The CCA indicates that hydrosoil organic matter is more important than water organic matter in affecting the growth and distribution of the submerged macrophyte communities. However, Caffrey (1985) demonstrated that alteration of the chemical composition of water, due to organic pollution, may affect both macrophyte species composition and abundance in rivers.

Some authors have claimed that the existence of submerged plants is determined mostly by the quality of the water (Agami, 1984). The present study suggests that hydrosoil pollutant factors are probably at least as important as water pollution factors, and sometimes may be more influential on submerged plants survival and community composition.

Acknowledgments

The authors thank Dr K. Murphy, Division of Environmental and Evolutionary Biology, University of Glasgow for his support and comments on this work. We also thank the Organising Committee of the European Weed Research Society, 9th International Symposium on Aquatic Weeds, 1994 for part-funding the presentation of this work in Dublin 1994.

References

Agami, M., 1984. Effects of water pollution on plant species composition along the Amal River, Israel. Arch. Hydrobiol. 100: 445–454.

Ali, M. M., 1992. Ecological Studies on Freshwater Macrophytes in Regulated Waterbodies in Egypt and UK. Ph.D. Thesis, Faculty of Science at Aswan, Assiut University: 263 pp.

Allen, S. E., H. M. Grimshaw & A. P. Rowland, 1986. Chemical analysis. In Methods in Plant Ecology. P. D. Moore & S. B. Chapman (eds), 285–344.

American Public Health Association, 1980. Standard Methods for the Examination of Water and Waste Water. Amer. Publ. Health Assoc., Wash. D.C. 874 pp.

Best, E. P. H., 1979. Growth substances and dormancy in Ceratophyllum demersum L. Physiol. Plant. 45: 339–406.

Caffrey, J., 1985. A scheme for the assessment of water quality using aquatic macrophytes as indicators. J. life Sci. R. Dubl. Soc. 5: 105–111.

Drew, M. C. & J. M. Lynch, 1980. Soil anaerobiosis, microorganisms, and root function. Annu. Rev. Phytopathol. 18: 37–66.

Grime, J. P., 1979. Plant strategies and vegetation processes. John Wiley & Sons Ltd. Publ., Chichester. 222 pp.

Jones, G. & D. R. Cullimore, 1973. Influences of macro-nutrients on the relative growth of water-plants in the Qu'Apelle lakes, Canada. Envir. Pollut. 4: 283–290.

Kadono, Y., 1982. Occurrence of aquatic macrophytes in relation to pH, alkalinity, Ca^{++}, Cl^{-} and conductivity. Jap. J. Ecol. 32: 39–44.

Kautsky, L., 1987. Life-cycles of three populations of Potamogeton pectinatus L. at different degrees of wave exposure in the Askö area, Northern Baltic proper. Aquat. Bot. 27: 177–186.

Kautsky, L., 1988. Life strategies of aquatic soft bottom macrophytes. Oikos. 53: 126–135.

Murphy, K. J., B. Rørslett & I. Springuel, 1990. Strategy analysis of submerged aquatic lake macrophyte communities: an international example. Aquat. Bot. 36: 303–323.

Raven, P. H., L. R. Berg & G. B. Johnson, 1993. Environment. Saunders College Publ. 569 pp.

Seager, J., F. Jones & G. Rutt, 1992. Assessment of control of farm pollution. Journal if the I.W.E.M., 6: 48–54.

Ter Braak, C. J. F.,1987. CANOCO – a FORTRAN Program for Canonical Communitiy Ordination By Parial Detrended Canonical Correspondence Analysis, Principal Components Analysis and Redundancy Analysis (Version 3.1). Agriculture Mathematics Group, Wageningen.

Van Wijk, R. J., 1986. Life cycle characteristics of *Potamogeton pectinatus* L. in relation to control. Proceedings EWRS/AAB 7th Symposium on Aquatic Weed 375–380.

Hydrobiologia **340**: 85–92, 1996.
J. M. Caffrey, P. R. F. Barrett, K. J. Murphy & P. M. Wade (eds), Management and Ecology of Freshwater Plants.

Effects of lake water level regulation on the dynamics of littoral vegetation in northern Finland

S. Hellsten & J. Riihimäki
VTT, Communities and Infrastructure, Water Engineering and Ecotechnology, P.O. Box 19042, FIN–90571 Oulu, Finland

Key words: regulated lakes, littoral vegetation, dynamics, dissimilarity

Abstract

The effects of water level regulation on the dynamics of littoral vegetation were studied in regulated Lake Ontojärvi and unregulated Lake Lentua in northern Finland by using permanent plots. The study was carried out during 1984–1988 and the abundance of plant species was measured yearly. The annual changes were measured by comparing the mean dissimilarities (D_m) in species abundance and the diversity (mean number of species, N_{sm}) between the different research years in the same square. The mean dissimilarities in Lake Ontojärvi and Lake Lentua were 0.238 and 0.297, respectively. The difference between the lakes was not significant. The effect of the increased ecological stress was observed on the littoral zone; the number of species was lower in Lake Ontojärvi than in Lake Lentua. In Lake Ontojärvi, the area of almost permanent submersion (sublittoral) showed higher dissimilarity (D_m) and lower diversity (N_{sm}) values compared to Lake Lentua. The vegetation in both lakes was well adapted to the disturbance caused by waves and penetrating ice. The diversity and D_m values were lower on the exposed shores in both lakes compared to sheltered shores. The peaty bottoms were the most stable environment, whereas the muddy bottoms were unstable in our research lakes. Generally speaking the vegetation in Lake Ontojärvi is equally stable as in Lake Lentua. Both diversity and dissimilarity values were slightly higher in Lake Lentua. The vegetation in regulated Lake Ontojärvi is well adapted to the ecological disturbance caused by the fluctuating water level.

Introduction

About 10% (33 522 km^2) of the total area of Finland is covered by lakes. Over one third of this area (11 900 km^2) is regulated, mainly for hydroelectric purposes. The regulation is usually achieved by raising the water level during the summer (0.5–3.5 m) and lowering the winter minimum by 2–7 m. During the first few years of regulation, the littoral area is subject to considerable erosion, depending among other things on the regulation height, exposure, steepness, quality of sediment and on the rate of water level uplift (Alasaarela et al., 1989). At the same time, the changes in vegetation are obvious, including a decrease of former macrophytic vegetation, as reported in several Scandinavian lakes (Nilsson, 1981; Rørslett, 1985a). The new vegetation on eroded shores consists of disturbance-tolerant species (e.g., *Ranunculus reptans, Eleocharis acicularis*) adapted to the altered ecological environment (e.g., Murphy et al., 1990), which is under succession for several decades (Koskenniemi, 1987; Nilsson & Keddy, 1990). The effects of lake water level regulation has been under intensive research from the beginning of eighties (e.g., Alasaarela et al., 1989), but the reports dealing with littoral vegetation are rare and published mainly in Finnish (e.g., Granberg & Hakkari, 1980; Hellsten & Joronen, 1984; Hellsten et al., 1989). This paper concentrates on describing the dynamics of littoral vegetation during a four-year study using permanent plots, while other part of the study assesses the general ecological environment and macrophytes in regulated lakes.

Table 1. Species composition, frequency (%) and abundance (%) observed at permanent plots of research lakes.

	ONTOJÄRVI		LENTUA	
	Freq. %.	Abun. %	Freq. %.	Abun. %
Agrostis canina L.	0.6	8.0	0.5	8.0
Agrostis capillaris L			3.0	12.0
Alisma plantago-aquatica L.	5.6	4.9	0.5	4.0
Alnus incana (L.) Moench			5.1	4.4
Alopecurus aequalis Sobol.	9.9	11.8	3.5	8.6
Andromeda polifolia L.			2.0	12.0
Aulacomnium palustre (Hedw.) Schwaegr.			0.5	8.0
Betula pubescens Ehrh.	0.6	8.0	0.5	8.0
Bryidae spp.	2.5	11.0	6.6	22.8
Bryum sp.	1.9	6.7	0.5	4.0
Calamagrostis sp.			1.0	32.0
Calliergon cordifolium (Hedw.) Kindb.	1.9	32.0	2.5	41.6
Calliergon megalophyllum Mikut.	1.9	18.7	0.5	24.0
Calliergon sp.	3.1	14.4	3.5	30.3
Callitriche palustris L.	8.1	38.6	2.5	44.4
Callitriche sp.	5.6	10.7	1.0	16.0
Calluna vulgaris (L.) Hull			4.0	62.0
Carex acuta L.			5.1	39.6
Carex lasiocarpa Ehrh.			6.6	54.5
Carex nigra subsp. *juncella* (Fries) Lemke			4.0	19.0
Carex nigra subsp. nigra (L.) Reichard			5.1	43.6
Carex panicea L.	1.9	20.0	1.0	16.0
Carex rostrata Stokes	8.7	52.0	8.6	28.5
Carex serotina Merat			1.5	25.3
Carex sp.			1.0	8.0
Climacium dendroides (Hedw.) Web. & Mohr	1.2	6.0	1.5	9.3
Drepanocladus sp.	18.6	43.3	13.1	30.6
Drosera anglica Hudson			2.5	69.6
Elatine hydropiper L.	2.5	28.0	2.5	24.0
Eleocharis acicularis (L.) Roemer & Schultes	29.8	62.3	17.7	64.0
Eleocharis palustris (L.) Roemer & Schultes	7.5	53.7	12.6	69.9
Equisetum fluviatile L.	8.7	67.4	10.1	32.6
Hieracium umbellatum L.			4.0	9.5
Hypnum lindbergii Mitt.			1.5	22.7
Isoetes echinospora Durieu	16.8	18.8	20.7	37.7
Isoetes lacustris L.			21.2	69.4
Juncus alpinoarticulatus Chaix	4.4	8.0	6.1	32.7
Juncus bulbosus L.	1.2	6.0	1.0	6.0
Juncus filiformis L.	16.8	25.6	18.7	38.0
Juncus sp.	0.6	12.0		
Lobelia dortmanna L.			14.7	43.0
Lysimachia thyrsiflora L.			1.0	6.0
Lythrum salicaria L.			1.5	26.7
Marcanthia polymorpha L.	0.6	4.0	14.7	51.0
Mentha arvensis L.			5.6	19.3
Menyanthes trifoliata L.			1.5	49.3
Molinia caerulea (L.) Moench			16.7	40.3

Table 1. Continued

	ONTOJÄRVI		LENTUA	
	Freq. %.	Abun. %	Freq. %.	Abun. %
Pedicularis palustris L.			0.5	4.0
Phragmites australis (Cav.) Trin. ex Steudel	0.6	8.0	6.1	10.7
Pinus sylvestris L.	1.2	6.0	5.6	5.5
Pohlia sp.	4.4	53.1	2.0	70.0
Pohlia nutans (Hedw.) Lindb.			2.0	38.0
Polytrichum commune Hedw.			0.5	4.0
Polytrichastrum longisetum (Brid.) G. L. Sm.	2.5	57.0	2.0	72.0
Polytrichum sp.	0.6	44.0		
Potamogeton gramineus L.	3.7	32.7	1.0	38.0
Potentilla palustris (L.) Scop.	2.5	10.0	19.7	22.4
Ranunculus reptans L.	61.5	63.7	28.8	46.4
Sagittaria natans Pallas	5.0	49.0	2.5	51.2
Salix sp.			2.0	21.0
Scorpidium scorpidioides (Hedw.) Lindb.	0.6	72.0	1.5	11.7
Selaginella selaginoides (L.) Beauv. ex Schrank & Mart	0.6	4.0	0.5	4.0
Sparganium sp.	9.3	18.4	9.6	37.2
Sphagnum sp.	2.5	18.0	10.1	25.0
Subularia aquatica L.	23.6	29.4	18.2	22.8
Trientalis europaea L.			3.0	9.3
Vaccinium uliginosum L.			3.5	13.7
Vaccinium vitis-idaea L.			2.0	36.0
Veronica scutellata L.	0.6	4.0	2.0	38.0
Warnstofia trichophylla (Warnst.) Tuom. & T. Kop	4.4	60.0	2.5	80.0

Description of sites studied

The two research lakes are situated on the Oulujoki water course (Figure 1). Lentua is a large (90 km^2) non-regulated lake with a water level fluctuation typical of Finnish lakes (Figure 2). Ontojärvi (102 km^2) has been regulated for hydroelectric purposes since 1951 with a maximum amplitude of 4.4 m and an average amplitude for 1960–88 of ca 3.4 m (Figure 2). At the beginning of the regulation, the mean summer water level was raised by one meter. The duration of the ice cover in the research lakes is usually 7–8 months. Both lakes are meso-oligotrophic, but the lake water in Ontojärvi contains slightly more phosphorous and humic substances compared to Lentua.

Material and methods

The field data covered the species frequency and the abundance of aquatic macrophytes and bryophytes, which were sampled yearly from August till the beginning of October in 1984–88. The sampling quadrats (0.5 by 0.5 m) were marked permanently by steel rods in 16 ecologically different shore areas (Figure 1) selected according to the results of transects studies (Hellsten et al., 1989). The number of analysed quadrats varied yearly according to the following schema:

	1984	1985	1986	1987	1988
Ontojärvi	13	53	33	30	32
Lentua	9	58	53	35	43

The observations were done by setting a 0.25 m^2 steel frame divided into twenty-five 1 dm^2 wire mesh squares on the area bordered by rods. The quadrats situated under the water level were analysed by SCUBA diving with the same method. All the species growing below the highest water level were recorded. The abundance was calculated from the presence of each species at the different subsquares (0, 4, 8,...,100 %) and the frequency as a mean frequency of all observed quadrats. Only the quadrats with continuous time series were included in analysis. Bottom quality

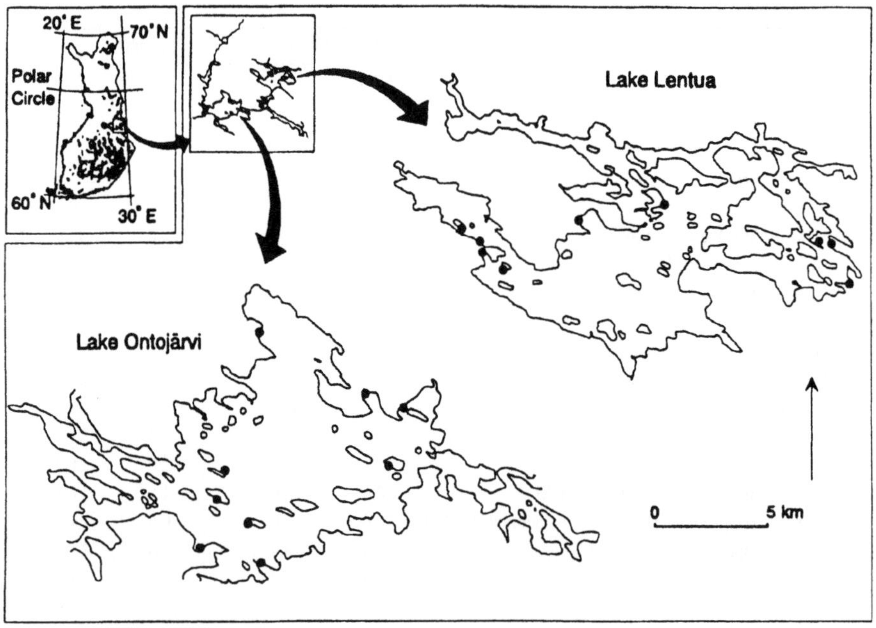

Figure 1. The location of Lake Ontojärvi and Lake Lentua in detail. Study sites indicated by filled circles.

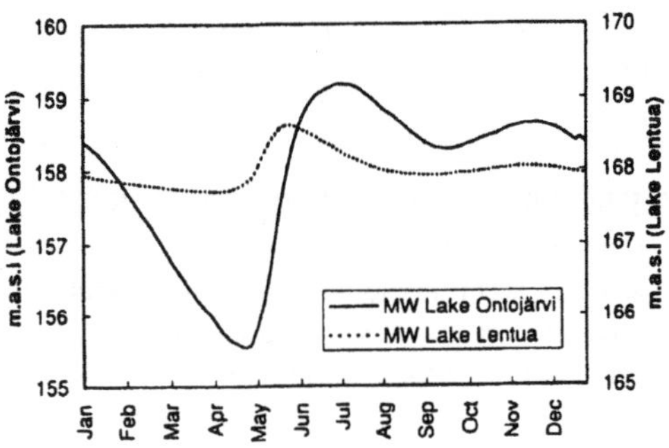

Figure 2. The mean water level (MW) calculated from the daily values during the years 1960–88.

of the quadrats was divided visually to four different classes (peat, sand, muddy sand and mud).

The water level data are obtained from the Hydrological Office (Figure 2). The vertical shore zonation was defined by the duration (d) of the water level during the iceless period (Hellsten et al., 1989). The eulittoral zone consisted of upper ($10\% < d \leq 25\%$), middle ($25\% < d \leq 50\%$) and lower eulittoral zones ($50\% < d \leq 75\%$). The sublittoral was divided to an upper ($75\% < d \leq 95\%$) and a lower zone ($d > \leq 95\%$). Effective fetch (L_f) was measured as proposed by Håkanson & Jansson (1983) and the shores were classified into open ($L_f > 1$ km) and sheltered (L_f 1 km) ones according to the Hellsten et al. (1989). The shores were also divided into steeply ($> 5\%$) and gently ($\leq 5\%$) sloping ones (Hellsten et al., 1989).

The annual changes in vegetation were measured by comparing the dissimilarities in species composition between the different research years at the same square. We used a calculation method modified from Nilsson & Keddy (1990):

$$D(x_1, x_2) = 1 - 1/2(W/A + W/B), \tag{1}$$

where A is the sum of species abundances in sample x_1, B is the corresponding value for sample x_2, and W is the sum of the minimum abundance values of each species observed in sample x_1 or x_2. The values varied from 0 (no changes in species composition and abundances) to 1 (no same species between the two observations). The mean values of all the observations on a given quadrat were used as an indicator of dynamics (D_0 or D_1). Dissimilarity values were also calculated between the lakes ($D_{0/1}$). Total number (N_{sto} or N_{stl}) of observed species in same lake and mean number (N_{sm}) of observed species on a given quadrat were used as an indicator of diversity.

Results

The littoral vegetation of Ontojärvi consisted of small growing isoetids (*Subularia aquatica, Eleocharis acicularis, Ranunculus reptans*) with a few (*Juncus filiformis, Eleocharis palustris, Equisetum fluviatile*) helophyte species (Table 1). In Lentua large isoetids (*Isoetes lacustris, I. echinospora, Lobelia dortmanna*) were typical, although the small isoetids were also common (Table 1). Elodeides and nymphaeides were quite rare in both lakes. In Lentua the vegetation at the uppermost littoral was dominated by terrestrial species (e.g., *Carex* species, *Potentilla palustris*, bryophytes), while their number was low in Ontojärvi. During the research period the diversity at the littoral of Ontojärvi was lower with 27 vascular plant species and 14 bryophytes compared to Lentua with 51 vascular plants and 17 bryophytes (Table 1). The mean number of species observed per plot in Ontojärvi was 5.8 (n = 36), which was sligthly less compared to Lentua (7.1, n = 53). The difference was statistically insignificant (Mann-Whitney U-test, p = 0.051). The mean abundance of vegetation at analysed quadrats was 59% in Lentua and 58% in Ontojärvi.

There were no significant differences in the dissimilarity values (D) between the research lakes. The mean dissimilarity (D_m) between the first and last observation was 0.269 in Ontojärvi and 0.297 in Lentua, the difference was not statistically significant (t-test, p = 0.170). When the dissimilarity values were calculated as a total mean between all continuous observations, the values were 0.238 in Ontojärvi and 0.297 in Lentua, but still without statistical significance (t-test, p = 0.630). The year-to-year dissimilarities in Ontojärvi (D_o) and Lentua (D_l) fluctuated between 0.17 and 0.25 (Figure 3). The values were slighly higher in Lentua, except in the first observation period with low number of observations (Figure 3). Also the total number of species (N_{st}) was higher in Lentua (Figure 3). Different species composition is also seen as a dissimilarity value between the lakes ($D_{o/l}$), which varied from 0.4 to 0.68 during the research period (Figure 3).

On the eulittoral zone, where most of the water level fluctuation takes place, both the dissimilarity values (D_m) and the mean number of species (N_{sm}) were higher in Lentua (Figure 4). N_{sm} was 2.5 times higher in Lentua compared to Ontojärvi on the sublittoral, whereas the differences in the dissimilarity values were small (Figure 4). The mean dissimilarities (D_m) between the different water level fluctuation zones in one lake and between the different lakes were statistically insignificant. The mean number of species (N_{sm}) decreased faster with increasing depth in Ontojärvi compared to Lentua (Figure 4).

The mean number of species (N_{sm}) and dissimilarities (D_m) varied quite slightly between the different exposure classes (Figure 5); in both cases the values were higher in Lentua without any statistical significance. The effect of the shore inclination varied; in Lentua dissimilarity values were higher at steeply sloping shores, while in Ontojärvi gently sloping shores were more unstable (Figure 5). The mean number of species was higher at gently sloping shores of both lakes, but again without statistically significant differences (Figure 5).

The D_m – and N_{sm} – values varied in different bottom quality classes with same trend observed in other ecological variables; both were higher in Lentua with the exception of D_m-value of peaty bottom (Figure 6). At peaty bottom N_{sm}-values were higher and the dissimilarity values were lower compared to other bottoms. Muddy bottom observed only in Lentua was unstable environment with low D_m-value (0.6). Statistically the differences were insignificant.

Discussion

Our results showed that during the four decades of water level regulation the vegetation of Ontojärvi has achieved a high stability. A different species composition shows, that the vegetation has obviously changed, but has now reached a stable state. The diversity of vegetation was higher in Lentua, which led also to higher dissimilarity values in species-poor littoral communities. Grime (1979) also pointed out that species richness was lower in an area of high ecological stress and disturbance (see also Rørslett, 1989). Nilsson (1981) investigated heavily regulated reservoirs in northern Sweden and found quite high dissimilarity values among the shore vegetation (see also Nilsson & Keddy, 1990). Koskenniemi (1987) pointed out that there was no notable stabilisation of vegetation on the open shores of several Finnish humic reservoirs during a six-year study. In our research lakes, the yearly fluctuation of dissimilarities was quite unpredictable, and no clear differences between the years and the lakes were observed. The high stability of the vegetation in Ontojärvi could be seen as a result of a moderate, aged water level regulation, compared to unstable heavily regulated reservoirs in northern Sweden (cf. Nilsson, 1981).

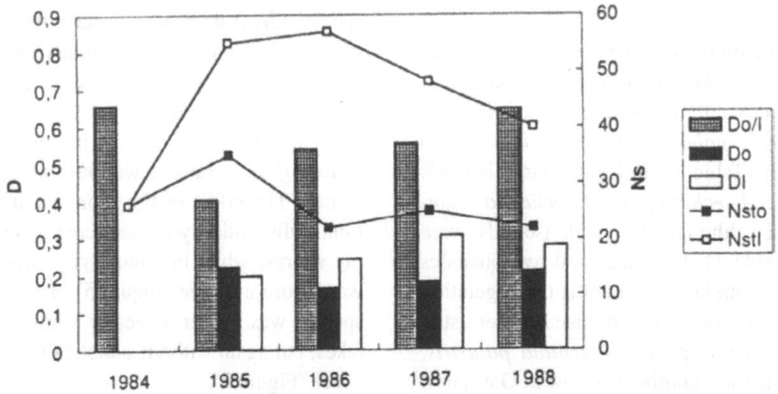

Figure 3. Dissimilarity values (D) and the number of species (Ns) during the research period. $D_{o/l}$ = yearly dissimilarity between Ontojärvi and Lentua, D_o, D_l = year-to-year dissimilarities in Ontojärvi and in Lentua, N_{sto}, N_{stl} = total number of species in Ontojärvi and Lentua.

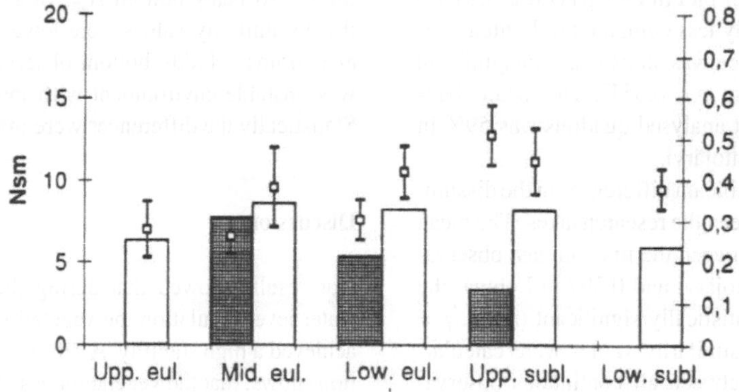

Figure 4. Mean number of species (N_{sm}) and mean dissimilarity (D_m) between the different water-level fluctuation zones. Symbols used: filled bars = N_{sm} in Lake Ontojärvi; white bars = N_{sm} in Lake Lentua; hi-low-lines = mean dissimilarity (D_m) with ±SE.

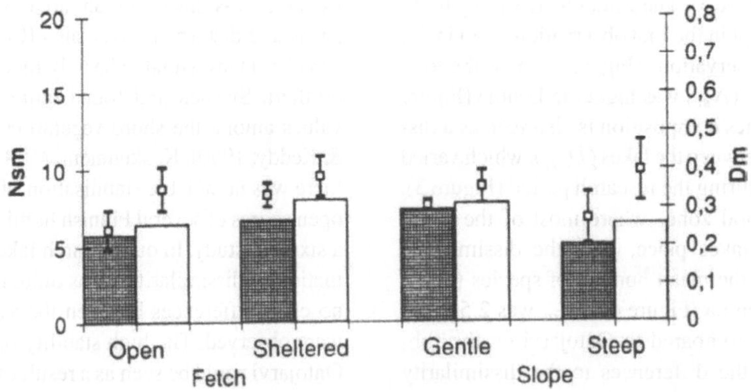

Figure 5. Mean number of species (N_{sm}) and mean dissimilarity (D_m) between the different exposure and inclination classes. Refer to Figure 4 for details.

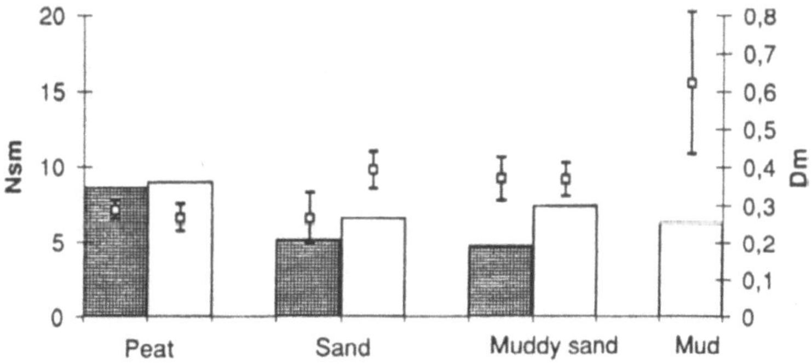

Figure 6. Mean number of species (N_{sm}) and mean dissimilarity (D_m) between the different bottom quality classes. Refer to Figure 4 for details.

The zonation of vegetation is mainly due to depth, but the range of water level fluctuation has an influence on width of the different zones (e.g., Hutchinson, 1975; Rørslett, 1985b). The dissimilarity values were higher on the eulittoral zone of Lentua compared to Ontojärvi; this Carex- dominated vegetation was rather stable in the driest part of the littoral and could probably be regarded as the most stable vegetation in Ontojärvi. In our research lakes, the water level is stable during the open water period, which tends to increase the stability of the vegetation. In the Swedish reservoirs the fluctuating water level during the summer affects negatively the abundance of vegetation (Nilsson & Keddy, 1990; Nilsson, 1981). Peat bottoms are also quite common on the eulittoral of sheltered shores, which increases the diversity and stability of vegetation as observed also by Nilsson (1981) and Koskenniemi (1987).

The littoral is also subject to penetrating ice during the winter. The vegetation in both lakes was well adapted to the disturbance of ice. In Ontojärvi the whole littoral was affected by ice, while in Lentua it affects only on the eulittoral (Hellsten et al., 1989). The effect of ice was also obvious on the sublittoral of Ontojärvi and caused slightly higher dissimilarity (D_m) and lower number of species (N_{sm}) compared to Lentua. The effects of ice on species composition and distribution of aquatic macrophytes are described in several studies (e.g., Erixon, 1981; Renman, 1986; Rørslett, 1985a) and they are clear also in our research lakes (Hellsten et al., 1989). The amphibious, short lived isoetids such as *Eleocharis acicularis*, *Subularia aquatica*, *Ranunculus reptans* and *Isoetes echinospora* have increased their distribution, whilst the long-lived isoetids *Isoetes lacustris* and *Lobelia dortmanna* have disappeared.

The sublittoral is usually quite a suitable environment for aquatic macrophytes. The positive relationship between macrophytes and sediment organic matter is also widely known (Duarte & Kalff, 1986; Wisheu & Keddy, 1989), but too soft sediments also limit the vegetation (Spence, 1982; Barko & Smart, 1983). In Lentua, the dissimilarity of vegetation is high at the area of muddy sediments. On the other hand, Rørslett (1985b) showed that the mortality of aquatic macrophytes was highest near the lowest limit of the macrophytes due to the reduced light intensity. In Ontojärvi, organic rich bottoms were situated in deeper areas due to the high rate of erosion, but the lower sublittoral was almost without vegetation due to reduced light penetration measured in studies of Hellsten et al., (1989).

Exposure affects the vegetation both directly through physical disturbance by waves and currents and indirectly by changes in sedimentation (e.g., Keddy, 1982). The diversity and D_m values on open shores were lower in both lakes compared to sheltered shores. It is generally believed that both the diversity and the abundance peak of vegetation occur on sheltered shores (Nilsson, 1981; Keddy, 1983). On the other hand, Rørslett (1987) pointed out in Lake Thyrifjord that exposure had no effect on floristic diversity, though its effects on the performance of the vegetation are clear. Nilsson & Keddy (1990) did not find any marked changes in the similarity values between weakly and moderately exposed shores. In our research lakes fewer species found on open shores are well adapted to the strong erosion.

Slope affects the organic matter content of the sediment (Håkanson & Jansson, 1983), and steep slopes are physically difficult places to be colonised by macrophytes. The effect of littoral slope was partly difficult to estimate. The diversity values were higher on gently sloping shores in both lakes as shown also by Duarte

& Kalff (1986), who found higher biomass and cover of macrophytes at gently sloping shores.

Conclusion

Littoral vegetation was relatively stable in both Ontojärvi and Lentua. The species observed in Ontojärvi were well adapted to the hostile ecological environment with several stress-tolerant and ruderal species, although the diversity was clearly lower. Surprisingly the dissimilarity was higher in Lentua, which could be related on higher diversity. The only remarkable exceptions were found at the sublittoral of Ontojärvi, where the effect of wave and ice-erosion caused higher dissimilarity with low number of species compared to Lentua.

From the viewpoint of lake management, the dissimilarity values observed at permanent plots do not alone give a true picture of the state of regulated lakes. Because vegetation is well adapted in a hard ecological environment, the often quite minimal yearly changes in species composition and abundance are rather difficult to analyse. On the other hand, research based on species composition and further to strategy analysis (e.g., Murphy et al., 1990) provides a better view of the state and succession of the lake ecosystem.

Acknowledgements

This work was financially supported by the VTT Building Laboratory, the Academy of Finland and the Foundation of Imatran Voima. We express our sincere gratitude to Dr Erkki Alasaarela of the National Board of Waters and Environment. We would also like to thank two anonymous referees for useful comments and Mrs Sirkka-Liisa Leinonen, M.Sc., for correcting the English of the original version of the manuscript.

References

Alasaarela, E., S. Hellsten & P. Tikkanen, 1989. Ecological aspects of lake regulation in northern Finland. In H. Laikari (ed.), River Basin Management - 5, Pergamon Press PLC. Oxford.

Barko, J. W. & R. M. Smart, 1983. Effects of organic matter additions to sediment on the growth of aquatic plants. J. Ecol. 71: 161–175.

Duarte, C. M. & J. Kalff, 1986. Littoral slope as a predictor of the maximum biomass of submerged macrophyte communities. Limnol. Oceanogr. 31: 1072–1080.

Erixon, G., 1981. Aquatic macrophytes and their environment in the Vindelälven river, northern Sweden. Wahlenbergia 7: 61–71.

Grime, J. P., 1979. Plant strategies and vegetation processes. Wiley, Chister, 222 pp.

Granberg, K. & L. Hakkari. 1980. Säännöstelyn vaikutuksista eräiden Kainuun järvien limnologiaan. Vesihallituksen tiedotuksia 255, 95 pp. (in Finnish).

Hellsten, S. & R. Joronen. 1986. Kemijärven litoraalin kasvisto ja kasvillisuus sekä niihin vaikuttavat ekologiset tekijät vuosina 1982–83. Rovaniemi, Kemijoen vesiensuojeluyhdistys ry, 39 pp. (in Finnish).

Hellsten, S., I. Neuvonen, R. Keränen, M. Nykänen & E. Alasaarela, 1989. Ekologiset näkökohdat joidenkin Pohjois-Suomen järvien säännöstelyssä. Osa 2. Rannan geomorfologia ja vesikasvillisuus. Valtion teknillinen tutkimuskeskus, tiedotteita 986, 131 pp. + 13 app. Helsinki 1989 (in Finnish).

Hutchinson, G. E., 1975. A treatise of limnology. 3. Limnological Botany. Wiley, New York, 660 pp.

Håkanson, L. & M. Jansson, 1983. Principles of lake sedimentology. Springer, Berlin, 316 pp.

Keddy, P. A., 1982. Quantifying within-lake gradients of wave energy: Interrelationships of wave energy, substrate particle size and shoreline plants in Axe Lake, Ontario. Aquat. Bot. 14: 41–58.

Keddy, P. A., 1983. Shoreline effects in Axe Lake, Ontario: Effects of exposure on zonation patterns. Ecology 64: 331–344.

Koskenniemi, E., 1987. Development of floating peat and macrophyte vegetation in a newly created, polyhumic reservoir, western Finland. Aqua Fenn. 17: 165–173.

Murphy, K. J., Rørslett, B. & I. Springuel. 1990. Strategy analysis of submerged lake macrophyte communities: an international example. Aquat. Bot. 36: 303–323.

Nilsson, C., 1981. Dynamics of the shore vegetation of a North Swedish hydro-electric reservoir during a 5-year period. Acta Phytogeographica Suecica 69.

Nilsson, C. & P. A. Keddy, 1990. Predictibility of change in shoreline vegetation in a hydroelectric reservoir, northern Sweden. Can. J. Fish. aquat. Sci. 45: 1896–1904.

Renman, G., 1986. Distribution of littoral macrophytes in a North Swedish river in relation to winter habitat conditions. Aquat. Bot. 33: 243–256.

Rørslett, B., 1985a. Regulation impact on submerged macrophytes in the oligotrophic lakes of Setesdal, South Norway. Verh. int. Ver. Limnol. 22: 2927–2936.

Rørslett, B., 1985b. Death of submerged macrophytes - actual field observations and some implications. Aquat. Bot. 22: 7–19.

Rørslett, B., 1987. Niche statistics of submerged macrophytes in Tyrifjord, a large oligotrophic Norwegian lake. Arch. Hydrobiol. 111: 283–308.

Rørslett, B., 1989. An integrated approach to hydropower impact assesment. II. Submerged macrophytes in some Norwegian hydroelectric lakes. Hydrobiologia 175: 65–82.

Spence, D. N. H., 1982. The zonation of plants in freshwater lakes. Adv. ecol. Res. 12: 37–125.

Wisheu, I. C. & P. A. Keddy, 1989. Species richness - standing crop relationships along four lakeshore gradients: constraints on the general model. Can. J. Bot. 67: 1609–1617.

Hydrobiologia **340**: 93–99, 1996.
J. M. Caffrey, P. R. F. Barrett, K. J. Murphy & P. M. Wade (eds), Management and Ecology of Freshwater Plants.

Influence of plants on redox potential and methane production in water-saturated soil

W. Grosse, K. Jovy & H. Tiebel
Botanical Institute, University of Cologne, Gyrhofstrasse 15–17, D-50923 Cologne, Germany

Key words: methane, pressurized ventilation, redox potential, soil anoxia, weeds, wetland vegetation

Abstract

Pressurized ventilation, which increases the oxygen supply of the roots and rhizomes, has been detected on three waterlilies (*Nymphaea capensis, N. lotus* var. *lotus, N. odorata*), two Japanese swamp grasses (*Ischaemum aristatum* var. *glaucum, Isachne globosa*), and three willow species (*Salix alba, S. cinerea,S. viminalis*). All of these plant species are able to generate sufficient convective gas flow to meet the oxygen demand of their organs buried in the anoxic soil. Excretion of surplus oxygen maintains higher redox potential in the tussock of *I. aristatum* and also in the rhizosphere of the waterlilies and willows, thereby protecting the root system from phytotoxin uptake. High methane production rates in reduced sediments contrast to the significantly lower rates of methane formation in the oxidized rhizosphere surrounding *N. lotus* roots. This is an example of how wetland plants use pressurized ventilation to alter microbial activities in their habitat. Pressurized ventilation seems to provide these plant species with a competetive advantage over species that rely on diffusive aeration of their submerged parts, thereby enabling them to become dominant weeds in their aquatic ecosystems or in wet meadows of nature reserves.

Introduction

Wetland soil and lake sediments are characterized by oxygen deficiency and redox potentials (E_h) below −150 mV. Plants growing in these areas require adequate internal aeration from the shoots and leaves for survival of roots and rhizomes. Aerenchyma formation and the associated increased rate of oxygen diffusion is a well known, long-term adaptation of plants to water-saturated soil or flooding.

Many emergent plant species including wetland trees have established a pumping system, powered by solar radiation, which superimposes a pressurized ventilation of their below-ground and submerged organs on the basic oxygen supply by gas diffusion. Differences in temperature and humidity across a porous partition separating the warmer, moister air inside the plant aerenchyma from the cooler, drier ambient atmosphere, are responsible for the pressurization of air in the aerenchyma. This pressurization drives convective gas flows through the coherent intercellular system of the plant (Dacey, 1980).

Pressurized ventilation aerates the submerged plant organs of floating-leaved plant species of the *Nelumbonacean* (Ohno, 1910; Dacey 1987; Mevi-Schutz & Grosse, 1988), *Nymphaeacean* (Dacey, 1980, 1981; Grosse & Bauch, 1991; Grosse et al., 1991; Schroeder et al., 1986) and *Menyanthacean* families (Grosse & Mevi-Schuetz, 1987), wetland grasses (Armstrong & Armstrong, 1991; Brix et al., 1992) and wetland trees (Grosse & Schroeder, 1984, 1985; Grosse et al., 1992, 1994; Schroeder, 1989). Some oxygen escapes from the roots into the rhizosphere as descibed for *Alnus* and *Salix* species (Armstrong, 1968, Grosse et al., 1993), *Phragmites australis* and *Oryza sativa* (Armstrong, 1969, 1971; Flessa & Fischer, 1992a, 1992b). This oxygen protects *Alnus glutinosa* from accumulation of toxic concentrations of heavy metals in the roots (Lattermann, 1994).

The present studies demonstrate the effects of pressurized ventilation on redox potential in the rhizosphere of waterlilies, two swamp grasses, and willow trees and on the microbial methane production which is a characteristic feature of anoxic freshwater sediments.

Materials and methods

Plant material

Waterlilies (*Nymphaea odorata* (Ait.), *Nymphaea capensis* (Thunb.), and *Nymphaea lotus* var. *lotus* (L.) were cultivated in 100 l plastic containers (60 cm in diameter and 35 cm in height) with the lower third filled with a mixture of potting-soil and sand in a ratio 1:1 and completely flooded with tap-water. The plants were grown in the greenhouse at 25°C/20°C and 16 h/8 h day/night.

Rhizomes of the Japanese swamp grasses *Isachne globosa* (Kuntze) and *Ischaemum aristatum* var. *glaucum* (Honda) were sampled during the summer of 1992 at a renaturalized paddy field of the Mobara-Yatsumi marsh, Central Japan, as described by Yabe & Numata (1984). Both grasses were cultivated in the greenhouse on water-saturated soil for one year. Temperature and light conditions in the greenhouse, soil, and the 100 l plastic containers were the same as used for the waterlilies. The *I. aristatum* formed a tussock, *I. globosa* has spread throughout the whole container.

Cuttings about 40 cm in length were taken from mature branches of willow trees (*Salix alba* (L.), *Salix cinerea* (L.), *Salix viminalis* (L.)) and rooted for 6 weeks in water-saturated Growing Medium Einheitserde Type P.

Induction of pressurized ventilation

A 150 W spotlight was used at a distance of 38 cm to generate pressurized ventilation through the willow trees (photon flux at the stem surface was 280 μmol m^{-2} s^{-1}) as described by Grosse et al. (1993). Pressurized ventilation in the *Nymphaea* species and grasses was induced by irradiation of the newly-emerged influx leaves of the waterlilies and the whole leaves of the grasses, respectively, using spotlight bulbs as above and a ventilating fan (model Multifan 3312, 12V = DC 2.4W, Papst-Motoren, St. Georgen, Germany) with wind speed of 40 m min^{-1}.

Gas flow measurements

For measuring gas flow through waterlilies (3-month-old seedlings of *N. odorata* with 3 floating leaves and 18-month-old *N. lotus* var. *lotus*) a soap bubble flowmeter was connected to the petiole of the oldest leaf after cutting off the leaf blade. For gas flow measurements on grasses (*I. aristatum* var. *glaucum* and *I. globosa*) the flowmeter was connected to the internal gas spaces of the basal end of the leafy shoots by a hypodermic needle.

Oxygen exchange measurements

Oxygen uptake and excretion by tree roots (cuttings of *S. alba, S. cinerea, S. viminalis*) were measured polarographically. A Clark-type oxygen electrode (Delieu & Walker, 1972) was installed in the bottom of a measurement chamber, and connected to a powered control box. The chamber and its lit was made from an opaque grey PVC material to prevent illumination of submerged parts of the stem. The chamber, filled with tap-water and surrounded by a cooling jacket, enclosed a tree's root system and was sealed gas-tight with Terostat IX. A syringe needle was introduced to the chamber's lid, and connected by silicon tubing to a water reservoir. The reservoir was covered with liquid paraffin to restrict gas exchange with the atmosphere. The system allows the water lost by transpiration to be replaced, while simultaneously restricting air entry into the chamber. A figure illustrating the experimental setup for oxygen exchange measurements have been published previously (Grosse et al., 1993). Efflux of oxygen from the sectional plane of the cuttings was blocked by liquified paraffin (<50 °C).

Redox potential measurements

Redox potentials in the rhizosphere and in water-saturated soil not affected by roots, were measured using platinum microelectrodes which were prepared from Pt wire (0.5 mm in diameter, 70 mm in length for helical-shaped and 30 mm in length for pin-shaped one, respectively) and were checked for identical readings. Ag/AgCl probes were used as reference electrodes. The E_h values were recorded using a computer as multiplexer and a multichannel data-logger. Roots of *N. capensis* (12-month-old with 8 floating leaves) were placed inside the helically-shaped Pt electrodes. For all other redox measurements pin-shaped Pt electrodes were placed directly into the soil and along the roots, respectively. Roots of *S. viminalis* were embedded in 1% agar prepared from a mixture of potting-soil and water in a ratio 1:1. The embedded roots and the submerged part of the stems were kept darkened during

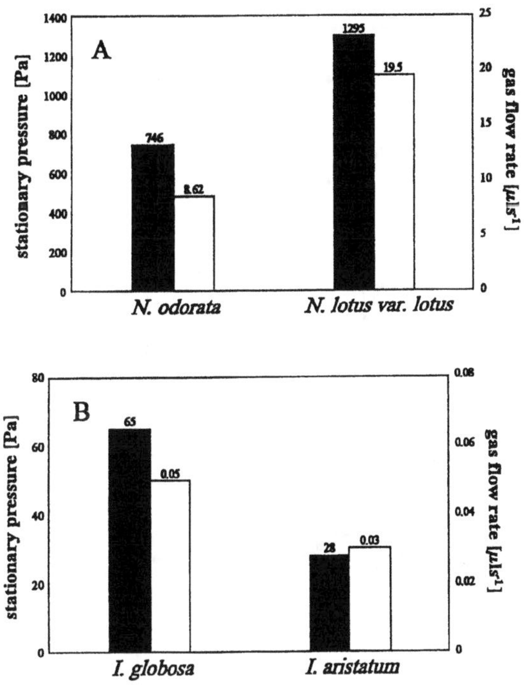

Figure 1. Internal pressurization (black columns) of the aerenchyma and convective gas through-flow (white columns) caused by pressurized ventilation were measured on the oldest leaf petioles of (A) two *Nymphaea* species and on cut culms of (B) two Japanese swamp grasses. The values are means of 16 replicates each.

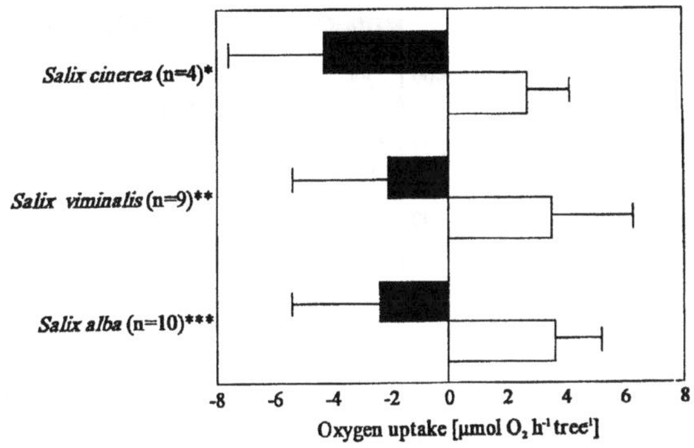

Figure 2. Respiratory oxygen uptake (white columns) by the roots from an aqueous medium at 8 °C and oxygen release (black columns) into this medium as a result of pressurized ventilation were assessed on 6-week-old rooted cuttings of three *Salix* species. Values are means of 4 to 10 replicates and are significantly different within species at the $P=0.001$ (∗ ∗ ∗), $P=0.01$ (∗∗), and $P=0.05$ (∗) levels of the *t*-test.

the whole period of measurements, so that the unsubmerged parts only were irradiated. Manufacture of the Pt microelectrodes and the redox measurements were adopted from Armstrong (1965).

Pressure, irradiation and rH measurements

Pressure differences between the aerenchyma and the atmosphere were measured with a highly sensitive pressure transducer and with a manometer containing Brodie solution as manometric liquid (density: 1033 g l^{-1}), respectively. Light intensity was measured with a quantum sensor (LI-190 SB) and recorded with data-logger A Hygro-Thermograph was used to measure relative humidity (rH).

Methane measurements

Tubes of stainless steel (5 mm inside diameter and 100 mm in length) were used as gas sampling probes for measuring the methane concentration in the sediment. Each tube was closed by a porous, water-tight PTFE membrane at the end which was inserted into the sediment, and by gas-tight septa at the end which was exposed to the water body. These gas sampling probes were flushed with nitrogen gas just before positioning for methane collection along a transect of the sediment from the root stock of an 18-month-old *N. lotus* var. *lotus* to the root-free soil. Following a three-day period of equilibration, the tubes were sampled once per day by a gas-tight Hamilton syringe. The methane content of the sample was measured on a Perkin Elmer Series 8500 gas chromatograph with a thermal conductivity detector.

Results and discussion

In the aerenchyma of young seedlings of *N. odorata* and mature of *N. lotus* stationary pressures of 746 Pa and 1295 Pa were recorded following irradiation. The pressurization of the lacunar system of young influx leaves results from absorption of light energy and, following an increase of temperature in leaf, a convective gas flow was created with flow rates of 8.62 ± 0.8 µl s^{-1} for *N. odorata* and 19.5 ± 1.7 µl s^{-1} for *N. lotus* (Figure 1A). These flow rates are low, but in the range of those reported for other *Nymphaea* species (Grosse et al., 1991).

Under the same conditions, internal gas pressures of up to 65 ± 11 Pa and 28 ± 5 Pa occur in the culms of *I. globosa* and *I. aristatum*. Gas flow rates of 0.05 ± 0.004 and 0.03 ± 0.003 µl s^{-1} were measured in

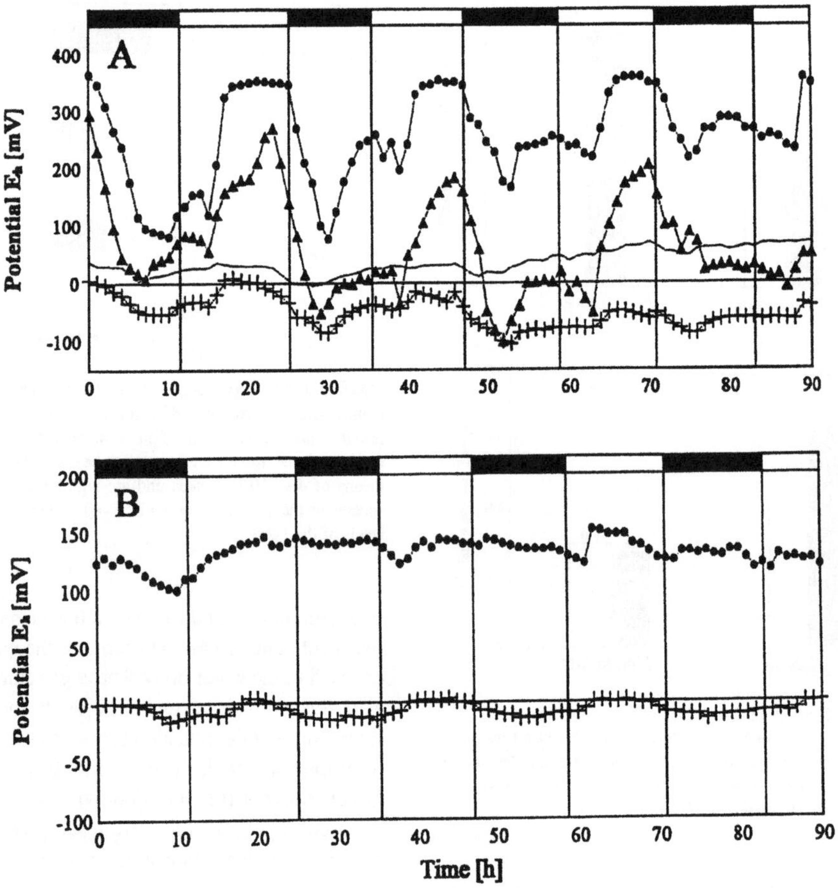

Figure 3. Changes in redox potential caused by oxygen release from the plants following pressurized ventilation is shown for (A) a *Nymphaea capensis* root (30 cm in length) close to the root tip (–▲–), in the middle of the root (- - -), at center of the root system (–●–), and in the soil about 10 cm far from the rhizosphere (-+-), and (B) in the tussock of *Ischaemum aristatum* var. *glaucum* (–●–) and in soil not affected by plant roots (-+-). Day and night time are indicated by white and black bars, respectively.

the swamp grasses (Figure 1B). Compared to the flow rate reported for *Phragmites australis* (Armstrong & Armstrong, 1991), these flow rates seem to be quite low, but the values might be reduced by the high relative humidity (83 rH) in the greenhouse and by the flow resistance of the lamella in the flowmeter, which is in the range of 17 Pa. When flow rates were measured with *I. globosa* in a dryer ambient atmosphere (60 rH) and 10 excised culms were working together into the same flowmeter, the mean flow rate was 0.25 ± 0.02 μl s^{-1} $culm^{-1}$ (n = 40) with a static pressure differential of 191 ± 7 Pa ($n = 30$) after blocking the gas flow. This flow rate is still significant lower than that 10.0 ± 0.3 μl s^{-1} $culm^{-1}$ ($n = 10$) of excised *P. australis* shoots which have been assessed under the same conditions (1440 Pa static pressure differential). But when the surface of the leaf sheaths are taken into account, which are the main entrance area of atmospheric gases, including oxygen, into leafy shoots of grasses (Armstrong & Armstrong, 1990), a rough calculation results in flow rates in the same range between both species. With a mean leaf sheath surface of 8.1 cm^2 $culm^{-1}$ in *I. globosa* and 101.1 cm^2 $culm^{-1}$ in *P. australis*, calculated from sheath lengths and shoot diameters, the flow rates are 0.035 and 0.1 μl s^{-1} cm^{-2} leaf sheath surface in *I. globosa* and *P. australis*, respectively.

Oxygen supply to roots by gas diffusion is only effective over short distances (Armstrong, 1979). Typically, tree roots depend on oxygen uptake from soil

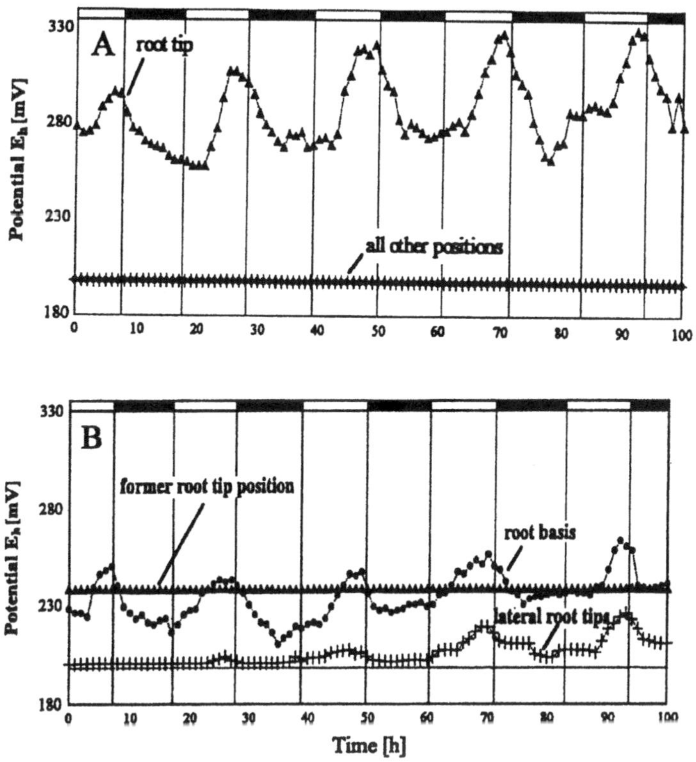

Figure 4. Changes in redox potential caused by oxygen release from a *Salix viminalis* root following pressurized ventilation. The E_h values are shown in the rhizosphere close to the root tip (–▲–), in the middle of the root (- - -), at the root basis (–●–), and in the soil 12 mm far from the root basis (-+-). Measurements were started (A) immediately after application of the roots in the experimental setup and (B) one week later. Day and night time are indicated by white and black bars, respectively.

gas phase. In the willow species *S. cinerea*, *S. viminalis*, and *S. alba* the oxygen uptake of roots from the rhizosphere was in the range of 2.7 to 3.6 μmol O_2 h^{-1} $tree^{-1}$ (Figure 2) when the internal oxygen supply through the intercellular spaces of the trees to the roots was diffusive. However, when gas flow to the roots was established by pressurized ventilation, the O_2 demand of the roots was covered completely by the internal aeration. Surplus oxygen at rates of 2.1 to 4.3 μmol O_2 h^{-1} $tree^{-1}$ was excreted into the rhizosphere.

Considerable differences in redox potential, up to 550 mV between soil in the tussock margin and of bare land near the tussock of *I. aristatum*, are reported by Yabe & Numata (1984). The observed high positive values of the redox potential at the tussock margin should be caused by radial loss of oxygen from roots, which is corroborated by the described increase of redox potential in the rhizosphere of rice (*O. sativa*) and common reed (*Ph. australis*) by Armstrong & Armstrong (1988), Beckett et al. (1988) and Flessa & Fischer (1992a, 1992b).

The important role of pressurized ventilation in creating such oxidized zones around the most susceptible root tips has been demonstrated by redox measurements on waterlily (Figure 3A), swamp grass (Figure 3B), and willow (Figure 4). For *N. capensis* (Figure 3A) the low redox potential changes to more positive values as soon as the plant is able to improve the oxygen supply to its roots by convective gas flow during the day. The oxidizing power is highest close to the root tips and in the center of the root system where new roots are developing continuously. Similar results were observed for the tussock forming *I. aristatum* (Figure 3B). In soil unaffected by roots, the E_h values were 100 mV lower than in soil surrounding roots. The oxidizing effect of pressurized ventilation is demonstrated by the daily oscillation of the redox potential in the soil.

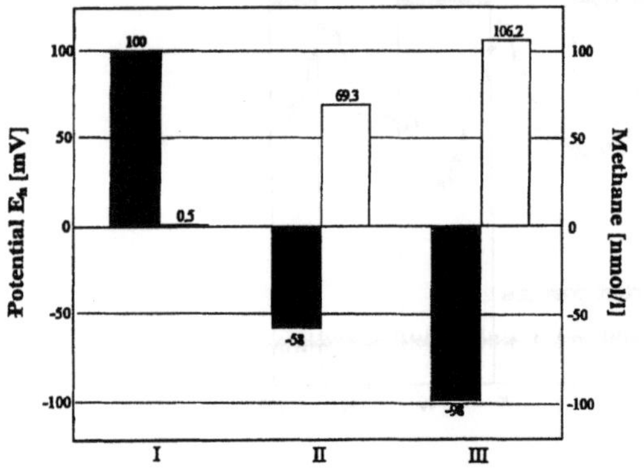

Figure 5. Effect of pressurized ventilation in *Nymphaea lotus* var. *lotus* on redox potential (black columns) and methane concentration (white columns) in pond sediments in the center of the root system (I), at marginal rhizosphere (II), and in soil 10 cm far from the rhizosphere (III).

The oxidizing effect of pressurized ventilation could be observed most clearly in the root tip area of *S. viminalis*, while relatively low but constant E_h values were recorded in the synthetic soil and in the rhizosphere close to older root zones (Figure 4A). When the measurements were started again following about one week of root elongation, the E_h values are constant, but oscillated at the root base due to growth of lateral roots (Figure 4B).

Pressurized ventilation is a most effective mechanism by which plants alter conditions in the rhizosphere. Recently, Lattermann (1994) using ESI (Electron Spectroscopy Imaging) and EELS (Electron Energy Loss Spectroscopy) techniques has shown that oxidation of rhizosphere by pressurized ventilation protects root tips of alder trees (*Alnus glutinosa*) from excess ferrous deposition close to the root meristems. While in soil which is contaminated with reduced ferrous compounds, the root tissues at low redox potentials are infiltrated by the highly soluble Fe(II)-ions, insoluble Fe(III)-compounds accumulate inside the roots. In contrast, the Fe(II) is precipitated at the root surface following oxidation to Fe(III), when the plant has generated high redox values in the rhizosphere.

Correlation between redox potential and methane production in vicinity of the root stock of *N. lotus* and root-free sediment is a further appropriate example of how plants are able to alter conditions in the rhizosphere (Figure 5). At the center of the root system a potential of 100 mV in the rhizosphere completely inhibited microbial methane production, while in the marginal rhizosphere (−58 mV) only partial inhibition occurred. In the absence of roots, sediment had redox potentials low enough (−98 mV) for optimal microbial methane production.

Acknowledgments

The authors are grateful to Jonathan Frye, McPherson College, McPherson, USA, for reading the manuscript, to William Armstrong, Department of Applied Biology, The University of Hull, Hull, UK, for assistance in preparing the equipment for redox potential measurements, to Takayoshi Tsuchiya, Department of Biology, Chiba University, Chiba, Japan, for assistance sampling the swamp grasses at the natural site in Central Japan, and to the Deutsche Forschungsgemeinschaft (DFG) for financial support.

References

Armstrong, J. & W. Armstrong, 1988. *Phragmites australis* – a preliminary study of soil-oxidizing sites and internal gas transport pathways. New Phytol. 108: 373–382.

Armstrong, J. & W. Armstrong, 1990. Pathways and mechanisms of oxygen transport in *Phragmites australis*. In P. F. Cooper & B. C. Findlater (eds), The use of constructed wetlands in water pollution control. Pergamon, Oxford: 529–533.

Armstrong, J. & W. Armstrong, 1991. A convective through-flow of gases in *Phragmites australis* (Cav.) Trin. ex Steud. Aquat. Bot. 39: 75–88.

Armstrong, W., 1965. Studies relating to the survival of plants in waterlogged soils. Ph.D. Thesis, University of Hull, UK.

Armstrong, W., 1968. Oxygen diffusion from roots of woody species. Physiol. Plant. 21: 539–543.

Armstrong, W., 1969. Rhizosphere oxidation in rice: An analysis of intervarietal differences in oxygen flux from the roots. Physiol. Plant. 22: 296–303.

Armstrong, W., 1971. Radial oxygen losses from intact rice roots as affected by distance from the apex, respiration and waterlogging. Physiol. Plant. 25: 192–197.

Armstrong, W., 1979. Aeration in higher plants. In H. W. Woolhouse (ed.), Advances in Botanical Research. Vol. 7. Academic Press, London: 225–332.

Beckett, P. M., W. Armstrong, S. H. F. W. Justin & J. Armstrong, 1988. On the relative importance of convective and diffusive gas flow in plant aeration. New. Phytol. 10: 463–468.

Brix, H., B. K. Sorrell & P. T. Orr, 1992. Internal pressurization and convective gas flow in some emergent freshwater macrophytes. Limnol. Oceanogr. 37; 1420–1433.

Dacey, J. W. H., 1980. Internal winds in water lilies: an adaptation for life in anaerobic sediments. Science 210: 1017–1019.

Dacey, J. W. H., 1981. Pressurized ventilation in the yellow waterlily. Ecology 62: 1137–1147.

Dacey, J. W. H., 1987. Knudsen-transitional flow and gas pressurization in leaves of *Nelumbo*. Plant Physiol. 85: 199–203.

Delieu, T. & D. A. Walker, 1972. An improved cathode for the measurement of photosynthetic oxygen evolution by isolated chloroplasts. New Phytol. 71: 201–225.

Flessa, H. & W. R. Fischer, 1992a. Redoxprozesse in der Rhizosphäre von Land- und Sumpfpflanzen. Z. Pflanzenernähr. Bodenk. 155: 373–378.

Flessa, H. & W. R. Fischer, 1992b. Plant-induced changes in the redox potentials of rice rhizospheres. Plant and soil 143: 55–60.

Grosse, W. & C. Bauch, 1991. Gas transfer in floating-leaved plants. Vegetatio 97: 85–192.

Grosse, W. & J. Mevi-Schuetz, 1987. A beneficial gas transport system in *Nymphoides peltata*. Am. J. Bot. 74: 947–952.

Grosse, W. & P. Schroeder, 1984. Oxygen supply of roots by gas transport in alder-trees. Z. Naturforsch. 39C: 1186–1188.

Grosse, W. & P. Schroeder, 1985. Aeration of roots and chloroplast free tissues of trees. Ber. Deutsch. Bot. Ges. 98: 311–318.

Grosse, W., H. B. Buechel & S. Lattermann, 1994. Root aeration in wetland trees and its ecophysiological significance. In A. D. Laderman (ed.), Coastally Restricted Forests, Oxford University Press, New York (in press).

Grosse, W., H. B. Buechel & H. Tiebel, 1991. Pressurized ventilation in wetland plants. Aquat. Bot. 39: 89–98.

Grosse, W., J. Frye & S. Lattermann, 1992. Root aeration in wetland trees by pressurized gas transport. Tree Physiol. 10: 285–295.

Grosse, W., A. Schulte & H. Fujita, 1993. Pressurized gas transport in two Japanese alder species in relation to their natural habitats. Ecol. Res. 8: 151–158.

Mevi-Schutz, J. & W. Grosse, 1988. A two-way gas transport system in *Nelumbo nucifera*. Plant, Cell and Envir. 11: 27–34.

Lattermann, S., 1994. Strukturelle und physiologische Anpassungen von *Alnus glutinosa* (L.) Gaertn. an Flutung und Bodenanaerobiose. Ph.D. Thesis. University of Cologne, Germany.

Ohno, N., 1910. Über lebhafte Gasausscheidung aus den Blättern von *Nelumbo nucifera* Gaertn. Z. Pflanzenphysiol. 2: 641–664.

Schroeder, P., 1989. Characterization of a thermo-osmotic gas transport mechanism in *Alnus glutinosa* (L.) Gaertn. Trees 3: 38–44.

Schroeder, P., W. Grosse & D. Woermann, 1986. Localization of thermo-osmotically active partitions in young leaves of *Nuphar lutea*. J. Exp. Bot. 37: 1450–1461.

Yabe, K. & M. Numata, 1984. Ecological studies of the Mobara-Yatsumi marsh. Main physical and chemical factors controlling the marsh ecosystem. Jap. J. Ecol. 34: 173–186.

Hydrobiologia **340:** 101–107, 1996.
J. M. Caffrey, P. R. F. Barrett, K. J. Murphy & P. M. Wade (eds), Management and Ecology of Freshwater Plants.

The aquatic microphytes and macrophytes of the Transvase Tajo-Segura irrigation system, southeastern Spain

M. Aboal, M. Prefasi & A. D. Asencio
Departamento de Biología Vegetal (Botánica), Facultad de Biología, Universidad de Murcia, Campus de Espinardo, E-30100 Murcia, Spain

Key words: algae, angiosperms, flora, irrigation channels, control, southeastern Spain

Abstract

The aquatic plant communities of the irrigation channels of the Transvase Tajo-Segura project are described in order to assess the various options for the control of nuisance vegetation. The low flow rate permits relatively complex planktonic assemblages in addition to epipelic cyanophytes, diatoms and chlorophyes. The only macrophytes to establish themselves in the bottom sediment and form well developed mats were *Cladophora glomerata* and *Potamogeton pectinatus*. The water quality of these channels is characterised by high temperature, pH, conductivity and sulphate and chloride concentrations. Moreover, the channels may remain dry for several months.

Introduction

Murcia, located in the southeastern Spain, is the most arid part of the country and of Europe. It has, for many centuries, been dedicated to agriculture because of the good quality of its soils. With the reduction of flow in the Segura river, it has been necessary to import water from an adjacent basin (M.O.P.U., 1977). The water from Tajo River arrives to Talave Reservoir, which belongs to the Segura basin. The water then flows through the Segura river to Ojós Reservoir, where two canals start, one of them finishing in the Campo de Cartagena, after passing through La Pedrera Reservoir (Figure 1). The total water resource available for irrigating the area (133 356 ha) is 65 000 m^3 conducted along 275 km of channels which are trapezoidal in cross-section being 3 m wide at the top of the bank.

Materials and methods

The aquatic plant communities of 50 km were described over the period October 1993 to September 1994 using seven sample points, four in the Campo de Cartagena Canal, one in La Pedrera Reservoir, one in a canal at Fortuna, and the seventh in the Ojós Reservoir. The first four sites are only temporary, drying out from September to January, the others (5–7) are permanent. Each site was visited every 7–14 days during the irrigation period, i.e. on average 16 samples per year, measurements being made of the aquatic plants (microphytes and macrophytes) and water quality. The different substrata present were sampled for algal material and both these and water samples were transported in a cold box to the laboratory. The algal material was fixed or isolated in liquid or solid Bold Basal Medium (Nichols, 1973). Water samples were frozen for later analysis by high resolution ionic spectrophotometry for sulphate, chloride, nitrate, ammonia, calcium, magnesium, sodium and potassium. Also included were field measurements of pH, channel discharge, temperature, conductivity, dissolved oxygen and photosynthetically active radiation (P.A.R).

Results

A total of 217 species were recorded from the seven samples sites and the most frequent benthic species are listed in Table 1. The predominant groups were diatoms, cyanophyceae and chlorophyceae (Figure 3a). The most significant component was the epipelon but

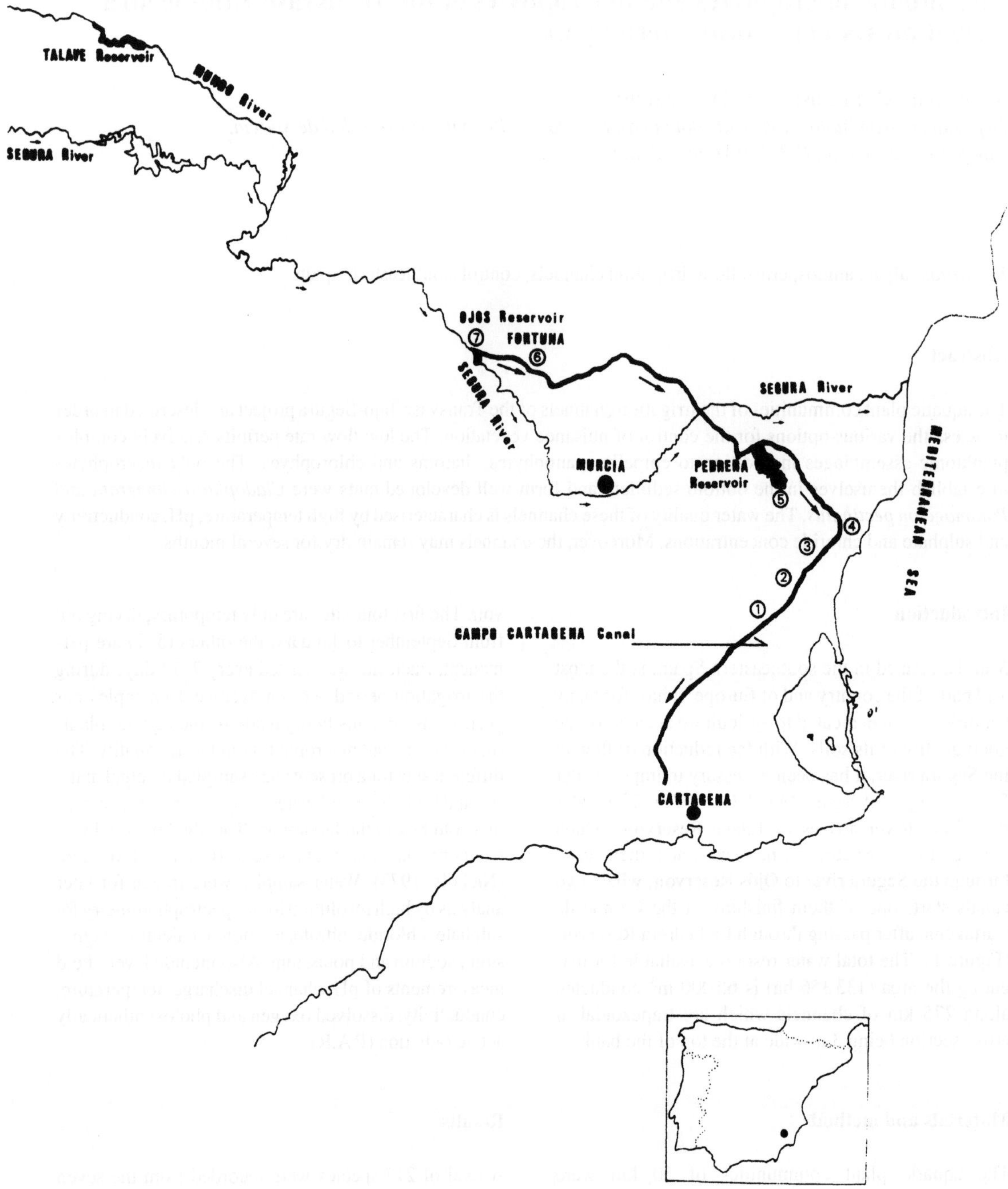

Figure 1. Geographical situation of studied area, southeastern Spain.

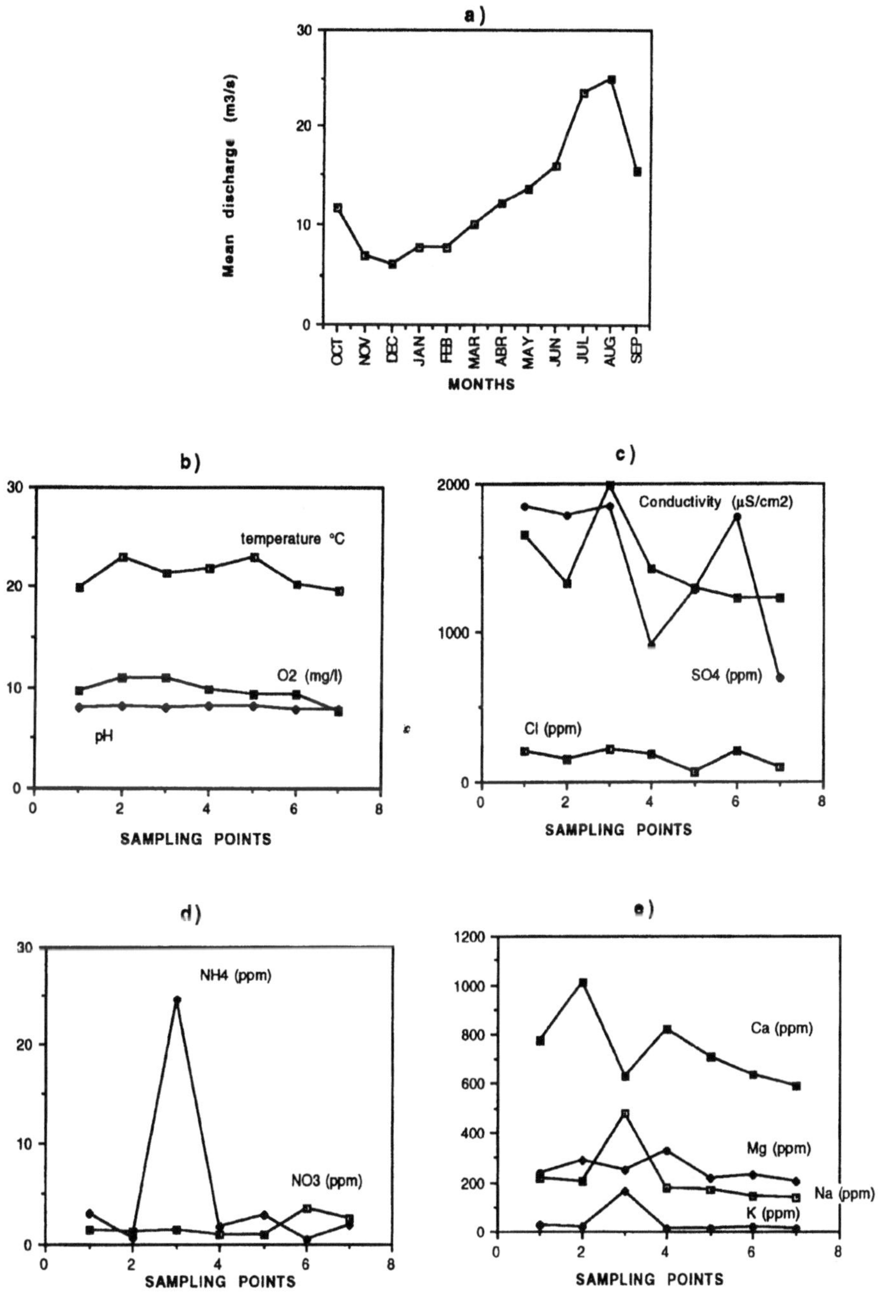

Figure 2. 1. Species number belonging to each taxonomical group, 2. species number belonging to each biological type, in the different sampling points.

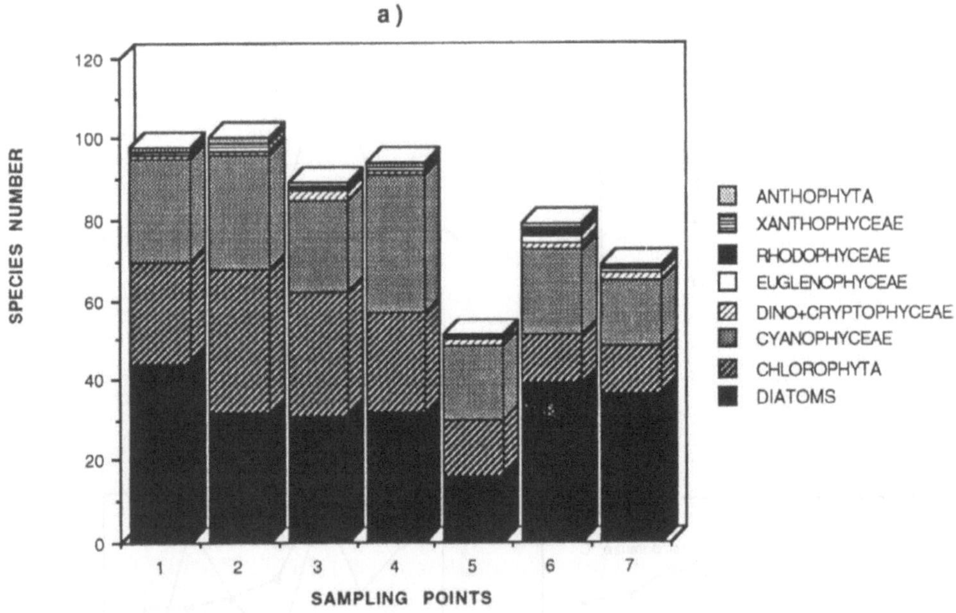

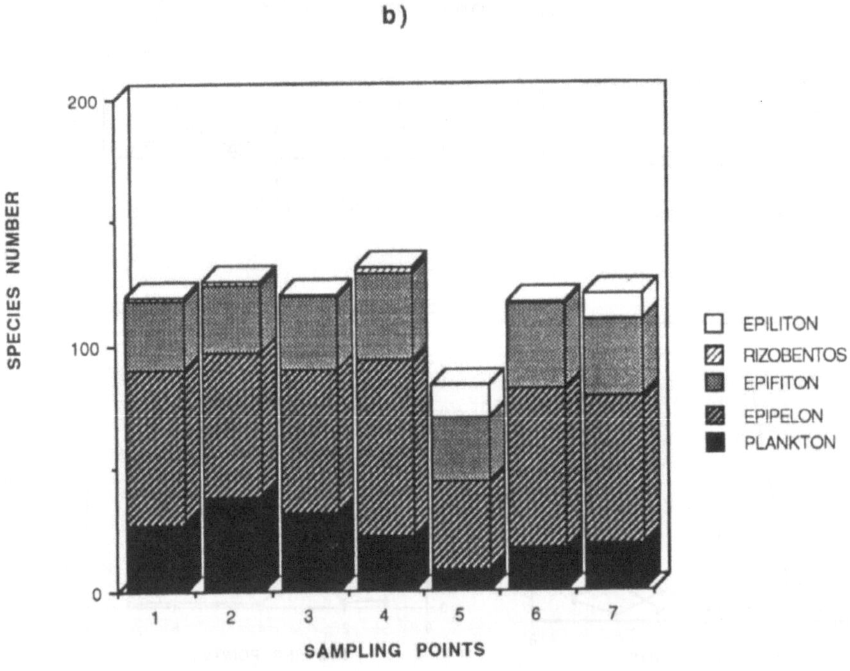

Figure 3. Fluctuations of the main physico-chemical characteristics of water from the canals during 1993–94: a. discharge, b. temperature, pH and oxygen, c. conductivity, chloride and sulphates, d. nitrate and amonium, e. sodium, magnesium, calcium (Each point is the mean calculated from 16 samples).

Table 1. Channels flora: benthic species present in more than one sampling point are only indicated.

Taxa	1	2	3	4	5	6	7
Bacillariophyceae							
Achnanthes minutissima	*	*	*	*	*	*	*
Caloneis permagna	*	*	*				*
Cocconeis pediculus	*	*	*	*	*	*	*
Cocconeis placentula	*	*	*	*	*	*	*
Cyclotella alvaniensis		*		*		*	*
Cyclotella kützingiana	*			*		*	
Cyclotella meneghiniana				*		*	
Cyclotella ocellata	*	*	*	*	*	*	*
Cymatopleura solea	*	*	*	*		*	*
Cymbella affinis	*		*	*	*		
Cymbella minuta	*	*		*		*	
Cymbella prostrata	*	*		*	*	*	
Denticula tenuis	*	*	*	*	*	*	*
Diatoma elongatum	*	*	*	*	*	*	*
Diatoma vulgare	*	*	*	*	*	*	*
Diploneis oblongella		*				*	*
Diploneis ovalis						*	*
Gomphonema angustatum	*	*	*	*			
Gomphonema constrictum	*	*	*	*	*		*
Gyrosigma acuminatum	*	*	*	*			
Navicula cryptocephala	*						*
Navicula lanceolata	*		*			*	*
Navicula pygmaea	*		*			*	*
Navicula radiosa	*	*	*	*		*	*
Navicula tripunctata	*	*	*	*		*	*
Nitzschia acicularis				*		*	*
Nitzschia cf. palea	*	*					
Nitzschia hungarica	*		*			*	*
Nitzschia linearis	*	*		*		*	*
Nitzschia sigmoides	*					*	
Nitzschia tryblionella	*	*		*		*	*
Nitzschia tubicola	*					*	*
Rhoicosphenia abbreviata	*		*	*	*	*	*
Surirella ovalis	*	*		*		*	
Surirella striatula	*		*				
Synedra acus		*	*	*	*	*	
Synedra ulna	*	*	*	*	*	*	*
Cyanophyceae							
Calothrix parietina	*	*		*	*	*	
Calothrix fusca			*	*			
Chamaesiphon cylindricus			*	*			
Lyngbya sp.	*	*	*	*	*	*	
Lyngbya articulata				*	*		
Lyngbya cf pusilla				*	*		
Lyngbya kützingiana		*	*	*	*		
Microcoleus sociatus	*	*	*	*			
Microcoleus vaginatus	*	*	*	*		*	
Myxosarcina coccinna	*	*	*	*	*	*	
Oscillatoria articulata	*	*	*	*		*	
Oscillatoria brevis	*	*	*	*			
Oscillatoria limnetica	*		*	*			
Oscillatoria quadripunctulata		*	*	*			
Phormidium autumnale	*		*	*		*	
Phormidium favosum	*	*	*	*		*	
Phormidum trealeasei		*		*		*	
Pleurocapsa minor		*	*	*	*	*	
Pseudanabaena catenata	*	*	*	*			
Pseudanabaena galeata	*			*			
Schizothrix symplocoides		*		*	*	*	
Schizothrix trealesii				*		*	
Schizothrix vaginata	*						
Schizothrix viguierii	*	*		*	*		
Spirulina major	*	*	*	*		*	
Staniera cyanosphaera	*	*	*	*	*	*	
Synechocystis sp.	*	*	*				
Xenococcus kernei				*	*	*	
Rhodophyceae							
Audouinella sp.					*		
Audouinella hermanii						*	
Xantophyceae							
Vaucheria sp.	*		*				
Chlorophyceae							
Aphanochaete repens			*	*	*		*
Bulbochaete sp.	*				*		
Cladophora glomerata	*	*	*	*	*	*	*
Closterium acerosum	*	*					
Closterium pronum	*	*	*	*	*		*
Cosmarium laeva	*	*	*	*			
Chara globularis				*			
Chara vulgaris		*					
Entocladia sp.				*	*		
Keratococcus raphidioides	*		*				
Klebsomidium faccidum	*	*	*				
Mougeotia sp.	*	*	*	*			
Oedogonium sp.	*	*	*	*		*	
Pediastrum boryanum				*			*
Pediastrum integrum		*	*			*	
Pediastrum simplex	*	*	*	*			
Pediastrum tetrax		*	*				
Protosiphon botryoides	*	*					
Rhizoclonium hieroglyphicum	*	*	*	*			
Scenedesmus acutus		*	*	*		*	
Scenedesmus bijugatus	*	*	*				
Scenedesmus disciformis		*					
Scenedesmus quadricauda	*	*		*			
Spirogyra sp.	*		*	*	*	*	*
Staurastrum sp.	*	*	*	*			
Ulothrix tenerrima					*		*
Volvocales indet.			*			*	
Bacteria							
Beggiatoa sp.	*		*				
Antophyta							
Potamogeton pectinatus	*	*	*	*		*	

there were also well structured assemblages of plankton and periphyton (Figure 3b). The planktonic assemblages are composed mainly of dinophytes, chlorophytes, diatoms, cyanophytes and euglenophytes. There is a spring bloom of *Aphanizomenon flos-aquae* (L.) Ralfs in La Pedrera Reservoir, though there have been no toxicity related problems. The rhizobenthos is composed primarily of characeae (*Chara vulgaris* (L.) var. *gymnophylla* Braun and *C. globularis* (Thuillier) and Anthophyta (*Potamogeton pectinatus* L.) and the epilithon is, strictly speaking, limited to the reservoirs sites (5–7); the only places where solid substrata are not covered by a broad layer of silt.

Most of the species are common in water systems in the Murcia area (Aboal, 1989) but a number of species were recorded which are new to the area: *Franceia javanica* (Bern.) Hortob., *Dictyosphaerium botrytella* Kom. & Perm, *Keratococcus raphidioides* (Hansg.) Pasch., *Gloeotila curta* Skuja, *Coelastrum polychordum* (Kors.) Hind., *Staniera cyanosphaera* (Kom. et Hind.) Kom. et Anag., *Oscillatoria acutissima* Kuff. Some of these taxa were only known from Central Europe and others from tropical regions. The bank of the channel above the water line supports only lichens and a conspicuous black layer of cyanophyceae, mainly *Calothrix parietina* Thuret. These two components are rarely submersed. The bed of the channel is typically dominated by mats of cyanophyceae which are normally covered by seston deposits. The crust formation is approximately 0.5 cm thick with an upper 1.2 mm which is a biofilm (the crusts collected were measured in the laboratory with a stereo microscope).

Cladophora is able to attach itself to these crusts even though it has a preference for hard substrata (Dodds & Gudder, 1992). These crusts change from blue-green to yellow due to the incorporation of different diatom species.

The rodophyte *Audouinella hermanii* (Roth) Des. was noted in one canal sample point (6) for the period of one week cohabiting with *Spirogyra* sp. Its rapid dissappearance was due to grazing: both species being especially palatable to fish.

At the driest period at the end of winter/beginning of spring, *Protosiphon botryoides* (Kütz.)Klebs, *Chlorhormidium flaccidum* (Kütz.)Fott and *Stichococcus bacillaris* Nägeli were recorded. These species, typically abundant in wet soils, are associated with mats of cyanophytes which penetrate the crevices in the channel bed to protect themselves from dessication. As ponds begin to develop in the channels, extended carpets of *Vaucheria terrestris* Götz develop at the edges. These shallow ponds are then colonised by *Chara vulgaris* L. *gymnophylla* Braun and *C. globularis* Thuillier, intermixed with *Cladophora glomerata* (L.) Kütz., *Rhizoclonium hieroglyphicum* (Agardh) Kütz., *Spirogyra* sp. and *Potamogeton pectinatus* L. The latter filamentous chlorophytes can develop mats of more than 1m in length and 10 cm thick. The cell wall of the filamentous algae supports a highly structured epiphytic community (Cambra & Aboal, 1992).

When the channel are completely full of water a very homogeneous structure of plant vegetation can be seen. *C. glomerata* rapidly grows from akinetes in the sediments and from other propagules in the water or air. It prefers the sloping walls to the floor of the channel in areas where a thin layer of sediment has accumulated, and where the incident light is more intense. It needs a high light intensity for its development and its prolific growth makes it one of the most significant contributors to biomass production in the flowing water sites (Aboal et al., 1994). It is found, for example, growing as a narrow belt near the surface in the turbid waters of the numerous canals in the Netherlands (Van den Hoek, 1963) and Den Hartog (1958) describes a relationship between turbidity of the water and the downward extension of the *Cladophora* belt. When filament fragmentation occurs the basal portion remains attached to the substrate and can regenerate or remain semidormant through the winter (Whitton, 1970) while the upright portions drift and allow colonization of downstream habitats (Dodds & Gudder, 1992).

P. pectinatus competes with the latter species for light (Ozimek et al., 1991) but is better adapted to low light intensities (Blindow, 1992) and, although *Cladophora* begins its growth first, *P. pectinatus* very quickly establishes monospecific prairies along the canals. The capacity of *Potamogeton* to take phosphate from the sediments represents an advantage in these channels, since the water analysis shows that the concentration of this anion is so low as to be undetectable.

P. pectinatus can survive adverse conditions due to underground tubers, and a great number of seeds can reach the channels because of the existence of well developed populations of this species in the Segura River and in all nearby natural and artificial ponds (Ríos, 1994). In the channels it flowers throughout the year preferring places where there is a heavy accumulation of sediments.

During the period of this study, discharge increased over the spring rising to a maximum of 26 $m^3\ s^{-1}$ in

July and August 1994 (Figure 2a). Light was rapidly absorbed in the top few centimetres of the water column: 43% of the radiation penetrated to 50 cm and 35% reached 1m depth. The mean water temperature varied between 19.6 and 22.9 °C, whereas the pH and conductivity were surprisingly stable for a predominantly limestone catchment at around pH 8 and 1200 to 2000 μS cm^{-2}. Dissolved oxygen was high at all sampling points (Figure 2b). The main concentration of chloride ranged between 65 and 216 ppm and given that, due to the abundance of marl and gypsum, the Segura basin has the highest sulphate concentrations of all Spanish basins, it is not surprising that sulphate concentrations reached 1849 ppm (minimum 696 ppm). Figure 2 summarises the variation in the chemical variables ammonium, sodium. magnesium, and calcium. Phosphate concentrations were negligible during the whole year for all sample points.

Prospects for aquatic weed control

Planning for aquatic weed control poses particular problems. The hardness of the water precludes the use of certain aquatic herbicides and even so the volumes of water involved would make chemical control prohibitively expensive. There is also a specific problem with *P. pectinatus* and tuber production. These organs in some species are known to be resistant to herbicide application (Langeland, 1990). In this area tubers can be collected from February (in very small quantities) to December.

In those sections of the Transvase Tajo-Segura which are permanent aquatic habitats, there is currently no problem of weed accumulation. It is considered that this is due to grazing by fish, notably the common carp (*Cyprinus carpio* L.) and the barbel (*Barbus schlateri* Steindachner). Both species are frequent in these sites. It is proposed that management is instigated to ensure a sustainable population of such grazing species in the system and work is underway to explore this possibility along with research into resistance of *P. pectinatus* tubers to herbicides and the potential for using antifouling paint to control algal species and *C. glomerata* in particular.

Acknowledgments

This paper was funded through a grant from the ARECES Foundation.

References

Aboal, M., 1989. Epilithic algal communities from River Segura Basin, Southeastern Spain. Arch. Hydrobiol. 116: 113–124.

Aboal, M., M. A. Puig, A. Sanchez-Godinez & G. Soler, 1994. Algal standing-crop in some Mediterranean temporary rivers in southeastern Spain. Verh. Int. Ver. Limnol. 25: 1746–1750.

Blindow, I., 1992. Long- and short-term dynamics of submerged macrophytes in two shallow eutrophic lakes. Freshwat. Biol. 28: 15–27.

Cambra, J. & M. Aboal, 1992. Filamentous green algae of Spain: distribution and ecology. In C. Montes & C. Duarte (eds), Limnology in Spain. Asociación Española de Limnología: 213–220.

Dodds, W. K. & D. A. Gudder, 1992. The ecology of Cladophora. J. Phycol. 28: 415–427.

Dudley, T. L., 1989. Interactions among algae, invertebrates and the physical environment in stream riffle communities. Dissertation. University of California. Santa Barbara. USA.

Hartog, C. Den, 1958. Epilitische algengemeenschappen in Nederland. Hand Hydrobiol. Ver. 10: 6–8.

Langeland, K. A., 1990. *Hydrilla (Hydrilla verticillata* (L. F.) Royle) a continuing problem in Florida waters. Institute of Food and Agricultural Sciences. University of Florida. Cooperative Extension Service. Circular 884: 20 pp.

M.O.P.U., 1977. El agua en España. Madrid. Dirección General de Obras Hidráulicas. Centro de Estudios Hidrográficos. 288 pp.

Nichols, H. W., 1973. Growth media-freshwater. In J. R. Stein (ed.) Handbook of Phycological Methods. Culture methods and Growth Measurements. Cambridge University Press. Cambridge: 25–51.

Nichols, S. A., 1991. The interaction between biology and the managemeny of aquatic macrophytes. Aquat. Bot. 41: 225–252.

Ozimek, T., E. Pieczynska & A. Hankiewicz, 1991. Effects of filamentous algae on submerged macrophyte growth: a laboratory experiment. Aquat. Bot. 41: 309–315.

Rios, S., 1994. El paisaje vegetal de las riberas del río Segura, SE de España. Tesis Doctoral. Universidad de Murcia. Murcia, Spain.

Van Den Hoek, C., 1963. Revision of the european species of *Cladophora*. Otto Koeltz Science Publishers. Koenigstein, 248 pp.

Whitton, B. A., 1970. Biology of *Cladophora* in freshwaters. Wat. Res. 4: 457–476.

Hydrobiologia **340**: 109–113, 1996.
J. M. Caffrey, P. R. F. Barrett, K. J. Murphy & P. M. Wade (eds), Management and Ecology of Freshwater Plants.

Aquatic vegetation of the Orinoco River Delta (Venezuela). An overview

Giuseppe Colonnello Bertoli
Museo de Historia Natural La Salle, Apdo. 1930. Caracas, 1010-A Venezuela

Key words: Aquatic plants, inventory, delta, Orinoco

Abstract

The Orinoco River Delta is a complex of relatively unknown ecosystems made up of aquatic and semiaquatic habitats. The region includes a Mariusa National Park and the Orinoco Delta River Biosphere Reserve. The 23 sites studied included lentic habitats (lagoons and creeks) and lotic habitats (streams of variable breadth and current) which varied range from black to white waters. A collection of aquatic vascular plants was made and the physical and chemical variables were measured. A total of 100 species were identified, 48% of which are new records for the Delta territory. A Canonical Correspondence Analysis divided the species and communities into two groups: The lentic habitats of lakes and lagoons, which show a greater species richness and the lotic habitats of the main river courses. This study forms part of an ongoing research project on the floristic and abiotic characteristics of the aquatic communities in the Orinoco River Delta.

Introduction

Until 1969, the region of the Orinoco River Delta remained relatively isolated from the development of Venezuela due to the lack of roadways. This explains the shortage of biological studies in this vast territory of approximately 40 000 km^2 which is dominated by aquatic habitats. Even in the few existing studies, the aquatic plants rarely receive major attention. The two most relevant inventories were produced by Delascio (1975) and Danielo (1976). Actually, the entire Delta region is under a high pressure of development, tourism, logging and an extensive petroleum exploitation, in consequence the environment is going to face radical ecological changes. The objective of the present paper is to present an oveview of the communities and species of aquatic macrophytes in this Delta region.

Site description

The Delta Region (Figure 1) was predominantly formed by alluvial materials deposited by the Orinoco River above marine clays. It can be divided according to Canales (1985) into the Upper Delta (between 7 to 2.5 meters above sea level) which is seasonally flooded for relatively short periods and predominantly covered by a deciduous forest; the Middle Delta (2.5 to 1 masl) seasonally flooded for longer periods, and composed of evergreen and palm forests; and the Lower Delta (1 to −1 masl) quite permanently flooded and dominated by evergreen shade forests, grasses and sedges. Mangroves are found along the coastal zone.

The land masses constitute islands surrounded by channels. A typical profile consists of a levee with sandy soils covered by forests and depressed areas in the middle comprised of clay and silty material covered by peat, grasses and sedges, and temporary as well as permanent lakes (Figure 2). Soils, due to previous environmental conditions, are generally heavy with very poor drainage properties, high acidity, and low fertility.

The sites selected for study included representative samples of each habitat type including the different types of waters. These were determined by their physical and chemical characteristics and defined by the topographic and lithological characteristics of the particular portion of the river basin drained (Vásquez & Wilbert, 1992).

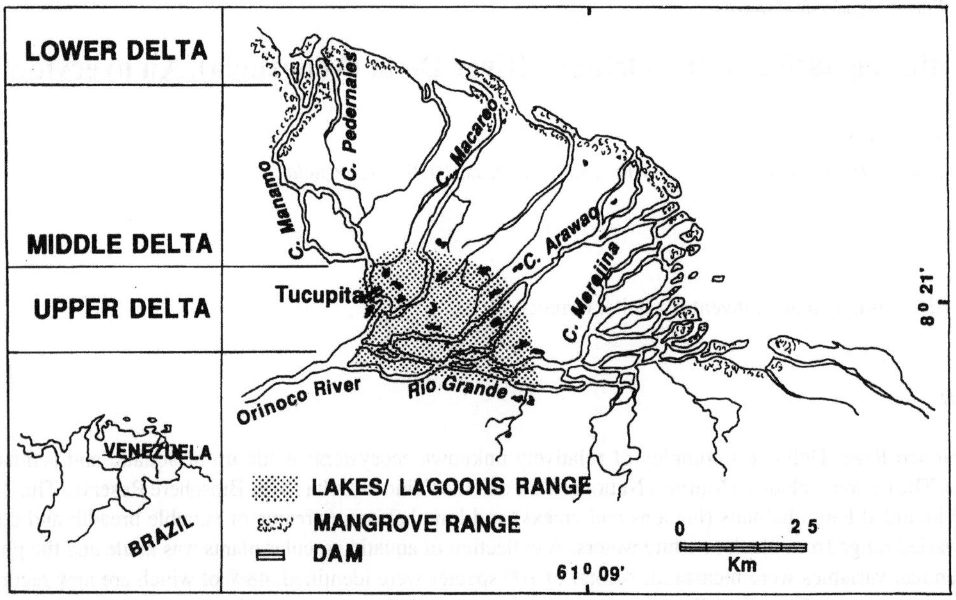

Figure 1. The Orinoco River Delta. The study area included distinct habitats in the upper, middle and lower Delta.

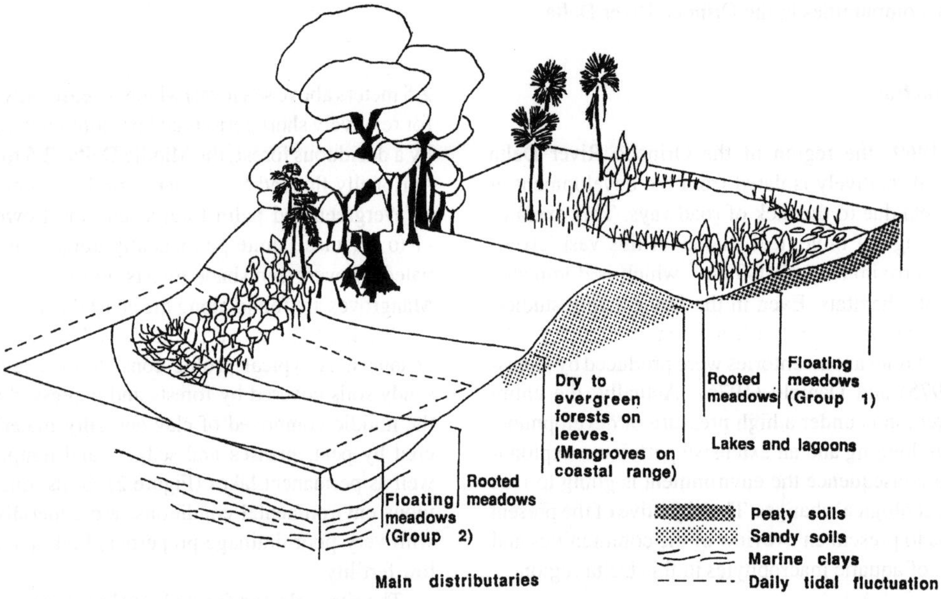

Figure 2. Typical profile through an area of the Upper Delta.

The Caño Manamo damming

An important event in the Delta was the damming, in the late sixties, of the Caño Manamo in order to prevent the flooding of the territories of the upper reaches of the river, and to create new lands for agriculture. This regulation reduced the water level oscillation from 2–3 m to 0.5–1 m and stopped the annual flooding of the upper reaches of the river altogether. This artificial drainage of the land promoted the oxidation of the underlying marine clays with the consequent impoverishment and acidification of part of the fluvial soils to pH 2.5.

The sedimentary processes caused new banks to emerge from the waters, promoting a colonization by mainly aquatic rooted species. There were also floristic variations, such as the disappearence from the main river course of *Echinochloa polystachya* and *Eichhornia azurea*, species adapted to high water level changes, and colonization by *Montrichardia arborescens* which dominates where water level fluctuations are small. Other effects included: (i) the modification of the current regime of tributaries, some of which developed extensive communities of floating plants obstructing the river courses, leading to the rapid sedimentation of the waterways. (ii) The increase of the mangrove community distribution (*Rhizophora* spp.) upstream of the Caños Manamo, Pedemales and Cocuina, due to the intrusion of the salt with the daily tides.

The emergent, free-floating and submerged species were systematically recorded. Plants were collected, preserved, and deposited in the herbarium of La Salle Natural History Museum (CAR).

At each site, 10 samples of 1 m^2 were taken. Vegetation cover was recorded during the survey and assigned a DAFOR rating: Dominant = 70–100%; Abundant = 30–70%; Frequent = 10–30%; Occasional = 3–10%; Rare = 0–3%. Water depth, color, transparency (Secchi disc), conductivity, dissolved oxygen (DO on surface and at 2 m depth) and flow rate (where appropriate) were measured. Water samples were refrigerated and sent for laboratory analysis to determine total P, Mg, K, Fe and pH. The resulting species and site data were analyzed using a Canonical Correspondence Analysis from the CANOCO package (Ter Braak, 1987–1992).

Table 1. Species belonging to the floating meadows ordered according to their constancy values.

Group 1(CCA), Lakes and lagoons
Salvinia auriculata Aubl.
Eichhornia crassipes (Mart.) Solms
Pistia stratiotes L.
Leersia hexandra Swartz.
Limnobium laevigatum (H&B exWilld) H.
Nymphaea connardii Wiersema
Nymphaea rudgeana G.F.W. Mey.
Salvinia sprucei Kuhn.
Hydrocotile umbellata L.
Utricularia sp 1
Paspalum repens Berg.
Echinochloa polystachya (H.B.K.) Hitch.
Eichhornia azurea (Sw.) Kunth.
Ludwigia helminthorrhiza (Mart.) Hara
Spirodela intermedia W. Koch
Nymphoides indica L. Kuntze
Utricularia foliosa L.
Ceratopteris pteridoides (Hook.) Hieron.
Marsilea polycarpa Hook. & Grew
Cabomba aquatica Aubl.
Lemna minor L.
Azolla filiculoides Lam.
Tonina fluviatilis Aubl.
Ludwigia sedioides (H. & B) Hara
Neptunia oleraceae Lour.
Urosphata sagittifolia (Rudge) Schott
Eichhornia heterosperma Alexander
Pontederia rotundifolia L.f.
Oryza latifolia Desv.
Utricularia inflata Walter
Utricularia sp 2
Sagittaria latifolia Willd.
Phyllanthus fluitans (Mull.) Arg.
Utricularia hydrocarpa Vahl.
Hydrocleis nymphoides (Willd.) Buch.
Heteranthera reniformis Ruiz & Pav.
Group 2 (CCA), Main distributaries
Eichhornia crassipes (Mart.) Solms
Salvinia auriculata Aubl.
Paspalum repens Berg.
Eichhornia azurea (Sw.) Kunth.
Echinochloa polystachya (H.B.K.) Hitch.

Results and discussion

Floristic inventory

More than 120 species were collected including free-floating, emergent and submerged plants. To date 100 species have been identified and are shown in Tables 1 and 2. According to the earlier inventories for the region (Delascio, 1975; Danielo, 1976; Velásques, 1987), the present survey increased the aquatic flora of the region by 48% (Colonnello et al., 1993). Additional collections have already added some 50 new records to this list (Colonnello, 1997).

The species richness and composition in the Orinoco delta region is comparable with similar areas

Table 2. Species belonging to the rooted meadows ordered according their constancy values.

ROOTED MEADOWS
Ludwigia octovalvis (Jacq.) Raven
Oxicaryum cubense (Poep & Kunth) K. Lye
Montrichardia arborescens (L.) Schott
Hymenachne amplexicaulis (Rudge) Nees.
Polygonum acuminatum H.B.K.
Mimosa pigra L.
Fuirena umbellata Rott.
Luziola subintegra Swallen
Cyperus distans L.f.
Ipomoea sp.
Sacciolepis striata (L.) Nash.
Alternanthera philoxeroides (Mart.) G.
Paspalum fasciculatum Willd.
Justicia laevilinguis (Nees) Lindau.
Eleocharis elegans (H & B.) Roem. & Shult.
Mikania congesta D.C.
Cuphea melvilla Lindl.
Bacopa aquatica Aubl.
Xyris caroliniana Walter
Eleocharis mutata (L.) Roem. & Shult.
Scleria microcarpa Nees ex. Kunth
Acroceras zizanoides (Kunth.) Dandy.
Acrostichum aureum L.
Begonia patula Haw.
Gynerium sagittatum (Aubl.) Beauv.
Rynchospora holoschoenoides (Rich.) Herter
Fimbristilis miliacea (L.) Vahl.
Eleocharis interstincta (Vahl.) R. & S.
Cyperus sp.
Cyperus sphacellatus Rottb.
Cyperus surinamensis Rottb.
Capraria biflora. L.
Hymenocallis tubiflora Salisb.
Blechnum serrulatum Rich.
Eclipta prostrata (L.)L.
Cyperus articulatus L.
Eleocharis filiculmis Kunth.
Leptochloa scabra Nees.
Panicum elefanthipes Nees. in Trin.
Panicum grande Hitch & Chase
Panicum parvifolium Lam.
Panicum pilosum Swartz
Panicum scabridum Doell. in Mart.
Paspalum wrightii Hitch. & Chase.
Thalia geniculata L.
Ludwigia decurrens Walt.
Ludwigia leptocarpa (Nutt.) Hara
Ludwigia torulosa (Arnott) Hara
Agalinis hispidula (Mart.) D'Arcy
Bacopa sp.
Sphenoclea zeylanica Gaertn.
Typha domingensis Pers.
Costus arabicus L.
Heliconia marginata (Greigg) Pittier.
Canna glauca L.
Habenaria repens Nutt.
Echinodorus sp.
Aeschynomene rudis Benth.
Chelonanthus alatus (Aubl.) Pulle

of the Amazonas River (Junk, 1990) and Parana River (Neiff, 1986). However, no further comparison can be made due to the different criteria used by the authors defining the status of the aquatic plants.

The aquatic communities and their environment

The species were preliminarily divided on the basis of those plants belonging to floating meadows (i.e. living in the water surface for the major part of their life cycle) and those belonging to rooted meadows (capable of living for long periods on soils which are not flooded).

The floating meadows (Table 1)

The results of the Canonical Correspondence Analysis showed a clear separation of two groups of floating meadow sites based on transparency and DO. The first group was represented by communities with relatively large number of species (more than 21) and characterized by lentic habitats. It was made up of a range of habitats from channels of black waters with very sinuous courses (the predominant source of water being percolation through the forest formations), to lagoons located in the Upper Delta. The sourse of water of these lagoons is rain water plus, during the middle of the rainy season, white water from the rivers. In the lakes (large permanent lagoons) the lack of current promoted the quick deposition of sediments, creating habitats on which submerged plants e.g. *Cabomba aquatica* and *Utricularia* ssp and floating plants e.g. *Eichhornia crassipes*, *Pistia stratiotes*, *Limnobium laevigatum*, *Nymphaea rudgeana* and *Ludwigia helminthorrhiza* developed together with many species of grasses and sedges e.g. *Leersia hexandra* and *Oxicaryum cubense*. The physical and chemical characteristics of these waters are low to medium pH values (3.8–4.9), medium conductivity (35–100 μS cm^{-2}) very low DO (0.1–1.6 mg l^{-1}), high transparency (50 to 150 cm) and low Ca and P levels.

Some sites like marshes and very polluted courses, near high population density areas had extreme values of conductivity and dissolved oxygen (max. 280 μS cm^{-2} and 0 mg l^{-1}) and restricted water current, leading to extensive populations of certain species such as *E. crassipes*, *P. stratiotes* and *Hydrocotile umbellata* which constituted a serious weed problem.

The second group of sites, contained communities with relatively few species (1–10). The habitats were associated with straight and wide channels with low sinuosity, high current speed, and low transparency which reduced the establishment of free-floating and submerged plants. Only few emergents were growing in occasional shallow banks. The species that can produce large communities in the shores are rooted plants adapted to this particular environment, e.g. *Paspalum repens*, *E. polystachya* and *E. azurea*, which have long floating branches. Some free floating species, e.g. *E. crassipes* and *Salvinia auriculata*, developed on areas protected from the currents, forming a dense community due to the intense vegetative reproduction, but during the major flooding events most of them were removed by the current. This is the case for the main distributaries of the Orinoco River (Río Grande, Caño Araguao, Macareo and Manamo) and those channels with a high daily fluctuation in the water level due to the tides, e.g. Pedernales and Guiniquina channels. The water at these sites has medium to high pH (5.3–6.6), low conductivity (10–35 μS cm^{-2}), very low transparency (8–50 cm), high DO (2.6–6.7 mg l^{-1}) and higher values of Ca and P. All these factors are partially influenced by the suspended sediments of the local waters which varied according to the season of the sampling.

The rooted meadows (Table 2)

The correspondence analysis did not further subdivide the rooted meadows sites, probably because a number of the rooted species are common to the different types of habitats. The environmental data collected were mostly refered to the water quality. Parameters such as thickness and type of the sediment, and the slope of the banks are currently under study. The most relevant species were *Ludwigia octovalvis*, *Polygonum acuminatum* and *O. cubensis*, which can grow rooted to the sediments or on floating mattresses. Species such as *M. arborescens*, *Hymenachne amplexicaulis* and *Mimosa pigra* occupy the highest positions on the shores.

Acknowledgments

This study is part of the project: Study of the aquatic vegetation of Mariusa National Park and its protected area was financially supported by The Educational Association for the Conservation of Nature (**EcoNatura**). I thank Lic. Maria Altagracia Sole and Sr. Jose Vicente Montoya for their help in the field and laboratory sample process, and Dr Max Wade for correcting the manuscript.

References

Canales, H., 1985. La cobertura vegetal y el potencial forestal del T.F.D.A. (Sector Norte del Río Orinoco). M.A.R.N.R. Divisiön del Ambiente. Secciön de Vegetaciön, 195 pp.

Colonnello, G., M. A. Solé & J. Velásquez, 1993. Inventario preliminar de las plantas acuáticas vasculares del delta del Río Orinoco, Venezuela. Memoria de la Sociedad de Ciencias Naturales La Salle 139: 147–159.

Colonnello, G., 1997. La vegetación acuática del Delta del Río Orinoco (Venezuela). Composición florística y aspectos ecológicos (I), Memoria de la Sociedad de Ciencias Naturales la Salle (in press).

Danielo, A., 1976. Végetation et sols dans delta de l'Orénoque. Annals de Géografie. Revue publiée, avec le concours de Centre Nationale de la Recherche Scientifique N° 471: 555–578.

Delascio, F., 1975. Aspectos biológicos del Delta del Orinoco. Instituto Nacional de Parques, Dirección de Investigaciones Biológicas. Divisiön de Vegetaciön, 64 pp.

Junk, W., 1990. Die Krautervegetation der Ubershwemmungsgebiete des Amazonas (Varzea) bei Manaus und ihre Bedeutung fur das Okosiystem, 349 pp.

Neiff, J. J., 1986. Aquatic plants of the Parana system. The Ecology of River System, Dr W. Junk Publishers, Dordrecht, 793 pp.

Ter Braak, C. J. F., 1987–1992. CANOCO – a FORTRAN program for Canonical Community Ordination. Microcomputer power, Ithaca, New York, USA, 95 pp.

Vásquez. E. & W. Willbert, 1992. The Orinoco: Physical, Biological and Cultural Diversity of a Major Tropical Alluvial River. In P. Calow & J. Petts (eds), The Rivers Handbook, Hydrological and Ecological Principles. Blackwell Scientific Publications, London, 526 pp.

Velásquez, J., 1987. Plantas Acuáticas Vasculares de Venezuela. Trabajo presentado para optar a la categoría de Profesor Titular en el Escalafón Docente y de Investigación. Escuela de Biología, Facultad de Ciencias, Universidad Central de Venezuela. Caracas, 1041 pp.

Hydrobiologia **340:** 115–120, 1996.
J. M. Caffrey, P. R. F. Barrett, K. J. Murphy & P. M. Wade (eds), Management and Ecology of Freshwater Plants.

Submerged vegetation development in two shallow, eutrophic lakes

Hugo Coops & Roel W. Doef
Institute for Inland Water Management and Waste Water Treatment (RIZA), P.O. Box 17, 8200 AA Lelystad, The Netherlands

Key words: macrophytes, *Potamogeton*, Charophytes, eutrophication, lake restoration

Abstract

Submerged macrophyte vegetation in two shallow lakes in the Netherlands, Lake Veluwemeer and Lake Wolderwijd, has been affected by eutrophication in the late 1960's and 1970's. Recent changes in the vegetation occurred in the period following lake restoration measures. Between 1987 and 1993, the dominance of *Potamogeton pectinatus* decreased, while Charophyte 'meadows' expanded over the same time interval. The pattern of change of the dominant macrophyte species might result from changes in the underwater light climate. Seasonally persistent clear water patches associated with the *Chara* meadows have been observed in the last few years. The interaction between submerged macrophyte vegetation succession and water transparency in the lakes is discussed.

Introduction

Submerged macrophyte vegetation plays a central role in the functioning of shallow lake ecosystems. Lake restoration measures (e.g., biomanipulation) often aim at recovery of macrophytes as a provision for phytoplankton grazers and piscivorous fish (Hosper, 1989; Moss, 1990). Furthermore, the role of submerged vegetation in lake nutrient dynamics is often stressed (Carpenter & Lodge, 1986). An increase of submerged vegetation may occur after improvement of the underwater light climate (Scheffer et al., 1993), but extensive studies documenting decline and recovery of macrophytes are scarce, because long-term monitoring results are hardly ever available. To evaluate vegetation recovery patterns, more should be known about fluctuations of the plant populations present among and within years.

In Lake Veluwemeer and Lake Wolderwijd, two shallow lakes bordering the Flevo Polders in the Netherlands, eutrophication has affected the submerged vegetation seriously since the late 1960's. Scattered patches of aquatic plants (mainly *Potamogeton pectinatus* L. and *P. perfoliatus* L.) occurred in the resulting turbid water. 'Meadows' of Charophytes, once an important component of the vegetation, disappeared in the late 1960's.

In order to assess the impact of lake restoration measures this paper describes changes in the submerged vegetation of both lakes, prior to and following nutrient reduction (Lake Veluwemeer) and fish removal (Lake Wolderwijd).

Description of the lakes

After disconnection from the sea in 1932, the formerly brackish Zuiderzee estuary in the Netherlands formed Lake IJsselmeer. Freshwater macrophytes (mostly *P. pectinatus*, *P. perfoliatus*, and Characeae) established in the late 1930's and 1940's, following salinity reduction. Lake Veluwemeer (3400 ha) and Lake Wolderwijd (2700 ha) were formed as marginal lakes after embankment of part of Lake IJsselmeer (the so-called Flevo Polders), in 1957 and 1969, respectively. Both lakes are shallow (mean depth ~ 1.5 m) and have been eutrophic since their origin. The water is in the range pH 8–9 and the sediments are predominantly sand and loam. Chloride content in Lake Veluwemeer is 150–200 mg Cl l^{-1} (due to flushing with relatively saline seepage water from the Flevo Polders), and in Lake Wolderwijd 80–150 mg Cl l^{-1}.

Originally, Lake Veluwemeer water was very clear and supported a dense mat of Characeae, while from 1965 onwards other macrophytes (*Potamogeton* spp. and *Myriophyllum spicatum* L.) gradually developed; an increase of turbidity and decrease of macrophytes occurred in the early 1970's. Lake Wolderwijd has been dominated by the blue-green alga *Oscillatoria agardhii* Gomont. since its origin, resulting in a sparse macrophyte vegetation (Berger, 1987). In Lake Veluwemeer, flushing with relatively phosphate-poor, chloride-rich seepage water from the Flevo Polders started in 1979 (Hosper & Meijer, 1986). There was a more or less immediate and marked reduction in total phosphorus from a mean of c. 0.4 mg l^{-1} to c. 0.1 mg l^{-1} which has been sustained ever since. Lake Wolderwijd was subject to a drastic reduction of the fish stock in the winter of 1991, aimed to promote zooplankton grazing and decrease the resuspension of sediments by benthivorous fish (Meijer et al., 1994).

Methods

Macrophytes have been monitored in Lake Wolderwijd and Lake Veluwemeer between 1987 and 1993. Each summer (July), the distributions of aquatic plant species were mapped. Species composition and cover were determined in quadrats located in a grid pattern over the lakes. Area estimates were made in GIS after preparing maps by interpolating gridpoints, following the procedure of Scheffer et al. (1992). The data were added to a previously constructed time series of the vegetation, which was reported in Scheffer et al. (1994).

The biomass of *P. pectinatus* was determined at a site in Lake Wolderwijd (52°20′N 5°35′E) during the vegetation periods of 1991, 1992 and 1993. At the site, eight to sixteen 40 × 40 cm quadrats were harvested at 2- to 3-weekly intervals. Aboveground parts, roots and tubers were separated before determination of ash free dry weight (AFDW) (48 hr at 550 °C).

Results and discussion

Vegetation development

Species composition of submerged vegetation in both lakes stayed relatively constant between 1987 and 1993, except Charophytes (principally *C. contraria* A. Braun ex Kützing and *C. aspera* Deth. ex Willd.) appeared for the first time in Lake Wolderwijd. Both lakes supported relatively abundant *P. pectinatus* and *P. perfoliatus*, with smaller quantities of *P. pusillus* L., *P. crispus* L., *Zannichellia palustris* L. and *Elodea nuttallii* (Planchon) St. John. Remarkable differences between the two lakes were the total absence from Lake Wolderwijd of *Myriophyllum spicatum* and *Alisma gramineum* Lej., which occurred abundantly in Lake Veluwemeer.

The areas occupied by the three dominant taxa, viz. *P. pectinatus*, *P. perfoliatus*, and *Chara* sp., between 1970 and 1993 are shown in Figure 1. The dynamics of the vegetation may be related to a variety of environmental factors, of which water depth and turbidity seem to be of major importance (Scheffer et al., 1992). In contrast to *P. pectinatus*, *P. perfoliatus* and Charophytes showed a strong reduction in Lake Veluwemeer in the early 1970's. Differences in response between the species may be explained by different growth forms, enabling plants to cope with reduced underwater light conditions to a variable degree. Plants with an erect architecture show less irradiance-controlled growth than bottom-dwelling species (Chambers & Kalff, 1987). *P. pectinatus* may tolerate increased turbidity by rapid shoot growth towards the water surface in spring, facilitated by carbohydrate reserves in buried tubers. The leaves form a canopy just below the water surface, thereby exploiting limited light resources efficiently (Van Wijk, 1988). *P. perfoliatus* has a much more pillar-like growth form, which potentially is less efficient in turbid water and is more prone to self shading. Furthermore, such a growth form is more susceptible to mechanical damage through water movement. Development of Characeae, which form dense mats of vegetation just above the soil surface, might strongly depend on high light availability as well as sufficiently high temperatures in spring and summer. Consequently, *Chara* spp. generally inhabited a shallower zone (30–75 cm deep) than *P. perfoliatus* (60–150 cm deep).

The seasonal cycle of *P. pectinatus* has a short duration in the lakes (Figure 2). The overwintering tuber biomass is a determinant of shoot emergence in the next year; its decrease during the winter period could be due to grazing. Several, probably interacting factors might cause the short duration of shoot growth: luxuriant epiphytic growth, high turbulence, as well as grazing by herbivorous birds (Van Wijk, 1988). A general decrease in above-ground biomass was evident between 1991–93, consistent with the increasing *Chara* cover. The area occupied by *Chara* meadows increased markedly after 1988 in Lake Veluwemeer

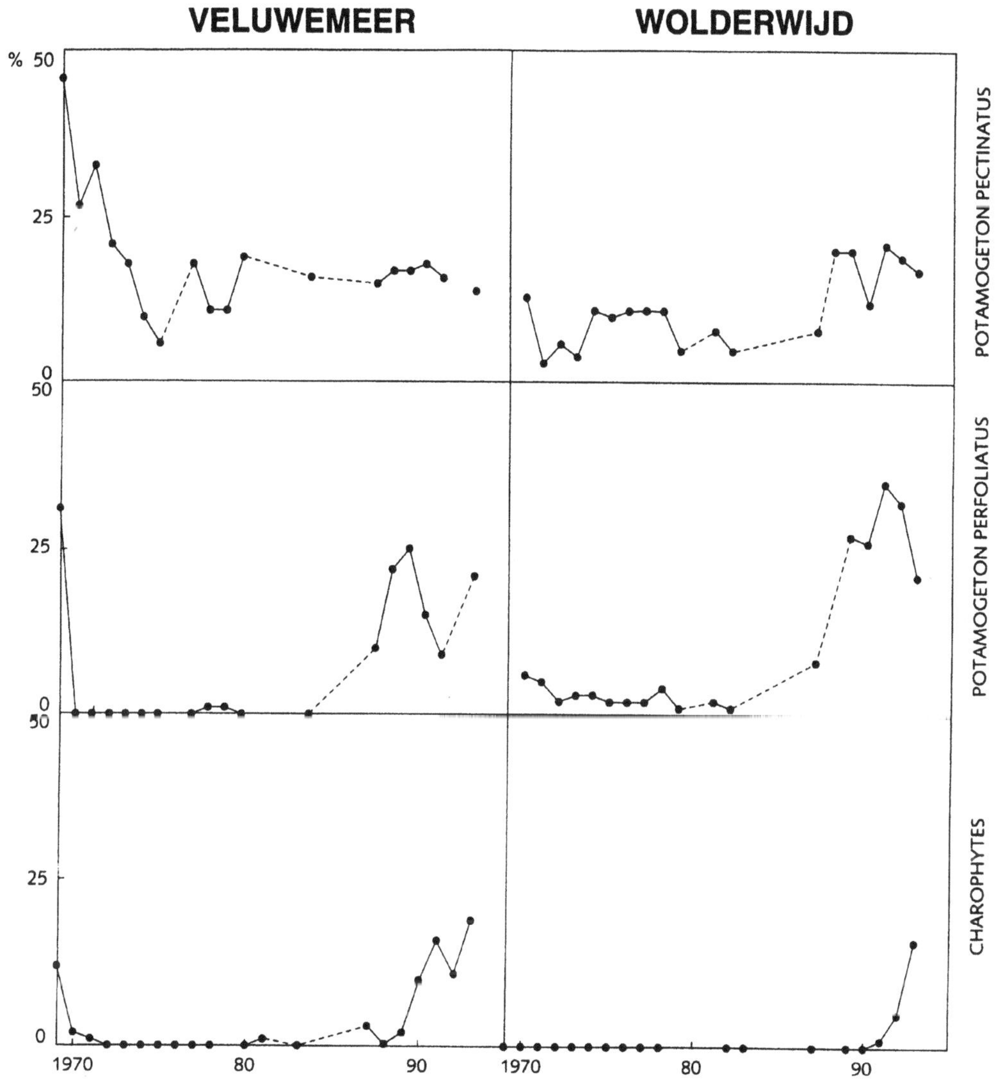

Figure 1. Presence (% of area) of *Potamogeton pectinatus*, *P. perfoliatus*, and *Chara* spp. in Lake Veluwemeer and Lake Wolderwijd between 1969 and 1993.

and after 1990 in Lake Wolderwijd. The recent colonization from 1990 onwards is demonstrated in Figure 3.

Management of the lakes

The recent increase of *P. perfoliatus*, and subsequently of *Chara* sp. indicates a gradual improvement of the water quality in both lakes from about 1985 onwards. Changed light conditions might cause competitive displacement of *P. pectinatus* by Charophytes, starting in the shallowest areas where light conditions are best (McCreary, 1991). Rapid spring growth of Charophytes may pre-empt the seasonal development of *P. pectinatus*.

Turbidity was reduced in the spring following biomanipulation in Lake Wolderwijd (Meijer et al., 1994). It is not clear, however, whether this triggered the simultaneous increase of Charophytes, since the first observations of Charophytes were made in 1990,

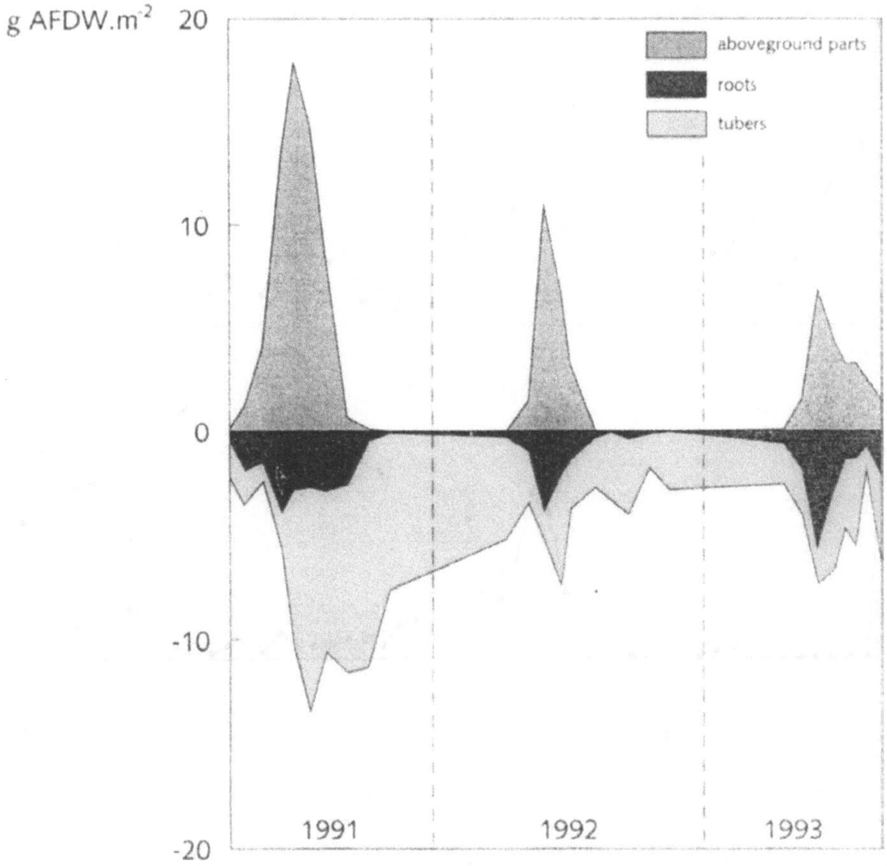

Figure 2. Biomass development and allocation between aboveground parts, roots and tubers of *Potamogeton pectinatus*, harvested from a site in Lake Wolderwijd in 1991, 1992 and 1993.

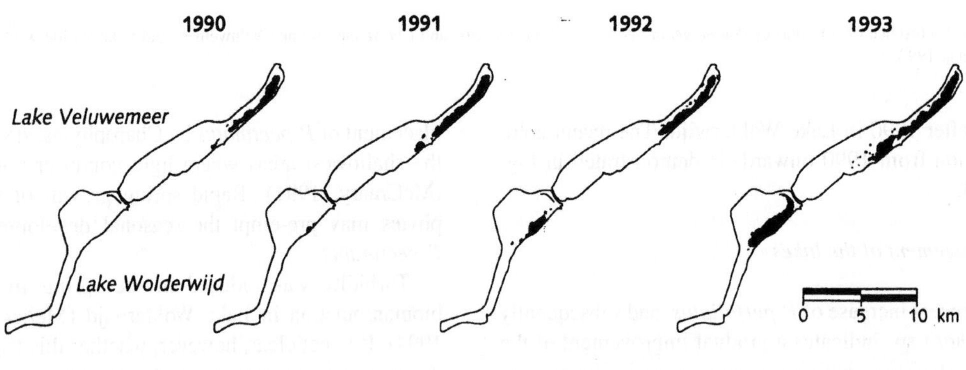

Figure 3. Pattern of distribution of *Chara*-dominated vegetation in Lake Veluwemeer and Lake Wolderwijd between 1990 and 1993.

before the biomanipulation started. Shallow, eutrophic lake ecosystems may potentially exist in one of two alternative states (Scheffer et al., 1992, Blindow et al., 1993): a clear, macrophyte-rich state characterized by cladocerans and predatory fish; and a turbid, phytoplankton-dominated state with benthivorous fish. Shifts from one state to the other due to management interference have been reported (Moss, 1990).

Vegetation has a large influence on ecosystem processes because of the effects of macrophytes on the physico-chemical environment and biota (Carpenter & Lodge, 1986). The coverage of the sediment surface may minimalize resuspension (James & Barko, 1990) Observations in both lakes during 1992 and 1993 of clear water patches above areas with dense Charophyte cover offer a demonstration of this. Imported suspended material may be trapped at the margins of the dense vegetation. Resuspension of sedimentated phytoplankton also might be prevented.

Phytoplankton growth around dense vegetation might also be limited by other mechanisms. Competition for nutrients may occur due to uptake from the water column. Charophytes may fixate relatively large amounts of P during the growing season (Blindow, 1992). However, as they take up nutrients from both water column and soil, no quantitative conclusion regarding their role in nutrient dynamics can be drawn at present. The excretion by Characeae of allelopathic substances that inhibit algal growth is often referred to (Hootsmans & Vermaat, 1991). However, any impact of allelopathy on water transparency in quite large wind-mixed lakes is unproven. Last but not least, dense Charophyte mats may act as a refugium for grazers (cladocerans, molluscs), which might be able to suppress both planktonic and epiphytic algal growth.

It seems likely that the mass emergence of *Chara* in the spring could dictate the appearance of clear water patches in summer. Once vegetation is established, clear water might be maintained, owing to the above mentioned mechanisms. However, their impact and relative importance require further study. The pathway of restoration of aquatic vegetation may involve more than one stage of succession (Figure 4). The colonization of deeper water by Charophytes might occur in interaction with the expansion of clear water patches over some years. Hence a clear lake might appear patchwise, interacting with the development of dense *Chara*-meadows. The influence of herbivorous birds, hydraulic and climatic fluctuations, edaphic factors, and nutrients in the water layer on the stability and future development of the 'partially clear state' of the lakes remains uncertain.

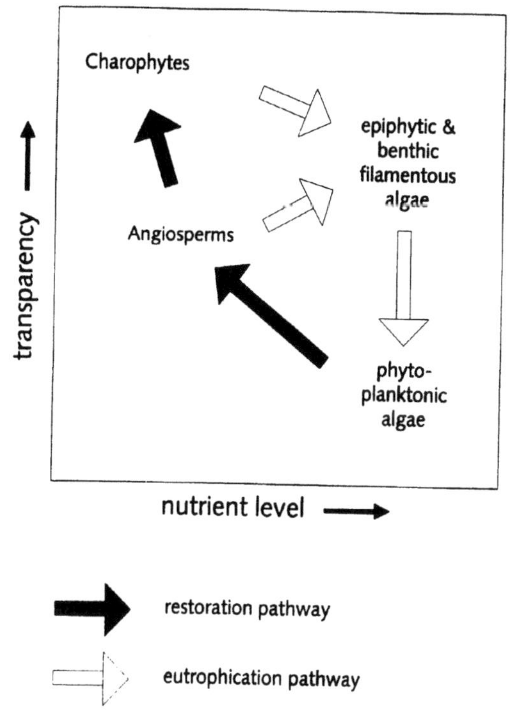

Figure 4. Schematic model of the pathways of submersed vegetation succession and regression in relation to water transparency and nutrient loading of the lakes. Starting from a phytoplankton-dominated system, lowering of the nutrient level and/or increasing water transparency results in the successive dominance of angiosperms and Charophytes. Eutrophication leads to a situation with poor macrophyte growth via enhancement of algal growth.

References

Berger, C., 1987. Habitat and ecology of *Oscillatoria agardhii* Gomont. Van Zee tot Land 55, 233 p. (In Dutch, with English summary).

Blindow, I., 1992. Long- and short term dynamics of submerged macrophytes in two shallow eutrophic lakes. Freshwat. Biol. 28: 15–27.

Blindow, I., G. Andersson, A. Hargeby & S. Johansson, 1993. Long-term pattern of alternative stable states in two shallow eutrophic lakes. Freshwat. Biol. 30: 159–167.

Carpenter, S. R. & D. M. Lodge, 1986. Effects of submersed macrophytes on ecosystem processes. Aquat. Bot. 26: 341–370.

Chambers, P. A. & J. Kalff, 1987. Light and nutrients in the control of aquatic plant community structure, I. In situ experiments. J. Ecol. 75: 611–619.

Hootsmans, M. J. M. & J. E. Vermaat, 1991. Macrophytes, a key to understanding changes caused by eutrophication in shallow freshwater ecosystems. Thesis, Wageningen, 412 pp.

Hosper, S. H., 1989. Biomanipulation, new perspectives for restoring shallow, eutrophic lakes in the Netherlands. Hydrobiol. Bull. 23: 5–10.

Hosper, S. H. & M. L. Meijer, 1986. Control of phosphorus loading and flushing as restoration methods for Lake Veluwe, the Netherlands. Hydrobiol. Bull. 20: 183–194.

James, W. F. & J. W. Barko, 1990. Macrophyte influences on the zonation of sediment accretion and composition in a north-temperate reservoir. Arch. Hydrobiol. 120: 129–142.

McCreary, N. J., 1991. Competition as a mechanism of submersed macrophyte community structure. Aquat. Bot. 41: 177–193.

Meijer, M. L., E. H. Van Nes, E. H. R. R. Lammens, R. D. Gulati, M. P. Grimm, J. Backx, P. Hollebeek, E. M. Blaauw & A. W. Breukelaar, 1994. The consequences of a drastic fish stock reduction in the large and shallow lake Wolderwijd, the Netherlands: can we understand what happened? Hydrobiologia 275/276: 31–42.

Moss, B., 1990. Engineering and biological approaches to the restoration from eutrophication of shallow lakes in which aquatic plant communities are important components. Hydrobiologia 200/201: 367–377.

Scheffer, M., M. R. De Redelijkheid & F. Noppert, 1992. Distribution and dynamics of submerged vegetation in a chain of shallow eutrophic lakes. Aquat. Bot. 42: 199–216.

Scheffer, M., H. Drost, M. R. De Redelijkheid & F. Noppert, 1994. Twenty years of dynamics and distribution of *Potamogeton pectinatus* L. in Lake Veluwe. In W. Van Vierssen et al. (eds), Lake Veluwe, a macrophyte-dominated system under eutrophication stress. Kluwer Acad. Publ: 20–25.

Scheffer, M., S. H. Hosper, M. L. Meijer, B. Moss & E. Jeppesen, 1993. Alternative equilibria in shallow lakes. Trends in Ecology and Evolution 8: 275–279.

Van Wijk, R. J., 1988. Ecological studies on *Potamogeton pectinatus* L. I. General characteristics, biomass production and life cycles under field conditions. Aquat. Bot. 31: 211–258.

Hydrobiologia **340:** 121–125, 1996.
J. M. Caffrey, P. R. F. Barrett, K. J. Murphy & P. M. Wade (eds), Management and Ecology of Freshwater Plants.

Noxious floating weeds of Malaysia

Mashhor Mansor
School of Biological Sciences, Universiti Sains Malaysia (USM) 11800, Minden, Penang, Malaysia

Key words: floating weeds, water hyacinth, aquatic, Malaysia

Abstract

After ten years of field surveys on various water bodies ranging from stagnant water ponds, pools and man-made lakes to flowing waters such as rivers, streams and canals, there is a clear evidence of four problematic weeds in Malaysia. These species are *Eichhornia crassipes*, *Salvinia molesta*, *Lemna perpusilla*, and *Pistia stratiotes*. Among these weeds, *E. crassipes* and *S. molesta* are widely distributed througout Malaysia. *E. crassipes* generally dominates canals and rivers although, recently, this species has spread to man-made lakes. The favourable tropical climate and conducive environmental factors help to trigger the massive growth of these weeds. The high nutrient concentrations, notably phosphate which has a soluble reactive concentration greater than 0.1 mg l^{-1}, initiate a high productivity. Manual control methods are generally used and several herbicides including 2,4-D and glyphosate are frequently employed to eradicate these weeds.

Introduction

Although Gopal (1990) has reviewed the aquatic weeds problems and management in Asia, few references were given on Malaysian aquatic weeds. Except for water hyacinth (*Eichhornia crassipes*), the research works on aquatic weeds, particularly on the floating species, are quite limited. Cheam (1974) was probably one of the first local scientists to highlight the aquatic weed problem in Malaysia.

Gopal (1987) stressed that the most widespread and obnoxious of all aquatic weeds in Asia are the two free floating weeds of south American origin, namely water hyacinth (*Eichhornia crassipes*) and salvinia (*Salvinia molesta*). Ho (1981) noted the problem of water hyacinth in the largest rice granary area (MUDA) in Malaysia since early 1970s. Then Baki (1982) stated the problem of water hyacinth in major drainage and irrigation canals in Malaysia. Mansor et al. (1983) also reported the massive occurrence of water hyacinth in Perak river which is one of the longest rivers in Malaysia.

Perhaps the second ranking floating weed in Malaysia is *Salvinia molesta*. Anwar (1978) has indicated the problem created by the species in ricefield areas. Apart from these two species, *Lemna perpusilla* and *Pistia stratiotes* could be regarded as noxious weeds. Although scattered information is available on various aspects of aquatic weeds in Malaysia, most works particularly on floating weed species are not well documented. Therefore the study was initiated in order to provide current information particularly on the status of floating weed species of Malaysia.

Decription of sites studied

Malaysia lies at the latitude 0 ° 60′–6 ° 40′N and longitude 99 °35′–119 °25′E. The country consists of 14 provincial states; 12 states are in the Peninisular Malaysia (West Malaysia, formerly known as Malaya) and two large states on the island of Borneo (East Malaysia), Figure 1. The country enjoys a warm and sunny equatorial climate (approximately 30 °C) with considerable rainfall throughout the year. Generally the highest values are recorded during the monsoon season. For example, the highest rainfall recorded in Taiping, Perak was 554.7 mm during the month of November 1992. On the other hand, the lowest rainfall recorded was 145.2 mm in March, 1992.

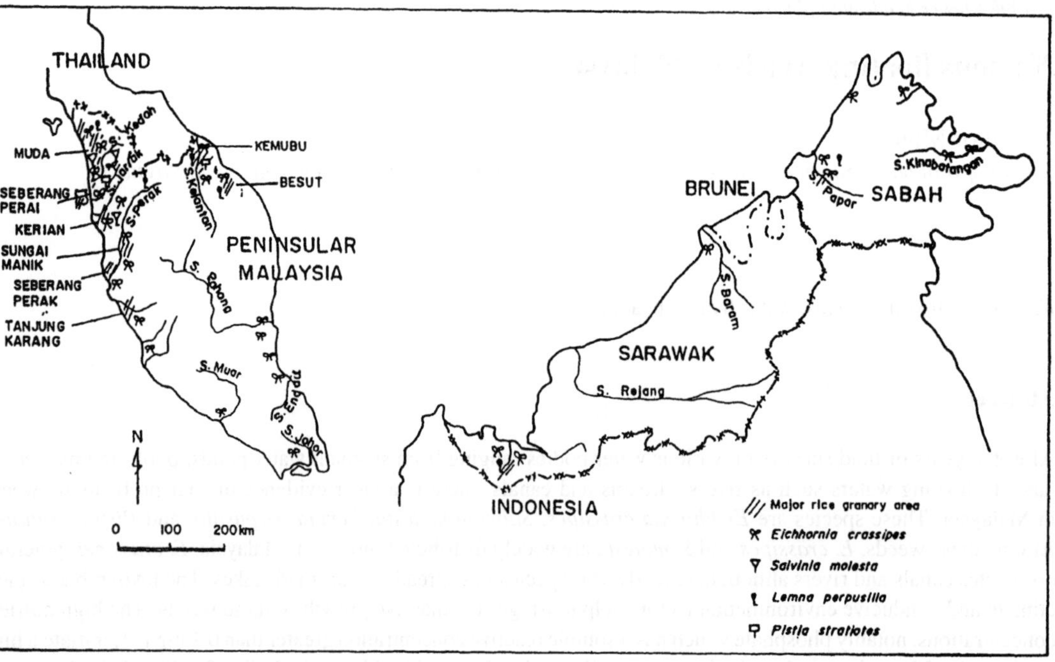

Figure 1. The map of Malaysia showing the major rice granary areas and the distribution of floating weed species.

Due to the relatively high rainfall, there are approximately 330 400 km^2 of land areas in Malaysia which are drained by more than one hundred river systems (Jaafar, 1986). However, most of them are relatively short and small. The largest river is Sungai Rejang in Sarawak with a total catchment area of 51 053 km^2 and a mean discharge of 4033 m^3 sec^{-1}. In addition, there are about 52 impoundments across the rivers which create man-made lakes. Tasik Kenyir with a surface area of 37 000 ha and Tasik Temenggor with a surface area of 15 000 ha are the two largest man-made lakes in Malaysia.

According to Kuan et al. (1991) rice granary areas cover 0.64 million hectares. Apart from these, there is a 1000 km network of drainage and irrigation canals. Monsoon drains and ditches are normally found in urban areas. Mining pools are also man-made lakes which resulted from the tin mining activities. Large fish ponds are gradually on the rise to cater for aquacultural industries.

Most of the aquatic habitats described above are conducive for good growth of floating weeds. Therefore, for the last ten years, the habitats were frequently visited in order to check the invasion of floating weed species.

Materials and methods

The survey of floating weeds was initiated in the year 1984. All the eight major rice granary areas in Peninsular Malaysia, namely Muda, Seberang Perai, Kerian, Seberang Perak, Sungai Manik, Tanjung Karang, Besut and Kemubu, were visited and the floating weed species were recorded. The largest rice granary area in Malaysia is the Muda ricefield which covers 96 000 ha of wet ricefields. Intensive monitoring of floating species was carried out at the area. The rivers like Sungai Perlis, Sungai Arau, Sungai Padang Terap, Sungai Kedah and Sungai Muda were frequently visited. Networks of drainage and irrigation canals generally connected to the river systems were also investigated. The Pedu and Muda dams which irrigated the Muda ricefields were also observed for the presence of aquatic weeds.

Intensive work was conducted on the 28 rivers (Table 1). Five sampling sites were established on each river and lake, based on presence and absence of the four floating species. Therefore, if the species were found in the five sampling sites of each river, then the status was categorized as abundant. For example, the weed species in Sungai Perak was abundant because in

Table 1. The rivers in Malaysia where floating weed have been recorded (E.c. = *E. crassipes,* S.m. = *S. molesta,* L.m. = *Lemna perpusilla* and P.s. = *P. stratiotes*)

No.	River	E. c.	S. m.	L. p.	P. s.	Status
1.	S. Perlis	+	+	+	–	abundant
2.	S. Aran	+	–	–	–	abundant
3.	S. Padang Terap	+	+	+	–	abundant
4.	S. Kedah	+	+	+	+	abundant
5.	S. Muda	+	–	–	–	moderate
6.	S. Jarak	+	–	+	–	abundant
7.	S. Pinang	+	–	–	–	abundant
8.	S. Kerian	+	+	+	–	abundant
9.	S. Kuran	+	+	+	–	abundant
10.	S. Beruas	+	–	–	–	moderate
11.	S. Perak	+	+	+	+	abundant
12.	S. Bernam	+	–	–	–	moderate
13.	S. Tinggi	+	–	–	–	moderate
14.	S. Selangor	+	–	–	–	moderate
15.	S. Kelang	+	–	–	–	moderate
16.	S. Langat	+	–	–	–	moderate
17.	S. Sepang	+	–	–	–	moderate
18.	S. Linggi	+	–	–	–	moderate
19.	S. Melaka	+	–	–	–	moderate
20.	S. Muar	+	–	–	–	moderate
21.	S. Benut	+	–	–	–	moderate
22.	S. Pulai	+	–	–	–	rare
23.	S. Johor	+	–	–	–	rare
24.	S. Pahang	+	–	–	–	moderate
25.	S. Besut	+	–	–	–	moderate
26.	S. Kelantan	+	–	–	–	abundant
27.	S. Sarawak	+	–	–	–	moderate
28.	S. Papar	+	–	–	–	moderate

Table 2. The lakes in Malaysia which have high population densities of foating plants. (E.c. = *E. crassipes,* S.m. = *S. molesta,* L.m. = *Lemna perpusilla* and P.s. = *P. stratiotes*)

No.	Lake	E. c.	S. m.	L. p.	P. s.	Status
1.	Bukit Merah	+	–	–	–	moderate
2.	Chenderoh	+	–	+	+	abundant
3.	Ringlet	+	–	–	–	abundant
4.	Temenggor	–	–	+	–	moderate
5.	Aman	+	–	–	–	abundant

all five sampling sites, the four floating weed species were found. On the other hand, at only one sampling site (Sungai Pulai) *Eichhornia crassipes* was found, therefore it is categorized as rare (Table 1). In addition to these, floating weed assessments were conducted at the five lakes including Lake Chenderoh and Lake Bukit Merah. At these locations, physico-chemical parameters like pH, conductivity, and phosphate (soluble reactive phosphate; SRP) were analysed.

An annual survey of Peninsular Malaysia was organised. During the survey, floating weeds species were recorded. In order to record weed species in East Malaysia, a two- week visit to Sarawak in 1991 and a ten-day visit to Sabah in 1992, were conducted. In addition to this, several Malaysian government agencies like MADA (Muda Agricultural Development Authority), MARDI (Malaysian Agricultural Research Development Institute) and DID (Drainage and Irrigation Deparment) provided some facilities during the field works.

Results

Figure 1 shows the distribution of floating weeds in Malaysia. Apparently most of the location are in the rice granary areas. The worst affected areas were Muda, Sebarang Perai and Kerian. At these places, not only the rivers, but the drainage and irrigation canals are also heavily infested with floating weeds particularly *Eichhornia crassipes*. Unlike the west coast states of Peninsular Malaysia, the east coast states and West Malaysia are relatively low in *Eichhornia crassipes* population.

Table 1 lists the rivers where floating species especially *Eichhornia crassipes* are abundant. Most of the rivers in rice granary areas like Sungai Perlis, Sungai Kedah, Sungai Jarak, Sungai Kerian, Sungai Perak and Sungai Kelantan are heavily infested with *Eichhornia crassipes*. The physico-chemical parameters of Sungai Perak taken at Teluk Intan showed that the pH values ranged from 6.64 to 6.91, conductivity readings were ranging from 30 to 49 μS cm^{-1} and SRP readings were ranging from 0.040 to 0.120 mg l^{-1}. While the physico-chemical parameters of Sungai Jarak taken at Tasik Gelugur showed that the pH values ranged from 6.20 to 7.30, conductivity readings were ranging from 45 to 59 μS cm^{-1} and SRP readings ranged from 0.09 to 0.23 mg l^{-1}.

Table 2 shows the five lakes in Malaysia which harbour relatively high density of floating weeds. Lake Ringlet in Cameron Highland, Pahang, and Lake Chenderoh in Perak have experienced a severe invasion of *Eichhornia crassipes*. Apparently, *Lemna perpusilla* thrives in Lake Temenggor, Perak. Lake Aman which was formerly a mining pool and converted into a recreational lake is situated in Petaling Jaya, one of the fast developing towns in Malaysia. The nutrient concentrations particularly phosphate (SRP) were relatively high (>0.10 mg l^{-1}) for Lake Ringlet, Lake Chenderoh and Lake Aman. On the other hand, the SRP concentrations for Lake Bukit Merah were relatively low (<0.02 mg l^{-1}). Although *Eichhornia crassipes* was recorded at this lake, the population was comparatively low.

Discussion

Soerjani et al. (1975) rated the floating weed species, *Eichhornia crassipes* and *Salvinia molesta* among the most noxious in Southeast Asia. They also singled out *Eichhornia crassipes* as the most problematic aquatic weed in Southeast Asia. In fact works on *Eichhornia crassipes* in Southeast Asia including Malaysia are well documented (Little, 1969; Nguyen, 1973; Soerjani, 1979; Lim & Salleh, 1983).

Apparently, the exact period the floating weed species of neotopic origin reached Malaysia is still unclear. Perhaps at the end of the nineteenth century or later. The species particularly *Echhornia crassipes* became noticeable only in 1970s.

At the time more drainage and irrigation canals were being built, dams being constructed across the rivers and changing of rice practice from the traditional transplanted method to direct seeding method was taking place particularly at Muda rice granary area. The increased usage of fertilizers resulted in the leaching of a certain amount of nutrient into the aquatic ecosystems. According to Carpenter & Adam (1978), and Mansor et al. (1985) a fertilizer that is generally rich in phosphate and nitrate could play an important role in triggering the tremendous growth of floating weeds.

When the massive invasion of floating weeds particularly water hyacinth clogged the irrigation canals and consequently blocked the water supply to the ricefields, there were few viable alternatives except to employ a team of labourers to clear the waterway by manual methods. However, in some areas herbicides such as 2,4-D and glyphosate were used to eradicate the weeds.

Generally the high population of *Salvinia molesta* is recorded in man-made canals and ricefields. Kerian ricefield in particular is one of the worst hit areas. Almost all the rice plots are covered with the species. Seemingly, it was quite successful for the insect to control salvinia in Malaysia. On the other hand, some of the biological control results are not that encouraging. As indicated by Salmah et al. (1991) a biological agent, *Neochatina eichhorniae*, was not doing much damage to water hyacinth.

According to Mansor & Sam (1990) a noticeable feature of ricefields particularly the ones that depend heavily on the use of fertilizer, is the occurrence of massive infestation of minute floating plants which are occasionally mistaken for algal bloom. *Lemna perpusilla* is by far the most widespread minute floating plant in Malaysia. Sometimes the high populations are observed to achieve 100 % coverage of stagnant parts of the canals and ditches. Perhaps the phosphate concentration at the locality is comparatively high and triggers the growth of this floating weed. Due to its small size and free-floating habit it is almost impossible to control the weed either manually or by mechanical methods. On the recent visit to Temenggor Lake (March 1994), *Lemna perpusilla* was observed in some parts of the lake. This is an unhealthy sign for the lake management. Prompt measures should be taken by the authority before it can cause a massive problem.

Among the floating weeds, *Pistia stratiotes* is the least problematic species. In Malaysia, the weed seldom grows well and unlike other floating species, it is generally attacked by several species of local insects. Kasno (1982) reported that several arthropods including *Proxenus hennia* and *Nymphula responsalis* were destroying the population of *Pistia stratiotes* in Malaysia.

Conclusion

By far the most noxious floating weed species in Malaysia is *Eichhornia crassipes*. The worst affected areas were rivers and canals in rice granary areas. Apparently not a single method could be considered effective enough to control this floating weed. Manual method is generally used. *Salvinia molesta* and *Lemna perpusilla* are the two floating species which could also be categorized as problematic weeds. Unlike other countries, *Pistia stratiotes* has not become much of a problem in Malaysia. Apparently its population is being controlled by various species of native insects.

Acknowledgments

I wish to thank Dr J. O. Rieley from the University of Nottingham for encouraging me to write this manuscript. The Universiti Sains Malaysia (USM) is gratefully acknowledged for providing facilities and research funds (R & D). To all my research students for helping me to complete the work.

References

Anwar, A. I., 1978. Some recent studies on Salvinia, an aquatic weed in rice. Information paper no. 6. Rice Research Branch, MARDI. 10 pp.

Baki, B. B., 1982. Aquatic weeds in major drainage and irrigation canals with special reference to water hyacinth in Malaysia. Ist Rice Advisory committee meeting, MARDI, Serdang, Selangor. 12 pp.

Carpenter, S. R. & M. S. Adam, 1978. Macrophyte control by harvesting and herbicides: implications for phosphorus cycling in Lake Wingra. J. Aquat. Plant Mgmt 16: 20–23.

Cheam, A. H., 1974. Current status of aquatic weed problems in Peninsular Malaysia. Southeast Asia Workshop on Aquatic Weeds, 25–29 June 1974, Malang, Indonesia, 18 pp.

Gopal, B., 1987. Water hyacinth. Aquatic Plant Studies 1. Elsevier, Amsterdam, 471 pp.

Gopal, B., 1990. Aquatic weed problems and mangement in Asia. In A. H. Pieterse & K. J. Murphy (ed.), Aquatic Weeds, the ecology and management of nuisance aquatic vegetation. Oxford Science Publications. 318–354.

Ho, N. K., 1981. A brief note on water hyacinth in MUDA area. Lembaga Kemajuan Pertanian Muda report, Alor Setar, Malaysia, 6 pp.

Jaafar, A. B., 1986. Quality and control of river discharges into Malaysian Coastal Waters. UNEP/ CDBSEA Singapore, 7 pp.

Kasno, 1982. A study on *Proxenus hennia* and *Nymphula responsalis* as potential biological agents of water lettuce (*Pistia stratiotes*). MSc thesis, Universiti Sains Malaysia (USM), 150 pp.

Kuan, C. Y., L. S. Ann, A. I. Anwar, T. Leong, C. G. Fee & K. Hashim, 1991. Crop loss by weeds in Malaysia. Proceedings of the third Tropical Weed Science Conference (ed. Lee, S. A. & K. F. Kon). Kuala Lumpur, Malaysia: 1–19.

Lim, W. C. & A. Salleh, 1983. *Eichhornia crassipes*: a serious weed in the Muda Irrigation Area. Proceeding of Weed Science in the Tropic Rajan A. & M. Rosli (eds): 12–18.

Little, E. C. S., 1969. The Floating Island of Rawa Pening.PAN, 15 pp.

Nguyen, V. V., 1973. Report on the aquatic weed problem of the Brantas River multipurpose project. Biotrop /WR/73/654. 10 pp.

Mansor, M., A. Ismail & M. N. Nordin, 1983. Factors governing the distribution of *Eichhornia crassipes* along Perak river. Proceeding of Weed Science in the Tropics Rajan, A. & M. Rosli (eds). Universiti Pertanian Malaysia: 36–41.

Mansor, M., A. P. Nordin & S. Kimi, 1985. Phosphate and the distribution of aquatic weeds in Northern Malaysia. Proceeding Asian-Pacific Weed Science Society; Tenth Conference. Chang Mai, Thailand: 438–451.

Mansor, M. & S. K. Sam, 1990. The competition between three species of small-leaf floating plants in rice growing areas of northern Peninsular Malaysia. Proceeding 3rd International Conference on Plant Protection in the Tropics K. Y. Lum (ed.): 242–246.

Salmah, M. R., M. Mansor & A. B. Ahmad, 1991. A prelimnary study of the distribution of *Neochatina eichhorniae*; a possible biological agent for water hyacinth hyacinth in Kerian District, Perak. Proceeding Asas dan Gunaan dalam Biologi A. A. Bidin (ed.). UKM: 166–171.

Soerjani, M., J. V. Pancho & N. V. Voung, 1975. Aquatic weed problems and control in Southeast Asia. Hyacinth Control Journal. 13: 2–3.

Soerjani, M., 1979. Recent trend in aquatic weed management in Indonesia. Proceeding Asian-Pacific Weed Science Society; Seventh Conference. Supp. Vol. 12 pp.

Hydrobiologia **340**: 127–135, 1996.
J. M. Caffrey, P. R. F. Barrett, K. J. Murphy & P. M. Wade (eds), Management and Ecology of Freshwater Plants.

Past and present distribution of stoneworts (Characeae) in The Netherlands

Jan Simons[1] & Emile Nat[2]
[1] *Department of Ecology and Ecotoxicology, Vrije Universiteit, De Boelelaan 1087, 1081 HV Amsterdam, The Netherlands*
[2] *Krayenhoffstraat 223, 1018 RJ Amsterdam, The Netherlands*

Key words: Characeae, The Netherlands, distribution, habitat factors, field ecology

Abstract

In The Netherlands 21 Characeae species occur. *Chara vulgaris, C. globularis,* and *Nitella flexilis* are common and widespread, occurring in at least 225 of the total of 1677 atlasblocks (5 × 5 km^2). *Chara aspera, C. contraria, C. major, Nitella mucronata* and *Tolypella prolifera,* occurring in at least 30 atlasblocks, are denoted as 'not uncommon'. Thirteen species are rare and recorded in less than 30 atlasblocks. Regarding the common species, the number of records significantly increased in recent time, presumably thanks to the recently increased flora inventory activities. The other species remained nevertheless rare, with a tendency of decrease. In the first half of this century mass occurrence of Characeae was a rather common phenomenon, especially in shallow lakes in the central western part of the country. Rich occurrence of Characeae is now restricted to localities with clear water which is low in nutrients. Important habitats are dune waters, polder ditches, shallow lakes and moorland pools. Physico-chemical factors in water and sediment, such as nutrients, salinity, $CaCO_3$, alkalinity, and in the sediment also the redox-value and organic matter, are suggested as important parameters for species composition. In recent years, at several sites where water quality has improved by restoration measures, Characeae reappeared or increased in species and biomass.

Introduction

In The Netherlands documented observations of Characeae date from about 1850 (Van den Bosch, 1853).

Rich occurrence and high biomasses are restricted to unpolluted sites with clear water and low nutrient status. Important habitats in The Netherlands are mostly shallow waters (depth < 4 m) dune pools, polder ditches, peat lakes in the central, western, and northern parts of the country, and moorland pools in the pleistocenic southern and eastern parts.

Recently, the interest of water management institutions for this group of water plants has grown. One of the reasons is that at many sites where water quality improved by restoration measures, Characeae rapidly reappeared or increased in species number and biomass (Simons et al., 1994). Apart from their well known value as indicators of good water quality (Krause, 1981), once established and stable Charophyte communities seem to improve and maintain good water quality systems with rich biotic components (Blindow, 1991).

The aim of this paper is to give a review of the pattern of distribution in space and time, in correlation with presumed important environmental parameters.

Past and present distribution

Distribution in the Netherlands relative to Europe

The 21 Dutch species are arranged with indications of frequency and geographic position in relation to Europe in Table 1. The total number of 21 species is comparable with the 25 species in Britain and Ireland (Moore, 1986), 20 species for Norway (Langangen, 1974), 23 for Belgium (Compère, 1992), 27 for Switzerland (Auderset Joye, 1993) but less than 32

Table 1. Recent occurrence and distribution of Characeae in The Netherlands.

	AFK 1975-1993	Area	Distribution over habitat groups 1a	1b	2	3	4	5	6	7
Chara aculeolata Kützing	7	c				+				
Chara aspera Detharding ex Willdenow	26	c	+	+,D		+	+,D			+
Chara baltica Bruzelius	8	m		+						
Chara canescens Desvaux & Loiseleur	11	s			+					
Chara connivens Salzmann ex A. Braun	7	c	+		+,D	+,D				
Chara contraria A. Braun ex Kützing	32	c	+	+		+	+,D	+	+	
Chara globularis Thuillier	246	c	+	+		+	+	+,D	+,D	+
var. virgata (Kützing) R.D. Wood	76	c		+		+	+	+	+	+
Chara major (Hartman) Hy*	56	c	+,D	+		+				
Chara vulgaris L.	307	c	+	+		+	+	+,D	+,D	
var. hispidula (A. Braun) J.A. Moore	69	c	+	+	+					
var. longibracteata (Kützing) J. Groves & Bullock-Webster	225	c	+	+	+	+	+	+	+	
Nitella capillaris (Krocker) J. Groves & Bullock-Webster	69	s	+			+	+	+		+
Nitella flexilis (L.) Agardh	225	c		+		+		+,D	+	+
Nitella hyalina (DC.) Agardh	22	c				+		+		
Nitella mucronata (A. Braun) Miquel	116	s				+		+,D	+	
Nitella opaca Agardh	3	c		+		+		+		
Nitella syncarpa (Thuillier) Chevalier	41	m						+	+	
Nitella translucens (Persoon) Agardh	8	c								+
Nitellopsis obtusa (Desvaux) J. Groves	10	s				+,D	+,D			
Tolypella glomerata (Desvaux) Leonhardi	5	c	+	+		+				
Tolypella intricata (Trentepohl ex Roth) Leonhardi	29	c		+				+	+	+
Tolypella prolifera (Ziz ex A. Braun) Leonhardi	43	m							+,D	

* Also known as Chara hispida var. major (Hartm.) R.D. Wood. Recently (1995), a new species: *Nitella tenuissima* (Desv.) Kütz. was recorded in a newly created wetland pond 'Dullaert' near Waalwijk, province N-Brabant. This renders the species number to 21.

AFK:

The number of atlasblocks (5 × 5 km) in which the species were recorded.

Area: (based on Van Raam & Maier, 1993)

c=central: the position of The Netherlands is completely inside the area and not at the border of the area.
s=subcentral: the position of The Netherlands is completely inside the area and more or less at the border.
m=marginal: the border of the uninterrupted area runs through The Netherlands or the finds in The Netherlands are completely outside the uninterrupted area but at a distance of less than 100 km.

Distribution over habitat groups:

1a: Dune waters in lime-rich area. Ib: Dune waters in the lime-poor Wadden area.
2: Coastal waters with brackish influence. 3: Hardwater shallow lakes.
4: Hardwater shallow lakes in the IJsselmeer area. 5: Hardwater polder ditches.
6: Clayish habitats in the river area. 7: Softwater moorland pools and ditches in the pleistocene area.
+: occurring. +,D: locally dominating the Charophyte vegetation.

Table 2. Some waterparameters of 17 Characeae.

	pH	Ca (mg/l)	Cl (mg/l)	SO_4 (mg/l)	Fe (mg/l)	total-P (mg/l)	total-N (mg/l)	NH_4-N (mg/l)	EC (mS/m)
Chara aspera	7.1–8.1–9.0	45.9–64–119	50–130–226	53–91–178	0.1–0.1–0.6	0.02–0.06–0.21	0.6–1.2–3.2	0.0–0.1–0.8	3.4–6.7–16
C. baltica	7.3–7.7–8.3	–	1800–4100–5770	–	–	0.38–0.95–2.0	2.2–3.9–5.6	0.0–0.1–1.9	–
C. canescens	7.1–7.6–8.3	–	91–104–3300	–	–	0.03–0.05–0.12	1.1–1.6–3.0	0.0–0.03–1.08	–
C. connivens	7.0–8.0–9.0	45.6–62.9–115	50–134.5–1070	41–79–134	0.1–0.1–0.6	0.02–0.06–0.21	0.6–1.2–3.0	0.0–0.1–0.8	3.4–7.2–30.7
C. contraria	6.7–8.1–9.7	40–63.6–100	39–130–700	9.5–66–134	0.03–0.2–0.88	0.02–0.08–0.49	0.6–1.3–6.3	0.03–0.1–0.8	2.6–6–22.5
C. globularis	5.9–7.9–10.1	0–61.3–534	12–146–1922	2.9–72–169	0.006–0.17–5.37	0.0–0.13–11.0	0.1–2.0–56.2	0.0–0.1–30.8	2.5–6.3–16
C. major	6.7–8.1–9.0	6.4–65–1130	30–115–1922	1–66.2–272	0.04–0.17–1.35	0.01–0.06–0.48	0.01–0.1–1.3	3.4–7.2–16	
C. vulgaris	5.6–7.4–10.1	9.6–75–534	9–98–3020	1–69.2–900	0.02–0.52–6.7	0.01–0.2–3.76	0.2–2.3–20.8	0.0–0.2–26.0	1.9–6.3–34.9
Nitella flexilis	5.6–7.7–9.0	12–57.3–94	14–61–226	3–75–210	0.07–0.2–3.8	0.0–0.08–1.4	0.1–1.3–23.5	0.0–0.2–5.6	2.4–5.6–10.8
N. hyalina	7.3–8.2–8.8	34–66.4–75.6	38–138–226	46–84–134	0.1–0.2–0.6	0.02–0.07–0.2	0.6–1.3–3.2	0.1–0.1–0.3	2.8–7.3–10.8
N. mucronata	6.6–8.0–9.0	28–64–75.6	32–129–226	39–77.5–137	0.1–0.1–2.2	0.02–0.08–1.3	0.6–1.4–17.6	0.01–0.14–1.0	2.5–7.1–16
N. opaca	7.4–8.1–9.0	45.9–62.9–78	76–153–226	53–91–134	0.1–0.1–0.6	0.02–0.04–0.12	0.6–1.1–2.8	0.1–0.2–0.3	340–760–10.8
N. translucens	5.8–7.4	4–17	12–24	–	–	0.013–0.189	–	0.04–0.09	0.56–2.3
Nitellopsis obtusa	6.4–8.1–9.0	40–64–95	40–137–688	39–71–134	0.02–0.1–0.6	0.02–0.07–0.21	0.6–1.2–3.2	0.0–0.1–0.84	3.2–6.7–22.5
Tolypella glomerata	7.1–8.1–9.0	48.9–64–925	83–131.5–189.6	75–94–118	0.1–0.1–0.3	0.02–0.05–0.21	0.7–1.2–3.0	0.0–0.15–0.8	4.1–7.5–1.6
T. intricata	6.2–6.9–7.6	–	22–37–41	0.5–0.5–0.6	0.11–0.98	0.05–0.08–0.28	–	0.16–0.4–3.0	–
T. prolifera	7.1–8.6	29.7–82	55.1–370	11.5–63.9	–	–	–	0.08–0.09	4.02–9.08

The figures represent the minimum level, median (if given) and the maximum level.

species recorded in Germany (Krause, 1984), and 29 for Sweden (Blindow, 1994). From other European countries no such complete reviews are available. The total number of species in Europe is 42.

The rarity of species

In Table 1 a quantification is given of the present frequency of records. *Chara vulgaris, C. globularis* and *Nitella flexilis* are most common and occur in at least 225 of the total number of 1677 atlasblocks (5 × 5 km^2). Each atlasblock is numbered and has a fixed position on the geographical map of The Netherlands. The past and present distribution of *Chara vulgaris* (Figure 1a) shows that the number of recent records (307) is far higher than in subrecent (64) and past (14) times. The group of *Chara aspera, C. contraria, C. major, Nitella mucronata* and *Tolypella prolifera* occurs in 26–55 atlasblocks, and could be denoted as 'not uncommon'. As an example, *Chara major* has been chosen (Figure 1b). The remaining 13 species can be denoted as rare, occurring only in 3–10 atlasblocks. Figure 1c shows the past and present distribution of *Nitella hyalina* in The Netherlands. *Chara tomentosa* has disappeared in the period before 1930.

Habitat distribution

Important habitats are dune waters, polder ditches, shallow peat lakes and moorland pools. Dune waters in the lime-rich southern area are often dominated by *Chara major,* and in the lime-poor northern Wadden area also by *C. aspera.* In polder ditches at places relatively free from agricultural contamination, *Chara vulgaris, C. globularis, Nitella flexilis* and *N. mucronata* are rather common. The rare species *Chara baltica* and *C. canescens* are restricted to brackish ditches and pools in the coastal area. In the clayish area along the rivers Rhine and Meuse, also some stoneworts can be found, and the relatively large *Tolypella prolifera* is more or less restricted to this area. Shallow hard water lakes in the central western part of The Netherlands can be rich in species, often in high biomasses. Examples are the lake Naardermeer with 12 species and the slightly brackish lake Botshol with 9 species. In these lakes the rare *Nitellopsis obtusa* may dominate at several sites. In the large lakes Veluwemeer, Wolderwijd, Gouwzee, rich charophyte vegetations developed after restore measures at the end of the 80's with *Chara contraria, C. aspera,* and *Nitellopsis obtusa* (Coops & Doef, 1996). In pools with macrophytes of the Littorellion community, the rare *Nitella translucens* may occur, besides the more common *Chara globularis* and *Nitella flexilis.* In Table 1 the distribution of the Characeae species over seven main habitat groups is indicated. Especially in some lakes as Botshol and Naardermeer, and locally in Ijsselmeerlakes as Veluwemeer dense stands with high biomass can be found. Small dune pools can wholly been filled by *Chara major* as in Oostvoorne and Egmond-Castricum (Simons, 1987).

Environmental parameters of water and sediment

Water quality

A survey of data within the last 10 years about nine water quality parameters at different sites of 17 species of Characeae (Table 2) will be evaluated.

pH: most species are alkaline or circumneutral, with median values between 7.4 and 8.2. *Tolypella intricata, Nitella translucens, N. flexilis, Chara vulgaris* and *C. globularis* can occur at relatively low pH values according to their minimum (5.6–6.2) or median values.

Ca: nearly all species occur at values above 60 mg l^{-1}, except *Nitella translucens* which occurs in the low range of 4.0–17.0 mg l^{-1}. Also *Nitella flexilis* has a relatively low median value of 57.3 mg l^{-1}.

Cl: 13 species, among which all Chara species are euryhaline and have median values above 100 (98) mg l^{-1}. *Chara baltica* and *C. canescens* have the highest salinity tolerance. Also the species *Chara connivens, C. globularis, C. major, C. vulgaris* and *Tolypella glomerata* have rather high salinity maxima, which agrees with their regular occurrence in coastal habitats. The remaining species of *Nitella* and *Tolypella*, except *Nitellopsis obtusa,* have a low salinity tolerance. *Nitella translucens* and *Tolypella intricata* with maximum values below 50 mg l^{-1} could be called halophobic.

SO_4: all median values are in a range of moderate to low concentrations (Lyon & Roelofs, 1986).

Fe: except a relatively high median value of *Chara vulgaris,* all values are at a low level (Lyon & Roelofs, 1986).

Total-P: *Chora globularis* and *C. vulgaris* have the highest median values of resp. 0.13 and 0.20 mg l^{-1} Median values of the other species are between 0.05 and 0.1. Total-N: *Chara globularis, C. vulgaris, Nitella*

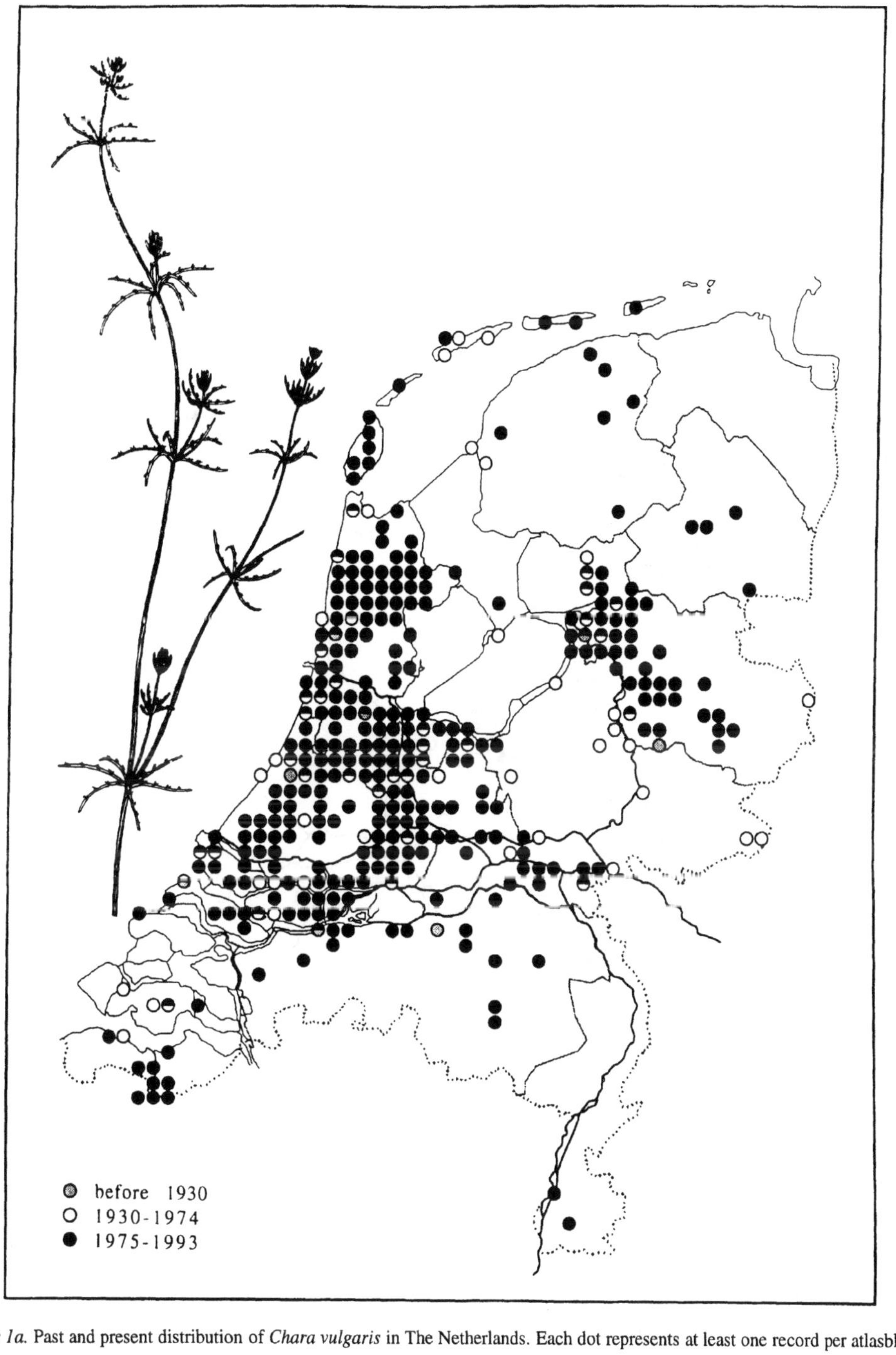

Figure 1a. Past and present distribution of *Chara vulgaris* in The Netherlands. Each dot represents at least one record per atlasblock.

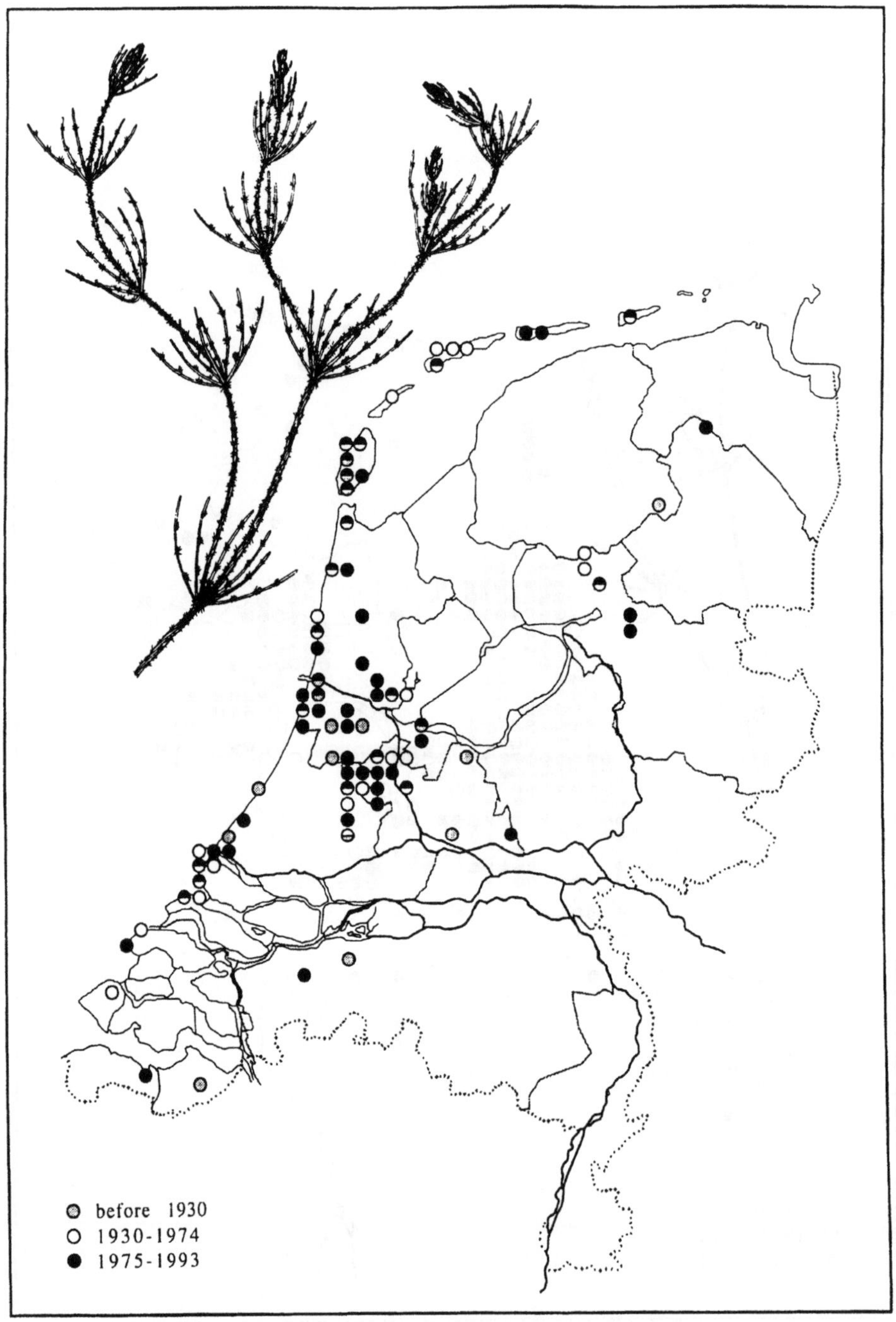

Figure 1b. Past and present distribution of *Chara major* in The Netherlands.

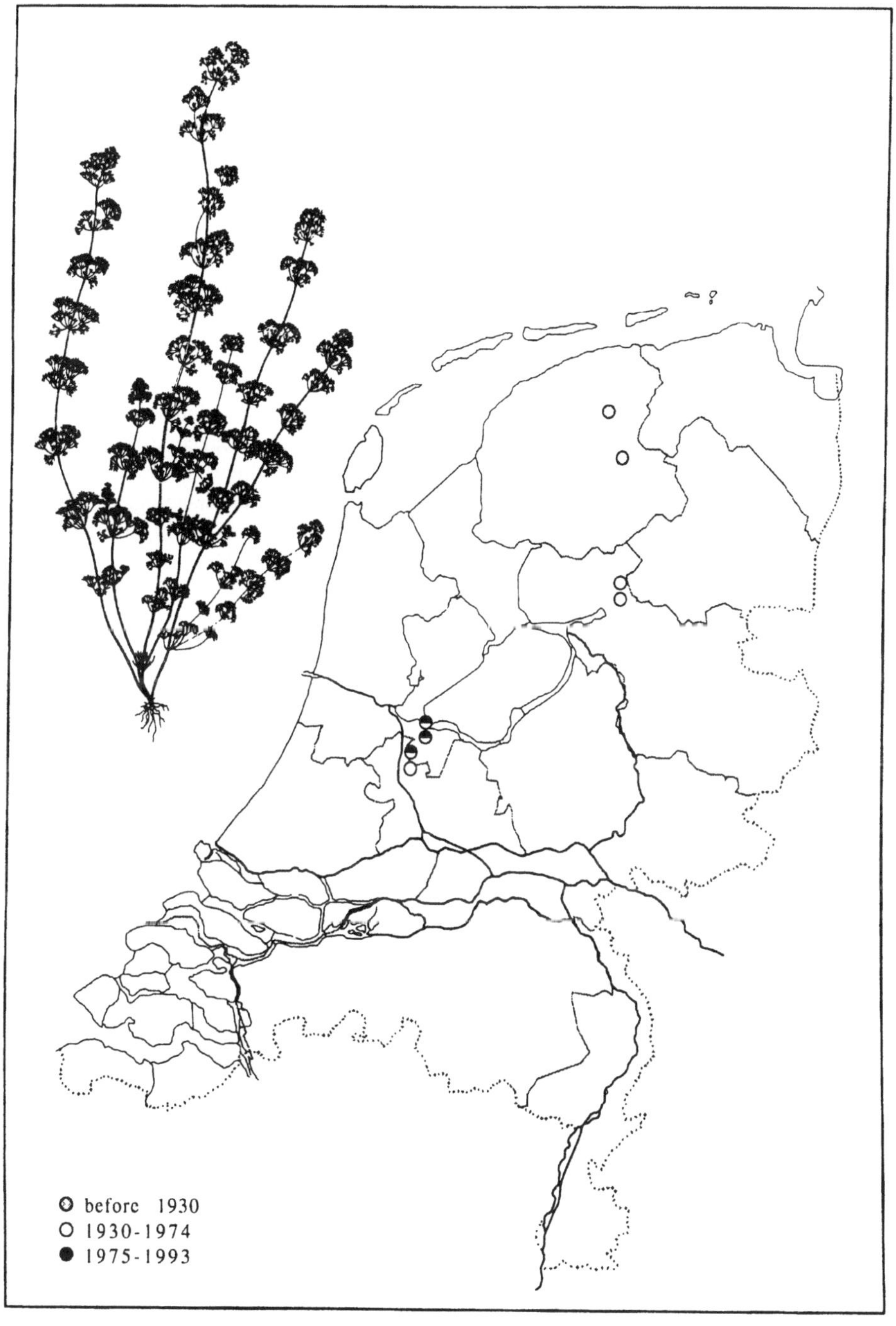

Figure 1c. Past and present distribution of *Nitella hyalina* in The Netherlands.

flexilis, and *N. mucronata* have the highest median and maximum values. Median values of most other species are between 1.0 and 1.6.

NH_4: all 17 species occur mostly at low values, except *Chara globularis* and *C. vulgaris* which occur at maximum values of 30.8 and 26.0 mg l^{-1}, respectively.

EC: Except *Nitella translucens,* all species are poly-ionic according to the scheme of Olsen (1950).

Discussion

Distribution and field ecology

Most Dutch inland waters are stagnant and shallow and are more or less strongly influenced by an input of Rhine water and many kinds of human contamination. Despite this influence, 21 species of Characeae, among which some are very rare, have maintained up to recent time, but mainly restricted to isolated areas.

Dune pools and polder ditches appear to be important habitats for Characeae. Nature managers in the dune area are aware of this, and one tries to preserve such waters by keeping them open, or even create new pools in which, in short time, a pioneering charophyte vegetation can develop, often with *Chara major* or *C. aspera* as dominant species. Polder ditches, widely and densely distributed in the alluvial peaty or clayish western area of The Netherlands, are strongly influenced by agricultural eutrophication. Yet some Characeae species, especially *Chara vulgaris, C. globularis, Nitella flexilis, N. mucronata,* are locally richly represented at sites relatively free from agricultural influences. This group is relatively eutrophication resistant (Hutchinson, 1975; Moore, 1986). But also some rare species can be found in ditches. Examples are *Nitella opaca, N. syncarpa, N. translucens, Tolypella intricata* and *T. prolifera.* The conditions for the occurrence of such species are not well known and more research is needed in this field.

Apparently, the many small and shallow water bodies with still and mostly hard waters, and presumably also features of the sandy, clayish or peaty substrates, are good habitats for Characeae. Moreover, the dynamic human influence may be useful in keeping or creating new and open habitats which are favorable to pioneering stoneworts.

Environmental parameters

The general statement of clear, nutrient poor, mostly alkaline hardwaters and calcium-carbonate rich sediments as prerequisits for Characeae occurrence (Hutchinson, 1975) is corrrect. Nevertheless there is hardly any knowledge about the exact role of each factor for individual species. Important species differentiating factors are salinity and factors related to eutrophication tolerance. Also, alkalinity plays an important role. Species occurring in soft waters, such as *Chara globularis* and *Nitella translucens,* are relatively good CO_2-users, whereas the strictly hardwater species, like *Chara major*, will be efficient HCO_3-users (Hutchinson, 1975).

Most knowledge about habitat factors concern parameters of the water column. The role of the sediment structure and chemical composition ought to be taken into account (Andrews, 1987; Hutchinson, 1975), as is the case with rooting phanerogamic water plants (Barko & Smart, 1981; Barko et al., 1991). For four species *(Chara globularis, C. major, C. vulgaris, Nitella flexilis)* a reasonable number of sediment parameters is available (Lyon & Roelofs, 1986). Mostly these values are in the same range as in the water column. All four species occur in moderately reductive (0 — –100 mvolt, *Nitella flexilis)* to reductive soils (–100 — –175 mvolt). The sediment parameters, point at a possible important role of redox state in relation to organic matter, availability of phosphorus, iron, and maybe other substances. Apart from *Chara major,* the sediments of the other three more common species are not very low in phosphate and iron. The role of iron and calcium-carbonate in the sediment may be important, among others in the sense of binding phosphorus input from the water column (Kufel & Ozimek, 1994). Also the competitive role of sulphur for iron may be important.

Rarity of species and recent developments

Despite the small size of The Netherlands, the number of 21 Characeae species is considerable, and not far lower than much larger countries as Britain, Norway, Sweden and Germany.

It can be stated that at least 13 of the 21 species are rare to very rare. It would be useful to incorporate such species in a red list system of endangered species and to preserve their habitats, which is not yet the case in the Netherlands. In Germany their exists a red list containing 28 of the 34 taxa of Characeae (Krause, 1984). The recent strong increase in records of the common species *Chara vulgaris, C. globularis, Nitella flexilis* is a consequence of the recently increased flora inventory activities of water board agencies and provincial institutions. Concerning the rare species, the observed tendency of decrease or in any case no increase, will be realistic, despite the risk of overlooking or undiscovering. Another point is that our data concern only records in one per surface area ('atlasblock') without taking into account the number of records per 'atlasblock' or biomass. Apart from these aspects, there are several indications of a general decrease, at least in biomass, in recent time, e.g. the Loosdrecht, Botshol and Nieuwkoop lakes. There is even a statement that stonewort masses were used as manure in cattle farming in the first half of this century. Recent cases of return of charophyte vegetation in lakes where restoration measures were taken (Coops & Doef, 1996; Simons et al., 1994), allow an optimistic view on future developments regarding this interesting and very old group of macrophytic algae.

Acknowledgements

The authors greatly acknowledge the cooperation with and financial support from the 'Rijksinstituut voor Integraal Zoetwaterbeheer en Afvalwaterbehandeling' (RIZA), Lelystad, especially Mr Hugo Coops, Mr Roel Doef, and Mrs Marita Cals. We express our thanks to many institutions and persons who made their Characeae data available for wider use, especially Mr Jan Roelofs for sediment data. Finally we thank Mrs Désirée Hoonhout for assistance with the make up of the text.

References

Andrews, M., 1987. Phosphate uptake by the component parts of *Chara hispida*. Br. Phycol. J. 22: 49–53.

Auderset Joye, D., 1993. Contribution à l'écologie des Characées de Suisse. Thesis, Université de Genève.

Barko, J. W. & R. M. Smart, 1981. Sediment based nutrition of submerged macrophytes. Aquat. Bot. 10: 339–352.

Barko, J. W., D. Gunnison & S. R. Carpenter, 1991. Sediment interactions with submerged macrophyte growth and community dynamics. Aquat. Bot. 41: 41–65.

Blindow, I., 1991. Interactions between submerged macrophytes and microalgae in shallow lakes. Doctoral dissertation, Lund University, 112 pp.

Blindow, I., 1994. Sällsynta och hotade kransalger i Sverige. (Rare and threatened charophytes in Sweden). Svensk Bot. Tidskr. 88: 65–73.

Coops, H. & R. W. Doef, 1996. Submerged vegetation development in two shallow, eutrophic lakes. Hydrobiologia 340 (Dev. Hydrobiol. 120): 115–120.

Compère, P., 1992. Charophytes - Flore pratique des algues d'eau douce de Belgique, 4. Jardin Botanique National de Belgique, Meise.

Hutchinson, G. E., 1975. A treatise on limnology, 3. Limnological Botany. J. Wiley & Sons, New York, 660 pp.

Krause, W., 1981. Characeen als Bioindikatoren für den Gewässerzustand. Arch. Hydrobiol. Suppl. 35: 305–317.

Krause, W., 1984. Rote Liste der Armleuchteralgen (Characeen). In: Blab et al.: Rote Liste der gefährdeten Tiere und Pflanzen in der Bundesrepublik Deutschland, 4 Aufl.: 184–187, Kilda, Greven.

Kufel, L. & T. Ozimek, 1994. Can Chara control phosphorus cycling in Lake Luknajno (Poland). Hydrobiologia 275/276: 277–283.

Langangen, A., 1974. Ecology and distribution of Norwegian charophytes. Norw. J. Bot. 21: 31–52.

Lyon, M. J. H. de & J. G. M. Roelofs, 1986. Waterplanten in relatie tot waterkwaliteit en bodemgesteldheid I, II. Catholic University, Nijmegen, 230 pp. (in Dutch).

Moore, J. A., 1986. Charophytes of Great Britain and Ireland. Botanical Society of the British Isles, London, 140 pp.

Olsen, S., 1950. Aquatic plants and hydrospheric factors. Svensk Bot. Tidskr. 44:1–34 & 332–373.

Simons, J., 1987. *Spirogyra* species and accompanying algae from dune waters in The Netherlands. Acta Bot. Neerl. 36: 13–31.

Simons, J., M. Ohm, R. Daalder, P. Boers & W. Rip, 1994. Restoration of Botshol (The Netherlands) by reduction of external nutrient load: recovery of a characean community, dominated by *Chara connivens*. Hydrobiologia 275/276: 243–253.

Van den Bosch, R. B., 1853. Characeae. Prodromus Florae Batavae 11(2): 186–189, Leiden. (in Latin).

Van Raam, J. C. & E. X. Maier, 1993. Overzicht van de Nederlandse kranswieren. Gorteria 18: 111–116 (in Dutch).

Hydrobiologia **340:** 137–140, 1996.
J. M. Caffrey, P. R. F. Barrett, K. J. Murphy & P. M. Wade (eds), Management and Ecology of Freshwater Plants.

Macrophytes and flood plain water dynamics in the River Danube ecotone research region (Austria)

G. A. Janauer & G. Kum
Institute of Plant Physiology, University of Vienna, Althanstrasse 14, A-1090 Vienna, Austria

Key words: ecotones, flow, macrophytes, mapping, species distribution, conservation, management

Abstract

In the ecotone research region of the Danube in Austria (Man-and-the-Biosphere (MaB)- project 5/21, Austrian Academy of Science) the macrophytes are one of the most important groups of organisms investigated. The species composition and the plant mass in hydrologically dynamic, and in predominantly stagnant sections of a system of lateral branches and sloughs were studied. This study showed that areas protected from frequent disturbance by floods had a greater number of species and higher biomass of aquatic macrophytes. Some species were shown to be evenly distributed throughout the branch system, whereas other species tended to form rare, but large singular patches. A set of new data elaboration techniques enabled us to describe the distribution pattern of the aquatic vegetation in this large branch system of the River Danube.

Introduction

The Austrian UNESCO 'Man-and-the-Biosphere' Project 'Aquatic- terrestrial Ecotones' aims at describing characteristic ecological processes to provide a scientific basis for state-of-the-art management of the fluvial corridor of the Danube river (Janauer, 1993). Hydraulics of riverine habitats have been proved by statistical methods to be among the most essential abiotic 'steering' factors (Janauer, 1994; Janauer & Hary, 1989) in this type of landscape.

Macrophytes are well studied in the ecotone project in Austria. They are important for conservational and ecological reasons: Many species are rather rare because of the lack of habitat and within the plant stands a highly diverse biocoenosis consisting of algae, invertebrates and different age-stages of fish is often found (Jacobsen & Sand-Jensen, 1994; Philips et al., 1993). Macrophytes even may indicate quality conditions of the water (Dennison et al., 1993).

In this study the aquatic vegetation of flood-prone and sheltered sections of river branch systems were compared. The results will later serve in the discrimination of backwater types essential for the description of characteristic ecotones of the fluvial corridor of the river Danube.

Study site

The river Danube, second largest river in Europe, stretches from the Schwarzwald (Germany) to the Black Sea (approx. 2850 km). After regulation (1870 to 1875) this originally braided mountain river was, in Austria, transformed to a single channel, serving shipping and flood protection. Most of the lateral stream branches were separated at the inflow by stone dams. Thus the hydrological dynamics of the floodplain were considerably reduced.

We studied the surface water system on the right bank of the Danube between Haslau (river km 1902) and Regelsbrunn (river km 1895). Different flood dynamics were distinguished using maps of flooding events (Danube Hydro Austria 1988, unpublished) and a characterization by Hein (1993). Moreover, we personally registered the flooding of individual water bodies *in situ* during major floods.

Flood-prone branches are up to 200 m wide, and more than two meters deep in the central parts. The

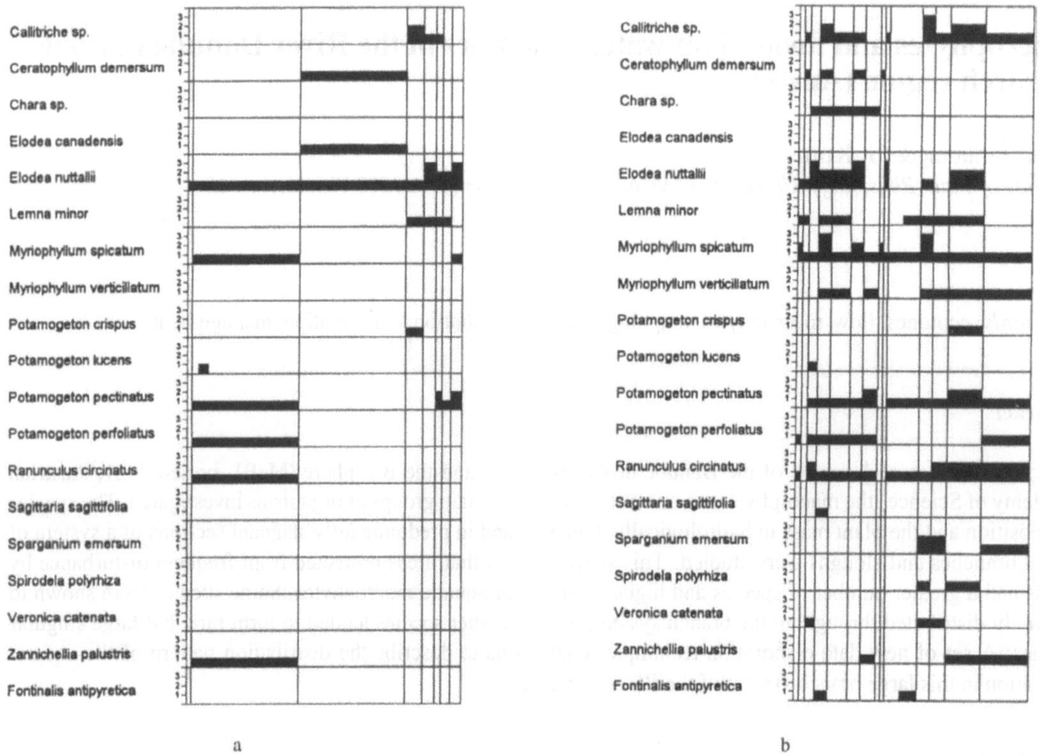

Figure 1. Plant distribution diagram: The length of the x-axis is proportional to the true length of the branch system. The y-axis represents the amount of plant mass. (a) Hydrologically dynamic sections, (b) Sheltered sections of the branch system.

sheltered type is small and frequently of a round shape, with a maximum depth of approximately 1.5 meters.

During low and mean discharges in the main river the lateral branch system mainly consists of still water bodies (Hein 1993). Flow >2 cm s^{-1} occurs usually only at the confluence with the main channel and at passages through small transverse dams which ensure a minimum water level during low run off. This situation changes when discharges in the Danube reach 3500 m^3 s^{-1} or more. The branch systems in the study area are then flooded. In this case flow velocities reach values high enough to influence species distribution.

For this study a contiguous branch system consisting of eight flood prone sections and 18 sheltered sections was selected. The system is part of one of the most species-rich habitats of aquatic vegetation in Austria.

Methods

The amount of vegetation was estimated by section mapping according to Kohler (1978). This is a standard method in central Europe using tested and reproducible observational estimates (Melzer, 1990; Heindl, in prep) and it avoids difficulties arising from Braun-Blanquet's (1964) approach. The amount of vegetation is referred to as 'mass index' (MI) in this study (Janauer et al., 1993).

All species found at the study site were identified following Casper & Krausch (1980, 1981).

Results

The distribution diagram (Figures 1a and b) shows the mass indices for each species in each section. The plant mass given on the y-axis is expressed using a linear scale (Melzer 1990, Janauer et al., 1993). The x-axis is proportional to the true length of the sections of the water bodies.

Figure 2 shows the two types of mean mass index, MMT and MMO (Janauer et al., 1993). In the sheltered branches (2b) the values are considerably higher (2.5 and more), whereas in the hydrologically dynamic systems (2a) the values are between 1 and 2.

Discussion

In Austria many systems of lateral stream branches still exist along the Danube, but separated from the main channel of the river at their former inflow by stone dams constructed in the course of river regulation 100 years ago. Studies by Janauer & Hary (1989) and Kum (1994) indicate disturbance by floods to be the dominating factor for the distribution of aquatic vegetation. The magnitude of the silt layer, which correlates positively with the mass of the aquatic species corresponds directly to the flood regime. Shading can be ruled out as a dominating factor when river branches are wider than five meters (Janauer & Kum, in press).

Only nine species occur in the hydrologically dynamic sections, but a total of 19 species are found in the sheltered branches (Figure 1). *Fontinalis antipyretica*, *Ranunculus fluitans* and *Sparganium emersum* (in its submersed form) have a wide flow tolerance, but all other species found are adapted to stagnant conditions (Janauer, 1981a, b; Haslam, 1978, 1987).

Only a single species, *Elodea canadensis*, was found to be restricted to the hydrologically dynamic sections, but this species is not considered to be resistant to high flow velocities (Haslam, 1978, 1987). Its appearance can be explained by the fact that most macrophyte stands in sections exposed to frequent disturbance by flooding are located close to the banks or within the riparian helophyte and sedge belts. However, in Austrian rivers *E. canadensis* is reported to tolerate at least moderate flow (Janauer & Pelikan, 1988; Janauer, 1981b). The central parts of the large branches are without any plant growth in most cases.

Figure 2 indicates the existence of different types of growth strategies for different species. In all cases where the MMT and MMO are of a similar size the species is rather evenly distributed throughout the whole branch system. If the MMT is much smaller, the species is restricted to a few sections, where it forms individual large patches.

In the sheltered sections all species reach higher values of MMT and MMO. This means that most of these sections are not only characterized by a high number of species, but additionally, by a substantial plant mass. In the flood-prone sections the frequent disturbance prevents the development of extensive macrophyte growth, directly by physical removal, or indirectly by influencing the nature of the substrate.

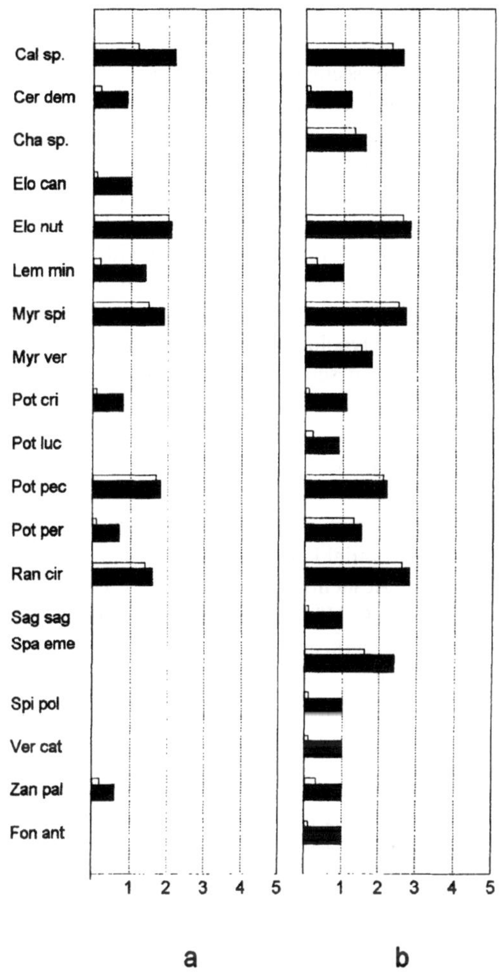

Figure 2. Mean mass indices, MMT and MMO. Outlined bar: Mean mass index of a species calculated on the total number of sections (MMT). Full bar: Mean mass index of a species calculated on the number of sections of occurrence (MMO). (a) Hydrologically dynamic sections, (b) Sheltered sections.

Conclusion

The method of section mapping the macrophyte vegetation enabled the establishment of a numerical basis for studying and comparing large branch systems of different hydrological regimes. In flood-prone sections only protected littoral habitats are refuges for the aquatic vegetation, whereas in sheltered sections large plant-patches reach high mass values and provide abundant structure for invertebrate communities and fish. Further studies on the distribution of the aquatic vegetation will lead towards a typology of branch systems needed for further detailed descriptions of the fluvial corridor.

Acknowledgments

This study is part of the UNESCO MaB-program 'Aquatic Terrestrial Ecotones' 5/21, funded in part by the Austrian Academy of Science (ÖAW). The authors are grateful to C. Hüpfel, who provided essential help with the field work, and to P. Christof-Dirry, who prepared the figures.

References

Braun-Blanquet, J. 1964. Pflanzensoziologie. Springer, Vienna.

Casper, S. J. & H. D. Krausch, 1980. Pteridophyta and AntophytaBd. 23. In H. Ettl, J. Gerloff & H. Heyning (eds), Die Süßwasserflora von Mitteleuropa, Frankfurt.

Casper, S. J. & H. D. Krausch, 1981. Pteridophyta and Antophyta Bd. 24. In Ettl, H. J. Gerloff & H. Heyning (eds), Die Süßwasserflora von Mitteleurops, Frankfurt.

Dennison, W. C., R. J. Orth, K. A. Moore, J. C. Stevenson & V. Carter, 1993. Assessing water quality with submersed aquatic vegetation – habitat requirements as barometers of Chesapeake Bay health. Bioscience 43: 86–94.

Haslam, S. M., 1978. River plants, Cambridge University Press, Cambridge, 396 pp.

Haslam, S. M., 1987. River plants of western Europe, Cambridge University Press, Cambridge, 512 pp.

Hein, T., 1993. Hydrologische Vernetzung – Schlüsselfaktor für Auensysteme? MSc-thesis, Vienna.

Jacobsen, D. & K. Sand-Jensen, 1994. Invertebrate herbivory on the submerged macrophyte *Potamogeton perfoliatus* in a Danish stream. Freshwater Biol. 31: 43–52.

Janauer, G. A., 1981a. Die Zonierung submerser Wasserpflanzen und ihre Beziehung zur
Gewässerbelastung am Beispiel der Fischa (Niederösterreich), Verh. Zool.-Bot. Ges. Österr. 120: 73–98.

Janauer, G. A., 1981b. Die Makrophytenvegetation als Indikator für Gewässerbelastung am Beispiel der Fischa und des Erlabaches. Proc. 22. Arbeitstagung Internat. Arbeitsgem. Donauforschung, Vienna, 201–204.

Janauer, G. A., 1994. An essential tool for assessing aquatic terrestrial ecotones. Proc. 1st. International Symp. on Habitat Hydraulics, Trondheim.

Janauer, G. A. & P. Pelikan, 1988. Kamp – Verkrautung. Abschlußbericht. Amt der NÖ Landesregierung und Bundesministerium für Land- und Forstwirtschaft (eds), Vienna, 39 pp.

Janauer, G. A. & N. Hary 1989. Interdisziplinäre Studie Donau. Österr. Wasserwirtschaftsverband, Vienna 362 pp.

Janauer, G. A., R. Zoufal, P. Christof-Dirry & P. Englmaier, 1993. Neue Aspekte der Charakterisierung und vergleichenden Beurteilung der Gewässervegetation, Ber. Inst. Landschafts-Pflanzenökologie Univ. Hohenheim 2: 59–70.

Kohler, A., H. Vollrath & E. Beisl, 1971. Zur Verbreitung, Vergesellschaftung und Ökologie der Gefäßmakrophyten im Fließgewässersystem Moosach (Münchner Ebene), Arch. Hydrobiol. 69: 333–365.

Kohler, A., 1978. Methoden der Kartierung von Flora und Vegetation von Süßwasserbiotopen. Landschaft + Stadt 10: 73–85.

Kum, G., 1994. Entwicklung der Makrophytenvegetation im Gießgang Greifenstein (Tullner Au) 1986–1992, M.Sc.-thesis, Vienna.

Melzer, A., 1990. Die Makrophytenvegetation des Tegern-, Schlier- und Riegsees. Informationsberichte Bayerisches Landesamt f. Wasserwirtschaft 2/90, München, 180 pp.

Philips, E. J., P. V. Zimba, M. S. Hopson & T. L. Crisman, 1993. Dynamics of the plankton community in submerged plant dominated during an annual cycle. Aquat. Bot. 43: 231–248.

Hydrobiologia **340:** 141–145, 1996.
J. M. Caffrey, P. R. F. Barrett, K. J. Murphy & P. M. Wade (eds), Management and Ecology of Freshwater Plants.

Stream vegetation in different landscape types

[1]Štěpán Husák & [2]Věra Vořechovská
[1]*Institute of Botany, Czech Academy of Sciences, Dukelska 145, 379 82 Trebon, Czech Republic*
[2]*Morava River Authority, Dřevařskà 11, 601 75 Brno, Czech Republic*

Key words: streams, vegetation, landscape

Abstract

A methodology is proposed for assessing the ecological value of streams in the catchment of the Moravská Dyje River. It is concluded that by using a wide range of parameters that a more objective assessment is achieved than if only one were used. The landscape of the catchment studied contains excessive amounts of nutrients and, in comparison to its natural state, has become too uniform.

Introduction

For the evaluation of streams in a prevailingly agricultural landscape, a small catchment area was selected, namely that of Moravská Dyje River (with tributaries): 237 sites were visited during the growing season of 1993 along these water courses (drains, brooks, rivulets as well as fishponds on these courses). At each site the plant species composition was described and dominance assessed by percentage cover per m^2. At 47 sites water samples were taken for determination of NO_3-, NH_4-, PO_{4-}^{3} and alkalinity. Further, the type of surrounding landscape including waterbodies was recorded (e.g., arable land, meadows). The species diversity and area or width of vegetation belts, type of forest or scattered trees and shrubs, intensity of erosion, type of settlements, nature of farming, intensity of fertilizer application to meadows and fields, and wastewater discharge were also noted.

Using local plant species, both negative and positive bioindicators were selected. After evaluation of all parameters obtained a 5-degree scale of quality was used for each stream or fishpond (1 = particularly important, 2 = important, 3 = average, 4 = negatively influenced, 5 = particularly poor quality).

Materials and methods

The bedrock of the area is Moldanubicum granite, a crystalline igneous rock poor in nutrients. Geomorphologically the area belongs to the Českomoravská vrchovina highlands, and is at an altitude of 404–734 m. This corresponds to the submontane vegetation tier.

Plant species were selected as 'positive' or 'negative' bioindicators at 237 sites (Figure 1) in various landscape types. Water samples were taken from 47 sites for the assessment of NO_3-, NH_4-, PO_{4-}^{3} and alkalinity, by methods recommended by Tecator (Sweden) (Tables 1 and 2).

Standards of evaluation

Any evaluation of a natural system with regard to the complexity of its structure and functioning is always relative and of limited value. Nevertheless, with the help of several criteria, an evaluation of each site studied was attempted, Table 2 lists the criteria used.

After a basic floristic analysis, aquatic and marsh plants species were selected, which could be used for the evaluation of different wetlands for this region. Selection of species as 'positive' and 'negative' indicators was derived from existing knowledge and experience of the autecology of the individual species (Bre-

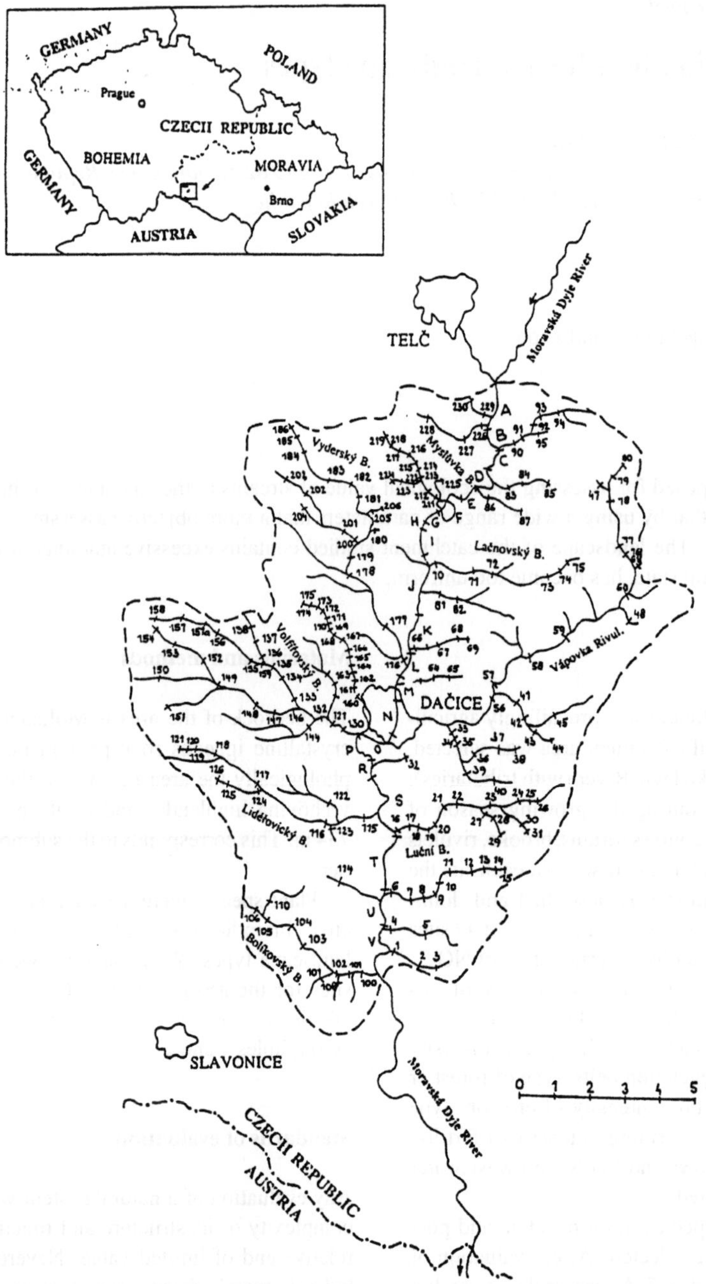

Figure 1. Location and catchment area of the Moravská Dyje River in the region of Dačice. Together 237 sites in tributaries and in this river (numbers + segments) were surveyed.

itig & Tümplich, 1982; Ellenberg et al., 1991; Frank & Klotz, 1988; Husák et al., 1989).

As positive indicators of waters the following were selected, as indicating waterbodies with predominantly soft water, poor in nutrients, unpolluted by either

Table 1. Water chemical analyses of different waterbodies (streams and fishponds) in catchment area of Moravská Dyje River sampled from October 18–20, 1993.
(Numbers or segments see map 1)

Sampling site	NO_3-N mg/l	NH_4-N mg/l	PO_4-P mg/l	Alkalinity mekv/l
Q	1.070	8.000	0.520	2.575
A1	1.680	0.830	0.190	2.240
A2	1.500	1.520	0.230	2.310
B	1.130	6.500	0.386	2.430
F	1.240	6.200	0.417	2.360
G	1.300	4.380	0.392	2.285
M	1.910	2.650	0.369	2.210
P	2.180	2.260	0.443	2.190
R	2.070	1.840	0.400	2.200
T	1.940	1.360	0.375	2.185
U	1.930	1.740	0.383	2.180
V	1.320	1.440	0.204	1.570
8	3.060	0.970	0.050	1.385
17	8.520	3.130	0.120	1.680
23	0.380	2.040	0.019	1.245
53	1.270	1.450	0.163	2.300
54	1.010	1.670	0.163	2.300
58a	2.030	1.090	0.106	2.020
59a	2.040	1.240	0.106	2.070
61	1.820	1.180	0.130	2.450
70	3.680	1.140	0.034	2.155
71a	1.550	1.030	0.035	2.705
73	0.050	1.110	0.026	3.055
75a	0.070	0.970	0.091	1.800
90a	1.660	2.170	0.019	2.600
94a	1.470	1.060	0.019	1.352
100	0.680	1.180	0.046	0.950
102	0.800	1.150	0.052	0.930
103a	1.520	1.000	0.059	0.930
103b	0.710	1.340	0.038	0.940
107	1.100	1.170	0.064	0.860
113	0.010	2.230	0.198	4.610
115a	0.000	12.500	0.491	4.620
130a	1.540	2.450	0.087	1.570
133	2.420	2.990	0.214	1.420
137	1.470	2.490	0.175	1.750
139	2.240	1.640	0.076	1.795
140a	2.220	2.510	0.048	1.305
144	6.460	2.070	0.095	0.960
146	1.210	2.210	0.045	1.415
155	2.110	2.210	0.044	0.985
160	0.120	4.790	0.784	2.850
173	0.000	1.090	0.019	2.325
210	2.260	1.850	0.372	1.665
214	2.960	2.870	0.400	1.710
216	2.790	3.160	0.372	1.760
220	2.710	3.960	0.428	2.035

organic or inorganic matter: *Chara fragilis, Riccia fluitans, Ricciocarpus natans, Fontinalis antipyretica, Callitriche hamulata, Potamogeton alpinus* and *P. obtusifolius*. Similarly the following marsh plants were selected as "positive" indicators: *Cardamine amara, Calamagrostis canescens, Calla palustris, Carex rostrata, Comarum palustre, Epilobium palustre, Equisetum fluviatile, Lysimachia thyrsiflora, Salix aurita, Sphagnum* sp. div. and *Veronica beccabunga*. All named species were evaluated according to their abundance by the symbols: P (positive) if present, PP if up to 20% abundance, PPP if over 20% abundance.

As 'negative' indicators of waterbodies, species were selected which are sometimes also common in waters with secondarily hard water: *Ceratophyllum demersum, Lemna gibba, Lemna minor, Myriophyllum spicatum, Potamogeton crispus, P. pectinatus, Spirodela polyrhiza* and *Zannichellia palustris*. Similarly the following marsh plants were selected as 'negative' indicators: *Glyceria maxima, Urtica dioica* and *Cirsium arvense*. They were evaluated as negative (N) if their abundance was more than 25%, as NN if 50–75% and NNN - 75–100% abundance. These criteria were used separately for water and banks or littorals (see Table 2).

Number of species per unit area is an important criterion of bitope quality. Low diversity is usually due to human impact, for example, eutrophication, ruderalisation or destruction of biotopes. In most cases high numbers of species are positively correlated with a higher quality of biotopes. The number of plant species from stream profiles fluctuated between 3 and 18 (Table 2). For standing waters, diversity was indicated by symbols DH = high diversity, if the numbers of plant species was more than 10 (in water and bank together) or LD = low diversity, if less than 10 species were present.

The results of the water analyses were used as an auxilliary criterion. Table 2 presents the symbols P = positive concentration, or N = negative concentration of nitrogen, ammonium, phosphorus and alkalinity, using the following key (Table 3).

The alkalinity reflects the fact that natural waters in the Dačice region are typically soft or medium-hard. Hard waters or very hard waters in this region are due to human impact.

Watercourses can have arable land extending to the river bank or fishpond shore, or they can have natural or seminatural banks with meadows, shrubs and trees or forest. Riparian vegetation is important as a buffer zone for the interception of nutrients and increases the total

Table 2. Results of assessment of 237 sites (here only first 30 sites) of streams and fishponds on these streams. 1. Numbers or letter of sites on tributaries of the Moravská Dyje R. or segments of this river, see map, Figure 1; 2. Bioindication, AP = aquatic plants. BP = bank or shore plants, P = positive indicators, N = negative indicators; 3. Biodiversity, numbers of species per unit area (usually 5 m^2), for fishpond are use symbols HD = high diversity or LD = low diversity; 4. Chemical composition , P = positive concentration, N = negative concentration, S = soft water, MH = medium hard water, H = hard water, VH = very hard water; 5. Landscape types, BB = broad belt, NB = narrow belt, AL = arable land; 6. Hydrotechnical and other criteria; 7. General assessment of streams or fishponds can be 1 = particularly important, 2 = important, 3 = average, 4 = negatively influenced, 5 = of a particularly poor quality, for details see text.

1	2. Bioindication		3. Bio	4. Chemical composition				5. Landscape	6. Hydrotechnical	7. General
	AP	BP	diversity	N	NH_4	P	Alk	types	and other criteria	assessment
1	–	–	8	–	–	–	–	–	–	3 stream average quality
2	–	–	10	–	–	–	–	–	–	3 stream average quality
3	N	–	–	–	–	–	–	AL	5	5 fishp. partic. poor quality
4	–	–	6	–	–	–	–	NB	–	3 stream average quality
5	–	–	10	–	–	–	–	AL	–	3 stream average quality
6	–	N	5	–	–	–	–	NB	–	4 stream negat. influenced
7	–	–	–	–	–	–	–	AL	5	5 fishp. partic. poor quality
8	–	P	10	P	P	P	MH	BB	–	2 stream important quality
9	–	–	7	–	–	–	–	–	5	3 stream average quality
10	–	–	11	–	–	–	–	AL	–	3 stream average quality
11	–	P	11	–	–	–	–	NB	7	3 stream average quality
12	–	N	8	–	–	–	–	–	–	4 stream negat. influenced
13	–	–	HD	–	–	–	–	–	–	3 fishp. average quality
14	–	–	–	–	–	–	–	NB	12, 13	4 fishp. negat. influenced
15	–	PP	HD	–	–	–	–	BB	–	1 fishp. particularly important
16	–	–	8	–	–	–	–	–	6	3 stream average quality
17	–	N	3	N	N	N	MH	NB	6	5 stream partic. poor quality
18	–	–	11	–	–	–	–	–	6	3 stream average quality
19	–	–	10	–	–	–	–	NB	–	3 stream average quality
20	–	P	14	–	–	–	–	BB	–	2 stream important quality
21	–	–	–	–	–	–	–	NB	16	4 fishp. negat. influenced
22	–	N, P	16	–	–	–	–	–	–	2 stream important quality
23	–	PPP	15	P	P	P	MH	–	–	1 stream particul. import.
24	–	–	13	–	–	–	–	–	–	3 stream average quality
25	–	P	HD	–	–	–	–	BB	–	2 fishp. important quality
26	–	P	HD	–	–	–	–	BB		2 fishp. imporatant quality
27	–	–	–	–	–	–	–	–	16	4 fishp. negat. influenced
28	–	–	–	–	–	–	–	–	16	4 fishp. negat. influenced
29	–	P	–	–	–	–	–	–	–	3 fishp. average quality
30	N	–	–	–	–	–	–	BB	–	2 fishp. important quality

value of the waterbody and landscape as a whole. Three levels were differentiated: streams with natural or seminatural vegetation on either bank in belts more than 30 m broad (indicated by the symbol BB = 'broad belt'); belt with natural or seminatural vegetation less than to 30 m broad (indicated by the symbol NB = 'narrow belt') and watercourses accompanied by arable land or intensively fertilized meadows (as well as streams or fishponds in settlements) (indicated by the symbol AL,=,'arable land').

From the field study, it was possible to use other criteria for the assessment of waterbodies. These criteria are divided into two groups. For streams and partly also for fishponds, the following parameters were used (numbered 1 to 11): 1 stream not regulated, 2 stream sign ificancly meandering, 3 stream with a natural river bed and with a rocky or stony bottom, 4 stream paved with prefabricated concrete panels, 5 stream or fishpond with concrete prefabricated panels in river banks or fishpond shores, 6 stream with bed paved with stones or cobbles, 7 stream or fishpond with stone walls along

Table 3. Key for evaluation of chemical analyses

	Concentration (mg l^{-1})	
	Positive (P)	Negative (N)
NO_3-N	below 3.0	over 3.0
NH_4-N	below 3.0	over 3.0
PO_4-P	below 0.1	over 0.1
	Alkalinity (mmol l^{-1})	
S	0.00–1.00 soft water	
MH	1.00–2.00 medium hard water	
H	2.00–4.00 hard water	
VH	more than 4.00 very hard water	

the banks or shores, 8 stream or fishpond filled with silt or sand, 9 stream strongly shaded, 10 stream banks mown, 11 crayfish present in stream. Special parameters for fishponds are listed below (numbers 12–16): 12 = deposits of bottom sediments along shores, 13 fishpond with deepened shores (without littoral vegetation zones), 14 extremely hypertrophic fishponds, often in the middle of villages or near farms, 15 rubbish heap on bank, or banks ruderalised or destroyed, 16 duck farms on the fishpond.

Using all these parameters an overall assessment of each waterbody was made (see Table 2 column no. 7). A 5-degree scale was used to describe the waters of each stream or fishpond: 1 = particularly important, 2 = important, 3 = average, 4 = negatively influenced, 5 = of particularly poor quality.

Discussion and conculsion

Evaluation of waterbodies in landscape using only one parameter provides limited results. By using more parameters it was hoped to obtain a more objective assessment (Wade & Husák, 1989). The present state of waterbodies within the Moravská Dyje River catchment is characterised by a heavy water pollution load persisting for the five years after a change from extremely intensive to extensive agriculture started in the Czech Republic. The data show that this landscape still contains excessive amounts of nutrients, and that the vegetation is too uniform for the landscape type given. Only in the upper sections of the tributaties are a few streams or fishponds preserved where the waters are not stressed by nutrients. These waterbodies have quite a rich vegetation, sometimes with rare and protected plant species. Practical experience will show whether the evaluation of streams and their surroundings is better accomplished in this elaborate way, or whether it should be confined to a smaller number parameters. We believe that a greater number of parameters will tell us more about the streams and their immediate surroundings.

Acknowledgments

We express our thanks to the Morava River Authority in Brno for financing this research, as well as to Dr J. Květ for linguistic assistance.

References

Breitig, G. & W. Tumpling, 1982. Ausgewahlte Methoden der Wasseruntersuchung. VEB Gustav Fishcher, Jena.

Ellenberg, H., H. E. Weber, R. Dull, V. Wirth, W. Werner & D. Pulissen, 1991. Zeigerwerte von Pflanzen in Mitteleuropa. Scripta Geobot. 18, Ver. E. Golze KG, Gotingen.

Frank, D. & S. Klotz, 1988. Biologisch-okologische Daten zur Flora der DDR. Martin-Luther-Univ. Halle-Wittenberg, Wissenschafl. Beitr. 1988/60, Halle (Saale).

Husák, S., V. Sládeček & A. Sládečková, 1989. Freshwater macrophytes as indicators of pollution. Acta hydrochim. hydrobiol. 17: 693–697.

Wade, P. M. & S. Husák, 1989. The restoration of the wetland flora of Hatfield Chase, England – a success for creative conservation. In Proc. Intern. Wetl. Conf. 19–23 Sept. 1988, Univ. of Rennes, France: 315–316.

Hydrobiologia **340**: 147–151, 1996.
J. M. Caffrey, P. R. F. Barrett, K. J. Murphy & P. M. Wade (eds), Management and Ecology of Freshwater Plants.

Coexistence of *Juncus articulatus* L. and *Glyceria australis* C. E. Hubb. in a temporary shallow wetland in Australia.

R. Geoffrey B. Smith & Margaret A. Brock
Botany Department, University of New England, Armidale, 2351 N.S.W., Australia

Key words: water regime, competition, emergent macrophytes, introduced species

Abstract

Juncus articulatus, a species introduced to Australia, is codominant over large areas of Mother of Ducks Lagoon but is rare in other lagoons. It occurs widely within the lagoon but is concentrated in lower, wetter areas that are more disturbed by birds and cattle. This suggests that *J. articulatus* may be separated over elevation (and therefore water regime) and disturbance gradients from the native grass *Glyceria australis*, the dominant species in the lagoon. This paper compares the growth and interaction of *G. australis* and *J. articulatus* under different water regimes. The species responded differently to water regime both in monoculture and in mixture. Above ground production of *J. articulatus* was greatest under fluctuating water levels, least under a damp water regime and intermediate under flooded conditions. *G. australis* production was greatest under the damp, least under the flooded and intermediate under the fluctuating water regime. The outcome of interaction is dependent on water regime and time. After one year *J. articulatus* was the superior competitor under all water regimes. At the end of two years *J. articulatus* was still the superior competitor under fluctuating and flooded water regimes but not under the damp regime. The change in outcome after two years was due to the competitive superiority of *G. australis* during the second year under all water regimes. The relative importance and the management implications for the invasive potential of *J. articulatus* are assessed in shallow Australian wetlands with fluctuating water regimes.

Introduction

The 'lagoons' on the northern tablelands of New South Wales, Australia, are shallow lakes which fill and dry at irregular intervals (Brock, 1991). *Glyceria australis* C. E. Hubb. (Australian sweetgrass) is the dominant species at Mother of Ducks Lagoon and is common in many others. It provides pasture for stock and habitat for water birds including *Gallinago hardwickii* (Gray). (Japanese snipe), the subject of an international agreement on migratory birds between Australia and Japan.

Juncus articulatus L., a species introduced to Australia (Sainty and Jacobs, 1981), is codominant over large areas of Mother of Ducks Lagoon, but is rare in other lagoons. Studies in Europe indicate *J. articulatus* has a long lived seed bank and is a colonising species. *J. articulatus* germinated from the seed bank of fens fifteen years after draining (Pfadenhauer and Maas, 1987) and is a characteristic species in the primary successional stage of wet heath (Neuhaus, 1990). *J.articulatus* is widespread in Australia where it also has a long lived seed bank. It germinated from Mother of Ducks Lagoon sediment stored for twelve years (Brock & Britton, 1995). In Western Australia, it reproduces readily from seed and rhizomes compared with other reproductively more specialised species (Chambers & McComb, 1992). It can be a minor weed of irrigation drains (Sainty & Jacobs, 1981).

G. australis and *J. articulatus* have similar emergent growth forms. *J. articulatus* occurs widely within Mother of Ducks Lagoon but is concentrated in lower, wetter areas which are more disturbed by birds and cattle. This suggests that the two species may be separated over elevation (and therefore water regime) and disturbance gradients. Several studies have focused on the effect of physical factors and competition on coexisting species (Mesléard et al., 1993., Zedler et al., 1990., Grace and Wetzel, 1981., McCreary et al., 1983). The role of species traits in competitive change over gradients has been highlighted as an area for more research (Bengtsson et al., 1994, Goldberg, 1990). This paper compares the growth and interaction of *G. australis* and *J. articulatus* under different water regimes. The relative importance of the invasive potential of *J. articulatus* for species management in Australian shallow wetlands with fluctuating water regime is assessed.

Description of study site

Mother of Ducks Lagoon (Latitude 30°23'5″S Longtitude 151°39'30″E, altitude 1,315 m asl) is approximately 3 km^2 and is unusual on the northern tablelands for its peat soil overlying a clay base. The township of Guyra on its eastern shore has affected the use of the lagoon. During this century the outlet of the lagoon was lowered approximately two metres to allow grazing and golf on the lagoon bed. About one quarter of the lagoon is now a nature reserve. Since 1988, the reserve has been leveed to restore a water regime closer to conditions prior to draining of the lagoon. Mother of Ducks Lagoon Nature Reserve has a maximum depth of just over one metre. As well as short term fluctuations, water levels have fluctuated between maximum depth and ground level for periods of up to several seasons duration. The remainder of the lagoon drains quickly and does not experience extended periods of higher water levels.

Methods

Interaction between *J. articulatus* and *G. australis* was measured in monoculture and in mixture (deWit, 1960). In September 1992, seedlings of the two species were transplanted into Mother of Ducks soil in 200 mm diameter pots at five different ratios (*J.articulatus*: *G australis*: 100%:0%, 75%:25%, 50%:50%, 25%:75%, 0%:100%) at a total density of 36 seedlings per pot. This density was representative of established stands of the species at Mother of Ducks Lagoon. Pots were grown in large tubs in a split plot design, under one of three water regimes (main plot factor); damp (bottom of the pot under water), flooded (0–10 cm water above soil surface) and fluctuating between these two relatively static levels. The tubs were blocked according to position and the pots allocated to blocks according to the size of the seedlings.

Performance of the species was measured as above ground dry weight. After the first growing season in February 1993, half the pots (one pot of each treatment combination from each tub) were clipped 2 cm above the soil and the harvested material was sorted to species and dried for 48 hours at 80 degrees C and weighed. The remaining pots (one from each treatment combination from each tub) were clipped, sorted, dried and weighed in May 1994, after the second growing season.

Differences in performance were assessed by comparing production per plant by analysis of variance and least significant difference. Normality of the data was checked using normal probability plots. Interactions between water, year and mixture were tested by three way analysis of variance for each species separately. The nature of the interaction between the two species was assessed by fitting the model of deWit (1960) using the maximum-likelihood estimation method of Machin & Sanderson (1977).

Results

The effect of water regime

The two species responded differently to water regime. *J. articulatus* production was always greatest under the fluctuating regime, least under the damp regime and intermediate under the flooded regime. The production of *G. australis* was always greatest under the damp regime, least under the flooded regime and intermediate under the fluctuating regime, except in monoculture in the first year. The production of *J. articulatus* in monoculture was significantly greater under the fluctuating and flooded treatments than under the damp treatment at the end of the first year ($P<0.05$), but by the end of the second year there was no significant difference between water regimes (Figure 1).

For *G. australis* after two years, production had increased significantly over all water regimes (Figure 1). For *J. articulatus* the increase was greater under the damp regime than under the fluctuating or flooded regimes ($P<0.05$). The interaction between water and year was significant for *J. articulatus* but not for *G. australis* (Table 1).

The effect of initial density

At the end of the first year *J. articulatus* production per plant decreased with increasing densities of *J. articu-*

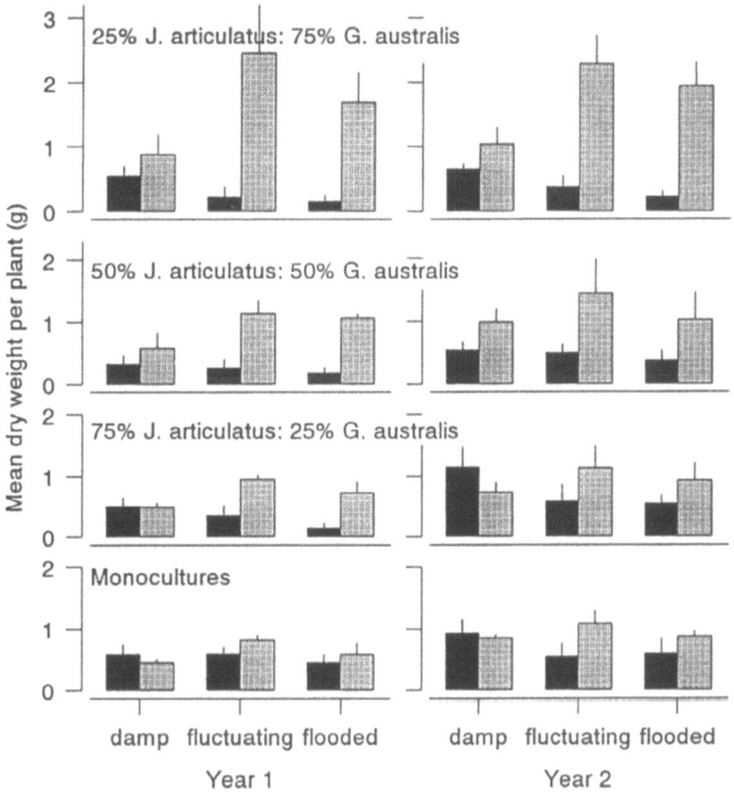

Figure 1. Dry weight per plant (+ se) of *J. articulatus* (shaded) and *G. australis* (black) grown for two years under three water regimes in three mixtures and in monoculture.

Table 1. Treatment interaction and analysis of variance for *J. articulatus* and *G. australis* grown in three mixtures and in monocultures over two years under three water regimes.

Source	*J. articulatus*		*G. australis*	
	F	*P*	*F*	*P*
Water	$F_{2,6} = 10.73$	<0.05	$F_{2,6} = 15.20$	<0.01
Year	$F_{1,63} = 39.71$	<0.000	$F_{1,63} = 74.45$	<0.000
Ratio	$F_{3,63} = 64.33$	<0.000	$F_{3,63} = 21.75$	<0.000
Ratio X water	$F_{6,63} = 3.010$	<0.05	$F_{6,63} = 3.959$	<0.01
Ratio X year	$F_{3,63} = 2.860$	<0.05	$F_{3,63} = 5.586$	<0.01
Water X year	$F_{2,63} = 5.321$	<0.01	$F_{2,63} = 1.648$	ns
Ratio X water x year	$F_{6,63} = 1.069$	ns	$F_{6,63} = 2.204$	ns

latus. In contrast, *G. australis* production per plant decreased with decreasing *G. australis* density under the fluctuating and flooded regimes, that is, the production of both species decreased with increasing *J. articulatus* density (Figure 1).

During the second growing season this trend reversed. For both species, the difference in production between year 2 and year 1 was greater with decreasing *G. australis* density. There was a significant year times ratio interaction term for both species (Table 1).

After two years, the decrease in *J. articulatus* production with increasing *J. articulatus* density had weakened. The production of *G. australis* was less at increasing densities of *G. australis* in mixture, but pro-

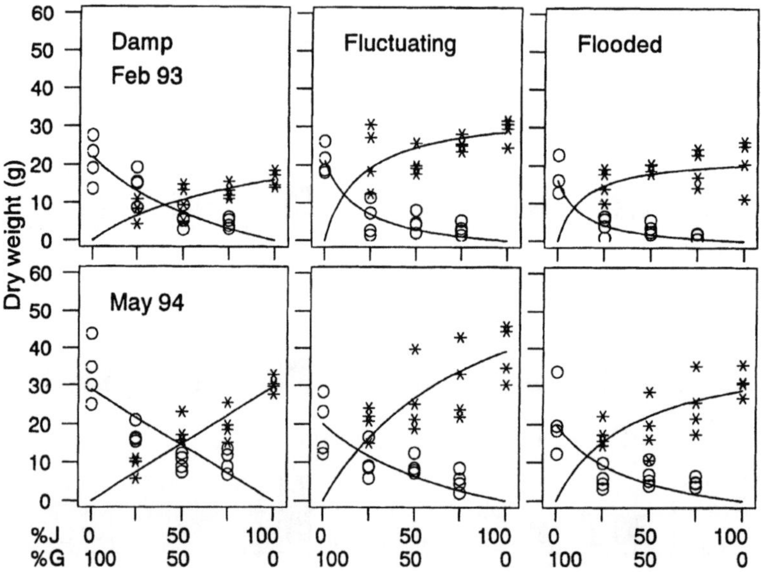

Figure 2. The model of deWit (1960) fitted to dry weight data of *J. articulatus* (J) and *G. australis* (G) for three water regimes and two years using the maximum likelihood estimation method of Machin and Sanderson, (1977). *J. articulatus* (J) ⋆. *G. australis* (G) ∘.

duction was still less in mixture than in monoculture (Figure 1).

Water regime affected the competitive outcome. Under the damp regime the patterns in production occurring with density changes were different. Both species were more successful at lower densities (Figure 1). The interaction between water and ratio was significant for both species (Table 1).

The nature of the interaction

The estimated parameters of the de Wit (1960) model describe interference of *G. australis* by *J. articulatus* for all treatments in the first year and for the fluctuating and flooded regimes after the second year (Figure 2). The reciprocal curvature of the two lines indicates interference of one species by the other. The downwards curvature indicates interference to growth. Under the damp regime after the second year the parameters describe mutual interference of equal magnitude (Figure 2).

Discussion

Competition between the two species varies with water regime. Under the fluctuating and flooded regimes in both years and the damp regime in the first year, the decreased production of *G. australis* per plant in mixture and the corresponding increase in production of *J. articulatus* per plant (Figures 1 and 2) indicates competition (as defined by Golberg and Barton, 1992, and described by Harper, 1977). The model of de Wit (1960) shows the relative intensity of the interaction (Figure 2). The influence of *J. articulatus* on *G. australis* is strongest under the flooded regime, even though the production of *J. articulatus* is greatest under the fluctuating regime. The competitve advantage of *J. articulatus* disappears during the second year. The effect of *G. australis* on individuals of the same species is greater than the effect of *J. articulatus* on *G. australis*. By the end of two years *G. australis* production in the lowest density mixture had caught up to the production in monoculture (Figure 1).

A change in interaction could occur where one of the two species (*J. articulatus*) has greater ability to deplete resource levels and the other species (*G. australis*) is better able to tolerate the low levels (Goldberg, 1990). After resource levels are depleted the tolerant species will have an advantage. In such a situation allocation of energy to high seed production would be an advantage for the species less tolerant of low resource levels. If this species was excluded it could have the potential to regenerate from seed at the next germination event.

Water regime manipulation could be used to favour one species over another. The variety of water regimes created by different management across Mother of Ducks Lagoon provides different niches for *J. articulatus* and *G. australis*. The fluctuating, wetter regimes inside the nature reserve may at times favour *J. articulatus* while the drier, more predictable regimes outside the lagoon may favour *G. australis*. Water levels outside the reserve generally remain below the soil surface and so do not favour *J. articulatus*. Inside the nature reserve water levels fluctuate from low to high for more extended periods. Fluctuations within two seasons favour *J. articulatus* over *G. australis* especially if nutrients are released by a drying event. The effect of longer term fluctuations is less certain. High water levels will favour *J. articulatus* over *G. australis*, but may favour deeper water species such as *Myriophyllum variifolium* J. Hook. and *Eleocharis sphacelata* R. Br. A decrease in performance of *J. articulatus* after water levels exceed the soil surface was demonstrated by Froend et al. (1993) in Western Australia.

J. articulatus prefers fluctuating and higher water levels and disturbance. *G. australis* is suited to vegetative growth over longer periods. If species reproductive and vegetative patterns under different water regimes are known, then manipulating water regime is a potential tool for management of invasive species. This can be achieved directly or indirectly: by changing the interaction of species by manipulating water levels or by influencing the pattern of disturbance caused by flooding, drought or water birds.

Acknowledgements

This work was made possible by a grant from the Land and Water Resources Research and Development Corporation and assistance of staff and students of the Botany Dept. UNE.

References

Bengtsson, J., T. Fagerström & H. Rydin. 1994. Competition and coexistence in plant communities. Trends in Evolution and Ecology 9: 246–250.

Brock, M. A., 1991. Mechanisms for maintaining persistent populations of *Myriophyllum variifolium* J. Hooker in a fluctuating shallow Australian lake. Aquat. Bot. 44: 211–219.

Brock, M. A. & D. L. Britton, 1995. The role of seed banks in the revegetation of Australian temporary wetlands. In B. D. Wheeler, S. C. Shaw, W. J. Fojt & R. A. Robertson, (eds), Restoration of Temperate Wetlands. Wiley, Chichester. pp 183–188.

Chambers, J. M. & A. J. McComb, 1992. Establishing wetland plants in artificial systems. In Wetland Systems in Water Pollution Control. Organising Committee of Wetlands Systems in Water Pollution Control Conference, Sydney. pp 20.1–20.6.

Froend, R. H., R. C. C. Farrell, C. F. Wilkins, C. C. Wilson & A. J. McComb, 1993. Wetlands of the Swan Coastal Plain, Volume 4: The Effect of Altered Water Regimes on Wetland Plants. Water Authority of Western Australia, Perth. 144 pp.

Goldberg, D. E., 1990. Components of resource competition in plant communities. In. J. B. Grace & D. Tilman (eds), Perspectives on Plant Competition. Academic Press, Inc., San Diego. 27–49.

Goldberg, D. E. & A. M. Barton, 1992. Patterns and consequences of interspecific competition in natural communities: a review of field experiments with plants. Am. Nat: 139: 771–801.

Grace, J. B. & R. G. Wetzel, 1981. Habitat partitioning and competitive displacement in cattails (Typha): experimental field studies. Am. Nat. 118: 463–474.

Harper, J. L., 1977. Population biology of plants. Academic Press, London. 892 pp.

McCreary, N. J., S. R. Carpenter & J. E. Chaney, 1983. Coexistence and interference in two submersed freshwater perennial plants. Oecologia, 59: 393–396.

Machin, D. & B. Sanderson, 1977. Computing maximum-likelihood estimates for the parameters of the de Wit competition model. App. Statist. 26: 1–8.

Mesléard, F., L. Tan Ham, V. Boy, C. van Wijck & P. Grillas, 1993. Competition between an introduced and an indigenous species: the case of *Paspalum paspalodes* (Michx) Schribner and *Aeluropus littoralis* (Gouan) in the Camargue (southern France). Oecologia 94: 204–209.

Neuhaus, R., 1990. Stadien und alter der primärsukzession von feuchtheiden in küstenden. Drosera, 90: 29–34.

Pfadenhauer, J. & D. Maas, 1987. Seed bank of fen soils of meadows under different management in the German Prealpes. Flora 179: 85–97.

Sainty, G. R. & S. W. L. Jacobs, 1981. Water plants of New South Wales. Water Resources Commission New South Wales, Sydney. 550 pp.

Wit, C. T. de, 1960. On competition. Versl. Landbouwk. Onderzoek. 66.8.

Zedler, J. P., E. I. Paling, A. J. McComb, 1990. Differential responses to salinity help explain ar the replacement of native *Juncus kraussii* by *Typha orientalis* in Western Australian salt marshes. Aust. J. Ecol. 15: 57–72.

Hydrobiologia **340**: 153–156, 1996.
J. M. Caffrey, P. R. F. Barrett, K. J. Murphy & P. M. Wade (eds), Management and Ecology of Freshwater Plants.

Interactions between national and local strategies for the management of aquatic weeds

D. S. Mitchell
The Murray-Darling Freshwater Research Centre, P.O. Box 921, Albury, New South Wales, Australia, 2640
Now at: The Johnstone Centre, Charles Sturt University, P.O.Box 789, Albury, New South Wales, Australia, 2640

Key words: water weeds, weed strategies, Australia

Abstract

Management of aquatic weeds is often handled primarily at the local level. However, both water and water weeds do not recognise political boundaries even when these coincide with rivers or catchment areas. Thus potentially effective management of a weed in one area may be undermined by absence of a complementary program of management in an adjacent area. Authorities in each of the eight States or Territories that make up Australia are separately responsible for managing water weeds in their own State or Territory. Originally there was little coordination of these programs, but during the 1980s a national strategy for control of Australian water weeds has been progressively devised and put into practice. This stresses prevention and includes policies on plant importation, nomination of noxious weeds, development of a research strategy, a public awareness campaign, guidelines on the use of herbicides in or near water, and a field guide. This strategy is currently being incorporated into a National Weeds Strategy.

Weed control strategies: some general principles

Locally-based strategies for the management of weeds are usually very effective because the community directly experiences the benefits of success or the consequences of failure. Both provide a powerful stimulus for the maintenance of an effective program, especially when the community is also responsible for the costs, as this ensures an acceptable balance between costs and benefits. However, it is possible for highly successful local programs to be undermined by failure to control the weed in question outside the local area. This may occur when the weed plant is assigned a different priority or not even regarded as a problem in the adjacent area. Furthermore, since many weeds have high reproductive capacities and efficient dispersal mechanisms, locally-based strategies for weed control are unlikely to succeed on their own and should be part of a broader regional strategy. The beneficiaries of such larger-scale strategies are more difficult to identify. The program is usually administered by a government agency on behalf of the whole community, who therefore pay for it with their taxes. This has the unfortunate effect of separating the people who pay for weed control from direct appreciation of the benefits of success or of the costs of failure, and the program comes to depend increasingly on the time-consuming and negative process of policing regulations.

The situation outlined above applies generally to all weeds. However, the multipurpose nature of many water bodies causes some aquatic plants to be regarded as weeds in terms of one use and as beneficial plants in terms of another. Such conflicts can be resolved in integrated regional strategies with multipurpose objectives. It is particularly important that native plant species that are regarded as weeds when they interfere with water use, are not treated as weeds in every situation. Thus native swamp plants, such as *Typha* spp, often inhibit flow in irrigation systems. Yet they are also an important component of many valuable wetlands and cannot be regarded as weeds in these situations. A different attitude however should be taken against plant species that are alien to an area and treated as potential, if not actual, weeds in every situation in which they occur.

The background to Australian weed control strategies

Australia has both the advantage and the disadvantage of being biogeographically isolated. Indeed, Good (1974) regarded Australia as the most isolated plant kingdom in the world. The disadvantage of this isolation is Australia's vulnerability to invasion by alien species, many of which were introduced in the 1800s with spectacular and long-lasting adverse consequences, exemplified by the introduction of rabbits (*Oryctolagus cuniculus*), foxes (*Vulpes vulpes*) and water hyacinth (*Eichhornia crassipes*). This process has been in progress for many centuries. For example, the Australian wild dog, *Canis familiaris* dingo, probably entered Australia from the north with a group of Aborigines between 5000 and 14 000 years ago (Newsome, 1983). However, the process has undoubtedly accelerated markedly since European colonisation commenced towards the end of the 1700s. At least 10% of the Australian flora now consists of introduced species (Groves, 1986).

The impact of several of these alien species has been so profound as to change the nature of the environment they have invaded. This has led to the concept of environmental weeds which Humphries et al. (1991) defined as those plants that 'cause major changes to species diversity, abundance or biomass of individual species of other plants, or of animals, as a consequence of modification of their habitats'. The first alien aquatic plant to be introduced to Australia was probably *Elodea canadensis* because it was in Tasmania in 1868 and was taken from there to New Zealand (Howard-Williams et al., 1987). Mitchell (1978) listed 32 plant taxa that he considered to be serious aquatic weeds in Australia. Ten of these are alien to Australia and there have been further introductions since then. Humphries et al. (1991) listed eight alien aquatic and semi-aquatic plants as serious environmental weeds. Together with probable dates of introduction, these are the free-floating plants *E. crassipes* (1884) and *Salvinia molesta* (1952); the wet pasture/wetland grasses *Brachiaria mutica* (early this century?), *Echinochloa polystachia* (1990), *Glyceria maxima* (early this century?), and *Hymenachne amplexicaulis* (1990); the floodplain shrub *Mimosa pigra* (late 19th century) and the marine kelp *Undaria pinnatifidia* (1982). Also some Australian native species can be invasive elsewhere in Australia. Thus *Typha orientalis*, which is native to the eastern States, is invading wetlands in Western Australia at the expense of species that are native to those systems (Zedler et al., 1990).

The advantage of Australia's biogeographic isolation is that the clear boundary that was responsible for the degree of isolation can be defended as a barrier against further invasion and, for a number of years, Australia has maintained a strict policy of preventing the introduction of unwanted plant and animal material.

Responsibilities for management of water weeds in Australia

The Commonwealth of Australia consists of eight States and Territories: Western Australia, Northern Territory, South Australia, Queensland, New South Wales, Victoria, Tasmania and the Australian Capital Territory. Each is responsible for the management of weeds in its area of jurisdiction. Each is also responsible for the management of its water resources. However, in some States, water weeds are managed by one or more water resource authorities, in others by weed officers in Departments of Agriculture and in some by both. Thus in Western Australia, the Waterways Commission is responsible for the management of weeds in estuaries (which can include whole catchments involving both fresh and brackish water), the Water Authority for inland waters and the Department of Conservation and Land Management for water bodies in national parks. The Agricultural Protection Board is responsible for the administration of the Noxious Weeds Act and will either require the landowner to carry out control measures against a particular plant or will carry out the work themselves and charge the cost to the landowner. Local Government is often involved in such weed management programs as a 'landowner'. In more complex situations a broadly based cooperative group may be established. For example a working group comprising State Government agencies, Local Government and community groups was set up by the Swan River Trust to develop a strategy for short-term control and long-term eradication of *Hydrocotyle ranunculoides* on that River (Klemm et al., 1993).

In New South Wales, by contrast, Local Government authorities are responsible for control of aquatic weeds under the overall direction of NSW Agriculture which operates on a regional basis within the State. The Department of Water Resources is responsible for weed control on its water storages and irrigation systems, while the Environment Protection Authority maintains jurisdiction over pollution of water bodies including pollution by herbicides. The other States and Territories operate on a broadly similar basis to New South Wales, though there are differences in detail. Within each State or Territory there is generally good interaction and good coordination of water weed management programs. However, Mitchell (1978), noted that, in the mid-1970s, there was comparatively little exchange of information between the States with respect to water weed management and that a more coordinated approach would be beneficial especially for troublesome alien plants, such as water hyacinth.

In 1979 a National Committee on Management of Aquatic Weeds was formed to recommend a coordinated national approach for the effective control of aquatic weeds in Australian waterways. This was succeeded in 1982 by the National Coordinating Committee on Aquatic Weeds (NCCAW), which is still in existence, although it has not formally met for several years and now functions as a corresponding network. Information is circulated when necessary to provide early warning of possible new problems that have the potential to have an Australia-wide impact and advice is provided when requested. An annual State-by-State assessment is made every year and these, together with appropriate recommendations, are compiled into an annual report on aquatic weeds in Australia for the Australian Water Resource Management Committee and the Australian Weed Committee.

Over the last 15 years better coordination of water weed management in Australia has progressively developed into a series of national activities, with particular emphasis on preventative measures. The National Committee on Management of Aquatic Weeds (1982) recommended that four species, *Alternanthera philoxeroides*, *E. crassipes*, *S. molesta* and *Lagarosiphon major* should be declared as noxious weeds for the whole of Australia. Eleven other species, *Pistia stratiotes*, *Egeria densa*, *E. canadensis*, *Brachiaria mutica*, *Glyceria maxima*, *Sagittaria graminea*, *Myriophyllum aquaticum*, *Trapa natans*, *Cabomba caroliniana*, *Myriophyllum spicatum* and *Stratiotes aloides* were nominated as serious or potentially serious weeds that would be more sensible to manage on a State or local level. It is noteworthy that *L. major* in the first list and the last four species in the second list were not known to be naturalised anywhere in Australia at that time. Also, all the plants listed were thought to be alien to Australia and thus could be barred from further introductions by the Australian Quarantine Service in support of this preventative approach. Subsequently, it has been established that *P. stratiotes* is almost certainly native to the Indo-Melanesian element of the Australian flora in the northern part of the continent. However, out of that element it behaves as an alien weed and quarantine measures against further introductions and efforts to prevent its spread are justified.

A nation-wide public awareness campaign was planned and progressively initiated during the 1980s. This included a poster illustrating the four plants that had been declared noxious for the whole country, a series of fact sheets, a video entitled 'The Good, Bad and Beautiful' and an illustrated field guide by Sainty & Jacobs (1988). NCCAW also recommended plants to be placed on quarantine lists, compiled a publication entitled 'Guidelines on the use of herbicides in and near water' (Australian Water Resources Council, 1985) and undertook a review of research and research needs. Research on *A. philoxeroides* was given top priority and research was funded on chemical control of this species and aspects of its physiology (Bowmer et al., 1993, Bowmer & Eberbach, 1993). This work proved to be of immense importance when the plant recently invaded wetland areas associated with the Murrumbidgee Irrigation Area in New South Wales (H. Milvain pers. comm.). In recent months the NCCAW has contributed to writing a National Strategy for all Australian weeds, for which its own actvities provided a useful model.

Interaction between local, regional and national strategies for management of water weeds

It is important to recognise that State boundaries have little or no ecological validity. It is therefore essential that weed management strategies are planned on a regional basis in relation to entire ecological units. An obvious example is the Murray-Darling Basin, an area of 1 058 800 km^2 (about 4.6 times the area of England, Scotland and Wales), which encompasses parts of Queensland, New South Wales, Victoria and South Australia and all of the Australian Capital Territory around Canberra. In 1975 Victoria and South Australia, which had successfully eliminated *E. crassipes* from their States, expressed serious concern about a 10 000 ha wetland infested with the weed in northern New South Wales in the Gingham Watercourse which drains into a tributary of the Darling River. A meeting of Commonwealth and State Ministers was held in July 1976 to approve an action plan proposed by the New South Wales Government. The work was carried out by New South Wales authorities and the costs shared among the governments concerned (Mitchell, 1978). This is a good example of successful coordination of water weed control at local and regional levels since the day-to-day weed control measures were mainly the responsibility of Local Government Officers, important drainage work was carried out by the State Department of Water Resources, general oversight was provided by the Department of Agriculture, and costs were shared among several States.

It is more difficult to establish a coordinated program when the objectives are less firmly held and only one Government is involved, as illustrated by the failure to establish an effective control program against *A. philoxeroides* in the vicinity of Newcastle on the north coast of New South Wales where it has infested wet pastures for a couple of decades. The weed is difficult and expensive to control in this situation and, even though concern was expressed that a very expensive problem would occur if the weed escaped and invaded irrigation systems in other parts of New South Wales, no serious attempt was made to reduce the extent of the infestation. However, the risk was difficult to assess, the cost of the control program was high, its outcome was uncertain and there was no external political pressure on State authorities. In 1993 the weed did invade water associated with one of the irrigation areas that were at risk and the potential for the plant to spread to other irrigation areas is now markedly increased. Fortunately the research promoted by the NCCAW meant that effective control measures were already established and could be applied immediately.

These two examples serve to illustrate the benefits and the difficulties of establishing cooperative weed management programs. In general, locally based units can be very effective in controlling weed infestation but often are not as effective in preventing problems. Preventative measures, including public awareness campaigns, etc. are best handled at the larger scale with the cooperation and support of local agencies. This approach has been adopted with the management of aquatic weeds in Australia with considerable, but not absolute, success. However, there is no doubt of the benefit of cooperation in the sharing of information and establishing uniformity in the naming of noxious weeds and in quarantine control. Also both local and larger-scale strategies benefit from well-planned public awareness campaigns and these provide further opportunity for positive interaction.

References

Australian Water Resources Council, 1985. Guidelines for the use of herbicides in or near water. Australian Government Publishing Service, Canberra.

Bowmer, K. H., P. L. Eberbach & G. McCorkelle, 1993. Uptake and translocation of ^{14}C-glyphosate in *Alternanthera philoxeroides* (Mart.) Griseb. (alligator weed). I. Rhizome concentrations required for inhibition. Weed Research 33: 53–57.

Bowmer, K. H. & P. L. Eberbach, 1993. Uptake and translocation of ^{14}C-glyphosate in *Alternanthera philoxeroides* (Mart.) Griseb. (alligator weed). II. Effect of plant size and photoperiod. Weed Research 33: 59–67.

Good, R., 1974. The geography of flowering plants. Longman Green & Co., London.

Groves, R. H., 1986. Plant invasions of Australia: an overview. In R. H. Groves & J. J. Burdon (eds), Ecology of biological invasions. Australian Academy of Science, Canberra: 137–149.

Howard-Williams, C., J. S. Clayton, B. T. Coffey & I. M. Johnston, 1987. Macrophyte invasions. In A. B. Viner (ed.), Inland Waters of New Zealand. DSIR, Wellington: 307–331.

Humphries, S. E., R. H. Groves & D. S. Mitchell, 1991. Plant invasions of Australian ecosystems. Kowari 2: 1–134 [National Parks and Wildlife Service, Canberra].

Klemm, V. V., N. L. Siemon & R. J. Ruiz-Avila, 1993. *Hydrocotyle ranunculoides*: a control strategy for the Canning River Regional Park. Swan River Trust Report No. 6, Perth, Australia.

Mitchell, D. S., 1978. Aquatic weeds in Australian inland waters. Australian Government Publishing Service, Canberra.

National Committee on Management of Aquatic Weeds, 1982. Water weeds in Australia: a national approach to management. Australian Water Resources Council, Water Management Series No. 3. Australian Government Publishing Service, Canberra.

Newsome, A. E., 1983. Dingo. In R. Strahan (ed.), The complete book of Australian mammals. Angus & Robertson, Sydney: 483–485.

Sainty, G. R. & S. W. L. Jacobs, 1988. Waterplants in Australia. Sainty & Associates, Sydney.

Zedler, J. B., E. Paling & A. McComb, 1990. Differential responses to salinity help explain the replacement of native Juncus kraussii by Typha orientalis in Western Australia salt marshes. Austral. J. Ecol. 15: 57–72.

Hydrobiologia **340**: 157–161, 1996.
J. M. Caffrey, P. R. F. Barrett, K. J. Murphy & P. M. Wade (eds), Management and Ecology of Freshwater Plants.

The economics of aquatic vegetation removal in rivers and land drainage systems

J. A. L. Dunderdale & J. Morris
Silsoe College, Cranfield University, Silsoe, Bedford MK45 4DT, United Kingdom

Key words: rivers, maintenance, vegetation, economics, longevity, agriculture

Abstract

One purpose of river maintenance within Britain is to deliver given standards of land drainage service relating to the control, within acceptable limits, of flooding and waterlogging on riparian, mainly agricultural land. Aquatic weed removal is a major maintenance activity. Authorities responsible for cost-effective river maintenance need to determine the extent and timing of vegetation removal in channels of various types. The impact of maintenance is being studied on 12 sites in five regions of the National Rivers Authority (NRA) in England and Wales. The impact of differing maintenance regimes on flooding and waterlogging and the consequences for agricultural performance are assessed. The longevity of maintenance in terms of the time taken for the 'without' maintenance watercourse condition to be reinstated following maintenance has been determined for gravel, sand and silt bed rivers on which vegetation cutting has been performed. The estimated benefits of river maintenance are set against costs to help formulate best maintenance strategies and prioritise and justify maintenance works.

Introduction

Increasingly, organisations responsible for river and drainage channel maintenance are required to justify and prioritise their maintenance activities in terms of costs and benefits. In the case of agriculture, the benefits of river maintenance are the avoidance of losses due to flooding and waterlogging associated with a lower standard of service in the absence of maintenance.

This paper reports on a study of the impacts of aquatic vegetation removal on standards of drainage service provided by lowland channels flowing within floodplains. Twelve sites throughout England and Wales provide the base information including vegetation and flow rates. Through morphological modelling, quantitative data on the effects of various vegetation densities on water levels and bankfull capacity are obtained.

Methods have been developed to determine financial benefits to agriculture of alternative standards of drainage service. These are combined with estimates of maintenance costs (and rates of deterioration in service following maintenance) to determine best strategies in terms of extent and frequency of maintenance.

Effect of maintenance of drainage standards of service

River maintenance activities such as desilting and weed cutting influence the profile and friction characteristics of the channel, which in turn influence the stage/discharge relationship. Thus, river maintenance influences river water levels and flood risk. Standards of drainage service for agriculture can be defined in terms of outfall facility for naturally or artificially drained land and flood risk. Hence the link between maintenance and standards of service.

Flood risk

River maintenance influences flood risk by increasing channel capacity and therefore bank full discharge (Qbf). Information regarding roughness and channel

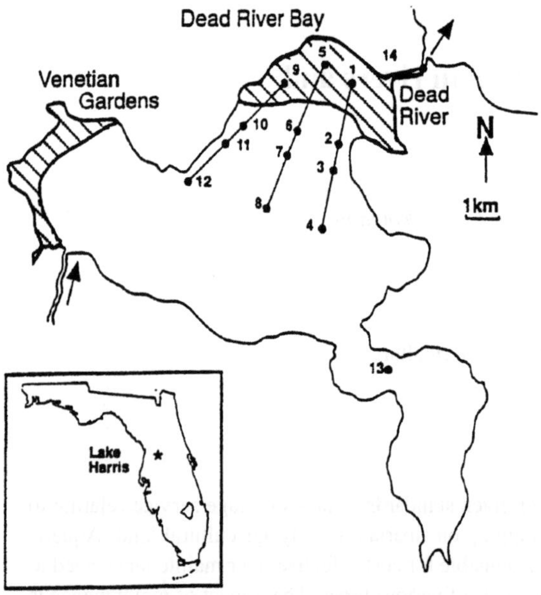

Figure 1. Impact of maintenance on flood risk and discharge, Woldgrift Drain

Table 1. Drainage status and financial returns.

% time Watertable at given depth	Watertable height (mm from surface)	Drainage status	Financial Return (£/ha 1994 prices)	
			Cereal/ oilseed	Intensive grass
>80	>500	Good	371	378
>50	<500 >300	Bad	271	315
>50	<300	Very bad	125	216

dimensions before and after maintenance are combined with data on flow frequency to assess changes in flood risk attributable to river maintenance. Figure 1 shows an example of the effect of maintenance on channel capacity and flood return period in the Woldgrift Drain, Lincolnshire.

Water table levels and drainage status

The influence of river and ditch water levels, and hence river maintenance, on field drainage conditions, taking into account local rainfall, evaportranspiration, soil type and drainage intensity was determined using a model developed at Silsoe College (Youngs et al., 1989). The model predicts water table depth in the

Table 2. Parameters for idealised channel models.

Channel bed type	Gravel	Sand	Silt
Depth (m)	1.5	1.25	1
Bed width (m)	12	9	6
Top width (m)	16	15	14
Bed slope	$4*10^{-3}$	$2.5*10^{-4}$	$1*10^{-4}$
Discharge range (m^3/s)	0.3 to 500	0.3 to 500	0.04 to 10
Bankful discharge (m^3/s)	30	9	3

adjacent floodplain (benefit area) against time for given river levels and weather conditions. These water table heights are then compared with those required for crop growth and trafficability of soils.

The model is run using water levels for the 'with' and 'without' maintenance situation. The same weather data are used for both cases to isolate the impact of maintenance on water table levels. It confirms circumstances where maintenance has a significant impact on drainage status and where neglect of maintenance may limit agricultural productivity.

The 'with' maintenance situation can be varied in order to determine the benefit of alternative levels of maintenance service in terms of water table levels, drainage status and flood risk.

Agricultural impacts

Drawing on research literature and farm surveys, three categories of drainage status have been defined which describe the level of agricultural productivity attainable under such conditions. Where the water table level lies within 0.3 m of the surface, productivity is under 'breakdown' conditions. Poor drainage imposes severe restrictions on land use. Levels between 0.3 m and 0.5 m command low levels of productivity as crop yields are depressed and access to the land by machinery and stock is restricted. Water table levels deeper than 0.5 m are not considered to impose restrictions on farming practices and productivity is 'normal'.

Drainage status has been defined in terms of the proportion of time the water table lies within these critical boundaries. A season with 80% of time with a water table deeper than 0.5 m from the surface is described as having 'good' drainage. A season which does not meet the latter criteria but has at least 50% of

Table 3. Effect of annual vegetation cutting on discharge capacity.

Channel type	Veg. cover*	Veg cut (%)	Year 1**	Year 5**	Year 10**
Gravel	10%	40%	0	0	0
		80%	+2	+1	+0.5
Sand	30%	40%	+6	+5	+4
		80%	+12	+10	+7
Sand	50%	40%	+10	+9	+8
		80%	+18	+18	+15
Silt	50%	40%	+20	+19	+19
		80%	+38	+37	+37
Silt	80%	40%	+17	+17	+17
		80%	+31	+31	+30

* Typical vegetation cover without maintenance.
** Values indicate change in discharge capacity as a % of original.

Table 4. Effect of annual vegetation cutting on water levels.

Channel type	Veg. cover*	Veg cut (%)	Year 1**	Year 5**	Year 10**
Gravel	10%	40%	0	0	0
		80%	–1	–0.6	–0.25
Sand	30%	40%	–3	–3	–2
		80%	–6	–5	–3
Sand	50%	40%	–5	–5	–4
		80%	–9	–9	–7
Silt	50%	40%	–6	–6	–6
		80%	–12	–12	–11
Silt	80%	40%	–6	–6	–6
		80%	–11	–11	–11

* Typical vegetation cover without maintenance.
** Values indicate change in water level as a % of original.

time with a water table deeper than 0.3 m commands 'bad' drainage and a season with at least 50% of time with a water table within 0.3 m of the surface results in 'very bad' drainage.

Information was derived from farm surveys and literature to compile agricultural productivity scenarios which describe the input-output relationships and resultant financial performance of land use and farming practice under the three categories of drainage status (Hess & Morris, 1985; Hess et al., 1989). For instance, drainage standards influence yields and crop options for arable farming and stocking rates, grazing seasons and the ability to make silage in the case of grassland. The estimates of benefits associated with the different productivity classes were used to determine the benefits attributable to the maintenance of standards of drainage service and thereby the benefit of river maintenance (Table 1). For example, if a good drainage status on land given to cereal and oilseed production generates a financial return of £371 ha^{-1} yr^{-1} and a bad drainage status generates a return of £271 ha^{-1} yr^{-1}, the net benefit of river maintenance which prevents a decline from good to bad drainage conditions is £100 ha^{-1} yr^{-1}. These estimates can be adjusted to 'economic' values which show benefits net of given subsidies. The latter are often used by public sector organisations to justify expenditure of public funds.

Recognising that the impact of maintenance varies with rainfall frequency and channel flows, the model was used to show the effect of different standards of maintenance on water table heights for different weather conditions. Using the methods described above, the drainage status and related benefits of maintenance schemes were estimated for wet, average and dry rainfall conditions and for 'with' and 'without' river maintenance. The benefits of maintenance were thereby identified by the change, if any, in drainage status for given climatic conditions. Drawing on historical weather data, the benefits were then weighted by the relative frequency of the three types of weather conditions to derive an average, expected annual benefit attributable to the maintenance activity.

A similar approach is adopted to estimate the benefits of flood alleviation associated with changes in maintenance standard. Estimates are derived of the cost of single or multiple flood events occurring on given land use (and productivity class) at different times of year and the change in flood risk associated with channel maintenance. For instance, if maintenance increases the time between floods from, for example, 2 to 3 years for which the average annual flood costs are £9 ha^{-1} and £6 ha^{-1}, respectively, the benefits of flood alleviation attributable to maintenance are £3 ha^{-1} over the area of flood risk.

Longevity of maintenance schemes

By definition, river maintenance is a repeat activity with intervals determined by rates of vegetation growth, siltation and erosion. The impacts of maintenance on sediment movement, deposition and erosion have been investigated using a numerical, morphological model based on typical channel parameters for British gravel, sand and silt bed rivers. These typical channel parameters have been derived from averaging the actual parameters of the 12 rivers in England and Wales which have been monitored for the purpose of this analysis. These 12 rivers are in themselves typical rivers of their type. The typical channel parameters used in the morphological modelling are presented in Table 2.

A model (Flumorph) developed by HR Wallingford, estimates changes in bankfull discharge and water level following maintenance and the longevity of these changes in terms of the number of years before the 'without' maintenance watercourse condition is reinstated. The longevity of aquatic vegetation management activities are referred to here.

Estimates have been obtained for typical gravel, sand and silt bed channels on the effect on discharge capacity and river water levels of removing 40% to 100% of the vegetation cover. This vegetation cover was set at levels which represented typical conditions according to levels which would occur in the absence of maintenance. All results presented here relate to the midpoint along the maintained reach and are based on the removal of submerged streaming weeds during the summer. The impact of removing emergent vegetation is simulated by using the results obtained from an analysis of the impact of channel widening.

The effect of maintenance on discharge and water levels over a period of 1 to 30 years has been studied in order to determine the longevity of various maintenance schemes in channels of different types. Flood return periods associated with these discharges are determined and hence the impact of maintenance on flooding over a period of 30 years is ascertained.

The effect of annual vegetation control on discharge capacity in gravel, sand and silt bed rivers varies according to the levels of vegetation 'without' maintenance and selected cutting intensity (Table 3). In a gravel bed channel where weed growth is usually low, the impact of cutting vegetation has a very small effect on discharge capacity. In sand rivers, improvements in discharge capacity due to annual weed control decline over time due to sedimentation. In sand bed channels therefore, weed cutting alone may not be sufficient to maintain standards of service. In silt rivers however, the additional discharge capacity provided by annual vegetation control is greater than for other river types and discharge capacities remain relatively constant over time.

Table 4 shows the impact of weed cutting on water levels. For typical vegetation cover and cutting regimes, the impact of annual cutting on water levels is greatest in silt rivers and least in gravel bed rivers. The longevity of the impact is also greatest in silt bed levels.

Justification of river maintenance

The benefits of flood alleviation when added to the benefits of drainage status provide an estimate of the total benefits attributable to river and drainage channel maintenance. These benefits, when aggregated within the benefit area of a maintenance scheme and set against the cost of providing the maintenance service, indicate the economic value of maintenance activities. Generally, the more intensive the land use, the better is the standard of field drainage (either due to piped underdrainage or freely draining soils) and the greater the sensitivity to standards of maintenance.

The results from modelling the longevity of various maintenance practices can be combined with flood risk assessment and water table modelling in order to determine the consequences for drainage status in the adjacent flood plain. Where agriculture is the dominant land use, estimates of financial benefits can be obtained and compared with the cost of maintenance in order to determine maintenance type, interval and justification.

Environmental impacts of river maintenance

Reconnaissance level physical monitoring (river corridor surveys) pre- and post- maintenance provide some data on the environmental impact of river maintenance. Given the greater importance now attached to environmental quality, environmental criteria such as invertebrates and fisheries need to be included more explicitly in the appraisal of river maintenance works both in terms of the design of and justification for maintenance. Circumstances where there are likely to be conflicts of interest in terms of the type of maintenance strategy to be adopted and where these different interests can be reconciled by modifications to maintenance regimes need to be identified. A separate study is being undertaken at present to address these issues. The results will be incorporated into guidelines for river maintenance appraisal and prioritisation.

Conclusions

The preceding discussion confirms that it is possible to formulate objective methods for the economic appraisal of weed control strategies which are justified in terms of providing standards of drainage service. The methods relate to lowland systems but can be adjusted to assess maintenance on highland carriers where the main service is the alleviation of flooding. The results presented here are applicable to the channel parameters examined but sensitivity analysis is being undertaken to determine circumstances in which the method and results can be applied with confidence.

References

Fisher K. R., J. A. L. Dunderdale, J. Morris & C. E. Reeve, 1994. The longevity of river maintenance. In CIGR., XII World Congress on Agricultural Engineering. CIGR General Secretariat, Belgium.

Hess, T. M. & J. Morris, 1985. A computer model for agricultural land drainage scheme appraisal. MAFF Conference of River Engineers, Cranfield, 16–18 July, 1985.

Hess, T. M., P. B. Leeds-Harrison & J. Morris, 1989. The evaluation of river maintenance in agricultural areas. In Agr. Eng.: 501–507.

Morris, J. & D. C. Sutherland, 1992. The evaluation of river maintenance. MAFF Conference of River and Coastal Engineers, Loughborough University, 6–8 July, 1992.

Morris, J., D. C. Sutherland & J. A. L. Dunderdale, 1994. River maintenance evaluation.

White, W. R. & J. Watts, River Flood Hydraulics. Wiley & Sons, Chichester, 491–500.

Youngs, E., P. B. Leeds-Harrison & J. M. Chapman, 1989. Modelling water table movement in flat low-lying lands. Hyd. Proc. 3: 301–315.

Hydrobiologia **340:** 163–172, 1996.
J. M. Caffrey, P. R. F. Barrett, K. J. Murphy & P. M. Wade (eds), Management and Ecology of Freshwater Plants.

The management of weeds in irrigation and drainage channels: integrating ecological, engineering and economic considerations

P. J. Barker, C. M. Ferguson, I. K. Smout & P. M. Wade
Water, Engineering and Development Centre, Loughborough University, Loughborough, Leicestershire LE11 3TU, United Kingdom

Key words: irrigation channel, drainage channel, aquatic weeds, hydraulic performance, cropping calendar, engineering economics

Abstract

The management of aquatic weeds in an irrigation scheme is constrained by the agro-economic system in relation to scheme layout, the nature and ecology of the aquatic weeds, agricultural practice, irrigation and drainage requirements, and the available resources for maintenance. The way in which the ecology, engineering and economics of irrigation and drainage channels interact to produce a pattern of management is investigated for the Mwea Irrigation Settlement Scheme, Central Province, Kenya. This is used to develop a simple model which enables the economic implications of varying the aquatic weed management practice to be identified. The model brings the selection of a weed control programme within the principles of engineering economy.

Introduction

Mwea Irrigation Settlement Scheme is located in the Kirinyaga District of Central Province, near Embu, Kenya. It extends over 12 140 ha and is the largest producer of rice in Kenya (JICA, 1988). The scheme is managed by the National Irrigation Board (NIB) but rice production is carried out by tenant farmers in accordance with a cropping schedule prepared by the NIB (Figure 1a).

The layout of the irrigation and drainage systems at Mwea is typical of irrigation schemes throughout the developing world (Kay, 1986). The scheme is divided into sections, each of which is administered by a NIB Irrigation Officer. Individual sections are sub-divided into units which, in turn, are split into fields. Irrigation water originates from the Thiba and Nyamindi Rivers and is distributed by gravity through a network of predominantly unlined open channels. Link and main canals (primary channels) and branch canals (secondary channels) convey water into the sections. Main or unit feeders (tertiary channels) carry water from the main and branch canals into the individual units, and feeders (quaternary channels) distribute water to the fields. Each feeder serves two lines of fields. The standard field measures 0.4 ha and is rectangular, with one short side abutting on the feeder, the other adjoining the field drain.

Drainage at Mwea is provided by networks of drains which discharge into the Kiruara, Thiba, Murubara or Nyamindi river. Field drains (quaternary channels) running almost parallel to the feeders on the opposite sides of the fields, discharge into collector drains (tertiary channels) which evacuate water from the units. Collector drains may deliver drainage water directly to a river; alternatively, they flow into main drains (primary or secondary channels) and thence into a river.

The growth of aquatic weeds in irrigation and drainage channels of schemes such as Mwea increases resistance to water flow (Chow, 1983; Brabben & Bolton, 1988), reducing the system efficiency. The NIB, in conjunction with the tenant farmers, has developed a channel maintenance programme integrated into the crop production cycle. This paper describes the management cycle for the scheme which takes into account both the crop and the weeds in the channels. Consideration is then given to the way in which the ecology, engineering and economics of the channels interact to produce the current pattern of management. A simple model is developed which enables the economic implications of varying the aquatic weed management practice to be identified.

Management programme

The management of weeds in irrigation and drainage channels at Mwea Irrigation Settlement Scheme is summarised in Table 1. Maintenance is apportioned between the NIB Works Department and the tenant farmers. The Works Department is wholly responsible for the primary and secondary channels and employs both mechanical and manual means of weed control. Mechanical control involves the use of hydraulic excavators to remove silt and weed from the channels (dredging), whilst manual control comprises clearance of weeds and some silt with simple hand-tools such as machetes.

The individual farmers at Mwea are obliged to maintain the irrigation and drainage facilities which directly serve their holdings (feeders and field drains, i.e., the quaternary channels). Channel clearance is carried out by hand, as described above. The management of weeds in main or unit feeders and collector drains is undertaken by the Works Department and the farmers. The Works Department generally shoulders the responsibility when the channels require dredging. Management on the part of farmers involves a communal effort by those individuals served by a particular watercourse.

The Works Department's weed management programme is largely dictated by climate and the cropping calendar (specifically the irrigation and drainage requirements) and the available resources of labour and hydraulic machinery. From December to mid-March, following harvest, rice fields are dry and free from any crop so any in-field maintenance which requires machinery is carried out at this time since plant can pass freely through the fields. During the same period management of the irrigation system commences with the primary and secondary canals and those tertiary canals serving the fields which are to be rotavated early in the year. Drainage is not an important function at this time; however, major drains are dredged during this period in preparation for the long rains in March/April. With the arrival of the long rains resources are focused on drainage maintenance to prevent water-logging (and bogging of tractors) in the fields, and to prevent water from over-topping drains and flooding in-field roads, restricting vehicular access.

Irrigation system maintenance recommences in May/June, canals being cleared systematically in advance of irrigation and rotavation of the fields they serve. September and October represent a critical period for water management. The demand for water is high since all the fields are under crop and high temperatures cause considerable evapo-transpiration (Figure 1b, c). Coincidentally, river flows are at their lowest during this period. It is imperative that the irrigation system has been maintained by this time.

During the period September–November the focus of the maintenance program reverts to the drainage system in preparation for the short rains in October/November and for the pre-harvesting drying-off period (Figure 1b, c). At this time main drains are maintained and flood protection works are carried out in the river channels.

The Works Department's management of weeds in irrigation and drainage channels is not confined to dredging. The recovery rate of vegetation is very rapid and the Works Department periodically deploys maintenance gangs to clear weeds from the channels by hand (Table 1).

Figures 2a, b illustrate the variability of clearance effort over an agricultural year at Mwea Irrigation Settlement Scheme. The maintenance records for 1992 indicate that the peak period for canal maintenance was May to July, and for drain maintenance, July to October. The records show that the allocation of labour and hydraulic machinery is consistent with the reported priorities of the management programme. The overwhelming requirement here is that rice-harvesting should commence in December and be completed as quickly as possible thereafter. All clearance effort is planned to secure this objective.

Ecology

The primary factor in managing weeds in the channels of irrigation schemes such as Mwea is the succession of the vegetation through seven clearly recognisable stages post-maintenance (Figure 3). The succession from one stage to the next is rapid due to both favourable light and temperature regimes, and the persistence of rhizomes, roots and other propagules in the channel bed beyond the reach of current maintenance techniques. The size of channel is also a significant factor in the successional process. The deeper and wider

Table 1. Weed management in irrigation and drainage channels at Mwea Irrigation Settlement Scheme.

Channel	Dimensions	Flow Regime	Principal Weeds	Maintenance Activity
Main Canal[a]	13330 m total length 2.00–6.50 m base width 0.80–1.50 m canal height 0.56–1.31 m water depth 1.95–6.35 m^3 s^{-1}	Flow year round – water supplied for domestic use as well as irrigation	*Acmella caulorhiza Commelina* sp. *Cyperus dives Polygonum senegalense*	Dredging to remove silt and weeds, once per year, in January/February Manual clearance of weeds, using pangas, hoes or spades, twice per year in June and September
Branch Canal[a]	45580 m total length 0.30–3.50 m base width 0.30–1.40 m canal height 0.12–1.23 m water depth 0.04–2.73 m^3 s^{-1}	Flow dependent on cropping program – February to November	*Acmella caulorhiza Ageratum conyzoides Commelina* sp. *Cyperus latifolia Eclipta alba Leersia hexandra Ludwigia abyssinica Panicum repens Polygonum senegalense Rhynchosia* sp.	Dredging to remove silt and weeds, once per year, before area served by canal is irrigated Manual clearance of weeds, using pangas, hoes or spades, twice per year
Main / Unit Feeder	*c.* 1.50–3.00 m bank top 0.028 m^3 s^{-1} / 20.25 ha (1 cusec / 50 acres)[b]	Flow dependent on cropping program – February to November	*Acmella caulorhiza Ageratum conyzoides Centella asiatica Commelina* sp. *Cynodon dactylon Leersia hexandra*	Dredging to remove silt and weeds, infrequently, as required Manual clearance of weeds, using pangas, hoes or spades, twice per year, during production period
Feeder	*c.* 0.50–2.00 m bank top 0.028 m^3 s^{-1} / 20.25 ha (1 cusec / 50 acres)[b]	Flow dependent on cropping program – February to November; period varies from 6–10 months	*Commelina* sp. *Cynodon dactylon Leersia hexandra*	Manual clearance of silt and weeds, using pangas and hoes, three times per year, before flooding, before transplanting and before top-dressing
Field Drain	*c.* 1.50–3.00 m bank top 0.003 m^3 s^{-1} / 1 ha) (0.05 cusec / acre)[b]	Flow dependent on cropping program – February to December; period varies from 6–10 months	*Commelina* sp. *Cynodon dactylon Echinochloa colona Fimbristylis* sp.*Leersia hexandra Ludwigia stolonifera*	Dredging to remove silt and weeds, infrequently, as required Manual clearance of weeds, using pangas and hoes, three times per year, before flooding, before transplanting and before draining for harvest
Collector Drain	*c.* 1.50–3.50 m bank top 0.003 m^3 s^{-1} / 1 ha) (0.05 cusec / acre)[b]	Flow dependent on cropping program – February to December	*Commelina* sp. *Cynodon dactylon Leersia hexandra Ludwigia stolonifera Panicum repens*	Dredging to remove silt and weeds, infrequently, as required Manual clearance of weeds, using pangas, hoes or spades, twice per year, in April/May and November/December
Main Drain[a]	32800 m total length 1.50–15.00 m base width 0.70–3.20 m canal height 0.42–2.83 m water depth 1.00–40.90 m^3 s^{-1}	Flow dependent on cropping program – February to December	*Commelina* sp. *Cynodon dactylon Echinochloa colona Echinochloa pyramidalis Leersia hexandra Marsilea* sp. *Polygonum senegalense Typha latifolia*	Dredging to remove silt and weeds, once per year Manual clearance of weeds, using pangas, hoes or spades, two or three times per year, before rains

[a] Design specifications JICA (1989)
[b] Design specifications Chambers and Moris (1973)

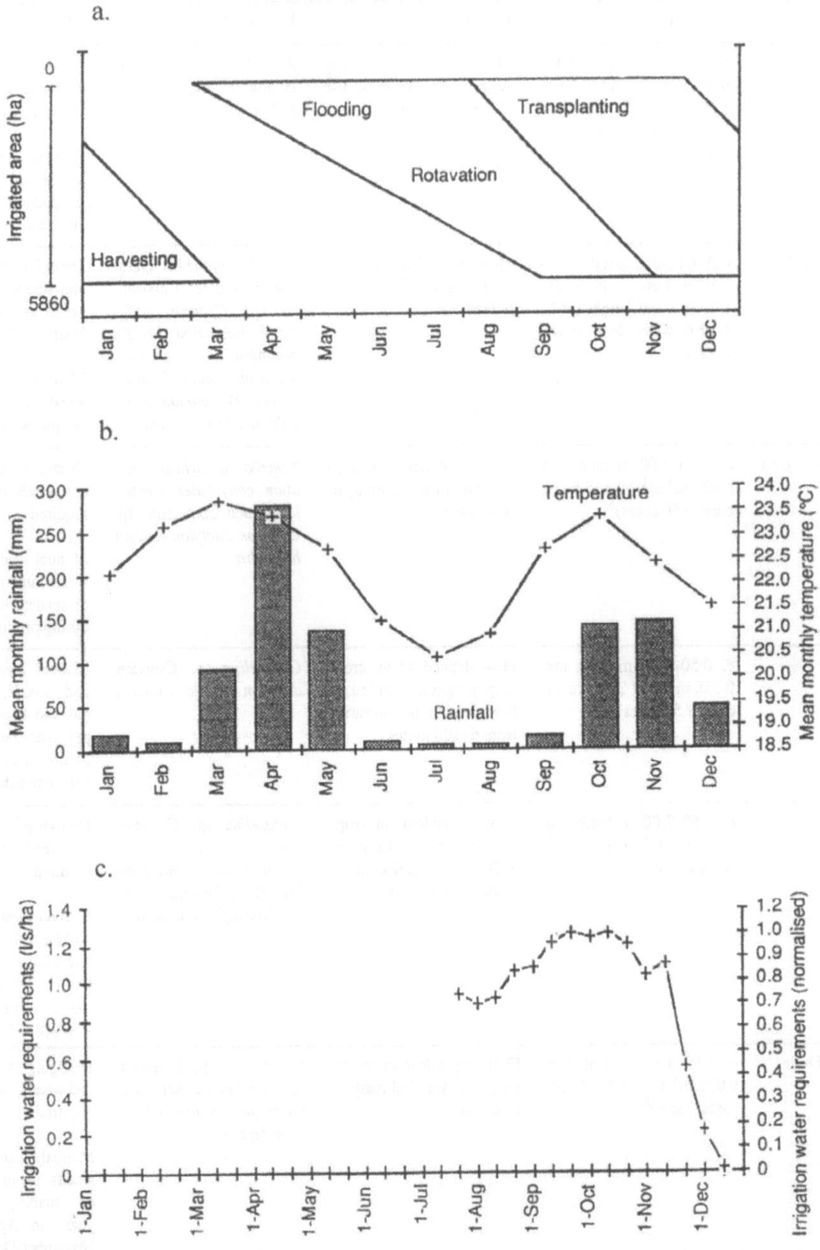

Figure 1. a. Rice cropping schedule at Mwea Irrigation Settlement Scheme. b. Mean monthly rainfall and temperature at Mwea Irrigation Settlement Scheme. c. Irrigation water requirements at Mwea Irrigation Settlement Scheme. (Data not available for January to August.

the channel, the slower the rate of change. Primary and secondary channels at Mwea exhibited all stages of succession, whereas smaller tertiary and quaternary channels characteristically passed from the open water stage directly to one with a high percentage cover of emergent grasses.

Certain species (*Cyperus latifolius* Poir., *Ludwigia abyssinica* A. Rich. and *Polygonum senegalense*

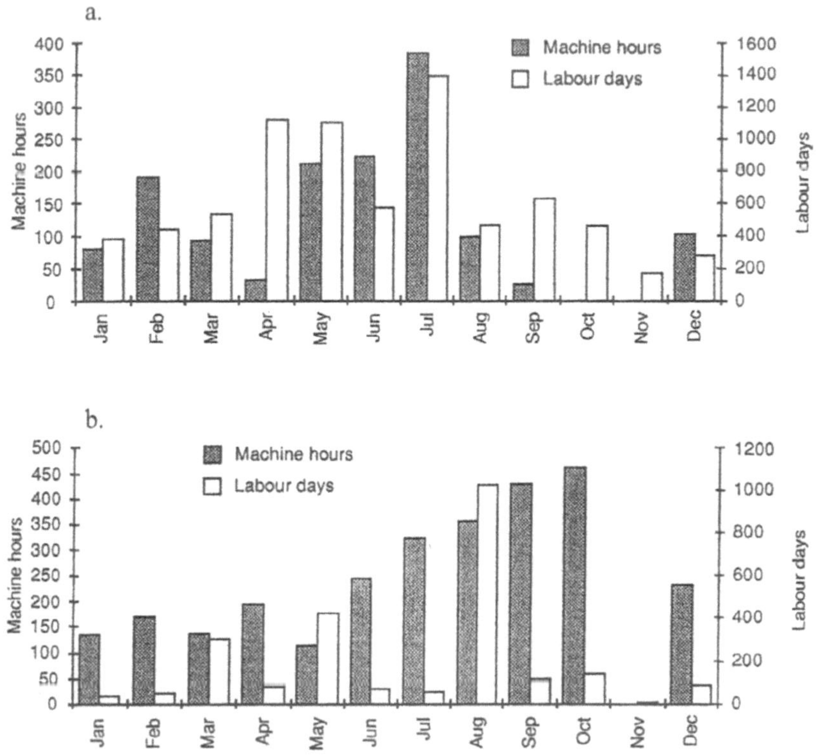

Figure 2. a. Canal maintenance at Mwea Irrigation Settlement Scheme. b. Drain maintenance at Mwea Irrigation Settlement Scheme.

Meisn.) were found to grow only in primary and secondary irrigation channels at Mwea: the depth of water combined with high turbidity preventing the growth of submerged species and the rate of flow inhibiting floating species. Flow and depth combined to slow the rate of encroachment of emergent vegetation. The rate of growth of emergent species in tertiary and quarternary channels was observed to be slower in those channels with a flow, i.e., during irrigation or drainage, than in those with still water conditions.

There are several differences between irrigation and drainage channels such that the ecology of each of these channel types is distinct. For example, flow is typically faster and turbidity usually higher in irrigation channels, particularly in primary and secondary canals. However, irrigation and drainage channels are both temporary aquatic habitats. At Mwea, tertiary and quaternary irrigation channels are without water for two to seven months of the year, and tertiary and quaternary drainage channels are dry for one to six months. Primary and secondary irrigation channels flow almost year-round because, in addition to their irrigation function, they supply water for domestic use. Similarly, primary and secondary drains flow almost year-round since they collect the tail waters from the primary and secondary irrigation channels and in some cases provide land drainage for areas outside the scheme.

Engineering

The management of irrigation and drainage channels can be analysed by using the concepts of condition and performance. The condition of a canal or drain at a particular time depends on the degree of structural and dimensional deterioration, weed infestation, and siltation. The condition worsens over time, but can be improved by maintenance operations.

The weed-related condition of the channel can be represented by its successional stage (Figure 3). Weed clearance changes a channel from a poorer to a better hydraulic condition by returning it to an earlier stage of

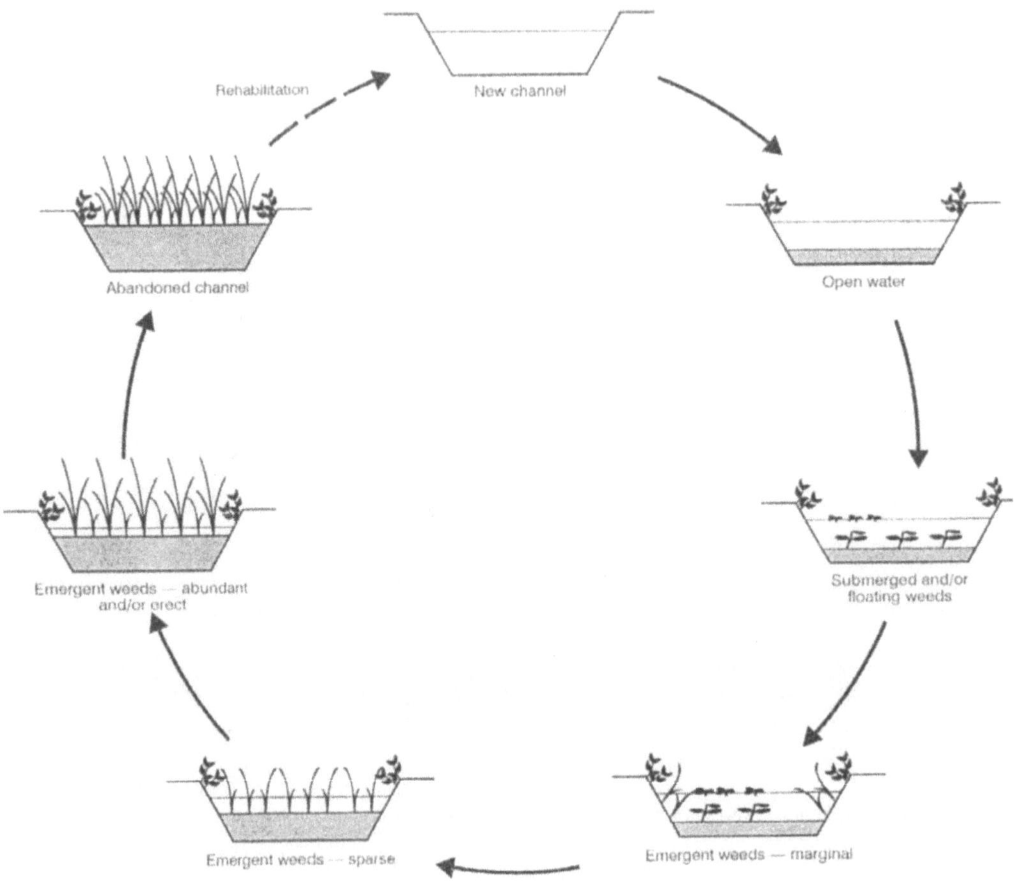

Figure 3. Stages in the succession of vegetation in irrigation and drainage channels.

succession. The silt-related condition can be represented similarly, but siltation normally occurs over a longer timescale, requiring less frequent clearance. Dredging operations remove weed, including root material, at the same time as silt, thereby returning the channel to an earlier stage of succession than do weed clearance operations.

The performance of a canal or drain at a particular time can be expressed by reference to its hydraulic objective: to pass a target discharge along the channel, while ensuring that the freeboard is not less than the target freeboard. The target discharge varies during the year with the irrigation requirements, depending on the crop calendar and climate (Figure 1c). In contrast, the target freeboard would normally be the same throughout the year to provide a safety margin against water over-topping the bank.

Performance can be represented quantitatively by the Delivery Performance Ratio (DPR) and the Freeboard Ratio (FBR), defined as follows:

$$\mathrm{DPR} = \frac{\text{Actual discharge}}{\text{Target discharge}};$$

$$\mathrm{FBR} = \frac{\text{Actual freeboard}}{\text{Target freeboard}}.$$

For optimum performance at a particular time: DPR = 1 and FBR = 1 or >1.

The actual freeboard at any time will depend on both the actual discharge, and the condition of the channel (Q, *n* and A in the Manning equation (Chow, 1983)). At those times of the year when the discharge is low, a poorer channel condition can be tolerated which will still pass the current target discharge at the target freeboard.

Economics

The central economic principle guiding irrigation and drainage maintenance (including weed management) is marginalist theory: maintenance is worthwhile only when its marginal benefit is greater than its marginal cost. Benefits may be thought of as additional crop values secured by improved yields, better quality produce, or both. They may also take the form of costs avoided, for example, costs attributable to bogged down machinery when drainage is inadequate.

Maintenance effort is governed by the need to convey water to and from the fields. Both of these imperatives require minimum levels of channel performance which vary according to season. At times when performance standards can be relaxed without jeopardising benefits, less effort and cost can be put into maintenance.

The Works Officer at Mwea prioritises the maintenance programme in accordance with the specific tasks required and the specific location of those tasks. Decisions in the formulation of the maintenance programme are largely determined by reference to system performance, but consideration is also given to equity amongst the tenant farmers. The need to disperse machinery to pursue equity occasionally conflicts with the aim of minimising costs.

Although the Works Officer formulates an efficient and fair maintenance programme which meets the requirements of the crop, the current pattern of management at Mwea is restricted to the achievement of short-term goals. It does not take account of the ecology of the succession of different weed communities which comprise the channel life-cycle (Figure 3) in that, in some instances, maintenance at an earlier stage in the cycle could slow down the succession. This could reduce the necessity for maintenance over the medium or even long term.

The current management strategy at Mwea is just one of a series of strategies which are potentially available to fulfil the programme. Other combinations of differing capital (hydraulic machinery) and labour intensity may be constructed to fulfil the maintenance programme. Alternatively, the input mix may be of machinery and herbicides, labour and herbicides, or include biological control. The viability of such a change to the maintenance regime would depend on how it might affect the crop cycle and whether or not there would be an economic gain.

The array of potential strategies could be filtered down to a small number of two or three by consideration of local economic and technical conditions. In developing countries some of the more important conditions might be:

- availability of labour, bearing in mind other labour-intensive demands (e.g., planting and harvesting crops);
- availability of hydraulic equipment and the need for maintenance facilities, and the need to optimise machine utilisation by spreading channel maintenance activities over time;
- availability of fuel, spares and skilled operatives for hydraulic equipment;
- availability of herbicides;
- public health and safety concerns (e.g., in the use of herbicides);
- weed type and growth characteristics which determine frequency of maintenance operations;
- severity of silting;
- variation in target discharge and hence permissible channel condition during the year.

Consideration of these factors will rule out some potential strategies. For example, at Mwea the use of irrigation water for drinking and bathing rules out certain types of herbicide application in irrigation channels and periodic labour shortages necessitate the use of machinery.

The identification of two or three feasible control strategies leads on to a more detailed specification of each maintenance programme and quantification of inputs (e.g., labour and machinery) required to accomplish it. Knowledge of input requirements and input costs allows unit costs to be calculated. Specification of a programme facilitates the breakdown of costs into capital (fixed) and operation and maintenance (variable) cost categories and, importantly, identification of their incidence through time (Table 2). A maintenance programme should be viewed over a planning period (e.g., 15 years) which allows for the inclusion of episodic components such as silt removal.

With costs classified and the years over which expenditure will occur identified, the selection of a single maintenance programme from the contenders can be accomplished by viewing each programme as an investment project with expenditures flowing through time. The flow of expenditures is likely to be uneven over the planning period because of the differing nature of maintenance tasks and their varied input requirements. No single year will be representative of resource expenditures and the whole programme should be viewed as an interdependent and sequential

Table 2. Maintenance expenditure on 90 km of primary and secondary canals at Mwea Irrigation Settlement Scheme.

Year	Inputs	Input Costs Per Unit	Number of Units	Annual Input Cost	Annual Total Input Cost	Discount Factor 20%	Present Value of Costs
1	Capital cost of excavator	9,000,000.00	5	11,250,000.00			
	Annual recurrent costs of excavator	701,108.63	5	876,385.79			
	Capital cost of hand tool (panga)	120.00	60	1,183.56			
	Annual cost of labour for cutting	33.42	3600	120,312.00	12,247,881.35	0.833	
2	Annual recurrent costs of excavator	701,108.63	5	876,385.79			
	Annual cost of labour for cutting	33.42	3600	120,312.00	996,697.79	0.694	
3	Annual recurrent costs of excavator	701,108.63	5	876,385.79			
	Annual cost of labour for cutting	33.42	3600	120,312.00	996,697.79	0.579	
4	Annual recurrent costs of excavator	701,108.63	5	876,385.79			
	Annual cost of labour for cutting	33.42	3600	120,312.00	996,697.79	0.482	
5	Annual recurrent costs of excavator	701,108.63	5	876,385.79			
	Annual cost of labour for cutting	33.42	3600	120,312.00	996,697.79	0.402	
6	Annual recurrent costs of excavator	701,108.63	5	876,385.79			
	Annual cost of labour for cutting	33.42	3600	120,312.00	996,697.79	0.335	
7	Annual recurrent costs of excavator	701,108.63	5	876,385.79			
	Annual cost of labour for cutting	33.42	3600	120,312.00	996,697.79	0.279	
8	Capital cost of excavator	9,000,000.00	5	11,250,000.00			
	Annual recurrent costs of excavator	701,108.63	5	876,385.79			
	Annual cost of labour for cutting	33.42	3600	120,312.00	12,246,697.79	0.233	
9	Annual recurrent costs of excavator	701,108.63	5	876,385.79			
	Annual cost of labour for cutting	33.42	3600	120,312.00	996,697.79	0.194	
10	Annual recurrent costs of excavator	701,108.63	5	876,385.79			
	Annual cost of labour for cutting	33.42	3600	120,312.00	996,697.79	0.162	
11	Annual recurrent costs of excavator	701,108.63	5	876,385.79			
	Capital cost of hand-tool (panga)	120.00	60	1,183.56			
	Annual cost of labour for cutting	33.42	3600	120,312.00	997,881.35	0.135	
12	Annual recurrent costs of excavator	701,108.63	5	876,385.79			
	Annual cost of labour for cutting	33.42	3600	120,312.00	996,697.79	0.112	
13	Annual recurrent costs of excavator	701,108.63	5	876,385.79			
	Annual cost of labour for cutting	33.42	3600	120,312.00	996,697.79	0.093	
14	Annual recurrent costs of excavator	701,108.63	5	876,385.79			
	Annual cost of labour for cutting	33.42	3600	120,312.00	996,697.79	0.078	
15	Capital cost of excavator	9,000,000.00	5	11,250,000.00			
	Annual recurrent costs of excavator	701,108.63	5	876,385.79			
	Annual cost of labour for cutting	33.42	3600	120,312.00	12,246,697.79	0.065	
	Sum of present value of costs						**17,385,454.54**
	Sum of present value of costs per kilometre	17,385,454.54 / 90					**193,171.72**
	Annualised cost per kilometre	193,171.72 × 0.214 (capital recovery factor)					**41,338.75**

The cost estimates above are in Kenyan shillings and based on operating conditions at Mwea Irrigation Settlement Scheme and, in this instance reflect operation of a Komatsu PC200–5 hydraulic excavator. All costs are measured in constant 1994 prices. No allowance for future inflation is included in the investment appraisals. The annual recurrent costs include insurance, road tax, operator wages and operation and maintenance costs.

series of activities through time. Some expenditure will be employed early in the planning period and some will be employed later. The former involves a larger sacrifice to the agency due to the loss of interest-earning potential.

To reflect the declining burden of later costs, decreasing weights (discount factors) are applied to annual costs in order to bring the series of costs through time to their present value. (Table 2 illustrates a calculation of the annualised costs in Kenyan shillings of dredging 90 km of primary and secondary canals once per year over a 15-year period). The discount rate is typically taken to be the interest rate that the agency has to pay on borrowed funds, or the interest rate that it might have earned on invested funds. Application of the discount rate through time allows the present values of the costs of alternative control programmes to be calculated and the selection becomes a matter of choosing the least cost programme.

The investing agency may find it useful to know the constant sum of money required on an annual basis to fund the selected programme. This may be readily achieved by multiplying the present value of costs by the appropriate capital recovery factor to determine the annualised cost (Table 2). For a specified number of years and at a specified interest rate, the capital recovery factor determines the constant annual sum that must be recovered in order to finance capital borrowed plus interest charges incurred to implement a control programme. This level sum of money has to be generated either through grants, loans or farmer payments to finance the selected least cost programme. It makes a valuable contribution to the agency in that it indicates the affordability of a programme over the entire planning period. Application of the model outlined above brings weed and silt control programme selection within the principles of engineering economy.

Potential for increase in efficiency

Economic efficiency requires that either output (maintenance contribution to system performance) is maximised for a given endowment of inputs, or a specified standard of system performance is achieved at the least cost of resources. As the proposed management objective for irrigation and drainage is the attainment of a standard of system performance, it is the second interpretation which is relevant in this context. To meet this objective, maintenance programmes should be formulated to fulfil performance targets as required to meet the water needs of the agricultural cycle. Feasible programmes should then be subjected to least cost analysis over a lengthy planning period.

Input availability should be inventoried and suitable measures of the productivity of maintenance inputs should be constructed. Measures such as distance or area cleared per worker or machine should be recorded. Field observations of the performance of different machines, classes of labour and chemicals should be made in order to measure the productivity of inputs under a variety of working conditions. At the same time, output indicators must be formulated and to this end channels should be classified according to their function and size.

The condition of channels should be assessed in terms of the extent of weeds and their significance for system performance. Because different channels have varying significance for system operation different standards of performance can be tolerated. Permissible minimum standards for each channel or network of channels need to be set allowing for variability over time. This exercise is set against the need to meet crop water requirements through irrigation and drainage at the appropriate times. The prerequisite to successfully accomplish these aims is the clear identification of the crop requirements over time. The agricultural cycle determines the permissible variation in channel performance over the year and consequently the intensity of clearance effort.

To achieve the specified performance objective at channel level, a feasible programme of maintenance needs to be designed taking account of the local constraints on input use. The necessary inputs to accomplish this programme are then identified and quantified. Recognition of the constraints is important because they mould the design of the feasible programme. This procedure is employed for each primary and secondary channel and at tertiary and quaternary level for networks of channels. In this way a series of programmes is designed and their input requirements recorded. The disaggregated system input requirements are then compared with the stock of available resources and, where necessary, adjustments in terms of amount or type of inputs made. The skill of the manager is in iteratively reallocating inputs to render compatible total requirements with the resource base whilst accomplishing the objectives of the system. Given the multiplicity of inputs and the size of irrigation systems, several overall feasible programmes capable of fulfilling system objectives may emerge. Each of these overall programmes can then be subjected to the least cost analysis as outlined above.

Irrigation managers report the importance of experience in the formulation and practice of maintenance programmes. Subjective evaluations of programmes can be greatly enhanced by systematic monitoring of individual programme performance. Realised input productivities can be recorded and compared with historical and expected performances. Targets can be set and in the wider context of system management, incentives and, where necessary, sanctions deployed to enhance system performance.

Acknowledgments

This paper was produced as part of a research project on Management of Weeds in Irrigation and Drainage Channels (R5500) funded by the Overseas Development Administration, UK. Thanks are also expressed to Dr S. K. Mutiso, University of Nairobi, Grace Thendi, John Honor, Welland and Deepings Internal Drainage Board and to the management and staff of the National Irrigation Board, Mwea Irrigation Settlement Scheme for their assistance.

References

Brabben, T. E. & P. Bolton, 1988. Hydraulic impacts of aquatic weeds in irrigation systems. Paper prepared for Joint TAA/ICID (British Section) Meeting on Weeds in Irrigated Agriculture, 14 November 1988. Overseas Development Unit, Hydraulics Research, Wallingford.

Chambers, R. & J. Moris, 1973. Mwea: An Irrigated Rice Settlement in Kenya. Welforum Verlag, Munich.

Chow, V. T., 1983. Open-channel Hydraulics. Twelfth Edition. McGraw-Hill, New York.

JICA, 1988. Feasibility Study on the Mwea Irrigation Development Project. Japan International Co-operation Agency.

JICA, 1989. Basic Design Study Report on the Project for Mwea Irrigation Settlement Scheme Development in the Republic of Kenya. Japan International Co-operation.

Kay, M., 1986. Surface Irrigation: Systems and Practice. Cranfield Press, Cranfield.

Hydrobiologia **340:** 173–179, 1996.
J. M. Caffrey, P. R. F. Barrett, K. J. Murphy & P. M. Wade (eds), Management and Ecology of Freshwater Plants.

Short- and long-term control of water lettuce (*Pistia stratiotes*) on seasonal water bodies and on a river system in the Kruger National Park, South Africa

Catharina J. Cilliers [1], D. Zeller & G. Strydom[2]
[1] *Plant Protection Research Institute, Private Bag X134, Pretoria 0001, South Africa*
[2] *Kruger National Park, Private Bag X402, Skukuza 1350, South Africa*

Key words: Water lettuce, *Pistia stratiotes, Neohydronomus affinis*, chemical control, biological control, aquatic weed, seasonal water bodies, perennial river

Abstract

Water lettuce (*Pistia stratiotes* L. (Araceae)) is alien to Africa and a declared weed in South Africa. In large perennial rivers it is effectively controlled by its biological control agent, *Neohydronomus affinis* (Coleoptera: Curculionidae). On those shallow isolated water bodies which are regularly subjected to alternate wet and dry regimes that become infested with water lettuce, chemical control is necessary to prevent further spread of the weed and to facilitate access to water. This paper discusses the short-term chemical control and the long-term biological control of water lettuce. The need for further research is outlined.

Introduction

Pistia stratiotes L. (water lettuce) is an alien plant to South Africa, the country of origin being South America and it is a declared weed in South Africa (Henderson et al., 1987). Water lettuce is one of three significant aquatic weeds in the Kruger National Park (KNP). It occurs in a number of areas within the KNP including shallow seasonal water bodies (pans) in the northern Pafuri area, on the Limpopo flood plain and in the southern area on a perennial river, the Sabie River. This river is known to have the highest species diversity for aquatic plants in South Africa.

An objective of the KNP is to manage and control alien plant invasions so as to prevent the disruption of the natural ecosystems. The term control encompasses actions aimed at eradication, limiting, maintaining or reducing infestations. (Zeller, 1993).

The pans in which water lettuce is a problem are Nhlangaluwe (22 °33′S 31 °16′E) and, in the same vicinity, Dakamila, Makwadsi and Mapimbi. These pans are seasonal but may contain water for several seasons depending on rainfall, then be dry for one or more seasons. The Sabie River runs through the southern part of the KNP where originally 12 km of the river were infested with water lettuce: a sparse infestation further downstream was followed by a dense infestation at Lower Sabie over approximately three kilometers (16–20 hectare). The control of alien plant invasions are of particular concern in these pans. When not controlled, further spread of the weed, because of its rapid uncontrolled growth, is possible and access to water is limited. The water becomes deoxygenated under dense infestations of the weed, evapotranspiration is increased, the indigenous fauna and flora are threatened and the whole ecological balance upset, contrary to the objectives of the KNP (Chikwenhere & Forno 1991; Deacon & Gagiano, 1992; Zeller, 1993).

Control measures for water lettuce are thus important and have to be ecologically acceptable. Biological control was successful in Australia (Harley et al., 1984) and was first used in 1985/86 with great success in Africa on seasonal pans in Nhlangaluwe and later Dakamila in the northern part of the KNP (Cilliers 1987, 1991). When water lettuce control programmes were started in the KNP in 1987/88 on the Sabie River, two options, chemical and biological control were followed. The study area of biological control on the

Sabie River was at Lower Sabie (25 °07′S 31 °53′E). Downstream of Lower Sabie there is a dam wall and water flow was not as rapid in this area as elsewhere on the river but usually there was a continuous overflow at the dam wall. The progress of the biological control agent *Neohydronommus affinis* (Hustache) (Coleptera: Curculionidae) was monitored to ascertain whether the eventual degree of control obtained would fall within ecologically acceptable levels, as biological control does not eradicate the target plant.

Chemical control on the Sabie river and later on the seasonal pans in the Pafuri area was undertaken to keep water lettuce levels as low as possible, to prevent further spread of the weed, and to provide access for wild animals to and in water.

For the purpose of this paper chemical and biological control on the Sabie River and chemical control of water lettuce on seasonal pans are described.

Methods

Biological control

A starter colony of the host specific beetle *N. affinis* was obtained and imported into South Africa from CSIRO, Brisbane, Australia in 1985. The beetle was first introduced onto a water lettuce infestation on Nhlangaluwe pan in December 1985 and the progress and effect on the plants was monitored (Cilliers, 1987). A population of 500 adult *N. affinis* was first released on the Sabie River at Lower Sabie in September 1987. Four further releases of between 100 and 1000 adults and larvae, totalling approximately 5000 beetles, took place over the next five years. The most important of these later releases were those beetles introduced at the source of the infestation in the Salitje River, upstream of Lower Sabie in July 1990 and again in January 1991. Various parameters were monitored every six weeks from August to May of each year in order to assess the progress and effect of *N. affinis* on water lettuce in this flowing river. The methods used were the same as described by Cilliers (1987). More parameters were included for the Lower Sabie monitoring than previously. For the purpose of this paper only the number of plants per m^2 and the number of those plants that were damaged by *N. affinis* as an index of beetle activity were analyzed. The samples were taken along two fixed transects, across the river and following the northern bank. The samples were taken from the left and right side of a boat where plants were present. Reference is made to the percentage of the total area covered by water lettuce, and stream flow was used to explain plant population fluctuations. The area covered by water lettuce at Lower Sabie was estimated from colour slides and photos taken from fixed points whenever sampling was undertaken and twice daily over the period September to October 1992. The total area at Lower Sabie where the water lettuce occured was 40 ha. The programme, Statgraphics Plus, Version 6, 1992, Manugistics Inc., USA was used to analyze the data.

Chemical control

The herbicide terbutryn was used for the chemical control of water lettuce, applied at a 3% mix with water either from a boat or from the river banks using CP15 backpack spray units. Aerial application of herbicide was by means of a helicopter using a micronair system giving 6 liters ha^{-1} at a 30% mix with water. Repeated follow-up operations were carried out. Chemical control was applied towards the end of the dry season when water levels were low, the plants more concentrated and access easiest. Areas under control were visually monitored for the presence of plants.

Results

Biological control

On Nhlangaluwe pan, in the Pafuri area, biological control was achieved within ten months (Cilliers, 1987). This pan then dried up and no water lettuce remained. Similar results were obtained at a nearby pan, Dakamila (Cilliers, 1991). When these two pans and two others, in the vicinity, Makwadsi and Mapimbi, again had water they became covered with water lettuce. Meanwhile it had been established that water lettuce produced viable seed in South Africa (Henderson & Cilliers, 1991). This provided an explanation as to why these isolated pans again became infested after a dry period. Although beetles were reintroduced onto the water lettuce these pans were also sprayed with herbicide. This decision was taken in order to try to deplete the seed reserve by preventing new seed reserves forming through continuous short term control.

At Lower Sabie the beetle population remained low and it was only a year after the initial release of beetles, in September 1987, that the beetle population and

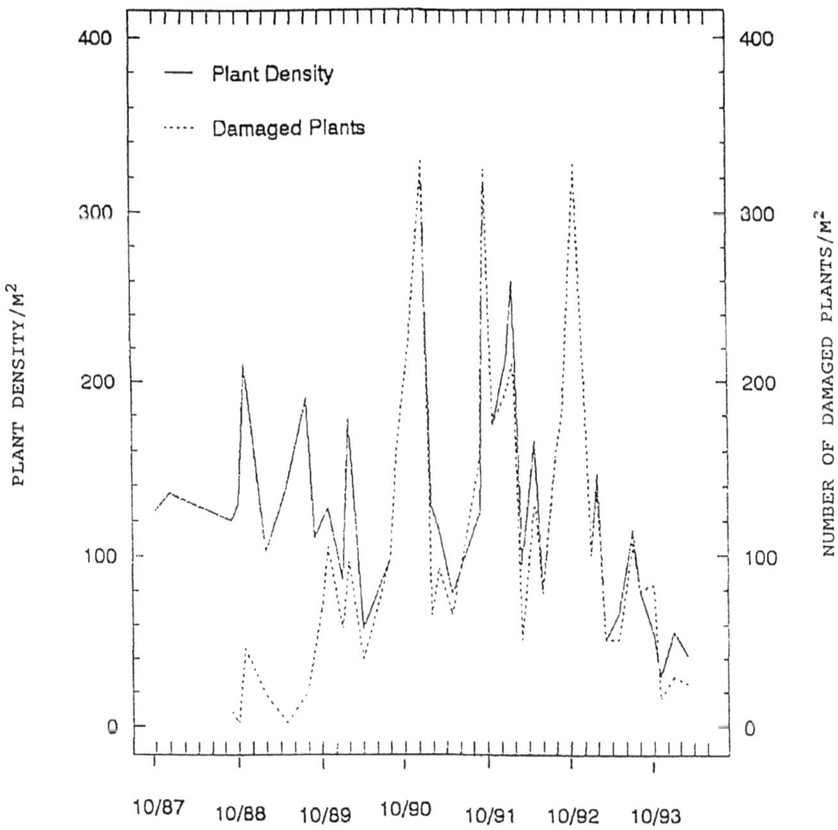

Figure 1. Time sequence plot of *Pistia stratiotes* (water lettuce) density per m^2 and the number of those plants damaged by the biological control agent *Neohydronomus affinis* at Lower Sabie. October 1987 to March 1994.

thus beetle damaged plants, could be readily observed during monitoring.

The density of plants in September 1988 was 120 m^{-2} of which only 8 plants showed insect damage (Figure 1). The total area covered by water lettuce was 60% (24 ha). Cyclical fluctuations occurred in the following years not only in plant population but also in number of beetle damaged plants. Plant populations peaked in November to February (summer) each year with a corresponding decline towards winter (Figure 1). By November 1990 and January 1991 all plants were damaged by *N. affinis*. However, large numbers of healthy plants were continually found at the top end of the study area where sampling was not done. In 1990 it was discovered that the Salitje river, a tributary of the Sabie River and upstream of Lower Sabie was a source of beetle free plants and beetles were therefore released on to plants in this river. In May and October 1991 and through to March 1992 those plants with insect damage were between 54–100%. In March 1992 not many plants were recorded per m^2 but the total area covered in water lettuce at Lower Sabie was 80% (32 ha) as opposed to between 10 to 15% (4 to 6 ha) cover in the previous year. In May 1992 the plant density again increased followed by a small decline in June 1992 and again a steady increase to November 1992. This was during a drought period in which temperatures were often above 40 °C and it was thought that more beetles were needed to curb the increase of plants. Booster colonies were released in April and May 1992. By September 1992 the surface covered by water lettuce had been reduced to less than 10% of the total area (less than 4 ha) with a continuous decline to 42 plants per m^2 in March 1994 (Figure 1). The positive correlation between plant density and number of damaged plants is illustrated by the high regression coefficient of $r^2 = 68.30\%$ (0.683), degrees of freedom = 37 (r^2 = the square of the correlation coefficient

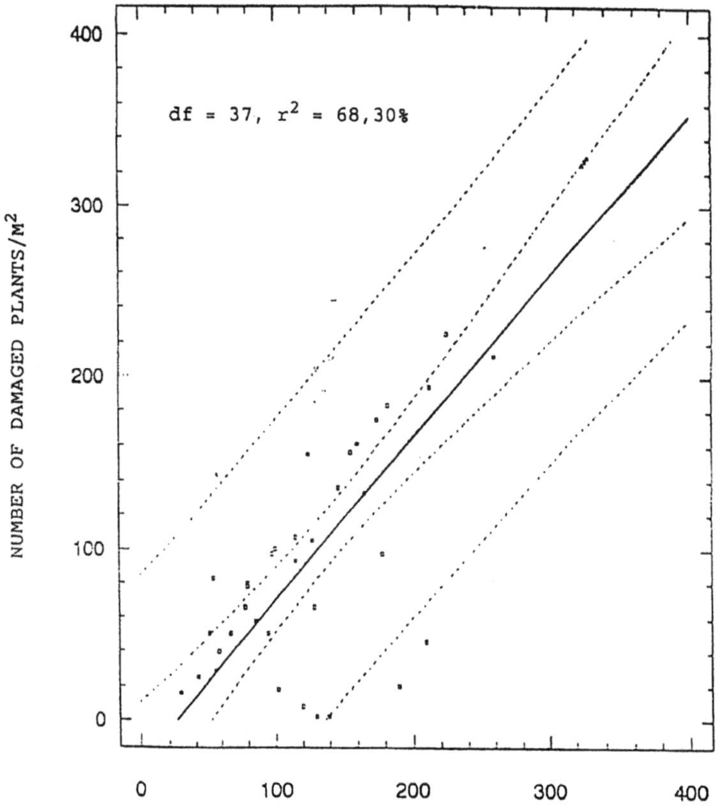

Number of plants damaged

Figure 2. Regression of *Pistia stratiotes* (water lettuce) density per m^2 and the number of those plants damaged by the biological control agent *Neohydronomus affinis* at Lower Sabie, October 1987 to March 1994. $r^2=68.3\%$, df=37. The solid line is the regression line. The pair of hatched lines closest to the regression line represent the 95% confidence limits. The pair of hatched lines furthest from the regression line represent the 95% confidence limits for future predictions.

as a percentage or 68.30% of 1 = 0.683) (Figure 2). A time series analysis of mean monthly river flow and the plant density showed an increase in plant numbers when there was a reduction in flow (Figure 3) and partly explained the increase in total area covered by water lettuce. If we restrict the prediction to the period after January 1990 when most of the plants were insect damaged the regression ($r^2=92.9\%$ (0.929) with degrees of freedom = 31) is further strengthened by restricting the regression to periods of low flow of less than 15 m^3 s^{-1} to give $r^2=95.88\%$ (0.9588) with degrees freedom = 21 (Figure 4).

By 1991 and the beginning of 1992 it was clear that *N. affinis* could control *P. stratiotes* on a flowing river.

Chemical control

Twelve kilometers of the Sabie River from Skukuza downstream were heavily infested with water lettuce and it was decided to chemically control this infestation. By the end of 1988 six kilometers of river below Skukuza were under control. During 1989 chemical control was continued and a further 6 km were brought under control and three follow-up operations were carried out on the total of 12 kilometers that year and again in 1990. At this time it was thought that water lettuce had been eradicated from this stretch of river and no controls were carried out during 1991. This proved to be wishful thinking as 49 man days, 11.5 litres herbicide, and a fortunate flooding of the river, were required to again clear this infestation during

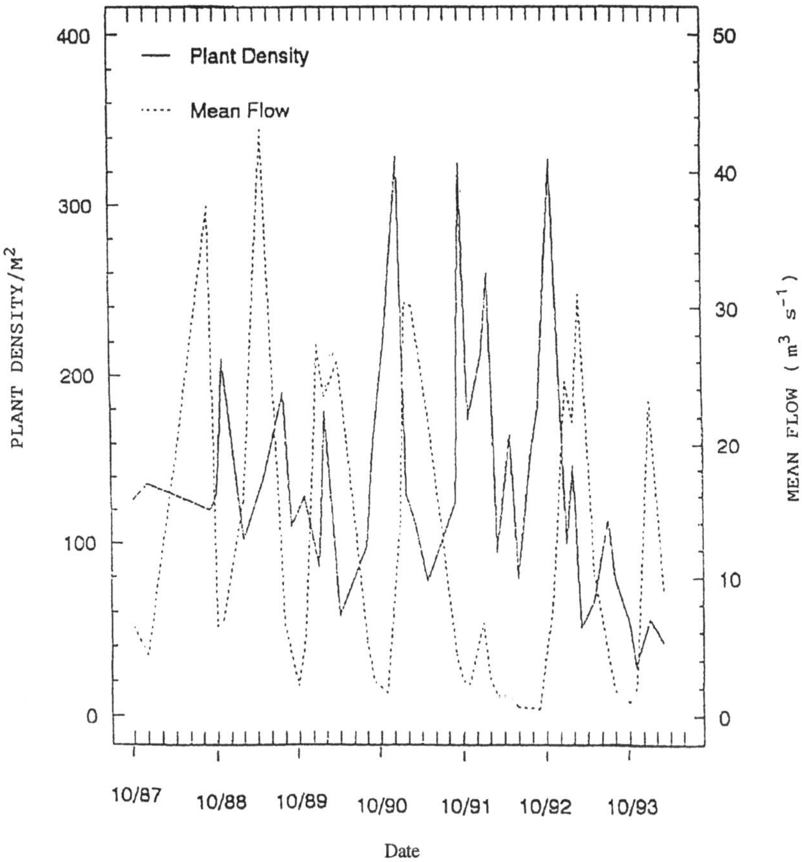

Figure 3. Time sequence plot of *Pistia stratiotes* (water lettuce) density per m^2 and mean monthly river flow at Lower Sabie, October 1987–March 1994.

1993. This section has remained clear of water lettuce since that time. A resurgence of water lettuce occurs on the pans in the north when they are recharged with water after rain and this is attributed to the high seed reserve. Herbicides are applied whenever water lettuce starts appearing, and before the plants are able to seed, in an attempt to reduce the seed reserve.

Discussion and conclusion

In biological weed control there is no eradication of the target plant but the aim is to bring the weed population down to an environmentally/ecologically acceptable level through the use of one or more natural enemies. The dramatic increase in the surface area of the study site covered in water lettuce during March to September 1992 was attributed to the extremely low river flow, when plants were not being washed downstream and over the dam wall but were able to accumulate. A stable population of plants developed on which the beetles could build up in numbers without being continuously thinned out. During this time plants that were beetle damaged varied between 80–100% (Figure 1). It was thus wrong to assume that the beetles may have succumbed to heat experienced during the excessive drought and booster colonies were unnecessary. A series of photographs taken in September and October 1992 showed that the water lettuce was being moved either towards the study area or away from it depending on the wind direction. Based on these facts it is concluded that *N. affinis* was able to control water lettuce on a flowing river and other natural enemies of water lettuce need not be considered further. Without the fixed point photography a wrong impression might have been formed of the area covered with water let-

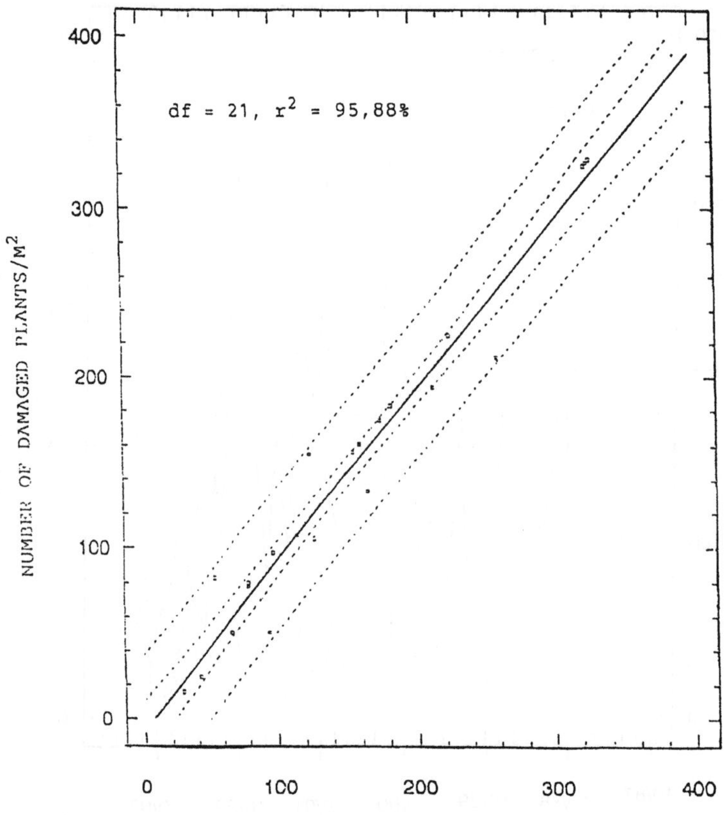

Figure 4. Regression of *Pistia stratiotes* (water lettuce) density per m^2 damaged by the biological control agent *Neohydronomus affinis* at Lower Sabie when mean river flow was below 15 m^3 s^{-1}, January 1990 to April 1994. r^2 =95,88%, df=21. The solid line is the regression line. The pair of hatched lines closest to the regression line represent the 95% confidence limits. The pair of hatched lines furthest from the regression line represent the 95% confidence limits for future predictions.

tuce at Lower Sabie as the samples were always taken where plants occurred along the transect lines. Future sampling needs to take this factor into consideration.

The Sabie River system, the pans in the Limpopo flood plains and elsewhere in the KNP are of priority concern in the control of water lettuce. Biological control is very successful on the pans and on the Sabie River. It will remain the main form of control of *P. stratiotes* in the KNP, but will be augmented by chemical control where necessary. On the Sabie River at Lower Sabie a cover of less than 10% of the water surface is presently regarded as the residual plant population that has to be tolerated. There is still further need for research on the influence of *P. stratiotes* on a sensitive section of the Sabie River in Sabie Poort 10 km downstream of Lower Sabie. In the Sabie Poort River braided channels in the dry season become isolated pools covered with water lettuce. Although the beetles also eventually control water lettuce here, the rotting plants cause eutrophication of the water. Herbicidal control would have the same effect. These pools are important habitats for many aquatic species, being home to more than 10 species of fish of which one species is endemic to the Sabie River. Two highly sensitive species have already become extinct in the Olifants River (Dr A. Deacon, personal communication, Kruger National Park, Skukuza, 1993). This problem needs further research.

Acknowledgments

In the early phases of this work part of the financial assistance was given by the South African Department of Water Affairs and Forestry. For the latter part of the project financial assistance was from the Water Research Commission, Pretoria and throughout by our respective Institutions. To be able to present this paper at the 9th International Symposium on Aquatic Weeds part-sponsorship was given by Zeneca Agrochemicals, South Africa and Amanzimtoti Municipality, Amanzimtoti. The KNP Management Committee and the following persons are thanked for assistance in one way or the other for longer or shorter periods: Drs H. Biggs, A. Deacon, F. Venter, P. Reid, Messrs K. Maggs, E. Pietersen and T. Mhalungane.

References

Cilliers, C. J., 1987. First attempt at and early results on the biological control of *Pistia stratiotes* L. in South Africa. Koedoe 30: 35–40.

Cilliers, C. J., 1991. Biological control of *Pistia stratiotes* (Araceae), in South Africa. Agric., Ecos. Envir. 37: 225–229.

Deacon, A. & C. Gagiano, 1992. Visvrekte: Sabierivier 17/07/92. Unp. rep. on file Kruger National Park, Rivers Research Programme, Skukuza.

Chikwenhere, G. P. & I. W. Forno, 1991. Introduction of *Neohydronomus affinis* for biological control of *Pistia stratiotes* in Zimbabwe. J. aquat. Plant. Mgmt 29: 53–55.

Harley, K. L. S., I. W. Forno, R. C. Kassulke & D. P. A. Sands, 1984. Biological control of water lettuce. J. aquat. Plant Mgmt 22: 101–102.

Henderson, L. & C. Cilliers, 1991. Water lettuce. Farm. in S. Africa. Weeds A.33/1991, 2 pp.

Henderson, M., D. M. C. Fourie, M. J. Wells & L. Henderson, 1987. Declared weeds and alien invader plants in South Africa. Dept. Agric. and For., Direct. Agric. Inf., Pretoria, 167 pp.

Zeller, D. A., 1993. An approach towards alien plants in the Kruger National Park. Unpublished rep. on file Kruger National Park, Skukuza.

Hydrobiologia **340:** 181–185, 1996.
J. M. Caffrey, P. R. F. Barrett, K. J. Murphy & P. M. Wade (eds), Management and Ecology of Freshwater Plants.

Strategies for waterhyacinth (*Eichhornia crassipes*) control in Mexico

Eric Gutiérrez, Rubén Huerto, Pilar Saldaña & Felipe Arreguín
Hydrobiology Laboratory, Instituto Mexicano de Tecnología del Agua, Apdo. P. 235, CIVAC, Morelos, Mexico 62500

Key words: Eichhornia crassipes, control, management, maintenance control.

Abstract

In Mexico, more than 40 000 ha of dams, lakes, canals and drains are infested with waterhyacinth (*Eichhornia crassipes*). To prevail over the problems resulting from this infestation, specific management programs are needed. Under a national program to control the waterhyacinth, guidelines to deal with the related ecological, social, technical and economic factors, and specific strategies to reduce coverage were developed. The ecological factors which were noted include the identification of the most affected areas and the consequences of proposed treatments. The social aspects embraced the stimulation of user awareness as to the importance of water quality, the creation of organizations to coordinate user-sponsored control activities, and the awakening of a community identity. Basic to all are the technical and economic aspects which make the activities feasible and operational. Examples are given of control by means of water level management, mechanical controls using trituration, and the application of chemical and biological agents, all of which may be combined in an integral program.

Introduction

Waterhyacinth (*Eichhornia crassipes* (Mart.) Solms) is successful owing to its life cycle and survival strategies which have given it a competitive edge over other species. Its capacity for vegetative reproduction allows the plant to quickly occupy any available space. Regrowth from relatively small plant fragments, floatability and the production of viable seeds are efficient mechanisms for the dispersion and colonization of other areas, especially when combined with a minimum of growth-limiting factors, resistance to drying, morphological variety, root-ability, lack of natural enemies and adaptability to little-competed ecological conditions make eradication of this plant virtually impossible and control extremely difficult (Perazza et al., 1979; Niño & Lot, 1983; Gopal, 1987; Luu & Getsinger, 1988).

The basic units of a management program are the complex variables related to plant growth and the relationships among them. The tactics and strategies will combine these units with greater or lesser efficiency. In Mexico, more than 62 000 ha of dams, lakes, canals and drains are infested with water weeds. Of this total, 40 000 ha are covered with waterhyacinth. To overcome the problems resulting from this infestation, specific management programs are needed to reclaim these bodies of water. Although many variables and factors interact under these conditions, the Mexican Institute of Water Technology (IMTA) has worked to single out those which may be built into strategies that are both technically and economically feasible.

The Aquatic Weed Control Program (AWCP) was created in 1993 to combat the excessive presence of weeds in the nation's water courses. The objectives of this work are to present the main characteristics of the AWCP, and describe the control program implemented in the Ayutla River watershed as an initial stage of a larger national program.

Study area

The Ayutla River watershed comprises three dams in series, the Miraplanes, Tacotán and Trigomil (Figure 1). The mean annual temperature in the area is 20.9 °C

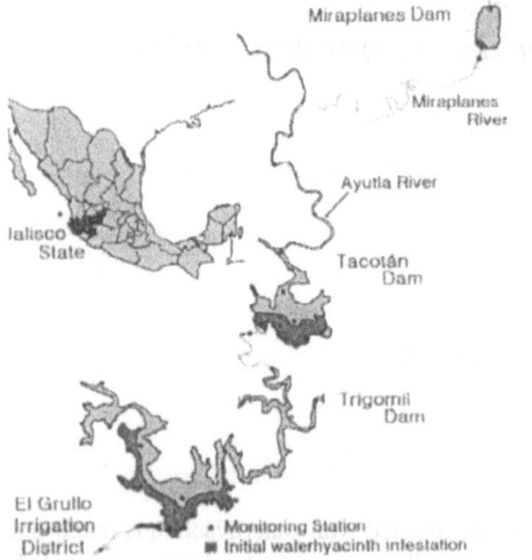

Figure 1. Water Hyacinth infested areas in the Ayutla watershed

Table 1. Characteristics of dams in the Ayutla Watershed

DAM Use	MIRAPLANES Irrigation	TACOTAN Irrigation, fishing	TRIGOMIL Irrigation, fishing
Volume (Mm^3)	0.73	149	324
Area (ha)	73	500	393
Mean depth (m)	1	20–30	60–80
Max. depth (m)	2	40	100
Weed present	W/T	W	W
Infested (ha)	73	204.8	211.4
Uninfested (ha)	0	257.5	181.8
Total surface (ha)	73	462.3	393.2

Satellite image (January 10, 1993). W = Waterhyacinth. T = Typha

with an average annual rainfall of 806.5 mm. Table 1 provides other related information.

Control program

The AWCP contemplates six phases.

Initial evaluation. The area was observed to evaluate the infestation and identify the users who are directly affected and would be interested in committing themselves to the maintenance phase of the program in conjunction with government authorities. The water uses, aquatic communities, location and types of crops surrounding the dam, weather conditions and possible control strategies were identified. The weed coverage was quantified by means of satellite images from LANDSAT TM. The resolution of these images was 25 m × 25 m per pixel (0.0625 ha).

Participation and communications. Meetings were convened with users to provide them with information concerning the proposed control strategy, establish user commitments, organize and define responsibilities for the short, medium and long term. Informative exhibitions were prepared to offer updates, and later training, to the users.

Economic feasibility study. Unit costs for labor, material, infrastructure and administration were calculated, as were costs for the monitoring and follow-up programs after the control efforts had ended.

Control. The control strategy was developed, based on the characteristics of each dam, the assigned budget, and the most appropriate control techniques available domestically. These latter included herbicidal, mechanical and biological procedures, and water level management (Table 2). Chemicals employed in control schemes have been mainly 2,4-D, diquat and glyphosate. The most commonly used mechanical method is a triturator placed on a raft with blades operating at 2000 rpm up to 30 cm below the water surface. The waterhyacinth weevil, *Neochetina eichhorniae*, has been observed in the three basins. This insect was introduced to Mexico toward the end of the seventies in an effort to establish a biological control (Bennet, 1984). During this phase, operations began. This included the programming of equipment and materials, supervision of tasks and quantification of the decrease in biomass.

Environmental monitoring. The elimination of aquatic weeds by any one of the means customarily used, modifies the preexisting conditions. In Gutiérrez et al. (1994) a water quality monitoring program is described in which analyses were made prior and subsequent to executing the control program. Studies were made of changes in the planktonic and benthonic communities in the affluent and effluent of the dam, and of herbicidal residues in water, sediment and tissues of edible fish species.

Maintenance. Tactics were developed wherein users and authorities were firmly committed to main-

Table 2. Evaluation of waterhyacinth control methods

Alternatives	Chemical	Extraction	Trituration	Biological	Water level management	Manual
A: Economic feasibility	Good	Poor	Average	Good	Good	Good
B: Technical feasibility	Good	Average	Average	Average	Average	Poor
– Availability	Good	Poor	Average	Average	Not applicable	Good
– Transport	Easy	Average	Average	Easy	Not applicable	Easy
– Access to water body	Good	Average	Average	Good	Good	Poor
– Efficiency	50–100% control in 20–80 days	240 m. tons/d 1.2 ha/8h*	716 m. tons/d 1.7 ha/16h	Poor to average	Average	2.5 m. tons/8h/man 0.01 ha/8h/man
– Short-term effects	Good	Good	Good	Poor	Good	Poor
– Long-term effects	Average	Good	Average	Good	Average	Poor
C: Environmental impact	Medium-high	Low	Medium-high	Low	Low	Low
D: Socio-economic conditions						
– Personnel training	Specialized	Intermediate-specialized	Intermediate-specialized	Not applicable	Not applicable	Non-specialized
– Foreign currency required	No	Yes	No	No	No	No
– Acceptation index	Low	High	Intermediate	High to intermediate	High	High

* Aquamarin H–10, manufacturer information

Table 3. Maintenance control guide for dams in the Ayutla watershed submitted to intensive control.

Infestation	Coverage %	Control	Opportunity
Serious	15–25	Chemical from airboat and/or trituration	Immediate
Dangerous	5–15	Chemical from airboat and/or trituration and/or mechanical extraction	7 days
Moderate to tolerable	0 to 5	Manual extraction from boat and/or shore, chemical from boat and/or shore	14 days

taining waterhyacinth levels below the problem threshold. Criteria were developed relating coverage with recommended control measures (Table 3). This information was provided to the user committee, together with training in the procedures listed. Routine inspections were made from previously-identified strategic points and the results compared with the criteria. The recommended control techniques for small scale application were those least likely to affect the ecosystem and water users. Often these measures were a part of an integral watershed management program. Maintenance control is considered essential in the reclamation process, as it is more cost efficient in the medium and long term, reduces the use of herbicides, lessens the environmental impact resulting from the destruction and decomposition of the aquatic weeds, and increases the efficiency of biological and mechanical control methods (Haller, 1981).

Management strategies

Three different control strategies were developed for three distinct bodies of water which had in common neither use, depth, size nor geographic location (Figure 1 and Table 1). Water level management was considered the most adequate for the Tacotán Dam. The water was released to the Trigomil Dam, downstream and 105 ha of waterhyacinth were left to dry on the shore and were burned by the users (fishermen). The remaining 100 ha were dusted, by helicopter, with 3.3 kg ha^{-1} of 2,4-D. The dam was then closed for 21 days. This first treatment was 60% effective. The remaining 40% was not sufficiently damaged to sink. However, with the combined effect of a reduction in population and a loss of turgidity, a greater surface area was made available. Diquat, a contact herbicide, was applied 55 days later at a rate of 1.7 kg ha^{-1} and provided 100% control. The dam was cleared after 110 days of operations.

For the Trigomil Dam, a combined chemical-mechanical program was prepared. The water from the dam is used in the El Grullo Irrigation District, restricting the time during which the dam may be closed and the chemical used. Glyphosate (Rodeo^{TM}), at 3.35 kg a.i./ha, was selected as there are no restrictions on its use in irrigation water. One-half of the infested area (104 ha) was treated initially; the remainder was sprayed 38 days later. The herbicide's action was irregular, with some areas showing excellent results while others reacted very slowly. The product was applied during the growing season based on results from small-scale tests performed that indicated greater effectiveness at that time (Gutiérrez, 1993). Though the plants sank slowly and inconsistently, there was a noticeable reduction in plant growth in most areas. There was also a marked change in the consistency from strong, healthy plants with an intense green color to yellow individuals which fragmented easily at the touch. Sinking was calculated in 20 to 40 hectares. The second dose was applied as scheduled and two triturators began a 15 day campaign to accelerate sinking, after which approximately 160 ha of waterhyacinth had been eliminated and 100% control was attained.

At the Miraplanes Dam (Figure 1), the presence of a large area of cattail (*Typha sp.*) affected the decision to use glyphosate. Westerdahal and Getsinger (1988) state that glyphosate is very effective against this species. Three treatments were programmed. The first was done from a small plane at 3.5 kg ha^{-1}. The second was 25 days later, from helicopter at 3.33 kg ha^{-1} and the third using the same method 207 days after the first. Fifteen days afterward, 70% of the dam was cleared. Three months later, the dam was totally weed-free.

The results of the analyses for residues of 2,4-D indicated that levels never exceeded 0.1 mg l^{-1}, the maximum accepted level for drinking water. Residues of 2,4-D, glyphosate and diquat were not detected in analyses of tissues of edible fish (tilapia, carp and catfish) and sediments. The low levels found in water may be explained by dilution and degradation, supporting claims of low persistence (Gutiérrez et al., 1994). The assimilation of the triturated or treated biomass into the water column modified its quality by incorporating nutrients and diminishing the dissolved oxygen through an increase in the COD. However, the change in quality was due mainly to an affluent in which high concentrations of organic material and other nutrients were detected. No dead fish were observed during or after the treatment period. Studies made of the biological communities (benthic and planktonic) in the Tacotán, Trigomil and Miraplanes dams indicated that they were unaffected (Gutiérrez et al., *op. cit.*). There was an increase in the number of weevils based on the observable foliar damage. As a second phase to the prevention program, *Neochetina bruchi* will be introduced to complement the maintenance program.

Suggestions and conclusions

The basis for a soundly-designed control program is early strategic planning. This means a timely evaluation of all environmental variables related to the process. Morphological characteristics, water use and quality, hydraulic operations, accessibility, relationships among users, and plant dynamics are just a few of these parameters. They must be correlated with the human, material and economic resources available. Finally, a cost-benefit analysis will define the most suitable alternative for control and maintenance under the conditions found at each site. To assure the fulfillment of the expectations of the project, in terms of scheduling, safety, goals and costs, constant on-site supervision is vital. Water quality analyses and aerial inspections are useful guides.

Most of the waterhyacinth control methods have been used in Mexico, harvesting by hand and machine, trituration, and treatment with herbicides and biological agents. Experience has taught us that the first phase of the control program must employ massive attack techniques for an important reduction in coverage, such as that seen with the use of chemical agents and triturators. The second phase should utilize all of the modern know-how combined into an integral management program to keep the population under the weed threshold. It is here that biological control can be an important component. A sustainable control program also requires a watershed-wide administration program in which all users are involved in the drafting of the alternatives. User involvement serves to stimulate an awareness of the causes and the magnitude of the problem, and to invite his direct participation in the cleaning process. In this context, user presence not only ensures the success of the restoration, but also reduces costs significantly.

Acknowledgements

Special recognition is given to Dr William T. Haller and Dr Alison Fox from the Center for Aquatic Plants of the University of Florida, for their contributions in the development of the AWCP. Thanks are given to Dianne

Hayward, for her valuable comments on the finished work and translation; to Ernesto Uribe, Eduardo Ruiz, Marco A. Mijangos and Ulises Bucio for their support in the fieldwork; to Alfredo Tapia for his assistance in the design and presentation of the manuscript.

References

Bennett, F. D., 1984. Biological control of aquatic weeds. In G. Thyagarajan (ed.) Proc. Int. Conf. Water Hyacinth. UNEP. Nairobi: 14–40.

Gopal, B., 1987. Water hyacinth. Aquatic plant studies. Elsevier Science Publishers, Amsterdam, 471 pp.

Gutiérrez, E., 1993. Effect of glyphosate on different densities of waterhyacinth. J. Aquat. Plant Manage. 31 (July): 255–257.

Gutiérrez, E., F. Arreguín, R. Huerto & P. Saldaña, 1994. Aquatic weed control. Int. J. Wat. Res. Devel. 10: 291–312.

Haller, W. T., 1981. Maintenance control of waterhyacinth. Aquatics 3(2):6–7, 11–12.

Luu, T. K. & D. K. Getsinger, 1988. Control points in the growth cycle of waterhyacinth. U. S. Army Corps of Engineers. Waterways Experimental Station. Envir. Lab. Vol. A-88-2: 1–5.

Niño, S. M. & A. Lot, 1983. Estudio demográfico del lirio acuático, Eichhornia crassipes (Mart) Solm. Dinámica de crecimiento en dos localidades selectas de México. Boletín de la Sociedad Botánica de México 45: 71–85.

Perazza, W. T., N. D. Pereida & T. M. Martins, 1979. Problematica de controle de plantas aquaticas. In Anais do 2° Simposio Nacional de Ecologia. Belem, Brasil, Nov. 19–23.

Westerdahl, H. E. & D. K. Getsinger (eds), 1988. Aquatic plant identification and herbicide use guide. Vol. 1 Aquatic herbicides and application equipment. Aquatic Plant Control Res. Prog. Tech. Rep A-88-9. U. S. Army Corps of Engineers. Vicksburg, Mississippi, USA, 222 pp.

Hydrobiologia **340:** 187–190, 1996.
J. M. Caffrey, P. R. F. Barrett, K. J. Murphy & P. M. Wade (eds), Management and Ecology of Freshwater Plants.

Management of *Hydrocotyle ranunculoides* L.f., an aquatic invasive weed of urban waterways in Western Australia

R. J. Ruiz-Avila[1] & V. V. Klemm[2]
[1]*Biospherics Pty Ltd (formerly of the Swan River Trust), P.O. Box 168, South Fremantle, Western Australia 6162*
[2]*Swan River Trust, P.O. Box 6892, Hay St East, East Perth, Western Australia 6892*

Key words: aquatic weeds, floating aquatic plants, *Hydrocotyle ranunculoides* L.f., integrated weed control, eutrophication

Abstract

Hydrocotyle ranunculoides L.f. is a stoloniferous perennial plant with floating and emergent leaves that is native to Europe. It is commonly used as an aquarium plant with little published information on its biology and natural range. In 1983 *H. ranunculoides* was first observed in the urban drainage network in the Canning River Regional Park, Western Australia. By 1991 the plant had extended throughout the drainage network into the river and adjacent wetlands. *H. ranunculoides* formed extensive mats, disrupting the ecology and recreational uses of the waterways, and posed a threat to other waterways. It is not known to be invasive in other Australian waterways. A group of state and local government and community members assessed environmental, technical and social interactions and developed an integrated management strategy for the weed, using a combination of physical, chemical and ecological techniques. The environmental significance of the affected waterways required the programme to be accompanied by appropriate ecological surveillance. The initial short-term control phase was completed successfully. An assessment of water quality and aquatic invertebrates during the initial phase showed only short-term disruption of river ecology following physical and chemical control. The long-term eradication phase is on-going.

Introduction

In Western Australia the Swan-Canning Estuarine system drains a catchment of approximately 140 000 km^2 before flowing though the capital city Perth and into the Indian Ocean (Figure 1). The Swan and Canning rivers are a scenic and recreational focus of Perth and a major portion of the Canning River is now managed as a Regional Park supporting 85 species of birds (State Planning Commission, 1988) and is used for a wide variety of recreational pursuits including canoeing, boating, swimming, fishing and walking. In the 1920s a weir was constructed to maintain a freshwater section of the river for riparian users in a section which was previously tidal. Water quality is affected by the clearing of native vegetation and resulting changes in land use. Land use in the Canning River catchment comprises urban areas, small scale agriculture, horticulture and native forest reserves and the freshwater section of the river is eutrophic and subject to annual blooms of the algae *Euglena* (Thurlow et al., 1986). Since the 1970s a number of introduced aquatic macrophytes have been recorded in the Canning River system, including an extensive outbreak of *Salvinia molesta* and lesser invasions of *Hydrilla verticillata*, *Eichornia crassipes* and *Nymphaea* spp.

Hydrocotyle ranunculoides was first observed in a minor tributary of the Canning River in 1983. A commonly used aquarium plant readily available in Western Australia, the initial infestation is believed to have arisen from release of aquarium wastes. Its distribution remained localised until early in 1991, when it entered the Canning River and spread rapidly throughout adjacent wetlands and the urban drainage network. It formed large floating rhizomatous mats which in some places covered the river surface completely.

It is not known if the major outbreak was due to the weed being introduced at one point or if a number

a)

b)

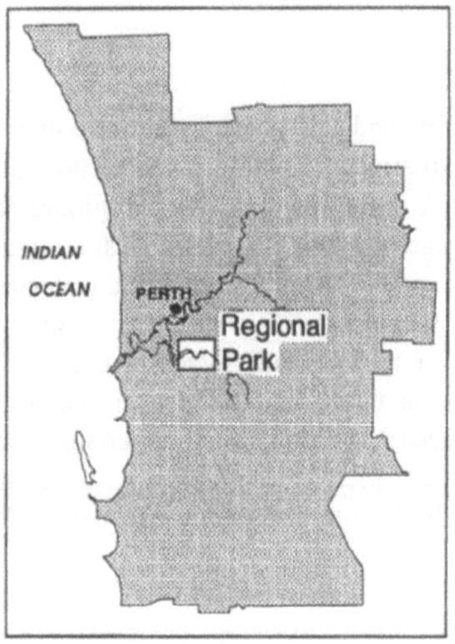

Figure 1. Location maps showing (a) the location of Perth in Western Australia and (b) the Canning River Regional Park where the invasion of *Hydrocotyle ranunculoides* occurred.

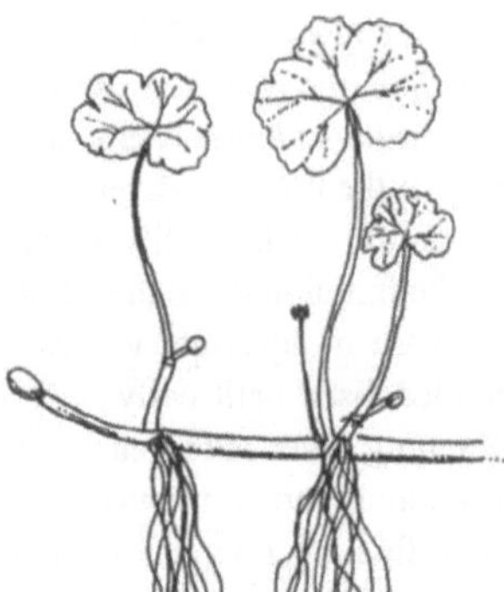

Figure 2. Growth form of *Hydrocotyle ranunculoides* L.f.

of simultaneous introductions occurred. The lag time between initial colonisation and the explosive phase of range expansion of *H. ranunculoides* in the Canning River was relatively short (eight years) compared with an average of up to 80 years for most major environmental weeds in Australia. The rapid growth and aggressive colonisation of *H. ranunculoides* threatened the ecological integrity of the Canning River system, interfered with recreational uses of the river and threatened economically important water resources.

Biology and ecology of *Hydrocotyle ranunculoides* L.f.

The genus *Hydrocotyle* is a widespread member of the Apiaceae family, containing 100 species (Marchant et al., 1987). It is found in both temperate and tropical regions (Hickey & King, 1988) with 55 species in Australia (Marchant et al., 1987). Most species are perennials, characterised by a creeping rhizomatous growth form, and generally occur in aquatic or moist terrestrial habitats. *Hydrocotyle ranunculoides* has been recorded as a component of the vegetation in a number of cool water bodies occurring in either high altitude tropical lakes (Gaudet, 1977; Denny, 1973; Wijninga et al., 1989) or low altitude coastal regions of the temperate zone (Steubing et al., 1980; Muenscher, 1944; Mason, 1957; Aulbac-Smith et al., 1990).

H. ranunculoides is an emergent aquatic macrophyte with a creeping stolon that bears profuse adventitious roots at the nodes (McChesney, 1994). The species is described by Marchant et al. (1987) and (Sainty & Jacobs, 1994). In the Canning River, *H. ranunculoides* flowered profusely during spring and summer months and formed large rhizomatous mats, covering the entire river for hundreds of metres. The mats were attached to the bank and grew, on average, half a metre above the water, with the root zone extending for half a metre below the water. *H. ranunculoides* also grew terrestrially amongst riparian vegetation without forming the same mat structure.

In a recent review of its biology based on observations from the Canning River and from available literature, McChesney (1994) concluded that *H. ranunculoides* possesses a number of typical weed characteristics. These include high growth rates (probably due to nutrient enriched environments), effective vegetative propagation (fragmentation and possible clonal integration), plasticity in growth response (e.g. overwintering, aquatic and terrestrial growth forms) and possible high resistance to herbivory (McChesney, 1994).

Development and implementation of an integrated management strategy

In late 1991 there was an estimated biomass of 175 tonnes of *H. ranunculoides* in the Canning River. Biomass was calculated by determining the surface area of the weed from aerial photographs and measured area/weight relationships. Early control attempts involved a two-week program of physical removal of *H. ranunculoides* by cutting the floating mats of *H. ranunculoides* with sickles from small boats. The mats were then pushed by small boats to an aquatic macroalgal conveyor harvester, floated to the bank and removed by a backhoe. Follow up maintenance control was continued until January 1992, when growth rates exceeded the rate of removal.

By September 1992, the estimated biomass in the river was 420 tonnes, mostly in the freshwater section of the river. During that summer mats in the freshwater section grew rapidly, covering the river from bank to bank. Downstream of the weir the rise in salinity destroyed all the *Hydrocotyle* mats.

While relatively successful at the time, the removal program of 1991 appeared to cause the spread of *H. ranunculoides*, by the dispersal of segments broken off mats during handling. New mats grew from these fragments over the following 12 months, greatly increasing the size of the infestation. However, the experience and knowledge gained during the first attempts at control proved invaluable in the development of the integrated strategy.

State and local government agencies, the local community and the Swan River Trust (the state government agency responsible for managing the Swan-Canning River system) assessed the complex technical, environmental, organisational, legislative, educational and social requirements in order to design an integrated management strategy (Klemm et al., 1993). The extensive consultation process was seen as a vital step in ensuring community support for the project. The aim of the strategy was in the long-term to eradicate the weed whilst minimizing adverse effects on water quality, the river ecosystem, recreational amenities and public health.

For the management strategy to be successful it needed to simultaneously address both the removal of the weed and the causes of the invasion. In 1992, *H. ranunculoides* was declared a noxious weed preventing importation, movement and trade, and ensuring that control and eradication measures were undertaken.

One of the key problems initially was the scarcity of information on the biology, autecology, synecology and management of the plant. A preliminary investigation on the biology of *H. ranunculoides* has since occurred (McChesney, 1994).

A two part management strategy was developed consisting of a short-term control program implemented in early 1993, and a long-term eradication program. Short-term control measures relied largely on physical removal techniques, similar to those used in 1991 with follow-up selective use of herbicides, removing 2000 tonnes (estimated from number of truckloads removed). The weed was then used for composting by the local council. Biological control was not considered to be an option for control of *H. ranunculoides*. Ecological control techniques were either unsuitable or unavailable for use in the short-term but were considered suitable for use in the long-term if eradication measures failed.

After most of the weed had been removed, herbicide was used along the banks to prevent new mats growing out. *H. ranunculoides* can grow up to 15 cm deep in the soil on the bank and glyphosate (Tradename: Round-up), a translocated herbicide, was selected on the basis of its success in glass-house experiments; its low toxicity to mammals, fish and microbes; and its low to medium toxicity to birds and other aquatic life. An application rate of 360 g ha^{-1} of active ingredient was initially used, with a stronger formulation used for the first time in 1994 (450 g ha^{-1}).

There were strong community concerns about the weed invasion and the techniques used in the management strategy, particularly with the use of herbicides adjacent to the river. It was essential for the Swan River Trust to obtain community support. This was done by an extensive process of public information and education through the media.

The long-term the aim to eradicate this weed from the river system, thereby preventing its spread through the State includes physical removal and use of herbicides, as in the short-term programme plus ecological control. Ecological techniques under investigation include the reduction of nutrient loads to the river and the removal of nutrient rich sediments. Currently a programme of integrated catchment is being developed to reduce nutrient loads entering the river via the drainage network. The reduction of nutrients to the Canning River would also reduce the opportunities for invasion by other aquatic weed species. Another option is to manipulate salinity in the freshwater section of the river by removal of the weir.

Ecological surveillance of management strategy

An ecological surveillance programme was established to identify and measure the environmental impacts of the management in response to the concern shown by the community that control techniques, particularly chemical control, would have a damaging effect on the ecology of the river system. Water quality and macro-invertebrates were monitored throughout the programme, the details of which are reported in Donohue (1994). Control techniques were found not to adversely affect either the water quality of the river or the benthic invertebrate populations (Donohue, 1994).

Conclusion and prospects

A number of points can be made from this exercise:

- The need for broad consultation.
- The need for integration of all appropriate measures.
- The importance of controlling the cause of the weed problem i.e. eutrophication.
- The need for commitment and long-term involvement to eradicate the weed.
- The desirability of early intervention when weed populations are still small i.e. in 1983.
- The risks of selling and dumping non-native aquatic plants.

Given the lack of available information on the biology and ecology of *H. ranunculoides*, and the rapidity of its invasion, it is considered that the integrated management strategy was successful in controlling the invasion with no long-term environmental impact. Further research on the biology of the plant is being conducted to aid in the efficacy of the long-term eradication strategy. The goal of eradication may be achieved in the next five years, although growth of the weed into the terrestrial environment is seen as a major setback. Focus should be made on autecological studies to determine life history attributes, population dynamics and habitat requirements. In addition, the Canning River system requires long-term rehabilitation to prevent recolonisation by *H. ranunculoides* or any other invasive weed.

Acknowledgments

The Swan River Trust provided funding for the writing of this paper. The authors would like to thank Dr Jon Dodd from the Department of Agriculture and Howard Willis for providing constructive comments on a draft of the manuscript. The invertebrate monitoring was undertaken by Robert Donohue of the Swan River Trust.

References

Aulbac-Smith, C. A., S. J. de Koslowski & L. A. Dyke, 1990. Aquatic and Wetland Plants of South Carolina. South Carolina Aquatic Plant Management Council, South Carolina.

Denny, P., 1973. Submerged and floating-leaved aquatic macrophytes (euhydrophytes). In P. Denny (ed.), The Ecology and Management of African Wetland Vegetation. Geobotany 6, Dr W. Junk Publishers, The Hague.

Donohue, R., 1994. Impacts of the *Hydrocotyle ranunculoides* removal program on the Canning River. Swan River Trust unpublished report, Perth, Western Australia.

Gaudet, J. J., 1977. The maintenance of plant and soil heterogeneity in dune grassland. J. Ecol. 76: 497–508.

Hickey, M., & C. King, 1988. One hundred families of flowering plants, 2nd ed. Cambridge University Press, Cambridge.

Klemm, V. V., N. L. Siemon & R. J. Ruiz-Avila, 1993. *Hydrocotyle ranunculoides*: A control strategy for the Canning River Regional Park. Swan River Trust Report No. 6, Perth, Western Australia.

Marchant, N. G., J. R. Wheeler, B. L. Rye, E. M. Bennett, N. S. Lander & T. D. McFarlane, 1987. Flora of the Perth Region. Western Australian Herbarium, Perth, Western Australia, 1080 pp.

Mason, H. L., 1957. A Flora of the marshes of California. University of California Press, Berkeley.

McChesney, C., 1994. Literature review of the Genus Hydrocotyle L. (Apiaceae), with particular emphasis on *Hydrocotyle ranunculoides* L.f. Swan River Trust Report No. 18, Perth, Western Australia.

Muenscher, W. C., 1944. Aquatic Plants of the United States. New York. Comstock Publishers, New York.

Sainty, G. R. & S. W. L. Jacobs, 1994. Waterplants in Australia. 3rd Edition. Sainty and Associates, Sydney, 327 pp.

State Planning Commission, 1988. Canning River Regional Park: Draft for Public comment. State Planning Commission, Perth, Western Australia.

Steubing, L. C., C. Ramirez & M. Alberdi, 1980. Energy content of water- and bog-plant associations in the region of Valdivia (Chile). Vegetatio 43: 153–161.

Thurlow, B. H., J. Chambers & V. V. Klemm, 1986. Swan-Canning estuarine system; environment, use and future. Waterways Commission Report No. 9, Perth, Western Australia, 463 pp.

Wijninga, V. M., O. Rangel & A. M. Cleef, 1989. Botanical Ecology and Conservation of the aguna de la Herrera (Sabana de Bogota, Colombia). Caldasia 16: 23–40.

Hydrobiologia **340:** 191–195, 1996.
J. M. Caffrey, P. R. F. Barrett, K. J. Murphy & P. M. Wade (eds), Management and Ecology of Freshwater Plants.

Submerged plant survival strategies in relation to management and environmental pressures in drainage channel habitats

M. R. Sabbatini* & K. J.Murphy
Institute of Biomedical and Life Sciences, Division of Environmental & Evolutionary Biology, University of Glasgow, Glasgow G12 8QQ, UK
**Author for correspondence: (present address: Departamento de Agronomía y CERZOS, Universidad Nacional del Sur, 8000 Bahia Blanca, Argentina)*

Key words: Submerged plants, survival strategies, drainage channels, aquatic weeds

Abstract

The abundance of submerged weeds, in relation to management regime and environmental factors, was surveyed during 1992 and 1993 in drainage channels located in four geographically-distinct areas of Britain. The aim of the study was to ascertain, using a multivariate approach, the degree to which species survival strategy and vegetation could be related to disturbance and stress pressures on plant survival. Indices of disturbance and stress were constructed from combined environmental data for each site. A species ordination using Canonical Correspondence Analysis showed that the combined disturbance variable explained more of the variability that did stress. Two main groups of species could be distinguished. The larger group scored low on the disturbance gradient and these species, with different tolerances to stress (especially light-limitation), appeared to be those better-adapted to habitats with low disturbance (e.g. *Potamogeton pectinatus* and *Potamogeton lucens*). The smaller group comprised species which tended to occur in sites with higher disturbance (e.g. regular cutting) such as *Callitriche stagnalis*. Using the terminology of strategy theory, most of the dominant species could be classed as 'competitive/disturbance tolerators (CD)' or variants of this established-phase strategy. The limitations are discussed of applying the strategy approach at species level in a defined habitat-type which shows a high degree of uniformity between sites, such as artificial drainage channels.

Introduction

Grime (1979) classified the external factors which affect vegetation into two broad categories: stress (phenomena which restrict photosynthetic production), and disturbance (pressures causing partial or total destruction of plant biomass). In the aquatic environment, factors such as light availability, water level fluctuation, desiccation and management regime are examples of such pressures acting to restrict the survival of submerged plants (Kautsky, 1988; Murphy et al., 1990; Nichols, 1991). Grime et al. (1988) included some submerged plants in describing the strategies of English plant species. For lake macrophytes, Murphy et al. (1990) attempted to produce a species-strategy classification using survival traits relevant to life cycle, morphology, regeneration and plant physiology.

A functional group of plant populations may be considered as a set of similarly-adapted species occurring together at one or more locations experiencing similar suites and intensities of stress and disturbance pressures (Hills et al., 1994). The application of strategy analysis to the submerged vegetation of drainage channels (which commonly experience widely differing aquatic weed control regimes) should lead to an improvement in knowledge of the relationship between functionally-defined groups of plants and the channel environment in which they occur.

The aim of this study was to ascertain the degree to which species survival strategy could be related to the disturbance and stress pressures present in British drainage channels, and to investigate the evidence for the existence of functional groups of channel plants which might be related to such pressures.

Materials and methods

Twenty four drainage channel stretches showing a broad range of environmental conditions and management regimes typical of this habitat in Britain were selected from four drainage areas: The Fens (10 sites, E England: 0°15′E, 52°40′N); Crossens (5 sites, W England: 3°00′W, 53°35′N); Solway (6 sites, NW England: 3°15′W, 54°50′N) and the Spynie Canal (3 sites, NE Scotland: 3°20′W, 57°40′N). In each stretch, a 50 m channel length (site) was sampled during 1992 and 1993 in early, middle and late summer. Sites were always sampled at least one month after the most recent weed control treatment. 10 random grapnel hauls were taken to sample submerged macrophyte species. Abundance of each species present per sample was rated as 1, scarce; 2, common or 3, abundant.

On-site electrometric measurement of underwater light (PAR) penetration was determined for each site on each sampling date using a twin-sensor SKYE SKP210 PAR linked to a SKYE Datahog SDL 2540 logger. Dissolved oxygen, pH, temperature and conductivity were measured using Hanna and WPA instruments. Additional water chemistry data were provided by the National River Authority (NRA Anglian Region, NW Region) and NE River Purification Board (Scotland). Information on management regime was provided by NRA (Ely, Crossens, Carlisle), Middle Level Commissioners (March) and the Maintenance Committee of the Spynie Canal.

Parameters likely to constrain the growth of submerged plant species were measured and transformed into a numerical index which increased with increasing pressure on plant survival, as follows:

(i) Water fluctuation: The coefficients of variation *cv* of the measures of water depth recorded during each visit, in both years, were transformed onto a 0–5 index, of equal-interval classes covering the full range of *cv* obtained. Water depth was correlated with water fluctuation and showed that shallow waters had maximum values. Water level fluctuation can affect both disturbance and stress (Kautsky, 1988), however we observed that high water fluctuation values occurred at sites that had very low water level during some weeks and a substantial loss of above-ground biomass. On these grounds, we considered that water level fluctuation was primarily acting as a disturbance in the channels studied.

(ii) Management: The management regime applied during the three years prior to sampling was subjectively rated on a scale of 0 to 5 as to the likely degree and timing of destruction of plant biomass (see Figure 1).

(iii) Light attenuation: From the PAR data determined for each site on each sampling date, the average extinction coefficient k was calculated (Moss, 1988). The euphotic depth z_{eu}, at which about 3% of the surface light still remains was then calculated as $z_{eu} = 3.51/k$, and the ratio of z_{eu} to average depth, d, determined. An increased value of this ratio suggested more light availability at the channel bed. The range of values calculated for the ratio was divided into a series of equal-interval classes to give an index on a scale of 0–5 for each site.

(iv) Other stress factors: These included saline intrusions, low oxygen, eutrophication and shade (from emergent plants or trees). Their likely effect on macrophyte survival was rated subjectively as low (1), medium (2) or high (3), and were summed for each site to give a 0–5 index range.

The disturbance index (I_D) was constructed as the sum of (i) plus (ii) and the stress index (I_S) as the sum of (iii) plus (iv).

Filamentous algae were included together, and treated as one taxon in the analysis: these included *Cladophora glomerata, Vaucheria dichotoma* and *Enteromorpha intestinalis*. Certain taxonomically-close vascular species, with marked similarities of morphological and reproductive survival traits were also combined as single taxa for the purposes of the analysis (see Figure 3).

Field data on plant abundance were analyzed using TWINSPAN (Two Way Indicator Species Analysis; Hill, 1979). The vegetation-environment dataset was analysed with CCA (Canonical Correspondence Analysis; ter Braak, 1989) using CANOCO (ter Braak, 1988).

Results

In total 40 euhydrophyte taxa were recorded from the target sites during 1992–93, including five filamentous algae, two bryophytes and one charophyte, the remainder being vascular plants. Of these, the 22 commonest, mainly submerged, taxa were included in the multivariate analysis. Excluded were species found at only one site and species with few occurrences located in

LOW DISTURBANCE	- Weed boat: every 1 - 2 years
	- Manual weed rake (from banks in deep channels): every year
	- Shallow dredge (Bradshaw bucket): every year
↓	- Shallow dredge and manual control: every year
	- Manual control (scythes, forks): 2 or more per year (shallow channels)
	- Herder bucket + manual control
HIGH	- Dredged every 3 - 4 years + annual manual clearance
DISTURBANCE	- Deeper dredging: every year

Figure 1. Management procedures in relation to the likely degree and timing of destruction of the plant biomass in the target sites.

sites with extreme conditions. We also excluded euhydrophytes with free-floating habit (e.g. *Lemna* spp.), or with most of their foliage above the water surface (e.g. *Hippuris*) because these were unlikely to be directly affected by the stress and disturbance pressures included in the I_D and I_S indices.

All the species recorded were included by Holmes & Newbold (1984) in the community group described as being typical of habitats which are 'either base-rich or nutrient-rich, and usually both'. The eutrophic condition of British drainage channels is shown by the nitrate concentrations in channel water recorded during summer 1993 at the study sites, which was in the range 1.4–13.5 mg l^{-1}. The study sites showed a wide range of physico-chemical characteristics: pH (5.6–8.6); oxygen concentration (2.3–11.0 mg l^{-1}). Conductivity (252–2088 μS cm^{-1}), with salinity intrusions at certain sites raising the maximum value to 3800 μS cm^{-1}; water depth (0.14–1.50 m); z_{eu}/d (0.85–8.70) and fluctuation of water depth (*cv*: 5–86%).

Figure 2 shows the CCA ordination plot displaying site scores and arrows for environmental variables (ter Braak, 1988). The eigenvalues, an indication of the amount of inherent variability within the data set accounted for along a given principal axis, were 0.40 and 0.11 for the first and second axes, respectively. In addition, Figure 2 indicates the sites included in both groups (I and II) in the TWINSPAN sample classification at level 1 (eigenvalue 0.44). In CCA, the first axis explained 79% of the variance in the weighted average of the species scores, and the inter- set correlation of environmental variables with axis 1 was 0.87. To investigate whether the observed differences could be accounted for by pure chance the Monte Carlo permutation test was used in CCA (ter Braak, 1988). The

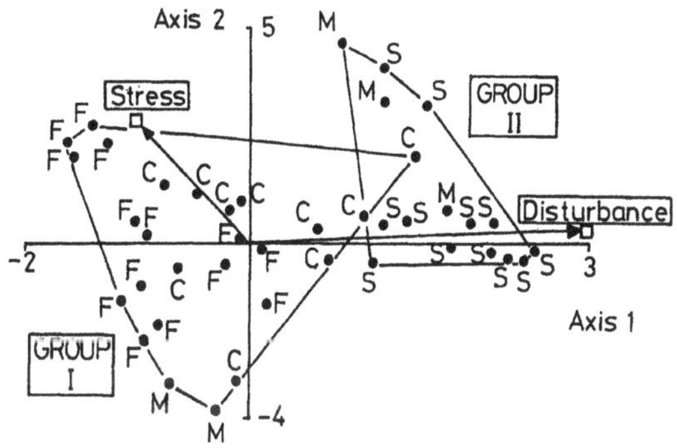

Figure 2. CCA ordination of sample scores (•) identified by region (F, Fens; C, Crossens; S, Solway; M, Spynie Canal) and arrowed biplot scores of environmental variables (□). The boundaries of two sample groups (I and II) identified at level 1 of TWINSPAN classification are overlaid on the ordination plot.

99-point random data set generated by random permutation all yielded a lower eigenvalue for the first axis and the overall analysis ($p \leq 0.01$).

Figure 3 shows the diagram of the same CCA analysis of above but displaying the species scores. Overlaid on Figure 3 are the strategy type for species given by Grime et al. (1988) and Murphy et al. (1990), together with the boundaries of TWINSPAN species groups (A and B) at level 1 (eigenvalue: 0.83).

Figure 2 shows that I_D clearly varies along axis 1. Sites supporting macrophyte communities more resistant to the disturbance produced by water fluctuations and management regime were located to the right of the diagram. The gradient also follows a geographical pattern, indicated by the drainage area location of each site. This may be explained by the fact that within an individual drainage area a relatively uniform set of weed control procedures is used, and that environ-

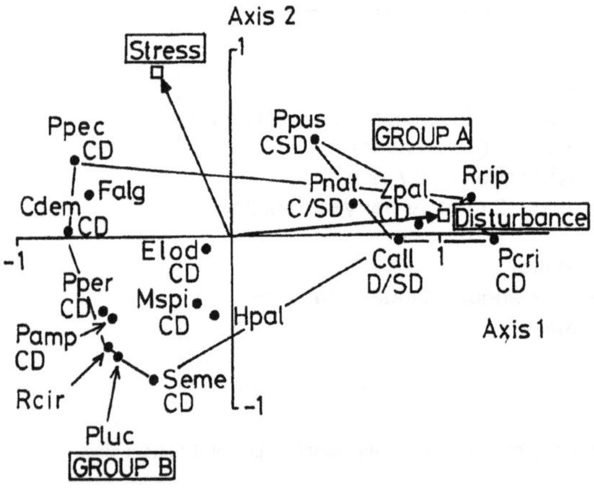

Figure 3. CCA ordination of species scores (•) and arrowed biplot scores of environmental variables (□) showing species name plus strategy type codes (where available): Call, *Callitriche stagnalis* and *C. platycarpa;* Cdem, *Ceratophyllum demersum*; Elod, *Elodea canadensis* and *E. nuttallii*; Falg, filamentous algae; Hpal, *Hottonia palustris*; Mspi, *Myriophyllum spicatum*; Pamp, *Persicaria amphibia*; Pcri, *Potamogeton crispus*; Pluc, *P. lucens*; Pnat, *P. natans*; Ppec, *P. pectinatus*; Pper, *P. perfoliatus*; Ppus, *P. pusillus* and *P. berchtoldii*; Rcir, *Ranunculus circinatus*; Rrip,*Rynchostegium riparioides* (moss); Seme, *Sparganium emersum*; Zpal, *Zannichellia palustris*. C, D and S indicate competition, disturbance and stress elements, respectively.

mental characteristics are likely more similar within than between geographically separated areas. Other environmental parameters that might also contribute to the observed gradient are average water salinity and pH. These parameters were not included in this study because the values we recorded were unlikely to have contributed significant stress or disturbance to plant growth.

TWINSPAN analysis can classify both samples and species into groups based on species assemblage data only and is not, as in canonical techniques, constrained by the environmental data. I_D was found to explain most of the variation shown by TWINSPAN at level 1. The indicator species for the large group (I) are *Elodea canadensis, Elodea nuttallii, Ceratophyllum demersum, Myriophyllum spicatum*, and *Potamogeton pectinatus*. All these species are considered to be high-nuisance submerged weeds (Pieterse & Murphy, 1990). The indicator species for the smaller group (II) are *Callitriche* species. These species are noted for their survival in channels occasionally dry for short periods or in temporary pools and they also have a strong tolerance of management based on disturbance, such as cutting (Haslam, 1978; Grillas & Duncan, 1986). Grime et al. (1988) allocated a strategy type with a strong element of disturbance-tolerance to *Callitriche stagnalis*.

In Figure 3, the small group A comprised *Potamogeton crispus, Potamogeton pusillus, Potamogeton berchtoldii, Zannichellia palustris*, and *Callitriche* spp. All these taxa scored high on the disturbance gradient, and are frequently mentioned in the literature as being disturbance-tolerant (e.g. Newbold et al., 1983). The larger species group B showed more varied tolerance of disturbance, although most species scored much lower on the disturbance gradient than group A species (Figure 3).

In Figure 2, the stress variable may permit differentiation of plant communities growing in sites with low disturbance, but affected by different intensities of stress, especially light availability. *Potamogeton pectinatus* scored highest on the stress-tolerant gradient (Figure 3); tolerance to shade and to high salinity is a well-known feature of this plant (e.g. van Wijk, 1988). A number of species appeared most characteristic of intermediate conditions of disturbance and stress: examples included *Potamogeton berchtoldii, Potamogeton pusillus* and *Potamogeton natans*.

Discussion

Most of the species included in this study probably have a rather similar established-phase survival strategy, particularly in relation to competitiveness and disturbance-tolerance. Disturbance-tolerance is likely to be a vital survival feature for channel plants in Britain, at least in part because of the widespread use of management based on disturbance-causing methods. All sites in this study experienced at least some disturbance from aquatic plant management: the lowest management index values were associated with a single annual weed-cut by boat; the highest with dredging (Figure 1).

The area of the ordination plot with low scores for disturbance and stress (i.e. the lower left corner) is were plants whose strategy incorporates a stronger competitive element would be expected to occur. In fact species with competitive strategies were located throughout the ordination diagram. This suggests that drainage channels habitats overall provide a rather productive environment for macrophyte growth.

The use of only two indices to summarize the complex influences of the environment on the growth of submerged plants is of value only up to point. Take, for example, water depth. This effects light availability for euhydrophytes (Kautsky, 1988; Blindow, 1992) and is therefore a stress factor, but changes in water level may provoke physical damage to plant tissues by partial or total desiccation, which is clearly a disturbance. Nevertheless, the use of integrated indices of stress and disturbance affecting channel sites proved successful in allowing us to distinguish two groups of plants showing differential disturbance tolerance. These may be considered as two separate functional vegetation groups, under the definition given in the introduction to this paper.

The study has identified a problem in the use of the strategy terminology put forward by Grime (1979) to classify and distinguish functionally defined groups of species of closely-similar established-phase strategy (here, mainly variants on competitors and disturbance-tolerators). If the functional analysis approach is to be successfully applied in these circumstances it would be useful to develop a terminology which would allow for more precise and detailed description of sub-categories of plant strategies. One such approach has recently been described by Hills et al. (1994) for wetland vegetation. Further work is needed to develop appropriate methods applicable to submerged plants.

Acknowledgments

We thank J. Hills, V. Abernethy and all other colleagues from the University of Glasgow who helped with data analysis and field-work. Also to the institutions mentioned in the text for allowing us access to site information. This study was part-funded by a CONICET (Argentina) grant to MRS.

References

Blindow, I., 1992. Long- and short-term dynamics of submerged macrophytes in two shallow eutrophic lakes. Freshwat. Biol. 28: 15–27.

Grillas, P. & P. Duncan, 1986. On the distribution and abundance of submerged macrophytes in temporary marshes in the Camargue (S. France). Proc. 7th Symp. Aquatic Weeds: 133–141. Loughborough, UK.

Grime, J. P., 1979. Plant strategies and vegetation processes. Wiley, Chichester, 222 pp.

Grime, J. P., J. G. Hodgson & R. Hunt, 1988. Comparative Plant Ecology. Unwin Hyman, London, 742 pp.

Haslam, S. M., 1978. River plants. Cambridge Univ. Press. 396 pp.

Hill, M. O., 1979. TWINSPAN: a FORTRAN program for arranging multivariate data in an ordered two-way table by classification of the individuals and attributes. Cornell University, Ithaca, NY, 90 pp.

Hills, J. M., K. J. Murphy, I. D. Pulford & T. H. Flowers, 1994. A method for classifying European riverine wetland ecosystems using functional vegetation groups. Funct. Ecol. 8: 242–252.

Holmes, N. T. H. & C. Newbold, 1984. River plant communities: reflectors of water and substrate chemistry. Focus on Nature Conservation 9. Nature Conservancy Council, Shrewsbury, UK. 71 pp.

Kautsky, L., 1988. Life strategies of aquatic soft bottom macrophytes. Oikos 53: 126–135.

Moss, B., 1988. Ecology of fresh waters. Blackwell Scientific Publ., 417 pp.

Murphy, K. J., B. Rørslett & I. Springuel, 1990. Strategy analysis of submerged lake macrophyte communities: an international example. Aquat. Bot. 36: 303–323.

Newbold, C., J. Purseglove & N. T. H. Holmes., 1983. Nature conservation and river engineering. Nature Conservancy Council, Shrewsbury, UK, 36 pp.

Nichols, S. A., 1991. The interaction between biology and the management of aquatic macrophytes. Aquat. Bot. 41: 225–252.

Pieterse, A. H. & K. J. Murphy (eds), 1990. Aquatic weeds. The ecology and management of nuisance aquatic vegetation. Oxford Univ. Press, UK, 593 pp.

ter Braak, C. J. F., 1988. CANOCO: a FORTRAN program for Canonical Correspondence Analysis. Tech. Rep.: LWA-88-02, Agricultural Mathematical Group, Wageningen, The Netherlands: 95 pp.

ter Braak, C. J. F., 1989. CANOCO. An extension of DECORANA to analyse species-environmental relationships. Hydrobiologia 184: 169–170.

van Wijk, R. J., 1988. Ecological studies on *Potamogeton pectinatus* L. I. General characteristics, biomass production and life cycles under field conditions. Aquat. Bot. 31: 211–258.

Hydrobiologia **340:** 197–203, 1996.
J. M. Caffrey, P. R. F. Barrett, K. J. Murphy & P. M. Wade (eds), Management and Ecology of Freshwater Plants.

The impact of drainage maintenance strategies on the flora of a low gradient, drained Irish salmonid river

James J. King
Central Fisheries Board, Mobhi Boreen, Glasnevin, Dublin 9, Republic of Ireland

Key words: Channel maintenance, cross-section modification, floral succession, *Sparganium erectum*, low gradient

Abstract

In 1990, the Central Fisheries Board initiated research on how drainage maintenance practices and strategies might be modified to enhance the salmonid carrying capacity of the affected water while maintaining an acceptable degree of conveyance. Much of the maintenance requirement was caused by dense in-channel weed beds impeding discharge and facilitating siltation. The impact of various maintenance regimes on the aquatic flora was examined in the course of pilot studies on channels of base width 3–9 m. The findings from one of these, the R. Tullamore Silver which chokes annually with *Sparganium erectum* L., are presented. Overdigging the centre of the channel and placement of spoil at the margins confined *S. erectum* to a narrow marginal zone and facilitated development of a submerged, open-water flora.

Introduction

In Ireland, under the 1945 Arterial Drainage Act, the Office of Public Works (OPW) was given responsibility for the design and implementation of drainage schemes on a catchment basis. The OPW was also required to 'permanently maintain arterial drainage works to an adequate standard' (Howard, 1980). Maintenance is usually required in those channels with a low longitudinal gradient and/or those which were over-widened at the works stage and thus rendered incapable of self-cleansing. Siltation in marginal or open water areas in such channel sections facilitates the development of various macrophyte species leading to further siltation. This process of sediment accretion and macrophyte expansion can lead to impedance of water, causing back-up in side channels carrying run off from land drains, reducing the conveyance in the channel and leading to demands from riparian owners for maintenance. While the impacts of channelization works on lotic ecosystems have been extensively documented (Swales, 1982), little study has been done on the ecological impacts of post-works maintenance. In 1990 the OPW requested the Central Fisheries Board to undertake research on alternative strategies which might be incorporated during routine mechanical maintenance programmes involving desilting and weed removal. It was envisaged that such strategies should have an enhancing effect on the channel's salmonid carrying capacity while maintaining an acceptable degree of conveyance. While particular attention was given to the status of salmonid fish stocks, the impact of the experimental works programme on aquatic plants was monitored where appropriate. The results below describe the vegetation changes observed in one such channel, the R. Tullamore Silver, in the four year period (1991 to 1994) since maintenance.

Study site

The R. Tullamore Silver, a tributary of the R. Clodiagh in the Brosna catchment, is an alkaline, moderately enriched channel of high conductivity. The study sites in this channel all lay in a 6 km section of uniformly low gradient (0.09%). Channel base width ranged from 5.5–6.5 m with lateral sediment deposition forming low shelf or berm areas, also known as secondary banks, close to the water surface. Tree/shrub cover occurred only on the bankfull line and accounted for less than 5% of channel length on any bank in the areas examined. While many bankside areas were unfenced, a combination of relatively high banks and steep bank slope restricted animal trampling and permitted a vigorous riparian cover of terrestrial grasses. The hydraulic regime was one of a continuous glide and the floral regime was characterised by two main types, an *Apium* sp.-dominated mixed flora in shallow (0.5–0.75 m) reaches on a hard stony-clay bed and a *Sparganium erectum* L.-dominated flora in deep (0.75–1.5 m) glides on silt overlying firm sandy clay. The channel held a good stock of brown trout (*Salmo trutta* L.) with angling until mid-May each year. Subsequently, the growth of *S. erectum* excessively impeded angling. High water levels and frost in late autumn generally cleared the channel of *S. erectum* and conveyance was unimpeded by macrophytes in the winter-spring period.

The R. Tullamore Silver was arterially drained in the early 1950's. From 1972 to 1994 it has been maintained on five occasions, on a *c.* 5-yearly basis. This work is currently executed by hydraulic excavators.

Materials and methods

Hydraulic machine works programme

The experimental maintenance programme stipulated that only material available on-site could be used in works and that all manipulation be done by hydraulic digging machines. This precluded strategies involving chemical control or mechanical cutting of nuisance weed.

Standard maintenance aims to restore the channel's design conveyance by re-profiling the cross-section, removing silt banks, instream macrophytes and other obstructions. Maintenance was carried out in an upstream direction, on a 9 km section of the R. Tullamore Silver between October 1990 and February 1991, beginning at the channel's confluence with the R. Clodiagh. Five sites, lying between 3.5 and 6 km of the channel's downstream end, were selected for detailed examination and specific maintenance treatment applied. One of the sites was an *Apium*-dominated zone and four were areas of heavy, full channel-width cover of *Sparganium erectum*. Qualitative assessment in autumn 1990, prior to maintenance, indicated an in-channel *Sparganium erectum* cover in excess of 75% at all four *Sparganium* sites chosen. The *Apium* site (site 1) was given standard maintenance as were two of the *Sparganium* sites (sites 2 and 3). A third *Sparganium* site had some degree of secondary bank area, colonized by *Glyceria maxima* (Hartman) Holmberg, along both banks. This site was left as an unmaintained control (site 4) to see if the vegetated marginal deposits might serve as a nucleus for further deposition with consequent channel narrowing and increased velocities leading to self-cleansing. The fourth *Sparganium* zone was maintained in an experimental manner (site 5). The centre of the channel was overdeepened, removing deposited silt, bed clay and nuisance weed material. The spoil produced was towed to the margins to form a secondary bank of clay material, giving the channel a V-shape in cross-section rather than the traditional trapezoidal cross-section. Sites 1, 2 and 4 were *c.* 100 m in length. Sites 3 and 5 were contiguous, the former being 25 m long and lying upstream of the latter, which was 75 m in length.

Monitoring of macrophytes

Monitoring of macrophytes took place annually from 1991 to 1994 in the September – early October period before any plant die-back had occurred. In the case of sites 1, 2, 3 and 5 this took the form of a scaled mapping (1:100) of the macrophyte cover within a representative 25 m sub-section of the test site. Marker pegs were set at 5 m-intervals along the bank in each mapping site. An engineering tape was stretched horizontally across the channel at each peg in turn and the width, from left-hand side, of each major vegetational cover form and of open-water recorded. These width values were transcribed onto metric graph paper for each 5-m transect and the points linked to show the extent of open water and major vegetative elements. Smaller stands of less prominent species or isolated stands of major elements were then drawn in on the mapping. The percentage vegetation cover and contribution to cover of individual species was compiled. The same 25 m sub-sections were mapped annually. Caffrey (1990) considered that a 25 m channel length was adequate to reflect the instream status of macrophytes present within longer, similar channel sections.

In the first series of monitorings after maintenance, in September 1991, it was impossible to carry out any instream monitoring in site 4 due to the dense nature of the marginal and instream cover. Instead, an annual photographic record was taken when other sites were being mapped. The natural development of a narrow open-water passage through this site in 1994 enabled

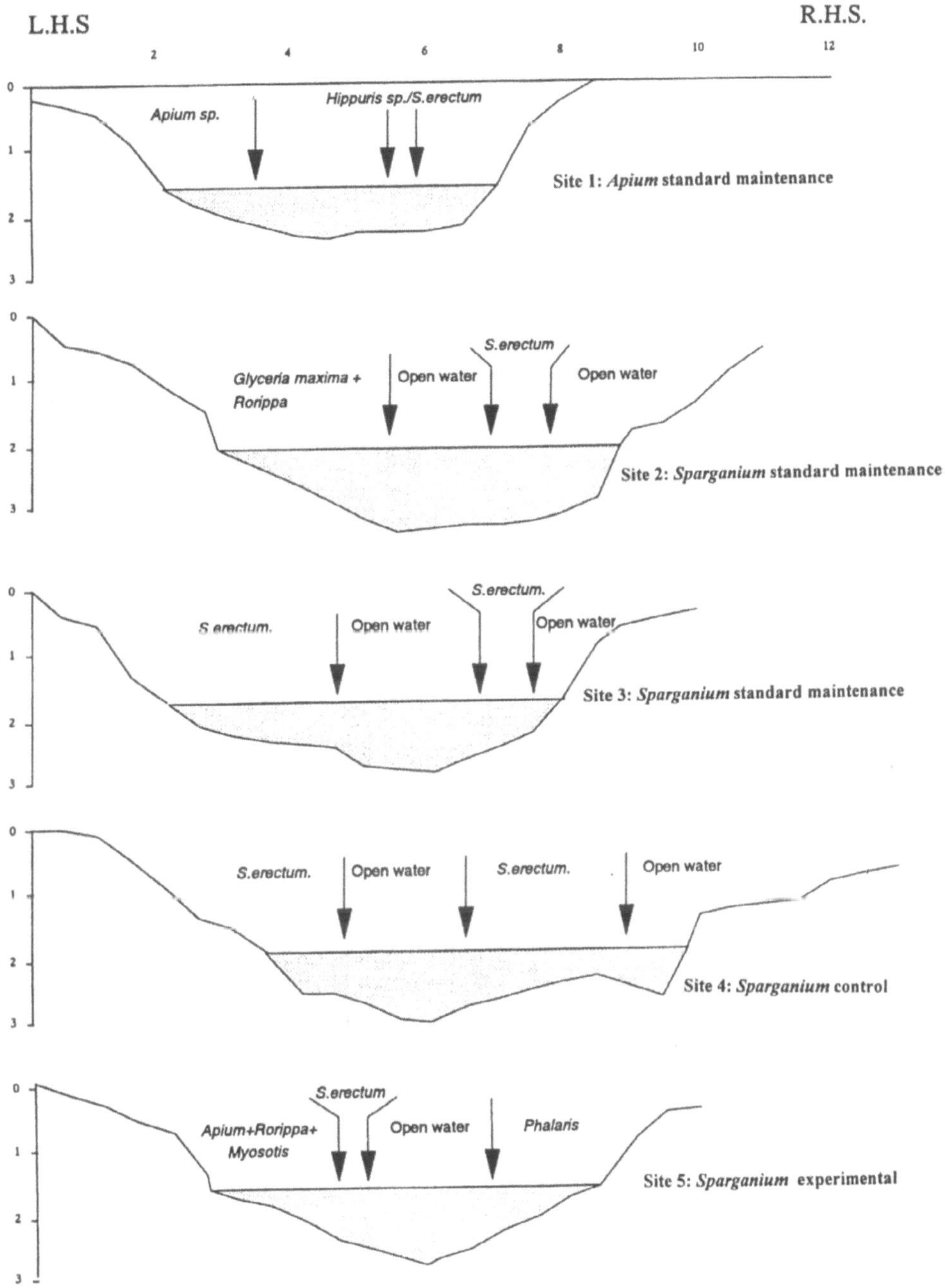

Figure 1. Characteristic channel cross-sections form sites 1 to 5 on R. Tullamore Silver, September 1993. Scales in metres. Left hand side (L.H.S.) and Right hand side (R.H.S.) with respect to observer facing downstream.

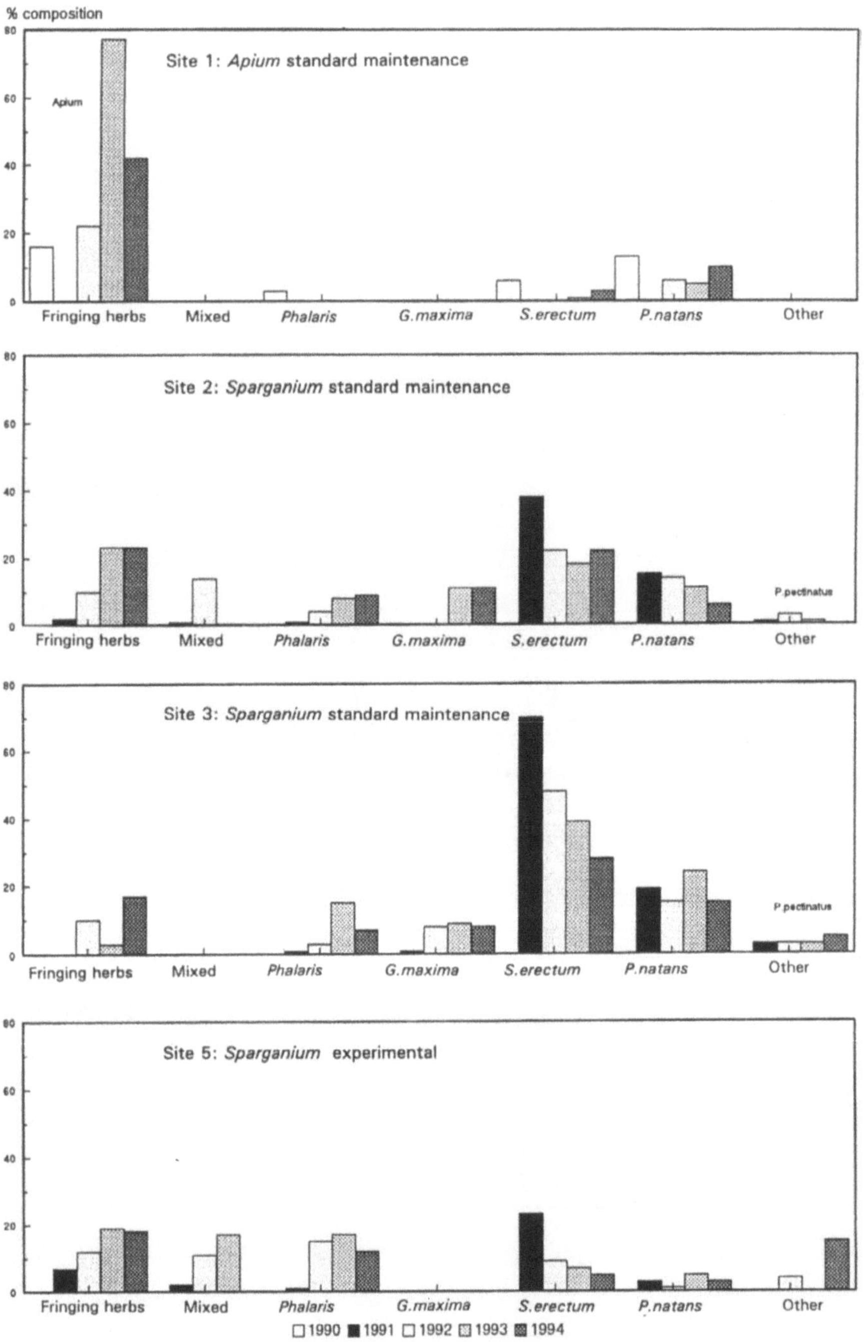

Figure 2. Changes in floral cover (% composition) at four test sites in the R. Tullamore Silver, 1990–1994.

boat access for electrofishing and instream monitoring. A floral mapping was also carried out in 1994.

Results

Site 1 – Apium standard maintenance

This was the shallowest site in cross-section with minimal maintenance required. The bed slope to the left bank from mid channel was a consistent feature of all cross-sections measured (Figure 1). Prior to maintenance the site had less than 50% floral cover composed mainly of *Apium* sp., *Potamogeton natans* L. and *Sparganium erectum*. The floral cover was substantially reduced following maintenance and remained at a level lower than that of the pre-maintenance regime for two years. However, a small increase in the *Apium* cover, a reduction of *P. natans* and loss of both *S. erectum* and *Phalaris arundinacea* was noted in the mapping site. Three years after maintenance a dramatic expansion of *Apium* occurred, with full-width coverage, in much of the site, of submerged and marginal emergent material (Figure 2). This emergent growth occurred almost entirely on the left-hand-side bed slope area where water depth was lowest. Half of this *Apium* cover was lost 12 months later and large areas of bare bed were observed. Increases in the cover of *P. natans*, *S. erectum* and *Hippuris* sp. were recorded.

Sites 2, 3 – Sparganium standard maintenance

Both sites showed evidence that maintenance, carried out from the left bank, had excavated material primarily from the right-of-centre of the channel leaving a uniform slope from mid-channel to the water surface on the left bank (Figure 1). Photographic evidence compiled during a fish stock survey three months after maintenance showed a continuous band of *Sparganium erectum*, up to 2.5 m in width, growing on this partially-maintained slope. Two similar vegetation changes occurred at both sites in the three-year period after maintenance. Firstly, each showed an annual decline in *Sparganium erectum* L. cover (Figure 2) and its replacement by various combinations of *Glyceria maxima*, *Phalaris arundinacea* or, where trampling by cattle had occurred, terrestrial grasses. In addition, low-growing fringing herbs, principally *Apium* sp., *Mentha aquatica* L., *Rorippa* sp., *Veronica anagallis-aquatica* L. and *Myosotis* sp. formed mixed or discrete stands either as an understorey to the tall emergent grasses or in open space. The emergent grasses grew on the landward side of the *Sparganium*. The more substantial cover of *S. erectum* at site 3 was related to the more extensive shelf of shallower water on the left side of the channel, relative to that at site 2 (Figure 1). Trampling by cattle was considered responsible for the greater initial expression of fringing herbs at site 2 than site 3, where cattle did not access the channel. Percentage cover of *S. erectum*, *P. arundinacea* and *G. maxima* remained stable between 1993 and 1994 at site 2 whereas the cover of *S. erectum* continued to decline at site 3, being replaced by fringing herbs, primarily *Rorippa* sp. and *Mentha aquatica* L.. On some occasions, it was not possible to identify individual floral elements due to heavy grazing pressure on some vegetated areas.. Such areas are referred to as 'Mixed Vegetation' in Figure 2. The status of the open water flora, comprising *Potamogeton natans* and *P. pectinatus* L., remained relatively constant in each site over the 1991–94 period.

Site 4 – Sparganium control

This site persistently showed a full channel width of *S. erectum*, fringed on both margins by *Glyceria maxima* and *Phalaris arundinacea*. Shallow water adjacent to cattle slips harboured stands of *Callitriche* sp. and *Veronica anagallis aquatica*. Photographic evidence taken annually in September confirmed the consistent, uniform status of the flora here and only one mapping was compiled in the study period. The cross-section (Figure 1) indicated the extent of the silt bar colonized by *S. erectum*. The small area of deeper water formed a narrow 'open-water' weed-free passage.

Site 5 – Sparganium experimental maintenance

The shallow marginal areas of bed clay created by maintenance were liable to inundation, depending on water level, (Figure 1) and were colonized by a range of fringing herbs within a year of maintenance. *Sparganium erectum* covered less than 25% of the site in this first year (Figure 2) and was confined to narrow marginal strips on the sides of the re-shaped channel. The contribution of *S. erectum* declined substantially in the following two years as colonization of the extensive secondary bank shelf on each side of the channel, by *Phalaris* and *Apium* (Figure 2), continued. The more extensive shelf on the left side also harboured the fringing herb species found at sites 2 and 3. A wide, open-channel area remained free of nuisance emergent vegetation at all times. *Potamogeton natans* was not a major component of the open-water flora, *S. erectum* was absent here and stands of submerged *Apium* were noted two years after maintenance. By the fourth year growth of *S. erectum* was noted in the open channel, although overall percentage cover of this species remained low. Grazing pressure on the left-hand berm had reduced the cover of *Phalaris* but facilitated an enhanced low-growing fringing herb assemblage and a large cover area of *Glyceria fluitans*, previously recorded in small amounts.

Discussion

Maintenance in the R.Tullamore Silver appeared to favour the initial proliferation, within six months, of *Sparganium erectum*. This differs from the findings of George (1976), Wade (1978) and Haslam (1978) in terms of the recovery rate and composition of the post-maintenance vegetation. George (1976) recorded the expansion of a submerged flora after maintenance with a recovery of floral diversity but low density three years after works. The 'main drains' studied by Wade (1978) were similar in width to the R. Tullamore Silver and also carried a submerged flora after maintenance, whereas in subsidiary drains (2.5 m wide) emergents, including *S. erectum*, dominated after works, to be replaced by an assemblage of *Phragmites* and *Carex*. Haslam (1978) reported that submerged forms (e.g. *Callitriche* spp., *Elodea* sp., and fringing herbs) grew best or only occurred in the first season after maintenance in clay channels. Species increasing to stable populations in the second and third year included emergent species such as *S. erectum* and *Glyceria maxima*. Haslam (1978) reported a recovery of vegetation in 2–3 years after dredging in streams and canals.

The subsequent changes in the composition of the instream flora in the R. Tullamore Silver are considered to be strongly linked to the form of the cross-section created in maintenance. The rapid development of *Phalaris* and negligible expression of *G. maxima* in the experimental site, compared to the more equal development of both in the standard maintenance sites, may be due to the substrate, depth regime and enhanced velocities in the experimental site. *Glyceria maxima* is considered to be favoured over *Phalaris* in low velocity channels (Krause, 1977) whereas *Phalaris* can colonize a range of sediment types and tolerate flooding for short or long periods (Conchou & Pautou, 1987). The more dramatic changes in the experimental site were considered to be due to the deliberate formation of the secondary bank, greater incline of bed side slope and stability of the secondary bank. The marginal floral growth at the experimental and standard maintenance sites was further enhanced by the impact of cattle trampling and grazing leading to the levelling out of clay at the foot of the bank and increasing areas for colonization by fringing herbs.

Excessive in-stream macrophyte growth is considered to increase channel roughness and increase impedence to flow. Maintenance is intended to reduce these effects and to increase conveyance. Increase in discharge generally reduces resistance to flow by drowning out roughness elements as submerged forms respond (Watson, 1987). However, Watson (1987) also recorded an increase in resistance to flow (as Mannings 'n') with increase in discharge at one site dominated by an emergent flora, the rigid stalks being resistant to bending with the flow. Hammill (1983) demonstrated the ability of *Sparganium erectum* to exert high retardation on discharge in a time-of-travel study in a reach of the R. Skerne, U.K., of similar width and gradient to the R. Tullamore Silver. These results would indicate a maintenance requirement in *S. erectum*-dominated channels. In the R. Tullamore Silver, velocities were substantially higher in the open-water experimental site, in fully weeded conditions, than in the partially weeded 'open-water' areas in the other *Sparganium* zones. Pitlo (1990) showed a relationship between resistance to flow and percentage of open water, pointing out that the bulk of flow, in a weeded channel, took place in the open channel area. Strategies designed to reduce maintenance requirements listed by Pitlo (1990) included oversizing the channel, in relation to depth or width, and partial maintenance of the cross section. Both strategies have had an impact on the flora in the R. Tullamore Silver. Overdeepening, in conjunction with a re-shaped cross-section, created a narrower, *Sparganium*-free central channel with an enhanced summer velocity regime. Partial maintenance led to the formation of a secondary bank dominated by *Phalaris* and *Glyceria maxima* and effectively narrowed the 'flowing water' area of the channel cross-section. Oversizing in relation to depth had the most desirable impact in engineering and fishery management terms. Absence of oversizing led to a more static *S. erectum*-dominated flora and dispersed the flow over a wider area of the cross-section.

George (1976) has indicated that time of maintenance is more likely to be set on a pre-determined time scale or maintenance cycle rather than in response to actual in-stream conditions. The remarkable development of *S. erectum*, in the R. Tullamore Silver, within 6 months of completion of maintenance, echoes the concerns of Krause (1977) and Wade (1978) as to the real benefits of maintenance in weeded, low-gradient channels. The plant succession reported here for maintained sites indicates a sequence of change, as opposed to the static regime in the *Sparganium* control site. The sequence of change led to the formation of a self-scouring area of flow with reduced or zero growth of *S. erectum* in the open channel. This developed over a period of 3–4 years but required a non-trapezoidal cross-section to achieve best effects. Any form of over-digging appeared to focus the flow along the line of maximum depth. It would be desirable that such a sequence of plant succession be monitored beyond the duration of the maintenance cycle and that over-digging strategies be designed and implemented over longer channel segments than the short treatment sites examined here.

Acknowledgments

The author gratefully acknowledges the financial support of the Office of Public Works for the Experimental Drainage Maintenance Programme, of which this study forms a part.

References

Caffrey, J., 1990. The classification, ecology and dynamics of aquatic plant communities in some Irish rivers. Ph.D. Thesis, the National University of Ireland, 254 pp.

Conchou, O. & G. Pautou, 1987. Modes of colonization of an heterogeneous alluvial area on the edge of the Garonne river by *Phalaris arundinacea* L. Reg. Rivers 1: 37–48.

George, M., 1976. Mechanical methods of weed control in watercourses – an ecologist's view. In Aquatic Herbicides B.C.P.C. monograph 16: 91–99.

Hamill, L., 1983. Some observations on the time of travel of waves in the River Skerne, England, and the effect of aquatic vegetation. J. Hydrol. 66: 291–304.

Haslam, S. M., 1978. River plants: The macrophytic vegetation of watercourses. Cambridge University Press, 396 pp.

Howard, J., 1980. Current practice in assessing drainage impacts. In Impacts of drainage in Ireland. National Board for Science and Technology, Dublin: Paper 7: 1–39.

Krause, A., 1977. On the effect of marginal tree rows with respect to the management of small lowland streams. Aquat. Bot. 3: 185–192.

Pitlo, R. H., 1990. Oversizing, and reduced maintenance in relation to aquatic plant growth and flow resistance. In Proc. EWRS 8th Symp. on Aquatic Weeds: 167–172.

Swales, S., 1982. Environmental effects of river channel works used in land drainage improvement. J. Envir. Mgmt 14: 103–126.

Wade, P. M., 1978. The effect of mechanical excavators on the drainage channel habitat. In Proc. EWRS 5th Symp. on Aquatic Weeds: 33 342.

Watson, D., 1987. Hydraulic effects of aquatic weeds in U.K. rivers. Reg. Rivers 1: 211–227.

Hydrobiologia **340:** 205–211, 1996.
J. M. Caffrey, P. R. F. Barrett, K. J. Murphy & P. M. Wade (eds), Management and Ecology of Freshwater Plants.

The effect of weed control practices on macroinvertebrate communities in Irish Canals

C. Monahan & J. M. Caffrey
Central Fisheries Board, Mobhi Road, Glasnevin, Dublin 9, Ireland

Key words: weed cutting, aquatic herbicides, dichlobenil, macrophyte-macroinvertebrate interrelationships

Abstract

Macroinvertebrates in aquatic habitats form an integral part of the diet of many freshwater fish. It is therefore important to understand the effects that weed control practices have on this community in canal fishery watercourses. The principal forms of weed control operated in the Grand and Royal Canals include mechanical cutting, using a variety of boat-mounted and land-based apparatus, and chemical treatment using dichlobenil. The community composition and relative abundance of macroinvertebrates in control, mechanically cut and dichlobenil treated canal sites was recorded on three to five occasions between 1993 and 1994. The results indicated that *Asellus aquaticus* was the dominant organism at all canal locations. The land-based Mowing Bucket effected the greatest reduction in macroinvertebrate numbers in the immediate aftermath of the cut. This reflects the capacity of the machine to cut vegetation to canal bed level, thereby removing any substrate for colonisation. At all eight sites examined, macroinvertebrate numbers increased relatively rapidly following treatment and no adverse effect on dependent fish life resulted. The Office of Public Works policy of removing obstructive vegetation from a central navigation channel, while preserving weeded marginal fringes, minimises the impact of weed control operations on the macroinvertebrate fauna.

Introduction

In 1990 the Office of Public Works (O.P.W.) commissioned the Central Fisheries Board to conduct a five-year environmental study on aspects of water quality status, aquatic plant management and recreational fisheries development in the Royal, Grand and Barrow Canals. These canals are currently being developed as multi-purpose recreational resources where navigation, angling and walking are the principal amenity pursuits. As part of this study, aquatic macroinvertebrate surveys in all three canals were conducted.

The canals are man-made, managed, linear bodies of water that have no direct counterpart in nature (Caffrey, 1988). They have been described as intermediate between flowing and static waterbodies (Murphy & Eaton, 1981) where, in long profile, they resemble lakes but in terms of width, depth and flow characteristics they are closer to lowland rivers. In the canals under examination the flow is slow (<5 cm s^{-1}) and the habitat is depositing. The deep mud substrate, relatively clear water and shallow (<2 m) channel enables submerged vegetation to occupy the full canal width (Caffrey, 1991). This provides an abundance of microhabitat niches for macroinvertebrate species. The presence of dense marginal vegetation provides further habitats for the canal fauna (Lillie & Budd, 1992). In Irish canals, weed control operations are conducted once or twice during the growing season in order to maintain navigation channels.

Macroinvertebrates in aquatic habitats form an integral part of the diet of many freshwater fish (Kennedy & Fitzmaurice, 1968, 1970, 1974). It is clear therefore that anything that influences the density, diversity and availability of these organisms will likewise impact on fish stocks. As aquatic plants can harbour large populations of invertebrates, providing food and shelter for them (Engel, 1988), it is probable that weed control procedures will significantly affect these populations (Pearson & Jones, 1978). For this reason, it is impor-

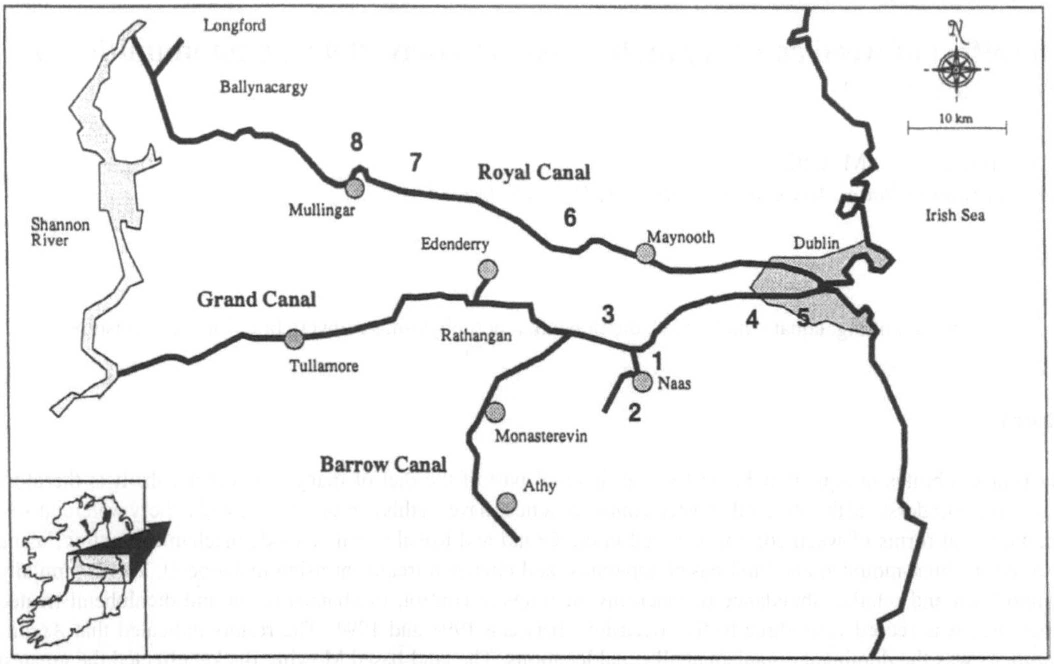

Figure 1. Map of the Grand and Royal Canals showing the location of the sites examined during 1993 and 1994. Sampling stations: 1 - Naas Branch, 2 - Corbally Line, 3 - Leinster Aqueduct, 4 - Lock 12, 5 - Harold's Cross, 6 - Cloncurry, 7 - Mc Neads Bridge and 8 - Mullingar.

tant to gain an insight into the community composition and relative abundance of macroinvertebrates in these canal habitats and to assess the effects that weed control procedures have on this component of the canal ecosystem.

Aquatic weed control in Ireland is essential for successful canal management. Research on these canals has revealed that the most efficient and environmentally safe control is achieved through the integrated deployment of mechanical and chemical methods (Caffrey & Monahan, 1995). For the purpose of this paper the effects on macroinvertebrate communities of spraying with dichlobenil, cutting with a mechanical harvester and a Mowing Bucket, and algal removal using a Wilder D-blade and Lifter were examined.

Study sites

The sites examined on the Royal, Grand and Barrow Canal systems are presented in Figure 1. The Naas Branch (Site 1) and Corbally Line (Site 2) were untreated and functioned as control sites. The land-based Mowing Bucket was operated west of the Leinster Aqueduct (Site 3), on the Grand Canal. The effects of mechanical cutting using a mechanical harvester were assessed east of Lock 12 (Site 4), also on the Grand Canal. The impact of spraying with dichlobenil on macroinvertebrate populations was examined at three sites. These were Harold's Cross (Site 5) on the Circular Line of the Grand Canal, and Cloncurry (Site 6) and McNeads Bridge (Site 7) on the Royal Canal. The effects of algae removal, using the Wilder boat, were studied at Mullingar (Site 8) on the Royal Canal. At all of the sites examined aquatic vegetation grew abundantly and, before treatment, presented in excess of 80% cover (as defined by Best, 1981).

Materials and methods

Sampling was conducted at each of the eight sites on at least three occasions during the sampling period (May 1993 to May 1994). Samples were collected using a stovepipe sampler (Weber, 1973; Caffrey, 1990). The stovepipe, made from heavy duty plastic material, measured 80 cm in height and 23 cm in internal diameter. It was thrust forcefully into the vegetation

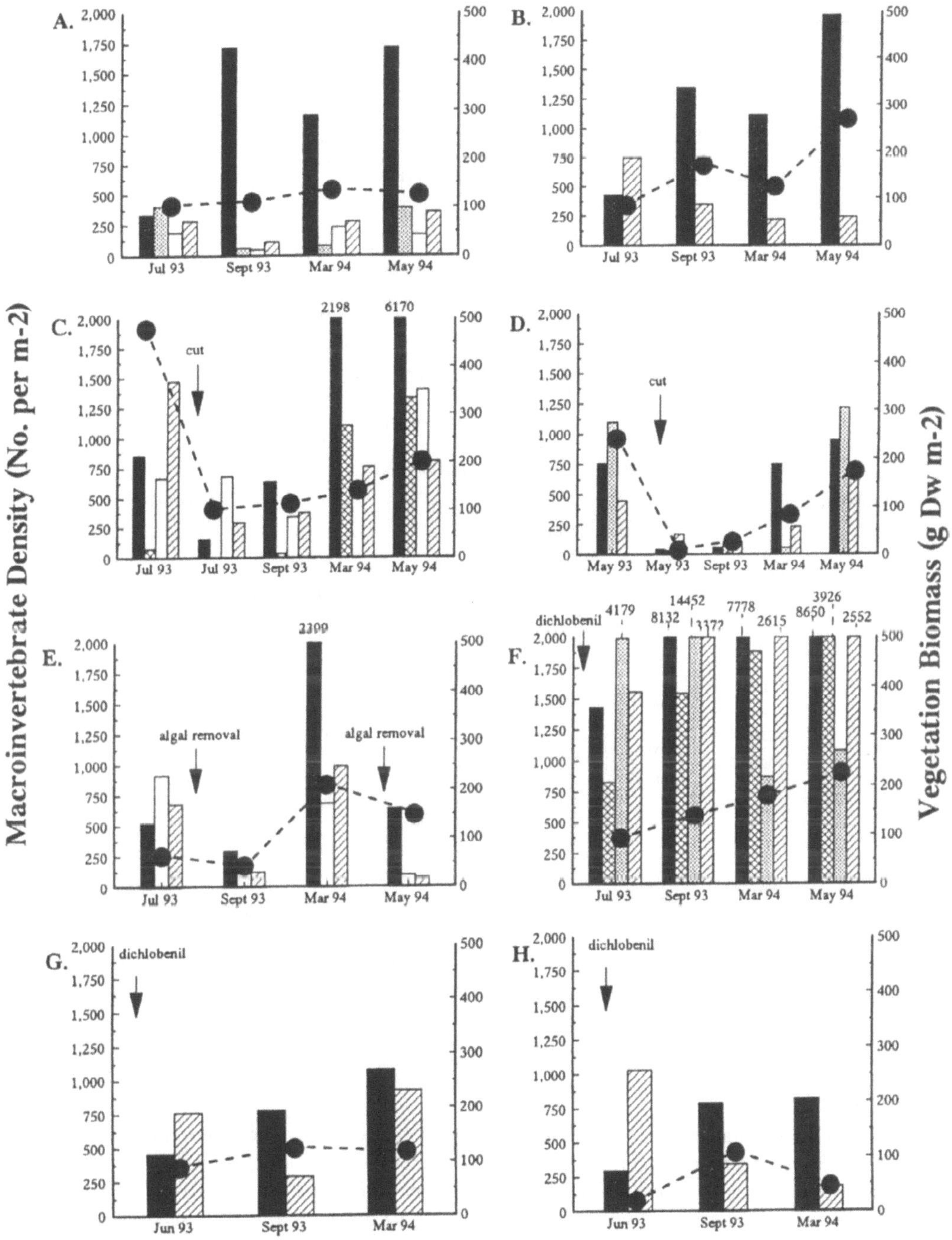

Figure 2. The effect of weed control practises on macroinvertebrate density and vegetation biomass in sections of canal during 1993 and 1994. Control sites; A - Naas Branch, B - Corbally Line; Harvester site: C - Lock 12; Mowing]3ucket site: D - Leinster Aqueduct; algal removal site: E - Mullingar; Dichlobenil treated sites: F- Harold's Cross, G - McNeads Bridge and H - Cloncurry. ■ *Asellus aquaticus* ▦ Mollusca □ Oligochaeta ▩ *Gammarus* ▨ Others ●—● vegetation biomass

Table 1. A list of macroinvertebrate taxa recorded in the canal sites examined during the investigation of the effects of weed control practices on macroinvertebrate communities in Irish canals.

TRICLADIDA			
	Planariidae		
		Planaria	*P. torva*
		Polycelis	*P. nigra*
			P. tenuis
		Dugesia	*D. lugubris*
	Dendrocoelidae		
		Dendrocoelum	*D. lacteum*
NEMATOMORPHA			
	Haplotaxidae		
		Haplotaxis	*H. gordioides*
OLIGOCHAETA			
	Lumbriculidae		
		Lumbriculus	*L. variegatus*
	Tubificidae		
		Tubifex	
HIRUDINEA			
	Piscicolidae		
		Piscicola	*P. geometra*
	Glossiphoniidae		
		Glossiphonia	*G. complanata*
			G. heteroclita
		Helobdella	*H. stagnalis*
		Hemiclepsis	*H. marginata*
		Theromyzon	*T. tessulatum*
	Hirudidae		
		Haemopis	*H. sanguisuga*
	Erpobdellidae		
		Erpobdella	*E. octoculata*
			E. testacea
GASTROPODA			
	Neritidae		
		Theodoxus	*T. fluviatilis*
	Valvatidae		
		Valvata	*V. piscinalis*
			V. cristata
			V. macrostoma
	Hydrobiidae		
		Bithmia	*B. tentaculata*
			B. leachu
		Potamopyrgus	
			P. jenkinsi
	Lymnaeidae		
		Lymnaea	*L. peregra*
			L. stagnalis
	Physidae		
		Physa	
	Planorbidae		
		Planorbis	*P. carinatus*
			P. complanata
			P. contonus
			P. corneus
			P. laevis
			P. planorbis
			P. spirorbis
	Ancylidae		
		Ancylus	*A. fluviatilis*
BIVALVIA			
	Sphaeriidae		
		Sphaerium	
	Pisidiidae		
		Pisidium	
ARACHNIDA			
	Araneae		
		Argyroneta	*A. aquatica*

MALACOSTRACA			
	Amphipoda		
		Gammarus	*G. lacustris*
	Isopoda		
		Asellus	*A. aquaticus*
			A. meridianus
EPHEMEROPTERA			
	Caenidae		
	Caenis	*C. luctuosa*	
ANISOPTERA			
Aeshnidae			
ZYGROPTERA			
	Lestidae		
	Coenagrionidae		
	Platycnemidae		
PLECOPTERA			
	Nemouridae		
HEMIPTERA			
	Gerridae		
		Gerris	
	Pleidae		
	Corixidae		
MEGALOPTERA			
	Sialidae		
		Sialis	
NEUROPTERA			
	Sisyridae		
		Sisyra	
TRICOPTERA			
	Limnephilidae		
		Limnephilis	
		Halesus	
		Stenophylax	
		Anabolia	
	Hydroptilidae		
		Agraylea	
	Leptoceridae		
		Athripsodes	
		Leptocerus	
	Phryganeidae		
		Phryganea	
	Polycentropidae		
		Cyrnus	
		Holocentropus	
		Polycentropus	
	Psychomyidae		
		Tinodes	
LEPIDOPTERA			
		Paraponyx	*P. stagnalis*
			P. stratiotata
DIPTERA			
	Tipulidae		
		Phalacrocera	
		Pedicia	
	Ceratopogonidae		
		Culicoides	
	Chironomidae		
	Tabanidae		
COLEOPTERA			
	Gyrinidae		
	Haliplidae		
	Dytiscidae		
	Hydrophiliidae		
	Helmidae		
	Chrysomelidae		
	Curculionidae		

and pushed into the substratum. The vegetation and attached macroinvertebrates were carefully removed from the stovepipe by hand and placed in labelled bags. When all the vegetation had been removed, a small fine-meshed (2 mm) aquarium net was swept through the water within the stovepipe to collect faunal organisms that may have fallen from the vegetation as it was being removed. Ten replicate samples were taken at each site. The samples were returned to the laboratory where all macroinvertebrates were washed from the vegetation onto a stack of sieves (mesh-width 2 cm, 1 cm and 0.5 mm) (Rietveld & Beltman, 1982). The invertebrates were then removed from the sieves and held in 40% alcohol for identification and enumeration.

The biomass (gm dry weight m^{-2}) of vegetation at all sites was recorded on each sampling occasion.

Table 2. Number of macroinvertebrate taxa recorded in the eight canal sites examined in 1993 and 1994. ( ) species diversity following cut.

Site	Site No.	1993 Jun/July	1993 Sept.	1994 March	1994 May
Naas Branch	1	18	14	12	15
Corbally Line	2	11	15	10	15
Leinster Aqueduc	3	10 (11)	18	14	15
Lock 12	4	18 (11)	16	17	17
Harold's Cross	5	16	14	17	16
Cloncurry	6	11	19	12	–
McNeads Bridge	7	15	15	23	–
Mullingar	8	12	17	22	8

Results

A rich diversity of macroinvertebrate organisms was present at all eight canal sites examined (Table 1). The aquatic flora, which was likewise abundant and diverse, comprised mixed assemblages of rooted and rootless plant species.

Asellus aquaticus was the most common macroinvertebrate found throughout the canal sites examined. This species was present in large numbers at all sites, although numbers varied considerably between the sites. For example, in May 1994 at Site 5 a total of 8650 individuals m^{-2} was recorded, while at Site 6 in March 1994, some 824 individuals m^{-2} were present. Other species that were commonly recorded were mud-dwelling Lumbriculid and Tubificid worms, *chironomid* larvae and Glossiphoniidae. Molluscs were commonly recorded, with *Bithynia* spp. and Sphaeriidae spp. being dominant. Limnephilid larvae, Haliplidae beetles and Zygopteran nymphs were also abundant.

The relative densities of predominating macroinvertebrate taxa at untreated control sites (Sites 1, 2) are presented in Figure 2A and 2B. At these sites mixed assemblages of *Oenanthe aquatica, Elodea canadensis, Myriophyllum verticillatum, Hippuris vulgaris, Sagittaria sagittifolia* and *Cladophora* spp. occupied these sections. Vegetation biomass values in excess of 250 g m^{-2} were recorded at these sites. *Asellus* numbers peaked in September 1993 and May 1994, with 1721 and 1716 individuals per m^2, respectively, being recorded at Site 1 and 1358 and 1976 individuals per m^2, respectively, recorded at Site 2.

The vegetation at Site 3 was mechanically cut in June, while that at Site 4 was cut in July 1993. *Oenanthe* spp., *Sparganium emersum, Elodea canadensis* and *Cladophora* spp. were the dominant species present at both sites. Immediately after cutting the numbers of macroinvertebrates decreased – from 2986 to 1135 per m^2 at Site 4 and from 2300 to 375 per m^2 at Site 3 (Figure 2C, D). At the latter site the Mowing Bucket cut the vegetation to the canal bed and practically no weed for colonisation by plant-dwelling insects remained. Natural recruitment among macroinvertebrates at this site was, consequently, slow (Figure 2D). The mechanical harvester removed vegetation to within 25 cm of the canal bed, leaving some plant material for insect colonisation. This accounts for the greater abundance of macroinvertebrates recorded following cutting with this machine at Site 4 (Figure 2C). The September peak in macroinvertebrate abundance observed at the control and dichlobenil treated sites in September 1993 was not recorded at either of the mechanically cut sites. Numbers did not fully recover until March 1994 at Site 4, where a total of 4436 invertebrates per m^2 was recorded, and until May 1994 at Site 3 where 2817 invertebrates per m^2 were counted.

Sites 5, 6, and 7 were treated with dichlobenil in May 1993. *Myriophyllum verticillatum* was the dominant aquatic plant present and was effectively controlled at all three sites. However, vegetation biomass at Site 5 remained relatively high due to the presence at this site of filamentous algae and *Ceratophyllum demersum,* which are both resistant to the herbicidal activity of dichlobenil (Caffrey, 1993a). Two month post-treatment *Asellus* number at all three sites decreased but by September numbers had increased significantly (Figure 2F, G, H).

Following algal removal at Site 8 there was a significant decrease in macroinvertebrate numbers, from 2137 per m^2 before cutting to 577 per m^2 after treatment (Figure 2E). *Asellus* numbers increased in March 1994 but decreased again in May, following further mechanical removal of the algae.

At the two control sites species diversity varied considerably over the seasons (Table 2). Although macroinvertebrate densities decreased at all sites after treatment, species diversity was not significantly affected. There was, however, a decrease in species diversity from 18 to 11 and 22 to 8 at Site 4 and Site 8, respectively, following mechanical weed removal (Table 2).

Discussion

The interrelations between aquatic plants and animals are well documented. Aquatic plants in slow-flowing or static situations provide a habitat of much greater physical and chemical complexity than is provided by the underlying and often homogenous silt/mud substrate (Rooke, 1984; Wright et al., 1992). In lakes, ponds and ditches, the habitat preference of macroinvertebrates seems to be influenced chiefly by the vegetation (Dvorak & Best, 1982; Scheffer et al., 1984). It is not surprising, therefore, that they support a greater variety of macroinvertebrates. Research on British lowland rivers conducted by Wright et al. (1992) also revealed a positive correlation between macrophytes and macroinvertebrates in lotic situations. Research further revealed that different growth forms among macrophyte species supported 9 significantly different numbers of taxa (Caffrey, 1993b).

In canal habitats plants provide shelter for macroinvertebrates from disturbance and predators. Plants also provide a large surface area for colonisation by epiphytic and periphytic algae which many macroinvertebrates, notably Mollusca, utilise as a source of food (Rooke, 1984). These plants also provide sites for the deposition of eggs by some macroinvertebrate species and emergence routes for species with an aerial stage in their life cycle.

Seasonal variations in diversity and abundance have been observed in benthic animal communities of various waterbodies (Holt & Stawn, 1983; Gargan, 1986). This seasonal trend was demonstrated at the two control sites (Sites 1 and 2). It is clear therefore that, while examining the impact that weed control procedures have on the density and diversity of macroinvertebrates in the canal habitat, seasonal trends and the timing of control operations must also be considered (Pearson & Jones, 1978).

Mechanical weed control is a destructive technique where vegetation and attached macroinvertebrates are removed from the habitat. A study conducted by National Rivers Authority staff showed that *circa* one million macroinvertebrates were removed with each tonne of *Ranunculus* harvested from the River Avon (K. Tibbett per. comm.). Treatment with dichlobenil is less destructive as the vegetation is killed *in situ*, permitting the attached macroinvertebrates to migrate and colonise adjacent untreated plants. In Irish canals dichlobenil is used for partial weed control only and the herbicide is selectively applied to a swath width of 6 to 8 m along the channel. This results in the creation of a relatively weed free central channel, fringed on both sides with vegetation (Caffrey, 1993a; Caffrey & Monahan, 1995). The use of dichlobenil at the canal sites examined did not significantly alter the normal seasonal trends where macroinvertebrate numbers peaked in May and September. Mechanical treatment, on the other hand, altered this trend and no significant increase in *A. aquaticus* numbers was recorded in the year of treatment. It is important therefore, in the interest of ecological sustainability, that mechanical cutting in this habitat is carefully controlled.

While large numbers of macroinvertebrates are removed from the canals during weed cutting and harvesting operations, it is rare that more than half of the channel width is targeted for treatment. The vegetation along both margins is normally left untouched. This provides a habitat for macroinvertebrates that are dislodged during harvesting, in addition to providing an ample reserve of individuals and taxa. Fox & Murphy (1990) revealed that the ecological impact of a non-selective cut in the River Windrush would have been minimised by the retention of a fringe of uncut plants at the channel edge, which would provide habitat for the macroinvertebrates and fish. Emergent macrophytes are generally macroinvertebrate-rich habitats, reflecting the fact that these perennial species provide year-round cover and food for the fauna (Jenkins et al., 1984; Omerod, 1988). The Mowing Bucket is employed to cut swims (weed free areas for unobstructed angling) in high profile angling areas along the canal. This limited cutting, combined with the fact that most macroinvertebrate species re-establish viable populations rapidly (Pearson & Jones, 1978), reduces the impact that this form of mechanical control has on the macroinvertebrate communities in treated canal habitats.

Asellus is the dominant macroinvertebrate in the canals studied. Observations to date reveal that this species is a major component in the diet of fish in Irish canals. While the weed control practices served to reduce the density of fish-food macroinvertebrates in the aftermath of treatment, the strategies adopted aim to ensure that no more than 50% of the vegetation in any one section is removed in any one period. This ensures the availability of a plentiful supply of food items for fish. The study sites examined are all important angling locations which regularly return large weights of fish to local and visiting anglers This supports the view that a sufficient macroinvertebrate density, to maintain the good fish stocks present, remains in the aftermath of judicious weed control.

Acknowledgments

The authors would like to express their gratitude to the Office of Public Works for funding this project.

References

Best, E. P. H., 1981. The submerged aquatic macrophytes in Lake Maarsseveen 1: the species composition, spatial distribution and productivity. Hydrobiol. Bull. 15: 72–81.

Caffrey, J. M., 1988. The status of aquatic plant communities in the Royal and Grand Canals, with reference to past and future weed management programmes. Office of Public Works commissioned report. Central Fisheries Board, Dublin. 107 pp.

Caffrey, J. M., 1990. The classification, ecology and dynamics of aquatic plant communities in some Irish rivers. Ph.D. Thesis, University College, Dublin. 254 pp.

Caffrey, J. M., 1991. Aquatic plants and plant mangement in the Inchicore area of the Grand Canal. In M. Connaghan, O. Gleeson & A. Maddock (eds), Inchicore and Kilmainham Development Project, Office of Public Works, Dublin: 66–68.

Caffrey, J. M., 1993a. Aquatic weed management practices using dichlobenil: an Irish experience. Pol. Arch. Hydrobiol. 40: 255–266.

Caffrey, J. M., 1993b. Aquatic plant management in relation to Irish recreational fisheries development. J. Aquat. Plant Manage. 31: 162–168.

Caffrey, J. M. & C. Monahan, 1995. Aquatic plant management in Irish canals, 1990–1995. Office of Public Works commissioned report. Central Fisheries Board, Dublin. 102 pp.

Dvorak, J. & P. H. Best, 1982. Macro-invertebrate communites associated with the macrophytes of Lake Vechten: structural and functional relationships. Hydrobiologia 95: 115–126.

Engel, S., 1988. The role and interactions of submersed macrophytes in a shallow Wisconsin lake. J. Freshwat. Ecol. 4: 329–341.

Fox, A. M. & K. J. Murphy, 1990. The efficacy and ecological impacts of herbicide and cutting regimes on the submerged plant communities of four British rivers. I. A comparison of the management efficacies. J. appl. Ecol. 27: 520–540.

Gargan, P. C., 1986. The biology of the fish and faunal communities in Lough Sheelin, Co.Cavan, a eutrophic lake in the Irish midlands. Ph.D. Thesis, University College, Dublin. 367 pp.

Holt, J. & K. Stawn, 1983. Community structure of macrozooplankton in Trinity and Galveston Bays. Estuaries 6 : 66–75.

Jenkins, R. A., K. R. Wade & E. Pugh, 1984. Macroinvertebrate habitat relationships in the Teifi catchment and the significance to conservation. Freshwat. Biol. 14: 23–42.

Kennedy, M. & P. Fitmaurice, 1968. The biology of the bream *Abramis brama* (L.) in Irish waters. Proc. Roy. Ir. Acad. 67B: 95–157.

Kennedy, M. & P. Fitzmaurice, 1970. The biology of the tench *Tinch tinca* (L.) in Irish waters. Proc. Roy. Ir. Acad. 69B: 31–82.

Kennedy, M. & P. Fitzmaurice, 1974. The biology of the rudd *Scardinius erythrophthalmus* (L.) in Irish waters. Proc. Roy. Ir. Acad. 74B: 245–303.

Lillie, R. A. & J. Budd, 1992. Habitat architecture of *Myriophyllum spicatum* L. as an index to habitat quality for fish and macroinvertebrates. J. Freshwat. Ecol. 7: 113–125.

Murphy, K. J. & J. W. Eaton, 1981. Waterplants, boat traffic and angling in canals. Proc. 2nd Brit. Freshwat. Fish. Conf.: 173–187.

Ormerod, S. J., 1988. The micro-distribution of aquatic macroinvertebrates in the Wye River system: the result of abiotic or biotic factors? Freshwat. Biol. 20: 241–247.

Pearson, R. G. & N. V. Jones, 1978. The effects of weed-cutting on the macro-invertebrate fauna of a canalised section of the River Hull, a Northern English chalk stream. J. Envir. Mgmt 7: 91–97.

Rietveld, W. & B. Beltman, 1982. A qualitative analysis of macrofauna sampling in ditches. Hydrobiol. Bull. 16: 149–157.

Rooke, J. B., 1984. The invertebrate fauna of four macrophytes in lotic system. Freshwat. Biol. 14: 507–513.

Scheffer, M., A. A. Achterberg & B. Beltman, 1984. Distribution of macroinvertebrates in a ditch in relation to the vegetation. Freshwat. Biol 14: 367–370.

Weber, C. I., 1973. Biological monitoring of the aquatic environment. Biological methods for the assessment of water quality. American Society for Testing and Materials, 46–60.

Wright, J. F., J. H. Blackburn, D. F. Westlake, M. T. Furse & P. D. Armitage, 1992. Anticipating the consequences of river mangement for the conservation of macroinvertebrates. In P. J. Boon, P. Calow & G. E. Petts (eds), River Conservation and Management, John Wiley and Sons Ltd.: 138–149.

Hydrobiologia **340**: 213–218, 1996.
J. M. Caffrey, P. R. F. Barrett, K. J. Murphy & P. M. Wade (eds), Management and Ecology of Freshwater Plants.

Physical control of Eurasian watermilfoil in an oligotrophic lake

Charles W. Boylen[1], Lawrence W. Eichler[1] & James W. Sutherland[2]
[1] *Rensselaer Darrin Fresh Water Institute, Rensselaer Polytechnic Institute, Troy, NY 12180-3590, USA*
[2] *New York State Department of Environmental Conservation, Lake Services Section, Albany, NY 12233-3502, USA*

Key words: Benthic barrier, milfoil, *Myriophyllum spicatum*, suction harvesting

Abstract

The introduction of Eurasian watermilfoil (*Myriophyllum spicatum*) into oligotrophic waters of high water clarity in temperate zones of North America has produced growth in excess of 6 m depth and yearly biomass approaching 1000 g m^{-2} dry weight. From its initial observation in Lake George, New York, USA in 1985, by 1993 milfoil had spread to 106 discrete locations within the lake. A 7-year study of one site having no management showed milfoil to grow expansively, suppressing native plant species from 20 in 1987 to 6 in 1993 with the average number of species m^{-2} quadrat declining from 5.5 in 1987 to less than 2 in 1993. Management of milfoil by means of hand harvesting, suction harvesting and benthic barrier has reduced the number of unmanaged sites from 106 in 1993 to 11. One year post-treatment at sites utilizing suction harvesting, showed a greater number of native species at all sites than pretreatment with a substantial reduction in milfoil biomass. At sites where benthic barrier was removed 1–2 years after installation, milfoil had recolonized 44% of grid squares within 30 days. Ninety days after barrier removal 74% of grid squares contained milfoil and one year later 71% of the grids supported milfoil. During the first year following mat removal, the average number of species m^{-2} peaked at 4.7 and stabilized at 4.5 during the second year. Hand harvesting by SCUBA in areas of limited milfoil growth (new sites of infestation and sites of former treatment) was found to reduce the number of milfoil plants present in subsequent years. Hand harvesting did not eliminate milfoil at any of the sites and regrowth/colonization necessitated reharvesting every 3 or more years. Results of evaluations of physical plant management techniques indicate that (1) an integrated program utilizing different techniques based on plant density reduced the growth of milfoil and (2) long term commitment to aquatic plant management is necessary since none of the techniques employed singly were found to eliminate milfoil.

Introduction

Oligotrophic waters of temperate North America support a diverse community of aquatic macrophytes. Lake George is characteristic with 48 species of primarily submersed types extending over a depth range of 0.5 to 8.0 m (Collins et al., 1987). Until recently thousands of these lakes have been spared the introduction of exotic species such as *Myriophyllum spicatum* (watermilfoil) from Eurasia. Beginning in the 1940s the Eurasian watermilfoil has spread from an initial nuisance-level infestation in the Potomac River/Chesapeake Bay region of the US to throughout much of North America, creating weedy growth and suppressing native plant populations (Couch & Nelson, 1985). Its impact has been particularly noticeable in nutrient enriched waters (Reed, 1977). However, since 1980 we have observed that Eurasian watermilfoil has increasingly invaded low nutrient waters which heretofore did not supported the establishment of this species. Many of these lakes are in pristine environments where water clarity allows for growth at depths in excess of 6 m. Aquatic weed management is often limited to early detection and selective removal of the milfoil.

Since 1987 the presence of Eurasian watermilfoil in Lake George has become an increasing aquatic weed problem. Lake George serves as a primary drink-

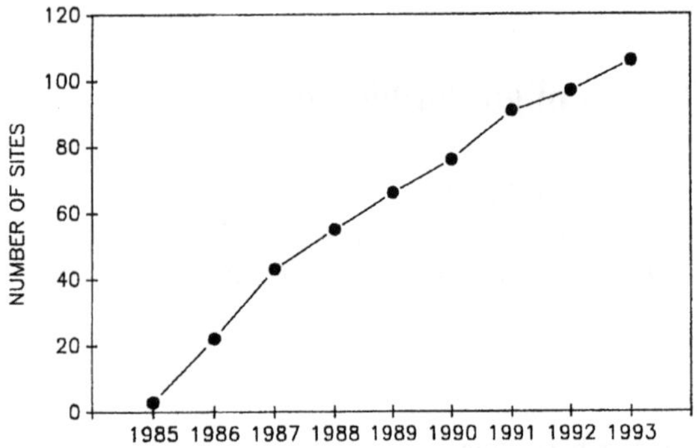

Figure 1. Yearly increase in the number of milfoil sites in Lake George.

ing water supply causing public concern over herbicide control of milfoil, and mechanical cutting would remove only the tops of plants thereby exacerbating its spread to other areas of the lake by fragmentation (Madsen et al., 1989). Consequently, a management strategy utilizing physical control techniques was initiated in 1989, which included hand harvesting, suction harvesting and benthic barrier installation. The early discovery of milfoil in the lake has provided an opportunity to systematically study the ecology of the establishment of this species and its impact on the native plant community while evaluating appropriate control strategies.

Study area

Lake George is situated on the eastern edge of the Adirondack mountains in upstate New York, USA. It has an overall length of 51 km, a mean width of 2.3 km and average depth of 20 m. The watershed is approximately 90% forested and development is confined largely to the south and southwest shore. Transparency by Secchi disk averages 7 m. The littoral zone can extend to 10 m depth if there is suitable sediment (Madsen et al., 1988). The waters of Lake George are classified as oligo-mesotrophic with productivity limited by phosphorus availability. The water is soft (alkalinity of 25–30 mg l^{-1} as $CaCO_3$), and characteristically low in nutrients (total phosphorus of 5–10 μg l^{-1} and nitrate and ammonia of <10–50 $\mu g\ l^{-1}$). Secchi disk transparency ranges from 6 to 12 m, with higher values generally reported in the north basin.

Material and methods

As part of the US Environmental Protection Agency (EPA) Phase II Clean Lakes Restoration Project (#S002287-01-3) of nuisance aquatic plant control targeted at Eurasian watermilfoil in Lake George, sites managed by hand harvesting were those usually having less than 1000 plants, sites managed by suction harvesting were limited to scattered plants of less than 50% cover, and sites managed with benthic barriers were areas where milfoil dominated the aquatic plant community (greater than 50% of total cover) and formed dense beds. Under this EPA program, management limitations were based on pronounced environmental impacts to the native plant community and cost considerations relative to each particular control technique.

For evaluation purposes, hand harvesting was carried out at 14 different sites by SCUBA due to the depth at which milfoil was growing, with care taken to remove both roots and shoots. Suction harvesting was evaluated at seven sites. The suction harvester was a diver-operated, hydraulic vacuum system created by a diesel-powered venturi pump mounted on a 9-m pontoon boat. The plant material, including vegetative stem and leaves plus roots, was pulled from the bottom by the diver and fed by hand into a vacuum hose (Eichler et al., 1993). The effectiveness of benthic barrier was determined at four sites. Benthic barrier material consisted of either Palco^{TM} (solid polyvinyl chloride (PVC) sheeting 20 mil thick) or Aquascreen^{TM} (open mesh sheeting). In each case 2.2 m by 30 m sections of the material was pulled from the boat by divers into the water and over the plants. Metal rods were used to secure the material to the lake bottom and to adjacent strips.

To evaluate the impact on sites managed by suction harvesting and benthic barrier, grid systems of contiguous 1-m^2 quadrats were constructed of PVC tubing and installed on the lake bottom. Two grids of 3 m by 6 m were installed at larger infestations while smaller areas of infestation received a single 3-m by 6-m grid. At sites suction harvested, species present in each grid and their relative abundance were recorded prior to harvest, shortly after harvest and 1 year later. At matted sites one or two 3-m by 6-m grids were put in place immediately following barrier removal based on the size of the original mat. Species presence and relative abundance were determined at the time of grid installation, 30 days and 1 year later.

Assessment of species present at each site and their relative abundance was recorded underwater by divers

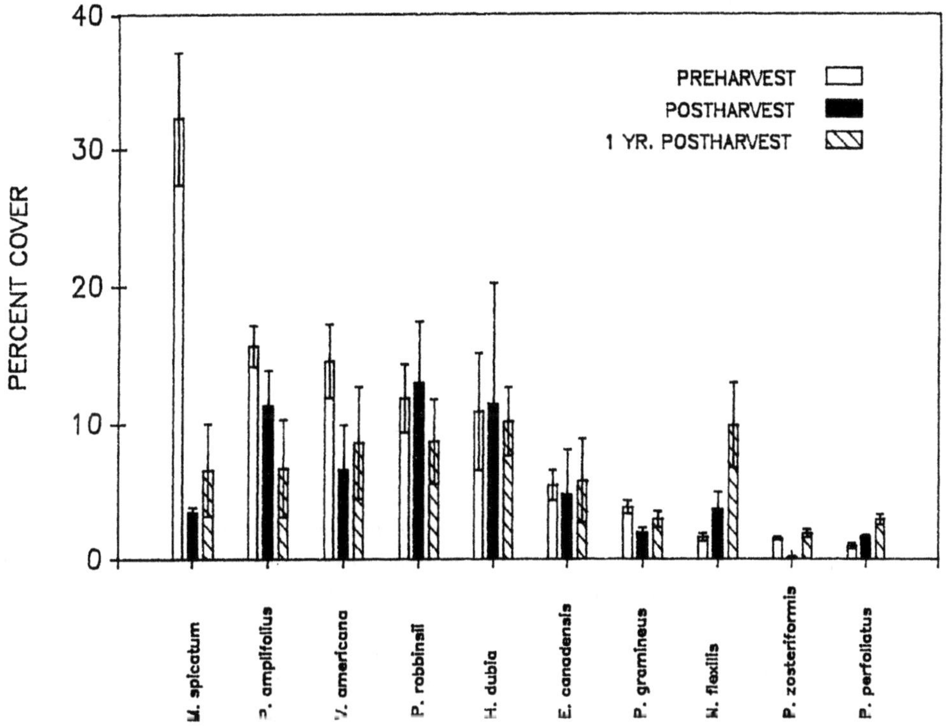

Figure 2. Comparison of relative per cent cover determined for the 10 most abundant species before (pretreatment), shortly after (postharvest) and 1 year following suction harvesting. Error bars represent mean values ± standard error, $n = 126$. Listed by decreasing abundance, they include: *Myriophyllum spicatum, Potamogeton amplifolius, Vallisneria americana, Potamogeton robbinsii, Heteranthera dubia, Elodea canadensis, Potamogeton gramineus, Najas flexilis, Potamogeton zosteriformis*, and *Potamogeton perfoliatus*.

using a modified Daubenmire scale (1968). Abundance classes were noted in % cover as: abundant (greater than 50%), common (25 to 50%), present (15 to 25%), occasional (5 to 15%), and rare (less than 5%). Person-days of effort is defined as individual effort in the field expressed in 8-h increments.

Results and discussion

As of the end of the growing season of 1993, 106 discrete lake locations were found to harbor milfoil (Eichler et al., 1994). Since 1985 a nearly linear yearly increase in the number of sites has been mapped around the lake perimeter (Figure 1). At the time of each discovery milfoil generally represented only scattered plants to small clusters of plants. Overall management of milfoil in Lake George has reduced the number of milfoil sites yet unmanaged to 11 (Eichler et al., 1994).

Impact of milfoil on native plant communities

In 1987 a small cluster of milfoil plants in a remote bay (Northwest Bay) was identified as a nonmanaged site to allow study of the impact of milfoil on native plant communities. Between 1987 and 1993 speciation and % cover estimates of milfoil and native species were determined throughout a grid system similar to that employed at managed sites. In 1987 20 species of native aquatics were found within the 6-m by 6-m grid; by 1993, only six species were found while the average number of species m^{-2} quadrat declined from 5.5 in 1987 to less than 2 in 1993. Key species lost included *Elodea canadensis, Najas flexilis, Potamogeton amplifolius*, and *Potamogeton gramineus*. In pristine environments where species diversity is high, there is concern that management of milfoil must take into account possible negative impacts on native plant communities. This site has provided the necessary control for the recolonization studies discussed below.

Table 1. Summary of hand harvesting efforts for 1989 and 1990.

	1989			1990		
	Total number of plants	Dry wt (kg) of plants	Effort in person-days	Total number of plants	Dry wt (kg) of plants	Effort in person-days
Total of all sites (n = 14)	21,200	17.1	26.4	3953	3.2	12

Table 2. Summary of suction harvesting efforts for 1990 and 1991.

	1990		1991	
	Dry wt (kg) of plants	Effort in person-days	Dry wt (kg) of plants	Effort in person-days
Total of all sites (n = 7)	710	28.0	49.6	5.7

Hand harvesting

A comparison of the hand harvest activities in 1989 and 1990 at 14 sites (Table 1) indicated an 81.4% reduction in the number of milfoil plants harvested. In other words, the initial year of hand harvesting was over 80% effective in removal of milfoil from these locations. Concomitantly there was a 56% reduction in the number of hours required to harvest these same sites. In situ hand harvesting has a minimal impact on native speciation because of the selective picking of individual plants. Since harvesting effort declined over time, once the number of individual plants has been sufficiently reduced, hand harvesting becomes a plausible maintenance tool for milfoil control.

Suction harvesting

Suction harvesting reduced both the biomass and per cent cover of milfoil. Milfoil was the dominant species by weight in the biomass samples prior to suction harvesting and declined to the fifth most abundant species after harvesting. A total of 28 person-days were spent suctioning milfoil in 1990 (Table 2). Species-by-species per cent cover (Figure 2) showed milfoil reduction to be greatest, as would be expected. From an average preharvest per cent cover of more than 30%, milfoil declined to less than 5% as a result of harvesting. One year later milfoil remained at an average of approximately 7% cover. Native species showed variable responses to suction harvesting. A decline in the per cent cover of *P. amplifolius* and *Vallisneria americana* was observed while *Potamogeton robbinsii, Heteranthera dubia, E. canadensis*, and *P. gramineus* reflected little change in per cent cover relative to harvesting. *N. flexilis*, however, showed substantial increases in per cent cover relative to harvesting. On a site-by-site basis, harvesting efforts for regrowth required between 64 and 89% fewer person-days than initial harvest efforts. Removal of regrowth by hand harvesting in 1991 required 5.7 person-days or 20% of the initial harvesting effort.

Benthic barrier

Within 30 days after barrier removal, all locations showed primary recolonization by native species. Six of the 7 grid areas had 10 or more species present; the seventh grid had 9 species. The number of species within the grids reached a maximum within 30 days after barrier removal (4.7 species m^{-2})and remained constant through year 2 (4.5 species m^{-2}). Average per cent cover was uniformly low 30 days after mat removal (Figure 3) with 44% of the grid squares having been recolonized by milfoil. One year after mat removal, regrowth of milfoil was apparent in 71% of the grid squares with several native species also increasing in average per cent cover, namely *P. robbinsii, H. dubia, E. canadensis, N. flexilis* and *Ranunculus longirostris*. Engel (1984) and Perkins et al. (1980) have shown similar usefulness in benthic barriers used for milfoil control.

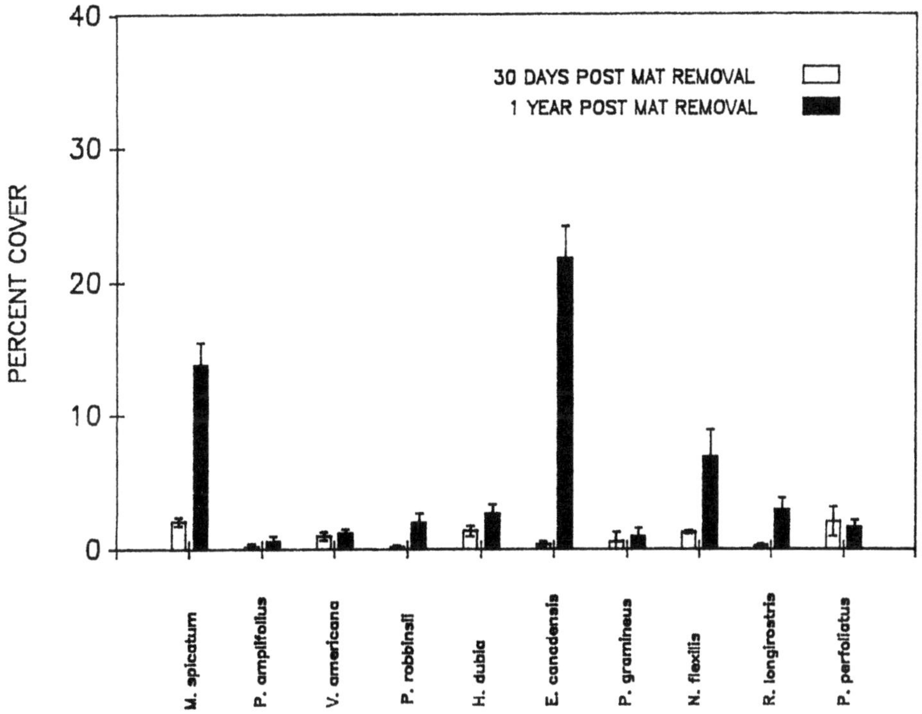

Figure 3. Comparison of relative per cent cover determined for the 10 most abundant species 30 days and 1 year following benthic barrier removal. Error bars represent mean values ± standard error, $n = 126$. Except for *Ranunculus longirostris*, refer to Figure 2 for species names.

Conclusions

Short term impacts by suction harvesting and benthic barrier on the native plant community are extreme; nonetheless, re-establishment of diverse native species occurs rapidly after treatment. Milfoil recolonizes the treated sites as well. Although short term management substantially reduces the amount of milfoil with negligible impact of restoration of a diverse native community, milfoil eradication is not achieved. Therefore a maintenance program of hand harvesting milfoil which has also recolonized must be carried out every 2–3 years to prevent milfoil dominance to reoccur.

Acknowledgments

We thank the Fund for Lake George for its financial support. Installation of benthic barrier material was supported by a USEPA Clean Lake Phase II project administered by NYSDEC. We thank J. D. Madsen and R. T. Bombard for field assistance and data interpretation. Contribution number 616 of the Rensselaer Fresh Water Institute and number 35 of the New York State Freshwater Institute.

References

Collins, C. D., R. B. Sheldon & C. W. Boylen, 1987. Littoral zone macrophyte community structure: Distribution and association of species along physical gradients in Lake George, New York, U.S.A. Aquat Bot. 29: 177–194.

Couch, R. & E. Nelson, 1985. *Myriophyllum spicatum* in North America, Proceedings of the First International Symposium on Watermilfoil (*Myriophyllum spicatum*) and Related Haloragaceae Species, 23–24 July 1985, Vancouver, British Columbia, Aquatic Plant Management Society, Vicksburg, MS.

Daubenmire, R., 1968. Plant Communities: A Textbook of Synecology. Harper and Row, New York. 300 pp.

Eichler, L. W., R. T. Bombard, J. W. Sutherland & C. W. Boylen, 1993. Suction harvesting of Euasian watermilfoil and its effect on native plant communities. J. aquat. Plant Mgmt 31: 144–148.

Eichler, L. W., R. T. Bombard & C. W. Boylen, 1994. Final Report on the Lake George Eurasian Watermilfoil Survey for 1993. Fresh Water Institute Technical report 94-1, Troy, New York 53 pp.

Engel, S., 1984. Evaluating stationary blankets and removable screens for macrophyte control in lakes. J. aquat. Plant Mgmt 22: 43–48.

Madsen, J. D., L. W. Eichler & C. W. Boylen, 1988. Vegetative spread of Eurasian watermilfoil in Lake George, New York. J. aquat. Plant Mgmt 26: 47–50.

Madsen, J. D., J. W. Sutherland, J. A. Bloomfield, K. M. Roy, L. W. Eichler & C. W. Boylen, 1989. Lake George Aquatic Plant Survey: Final Report. New York State Department of Environmental Conservation, Albany, New York. May 1989.

Madsen, J. D., J. W. Sutherland, J. A. Bloomfield, L. W. Eicher & C. W. Boylen, 1991. The decline of native vegetation under dense Eurasian watermilfoil canopies. J. aquat. Plant Mgmt 29: 94–99.

Perkins, M. A., H. L. Boston & E. F. Curren, 1980. The use of fiberglass screens for control of Eurasian watermilfoil. J. aquat. Plant Mgmt 18: 13–19.

Reed, C. F., 1977. History and distribution of Eurasian watermilfoil in the United States and Canada. Phytologia 36: 417–436.

Hydrobiologia **340:** 219–224, 1996.
J. M. Caffrey, P. R. F. Barrett, K. J. Murphy & P. M. Wade (eds), Management and Ecology of Freshwater Plants.

Response of *Elodea canadensis* Michx. and *Myriophyllum spicatum* L. to shade, cutting and competition in experimental culture

V. J. Abernethy, M. R. Sabbatini & K. J. Murphy
University of Glasgow, IBLS Division of Environmental and Evolutionary Biology, Brian Laboratory (Garscube), Glasgow G12 8QQ, UK

Key words: weed control, competition, disturbance, *Elodea canadensis*, *Myriophyllum spicatum*

Abstract

Elodea canadensis Michx. and *Myriophyllum spicatum* L. are widespread nuisance aquatic plant species. Their ecology is regarded as similar. Both species have been previously classified in terms of established-phase survival strategy as 'competitive disturbance-tolerant' species. Experimental data are presented to show that although this broad categorisation of strategy is probably correct for the two species, it is possible to demonstrate significant differences in terms of response to disturbance and competition. Less difference was discernible in their comparative response to stress. The drawbacks of applying broad descriptive terminology when dealing with two species of similar strategy are addressed. The results help explain reports of variable success in attempting to manage these two species using disturbance-based weed control measures, and suggest that *Elodea* is even less susceptible to such measures than *Myriophyllum*.

Introduction

Elodea canadensis Michx. and *Myriophyllum spicatum* L. are two submerged macrophyte species, which have successfully crossed the Atlantic during the past century, in the former case from North America to Europe, and in the latter from Europe to North America, to cause weed problems in a range of freshwater systems (Murphy et al., 1990a; Anderson, 1990; Steward, 1990; Simpson, 1984). Despite their differing provenances, both species are currently problem aquatic weeds in Europe.

The ecology of the two species is usually considered to be quite similar. Their established phase strategies both show strong elements of competitiveness and disturbance-tolerance (Grime et al., 1988; Murphy et al., 1990b). The two species tend to occur in similar freshwater habitats, and occur under broadly similar sets of physico-chemical environmental conditions (Simpson, 1984; Smith & Barko, 1990). The available evidence (as, for example, reviewed by Nichols & Shaw, 1986) therefore suggests that populations of the two species exhibit rather similar sets of phenotypically-expressed traits for tolerance of stress, disturbance and competition from other species (*sensu* Grime, 1979). When in direct competition there is some evidence that one species may successfully displace the other, but field observations are far from consistent (e.g. Madsen et al., 1991).

The question arises whether the application of management measures (which impose artificial stress or disturbance on weed populations) is likely to have similar effects on *E. canadensis* and *M. spicatum*, and whether such effects are modified in the presence of competitor plant populations.

The aims of the study were:

(i) to determine, under standardised experimental glasshouse conditions, the response of *Elodea canadensis* and *Myriophyllum spicatum* to artificially-imposed stress, disturbance, and interspecific competition; and
(ii) to use the information gained to refine knowledge of the established-phase survival strategy of the two species.

Methods

In all experiments plants were grown in Glasgow (RG19 Kelvinside & Maryhill water supply zone) tapwater (mean chemical characteristics: conductivity: 54.9 μS cm^{-1}; pH 8.26; nitrate 0.63 mg l^{-1}; calcium 4.6 mg l^{-1}; reactive phosphate 0.53 mg l^{-1}; chlorine 0.31 mg l^{-1}) in aerated 30 l black polypropylene tanks, under 16 hr light regime (Navilux 400 W sodium floodlights) augmenting natural daylight. Mean and standard error of incident PAR just above tanks varied seasonally, but during February–July 1993 was in the range 132.8 ± 1.11 to 222.1 ± 9.68 μmol m^{-2} s^{-1}), in a heated glasshouse (20 °C). The rooting medium was well-mixed natural river sediment, collected from the River Kelvin within the Garscube Estate of the University of Glasgow. Plants were established as 12 cm stem sections, each with a viable bud, and subjected in a series of experiments to varying intensities of stress, competition, and disturbance. A random-block design was used as standard, with 3 blocks; except in Experiment 4 where an incomplete factorial design was used, with 4 blocks. Variables measured were plant length, biomass per plant, and resource allocation (as biomass per stem, leaves and roots: Experiment 1 only). For each variable, and each species, % changes compared to untreated controls were calculated. Four experiments were conducted:

Experiment 1. Effects of stress caused by shade

Plants were grown in individual pots (1 plant/pot), with 2 plants of each species per tank. Individual tanks were shaded with one or more layers of white geotextile shade material, or left unshaded (9 tanks used), to give a design with 3 levels of the treatment factor: UNSHADED, LOW (23% reduction in photosynthetically-active radiation, measured using a Skye PAR meter at water level in the tank), and HIGH shade (40% reduction in PAR). Taking into account seasonal variation in PAR noted above, the PAR at water level in LOW shade tanks was in the range 102.3–171.0 μmol m^{-2}, s^{-1}; in HIGH shade tanks the corresponding values were 79.7–133.3 μmol m^{-2} s^{-1}).

Experiment 2. Effects of disturbance caused by cutting

Plants were grown in individual pots (1 plant/pot), with 11 pots/tank (18 tanks, each containing a random mix of treatment units). Cutting treatments were standardised to reduce individual plant length to 5 cm after each treatment. Two frequencies of cutting were used, to give a design with 3 levels of the treatment factor: UNCUT, LOW (cut 35 days after start of experiment) and HIGH cutting frequency (cut both 35 and 66 days after start).

Experiment 3. Effects of interspecific competition

An additive approach (Martin & Snaydon, 1982) was used to compare MIXED v. PURE stands of *Elodea canadensis* and *Myriophyllum spicatum.* Either 25 plants of each species in monoculture, or 25 + 25 plants of each species in mixed culture, were planted in trays (360 × 220 mm), with 1 tray/tank.

Experiment 4. Combined effects of shade stress and disturbance caused by cutting

The experiment was set up with plants grown in individual pots at a density of 10 plants per tank, of which 2 replicates per tank of each species were harvested. In total there were 6 treatment-combinations: untreated (UNTR), low shade (LS), high shade (HS), single cut (C1), two cuts (C2), and low shade + single cut (LS/C1). Shade treatments were as in Experiment 1.

Statistical treatments

Data were analyzed using GENSTAT, as follows: Experiments 1–3: ANOVA followed by orthogonal mean separation using Tukey's LSD test; Experiment 4: two-way ANOVA with orthogonal contrasts (UNTR v. LS; UNTR v. C1; LS v. HS; C1 v. C2; LS/C1 interaction). In the results outcomes are treated as significant at $P<0.05$ throughout, and the term 'significant' is used in text only where results of statistical testing produced this outcome. Within-treatment variation (measured as standard error) was always $<10\%$ of mean values.

Results

Experiment 1. Effects of stress caused by shade

Data shown in Figure 1 are 77 days from start of the experiment. Shade stress produced little significant response by either species. Both showed no significant change in length per plant (except for *Myriophyllum* under high shade: 19% increase) in response to reduced light availability. There were no significant effects on biomass per plant for either species. In terms of resource allocation no significant response was observed by either *Elodea* or *Myriophyllum* in biomass allocation to stem or leaves as a result of shading. For *Elodea* there was no change in root biomass either, but *Myriophyllum* showed a significant (71%) reduction in root biomass at high shade, compared with unshaded controls. The results for *Myriophyllum spicatum* mirror the findings of previous work, for example by Barko & Smart (1981).

Experiment 2. Effects of cutting (augmented disturbance)

Data shown in Figure 2 are 123 days from start of the experiment. Compared with untreated controls *Myriophyllum* showed a significant response to both single and double cut treatments: for both biomass (45 and 90% reduction after 1 and 2 cuts respectively), and length per plant: (22 and 70% reduction). For *Elodea* the effect was less, especially for length response, where there was no significant change after 1 cut, and only 44% reduction after two cuts. The biomass response of *Elodea* was more marked, with reductions of 41 and 59% after 1 and 2 cuts respectively.

Experiment 3. Effects of interspecific competition

Data shown in Figure 3 are 84 days post-treatment. The two species responded differently to interspecific competition. Compared with growth in monoculture, there was a significant reduction (25%) in plant length of *Elodea*, but no significant reduction in plant biomass, when grown with *Myriophyllum*. The converse was seen for the two variables measured in *Myriophyllum*: a significant reduction (33%) in plant biomass, but with no significant reduction in plant length, when grown with *Elodea*.

Experiment 4. Shade plus cutting (interacting stress and disturbance)

The results shown in Table 1 are % changes, for the orthogonal comparisons shown, two for variables measured 74 days after treatment. A stronger response to shade stress was seen in *Myriophyllum*, with biomass being significantly reduced by LS treatment, whereas no significant response was observed under low shade conditions for *Elodea*. Adding in cutting disturbance to low shade stress produced a greater effect on *Myriophyllum* than on *Elodea*. The effect on *Myriophyllum* was similar to that of high-disturbance treatment; much less for *Elodea*. The effects of cutting disturbance alone were similar for both species.

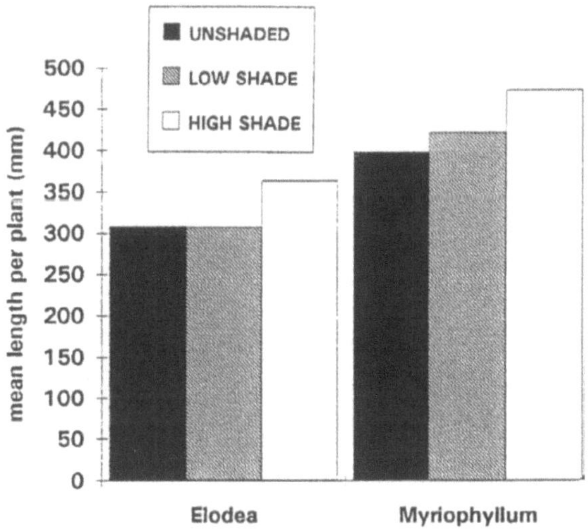

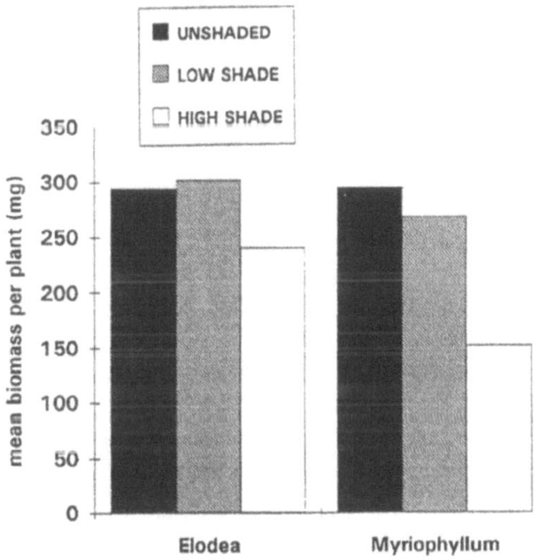

Figure 1A, B. Effects of stress caused by shade (UNSHADED, LOW SHADE, HIGH SHADE) on (a) length, and (b) biomass per plant of *Elodea canadensis* and *Myriophyllum spicatum*.

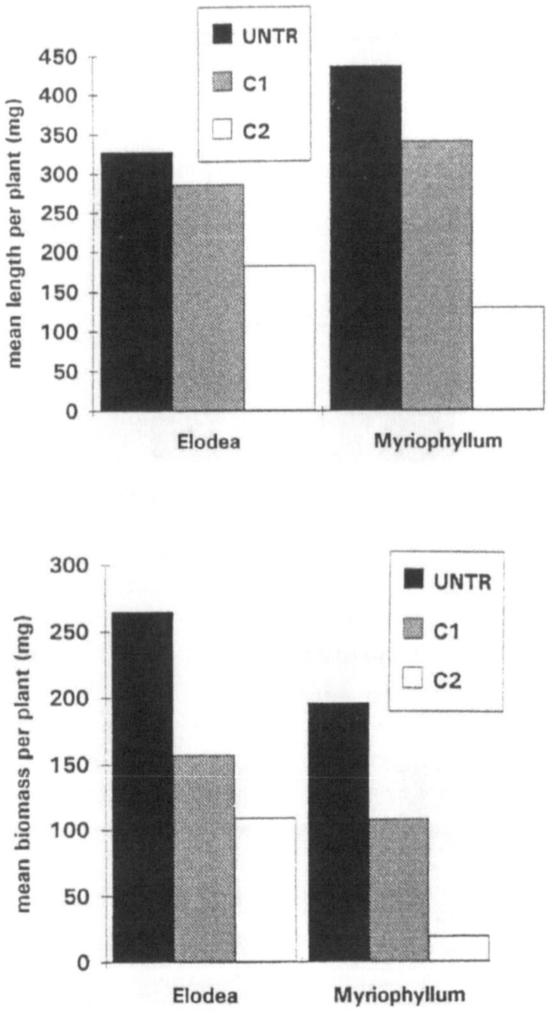

Figure 2A, B. Effects of disturbance caused by cutting on (a) length, and (b) biomass per plant of *Elodea canadensis* and *Myriophyllum spicatum.* UNTR = untreated; C1 = LOW cutting frequency; C2 = HIGH cutting frequency.

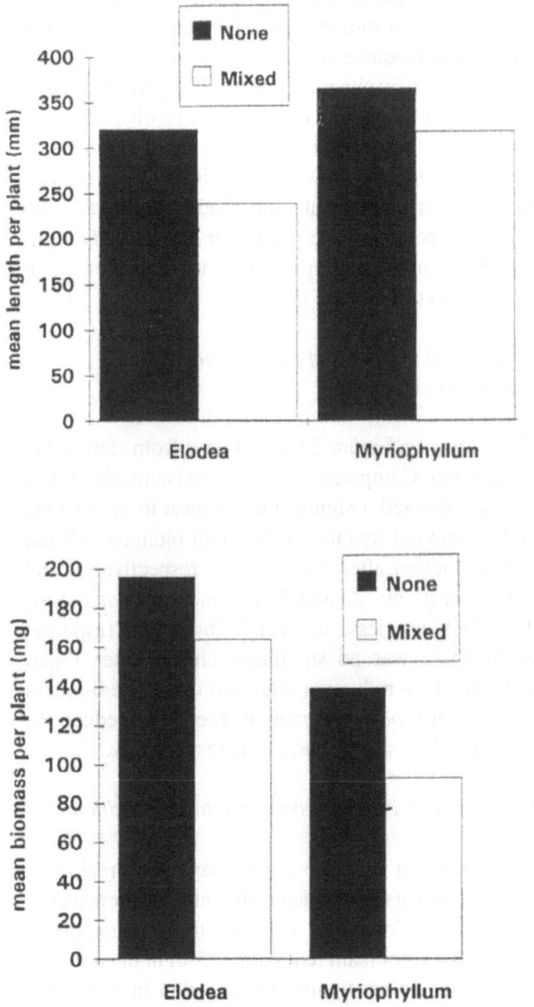

Figure 3A, B. Effects of interspecific competition on (a) length, and (b) biomass per plant of *Elodea canadensis* and *Myriophyllum spicatum*, grown in pure (NONE) and MIXED culture.

Discussion

Tolerance of stress and disturbance

Of the two species compared, *Myriophyllum* showed a more plastic growth response to shade stress: by reducing resource allocation to roots, and increasing its length. These results are suggestive of a rather low tolerance of stress (Grime, 1979). The results of Experiment 4 also suggested that *Elodea* was slightly more tolerant of shade stress than *Myriophyllum.*

Elodea was slightly more disturbance-tolerant than *Myriophyllum.* In both Experiments 1 and 4 the responses of *Myriophyllum*, in terms of biomass-reduction, and reduced plant length, were usually similar to, or greater than for *Elodea*. *Elodea* was more tolerant than *Myriophyllum* of combined stress and disturbance, at moderate intensities of both pressures.

These results are of relevance when considering the response of the two species to weed control mea-

Table 1. Percentage response of *Elodea canadensis* and *Myriophyllum spicatum* length and biomass per plant for 5 orthogonal comparisons. Treatment codes are given in text. NS: not significant (P>0.05); other values are significant at P<0.05 for comparison)

Treatment comparison	Reduction (%)			
	Elodea		*Myriophyllum*	
	Length	Biomass	Length	Biomass
UNTR v. LS	NS	NS	NS	53
LS v. HS	NS	77	NS	NS
UNTR v. LS/C1	NS	49	62	85
UNTR v. C1	37	38	NS	38
C1 v C2	43	88	66	83

sures based on stress and disturbance. *M. spicatum* has frequently been observed to respond positively to disturbance produced by cutting or harvesting (Smith & Barko, 1990). The results of our study suggest that applying disturbance-based weed control to *Elodea canadensis* is likely to produce an even worse result in weed control terms.

Competitiveness

From the results of Experiment 3, *Elodea* was the more competitive of the two species when grown in mixed culture with each other under standard glasshouse conditions. Although *Elodea* produced shorter plants in competition with *Myriophyllum*, *Elodea* showed no significant loss of biomass compared with monoculture controls. In contrast, *Myriophyllum* plants competing with *Elodea* showed significant biomass loss.

Separation of strategies of Elodea *and* Myriophyllum

The two freshwater plant species studied here, both of which act as opportunistic weeds, and which tend to occur in similar habitats (Nichols & Shaw 1986), had measurably different responses to stress, disturbance and competition, under standardized experimental conditions.

Field evidence from comparison of drainage channel habitats of the two species in Britain (Sabbatini & Murphy, 1996) has suggested that there is a tendency for *Elodea* to occur in slightly higher-stress conditions than *Myriophyllum*. Sheldon & Boylen (1977) found that *E. canadensis* had the deepest maximum depth (compared with *M. spicatum* and *Potamogeton crispus*) in US lakes. Nichols & Shaw (1986) considered that *E. canadensis* is the 'most efficient' of these three submerged macrophyte species in surviving low light conditions. There is further evidence in the literature that *M. spicatum* is only poorly-tolerant of shade stress (e.g. Chambers & Kalff, 1985). In neither species, however, does stress-tolerance seem to play a major role in established-phase strategy. Much more important are traits for disturbance-tolerance and competitiveness.

The established-phase strategies of these two species are certainly close (for most populations of the two species, probably competitive disturbance-tolerant CD), but there are interspecific differences in response to environmental pressures on survival, which indicate that their strategies can be separated. This highlights the problem of relying on a descriptive terminology for plant strategy, such as that put forward by Grime (1979). When two species have closely-similar strategies, as in the case of *Elodea canadensis* and *Myriophyllum spicatum*, classification into broad categories such as 'competitive disturbance tolerator' do not adequately reflect the functional differences between the species. What is needed is a numerically-based methodology to describe strategy and functional type of plant species, which would allow better quantification of the differential responses of plants to pressures on their survival and reproduction. An increasing amount of work is currently being devoted to developing functional analysis methods along these lines, for aquatic and wetland vegetation as well as terrestrial plants (e.g. Hills et al., 1994; Abernethy, 1994.; Pautou & Arens, 1994; Bornette et al., 1994; Hendry & Grime, 1993). The appropriate application of approaches such as these may lead to a substantially enhanced understanding of both the invasive potential, and susceptibility to control measures, of nuisance species such as *Elodea canadensis* and *Myriophyllum spicatum*.

Acknowledgments

This work was part-funded by a UK NERC studentship to VJA, and an Argentine CONICET grant to MRS. We thank Strathclyde Regional Council Water Laboratory for providing chemical analysis data for Glasgow tapwater.

References

Abernethy, V. J., 1994. The functional ecology of euhydrophyte communities of European riverine wetland ecosystems. Ph. D. Thesis, University of Glasgow. 265 pp.

Anderson, L. W. J., 1990. Aquatic weed problems and management in the western United States and Canada. In A. H. Pieterse & K. J. Murphy (eds), Aquatic Weeds. Oxford University Press, Oxford: 371–391.

Barko, S. W. & R. M. Smart, 1981. Comparative influences of light and temperature on the growth and metabolism of selected submersed freshwater macrophytes. Ecol. Monogr. 51: 219–235.

Bornette, G., C. Henry, M.-H. Barrat & C. Amoros, 1994. Theoretical habitat templets, species traits and species richness: aquatic macrophytes in the Upper Rhône River and its floodplain. Freshwat. Biol. 31: 487–505.

Chambers, P. A. & J. Kalff, 1985. The influence of sediment composition and irradiance on the growth and morphology of *Myriophyllum spicatum* L. Aquat. Bot. 22: 253–263.

Grime, J. P., 1979. Plant strategies and vegetation processes. Wiley, Chichester. 222 pp.

Grime, J. P., J. G. Hodgson & R. Hunt, 1988. Comparative plant ecology. A functional approach to British species. Unwin & Hyman, London. 742 pp.

Hendry, G. A. F. & J. P. Grime, 1993. Methods in comparative plant ecology. Chapman & Hall, London. 252 pp.

Hills, J. M., K. J. Murphy, I. D. Pulford & T. H. Flowers, 1994. A method for classifying European riverine wetland ecosystems using functional plant groups. Funct. Ecol. 8: 242–252.

Madsen, J. D., J. W. Sutherland, J. A. Bloomfield, L. W. Eichler & C. W. Boylen, 1991. The decline of native vegetation under dense Eurasian watermilfoil canopies. J. Aquat. Plant Mgmt 29: 94–99.

Martin, M. P. L. D. & R. W. Snaydon, 1982. Analysis of competition experiments. J. Appl. Ecol. 19: 263–272.

Murphy, K. J., T. O. Robson, M. Arsenovic & W. van der Zweerde, 1990a. Aquatic weed problems and management in Europe. In A. H. Pieterse & K. J. Murphy (eds), Aquatic Weeds. Oxford University Press: Oxford: 295–317.

Murphy, K. J., B. Rørslett & I. Springuel, 1990b. Strategy analysis of submerged lake macrophyte communities: an international example. Aquat. Bot. 36: 303–323.

Nichols, S. A. & B. H. Shaw, 1986. Ecological life histories of three aquatic nuisance plants, *Myriophyllum spicatum, Potamogeton crispus* and *Elodea canadensis*. Hydrobiologia 131: 3–21.

Pautou, G. & M.-F. Arens, 1994. Theoretical habitat templets, species traits and species richness: floodplain vegetation in the Upper Rhône valley. Freshwat. Biol. 31: 507–522.

Sabbatini, M. R. & K. J. Murphy, 1996. Submerged plant survival strategy in relation to management and environmental pressures in drainage channel habitats. Hydrobiologia 340 (Dev. Hydrobiol. 120): 191–195.

Sheldon, R. B. & C. W. Boylen, 1979. Maximum depth inhabited by aquatic vascular plants. Am. Midl. Nat. 97: 248–254.

Simpson, D. A., 1984. A short history of the introduction and spread of *Elodea* Michx. in the British Isles. Watsonia 15: 1–9.

Smith, C. S. & J. W. Barko, 1990. Ecology of Eurasian watermilfoil. J. Aquat. Plant Mgmt 28: 55–64.

Steward, K. K., 1990. Aquatic weed problems and management in the eastern United States. In A. H. Pieterse & K. J. Murphy (eds), Aquatic Weeds. Oxford University Press: Oxford: 391–405.

Hydrobiologia **340**: 225–228, 1996.
J. M. Caffrey, P. R. F. Barrett, K. J. Murphy & P. M. Wade (eds), Management and Ecology of Freshwater Plants.

Mechanical aquatic weed management in the lower valley of the Río Negro, Argentina

Dall'Armellina, A[1], A. Gajardo[1], C. Bezic[1], E. Luna[2], A. Britto[3] & V. Dall'Armellina[4]
[1]*Centro Universitario Regional Zona Atlantica-Universidad Nacional del Comahue, Ayacucho y Esandi, 8500 Viedma, Río Negro, Argentina*
[2]*Departamento Provincial de Agua, San Martin 342, 8500 Viedma, Río Negro, Argentina*
[3]*IDEVI Instituto de Desarrollo del Valle Inferior del Río Negro, Belgrano 536 (4to piso), 8500 Viedma, Río Negro, Argentina*
[4]*Secretaría General de la Gobernación de Río Negro, Belgrano y Laprida, 8500 Viedma, Río Negro, Argentina*

Key words: physical weed control, submerged macrophytes, irrigation channels

Abstract

A major irrigation system in the Lower Valley of the Río Negro, Argentina, has been invaded by aquatic plants, with *Potamogeton illinoensis* Morong dominant in irrigation channels and *Potamogeton pectinatus* L. dominant in drainage channels. Although several other macrophytes are present, problems are largely caused by the dominant species. Results are presented for plant biomass response to weed control treatments using a chain-cutting method in the principal irrigation channel of the system. Peak above-ground biomass of *Potamogeton illinoensis* was reduced by about 38% by this physical control regime. The treated populations regrew rapidly after spring clearance, but did not regrow after subsequent mid- and late-season clearance operations, even though untreated population biomass remained high during this period. The highest density of *Potamogeton illinoensis* ramets was found in treated areas. Chain- cutting produced no discernible effect on dissolved oxygen, water temperature, water conductivity, pH or light extinction coefficient compared with untreated check sectors of the channel.

Introduction

In arid and semi-arid regions, irrigation is essential for agriculture. The water for irrigation systems flows through supply and drainage channels which are often seriously affected by excessive aquatic plant growth causing blockage of water flow (Murphy, 1988a).

In the semi-arid lower valley of the Río Negro, Argentina management costs in irrigation systems are increased by the need for weed control measures. Aquatic weed growth in both irrigation and drainage channels increases the risks of crop production loss, salinization of soils, and flooding, because of the reduced efficacy of water movement through channels and irrigated soils. The risk is so great that the costs of channel weed control must be included, as an integral part of the crop management regime, and a range of physical weed control measures is available for this purpose (Murphy, 1988b; Fernández et al., 1978). The objectives of this study were: (i) to identify the weed species growing in the channels of a major irrigation system in the lower valley of the Río Negro, in southern Argentina; (ii) to determine the seasonal pattern of biomass production for each species; and (iii) to examine the impacts of physical weed control on these species.

Area of study

Study areas are located in Viedma, Río Negro province in the south of Argentina (40 ° 48′ S; 63 ° 05′ W). The climate here is arid to semi-arid. Water from the Río Negro supplies the IDEVI (Instituto de Desarrollo del Valle Inferior del Río Negro) irrigation system, comprising in total 235 km of irrigation channels, with an associated 550 km of drainage channels. The length of the main supply channel is 95 km, of which 12 km is concrete-lined, the remainder being unlined. Unlined sectors of the main channel were used for the work reported here. The width of the main channel reduces from 18 m at the river, down to 12 m at the far end; with water depth fluctuating in the range 1.8–3.2 m. The mean flow taken from the river during the 1993/94 growing season was between 9.5 m^3 s^{-1} in August, when the irrigation period started, and 20 m^3 s^{-1} in January when the irrigation demand was greatest.

Materials and methods

Two stations in the main channel were used, at Km 80 and Km 30. In total, 15 samples were taken at random, from treated and untreated sectors of each station every 21 days during the growth period, from September to April. Samples of above and belowground biomass (Madsen, 1993) were collected from a tethered raft using a 0.028 m^2 core sampler, and taken to the laboratory for processing.

Above and below ground parts of the plant samples were separated in the laboratory. Aboveground biomass was separated into different species, and dried in an electric oven at 105 °C for 24 hours.

Enviromental parameters measured at each site were water temperature, dissolved oxygen concentration (mg l^{-1}), light intensity (μE m^2 s^{-1}), pH and electrical conductivity (μS cm^{-1}).

At the sites where biomass was recorded a simple submerged aquatic weed control experiment was undertaken, using the chain cutting method, with treatments applied on 15 November when *Potamogeton illinoensis* was approximately 50 cm long on average, the second treatment on 3 January, and the third treatment on 22 February.

Results

Potamogeton illinoensis was present in 100% of samples taken from the two main stations, with other species present in different percentages. Development of aboveground biomass of all macrophytes growing in the main channel is shown in Figure 1. The growing period started in September, with substantial growth increase from the last week of October to a peak at the end of December.

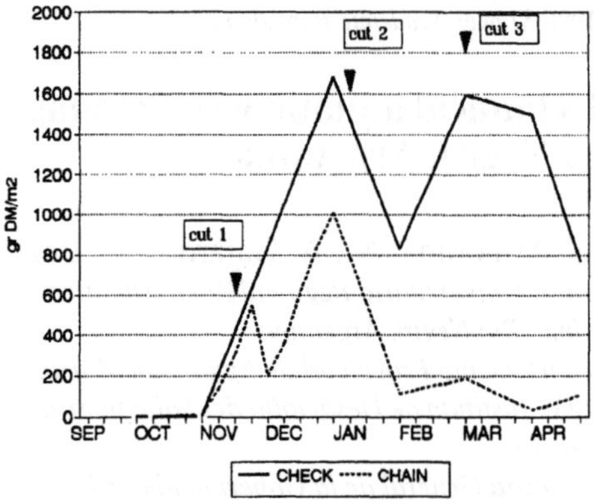

Figure 1. Development of aboveground biomass of macrophytes growing in the main channel (Km 80) in both treated and untreated areas.

Table 1. Belowground biomass (g DW.m^{-2}) of *Potamogeton illinoensis* in the main irrigation channel during the 1993/94 growing period in both, treated (chain) and untreated (check) sectors.

Date	Treated	Untreated
September	25.75 ± 2.57	32.30 ± 2.85
October	21.20 ± 4.25	12.62 ± 2.18
January	16.88 ± 2.34	47.31 ± 4.08
February	29.22 ± 2.34	40.38 ± 3.67
March	13.29 ± 1.63	50.89 ± 2.60
April	23.10 ± 2.56	77.19 ± 4.86

Note: the growing period mean values of belowgroun biomass in treated and untreated sectors were 21.57 ± 2.37 and 43.45 ± 8.74 gr DW.m^{--2} respectively.

Table 2. Environmental data registered in the irrigation main channel during the 1993/94 growing period in both, treated (chain) and untreated (check) sectors. (*T* = temperature; EC = electrical conductivity; DO = dissolved oxygen; *K* = underwater light attenuation coefficient; WV = water velocity).

	Treated	Untreated
T (°C)	14.57 ± 1.50	14.55 ± 1.50
EC (μS.cm^{-1})	171.25 ± 3.92	171.16 ± 3.96
DO (g.l^{-1})	10.38 ± 0.13	10.45 ± 0.15
K (m^{-1})	1.75 ± 0.24	1.63 ± 0.18
WV (m.s^{-1})	0.42	0.42

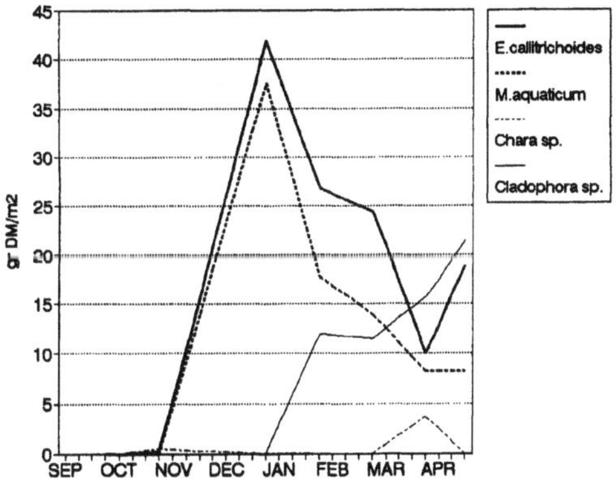

Figure 2. Development of aboveground biomass of several macrophytes growing in the main channel (Km 80) during the 93/94 season.

Potamogeton pectinatus was found in a high percentage of the samples taken in the main irrigation channel but its biomass production was always lower than *P. illinoensis*, which represented more than 93% of the total biomass produced in the channel. Below-ground data for *P. illinoensis* at treated and untreated sites are shown in Table 1. Minor species like *Elodea callitrichoides* and *Myriophyllum aquaticum* reached maximum development in February. Growth of *Cladophora* sp. and *Chara* sp. began late in the season, and produced only very low biomass (Figure 2).

Biomass changes after cutting showed that the first treatment had little effect on macrophyte growth. The second treatment was more effective, decreasing total biomass substantially compared with the untreated control data. The third treatment maintained this low biomass. There were no discernible differences in any water quality data between treated and untreated sites (Table 2).

Discussion and conclusions

The irrigation channels of the IDEVI area are clearly very favourable habitats for submerged plant growth, and production is high (approximating to some 18–20 t ha^{-1} yr^{-1}). This may be compared with productivity values of 25 t ha^{-1} yr^{-1} recorded by Howard-Williams (1978) for *Potamogeton pectinatus* at similar latitudes in New Zealand; and up to 40 t ha^{-1} yr^{-1} for *Potamogeton schweinfurthii* in tropical Africa (Wade, 1990).

The physical weed control regime used in the IDEVI irrigation channels appears to produce results comparable to those seen with mechanical clearance elsewhere in the world. Early-season clearance produces a brief check to growth, postponing the spring increase, but producing virtually no effect on rate of increase (as is clearly visible after the first cut in Figure 1). Later season clearances are more effective in reducing biomass. Very similar results have been reported from navigable canals in Britain (Eaton et al., 1981). Timing of weed control operations is clearly an important factor in managing this system effectively. Timing is also crucial with regard to reducing vegetative propagule formation for the main *Potamogeton* species present: rhizomes in the case of *P. illinoensis* (Bezic et al., 1996) and tubers in the case of *P. pectinatus*: Yeo, 1965). There is plenty of evidence to suggest that control is most effective against aquatic weeds dependent on vegetative propagules, when used before the formation of these structures (Wade, 1990). For rhizomatous species, regular defoliation of the plants can be an effective means of depleting carbohydrate reserves in the below-ground structures, producing longer-lasting control (e.g. Wallsten, 1983).

Although it does decrease below-ground biomass, the control method (chaining) currently favoured in the IDEVI area probably produces insufficient destruction of submerged foliage, for long enough, to produce significant effects on rhizomes. Indeed, the method probably favours the spread of species like *P. illinoensis*, by breaking up the plants, and allowing viable rhizome and stem fragments to move downstream. It is probable that a control regime which produced longer-term suppression of above-ground biomass would give better control of the main weeds present in the system. Possibilities might include drip-feed herbicide control, for example using acrolein (e.g. Bowmer & Sainty, 1977), or the use of biological measures such as grass carp. An integrated control regime utilising grass carp and physical control may well be the optimal solution to the problem in the IDEVI system, and in similar irrigation systems elsewhere in Argentina.

Acknowledgments

We thank Dr K. Murphy (Glasgow University, U.K.) for assistance in preparing the English language version of this paper.

References

Bezic, C., A. Dall'Armellina & O. Gajardo, 1996. Studies on vegetative production of *Potamogeton illinoensis* Morong in southern Argentina. Hydrobiologia 340 (Dev. Hydrobiol. 120): 7–10.

Bowmer, K. H. & G. R. Sainty, 1977. Management of aquatic plants with acrolein. J. aquat. Plant Mgmt 15: 40–46.

Eaton, J. W., K. J. Murphy & T. M. Hyde, 1981. Comparative trials of herbicidal and mechanical control of aquatic weeds in canals. Proc. Assoc. Appl. Biol. Conf. Aquatic weeds and their Control 1981, AAB, Wellesbourne, UK: 105–116.

Fernández, O. A., M. R. Sabbatini & J. Irigoyen, 1978. Aquatic management in drainage canals of southern Argentina. J. aquat. Plant Mgmt 25: 65–67.

Fernández, O. A., D. L. Sutton, V. H. Lallana, M. R. Sabbatini & J. H. Irigoyen, 1990. Aquatic weed problems and management in South and Central America. In A. H. Pieterse & K. J. Murphy (eds), Aquatic Weeds, Chapter 20. Oxford University Press, Oxford: 17–30.

Howard-Williams, C., 1978. Growth and production of aquatic macrophytes in a south temperate saline lake. Verh. int. Ver. Limnol. 20: 1153–1158.

Madsen, S. J., 1993. Biomass techniques for monitoring and assessing control of aquatic vegetation. Lake Reserv. Mgmt 7: 141–154.

Murphy, K. J., 1988a. Aquatic weed problems and their management. A review of worldwide scale of aquatic weed problems. Crop Protection 7: 232–248.

Murphy, K. J., 1988b. Aquatic weed problems and their management. Physical control measures. Crop Protection 7: 283–302.

Steward K. K., 1990. Aquatic weed problems and management in the eastern United States. In A. H. Pieterse & K. J. Murphy (eds), Aquatic Weeds. Chapter 19b. Oxford University Press, Oxford: 391–405.

Wade, P. M., 1990. General biology and ecology of aquatic weeds. In A. H. Pieterse & K. J. Murphy (eds), Aquatic Weeds. Oxford University Press, Oxford: 17–30.

Wallsten, M., 1983. Starch content in roots in relation to vegetative growth in natural and treated aquatic macrophyte areas. Proc. Int. Symp. Aquatic macrophytes 1983, Nijmegen, 292–297.

Yeo, R. R., 1965. Life history of sago pondweed. Weeds 13: 314–321.

Hydrobiologia **340**: 229–234, 1996.
J. M. Caffrey, P. R. F. Barrett, K. J. Murphy & P. M. Wade (eds), Management and Ecology of Freshwater Plants.

Patterns of aquatic weed regrowth following mechanical harvesting in New Zealand hydro-lakes

Clive Howard-Williams, Anne-Maree Schwarz & Virginia Reid
NIWA, P.O. Box 8602, Riccarton, Christchurch, New Zealand

Key words: mechanical harvesting, management, Hydrocharitaceae, *Lagarosiphon*, *Nitella*, *Ceratophyllum*, *Egeria*

Abstract

Mechanical harvesting is used to control submerged aquatic weeds in parts of the hydro-lakes in New Zealand's North Island. Problem species are *Egeria densa* and *Lagarosiphon major* (Hydrocharitaceae), and *Ceratophyllum demersum*. Experiments were conducted in two contrasting hydro-lakes. Lake Aratiatia; clear water (K_o 0.2 m^{-1}) and a low residence time (<8 h), and Lake Ohakuri; turbid water (K_o = 0.6) and a longer residence time (> 5 days). Growth rates were measured underwater in harvested and control (unharvested) plots. Regrowth of *C. demersum* was dependent on the prior establishment of the rooted Hydrocharitaceae. Regrowth of the Hydrocharitaceae was inhibited where significant water movement occurred. Regrowth declined after 3 six-monthly harvests allowing the establishment of low growing native *Nitella* spp. beds in the smaller clear water lake. In Lake Ohakuri there was a change in species dominance from *Ceratophyllum* to *Elodea canadensis* in shallow (1–2 m) water. No change in species dominance was observed in deeper (>2 m) water and native species were not able to re-establish. The recommended cutting frequency for management of surface weed growths was only once per year in Lake Aratiatia, but twice per year in Lake Ohakuri.

Introduction

During the last century lakes and rivers in New Zealand have experienced spectacular invasions by introduced submerged aquatic weed species. The displacement of native, non-weedy species by introduced weeds which form dense tall (4 m high) often monospecific stands, has led to the study and implementation of a wide variety of control mechanisms. The most problematic weeds are coontail (*Ceratophyllum demersum* L.) and members of the Hydrocharitaceae, notably lagarosiphon (*Lagarosiphon major* (Ridl.) Moss), elodea (*Elodea canadensis* L.) and egeria (*Egeria densa* Planch.). Aquatic weed problems in New Zealand lakes are a particular threat to hydropower stations in terms of commercial losses (Johnstone, 1981; Howard-Williams, 1993) and mechanical harvesting is now used to control submerged aquatic weeds in parts of hydro- lakes in New Zealands North Island.

The often rapid regrowth of aquatic plants following mechanical cutting and harvesting techniques is well known (Wade & Edwards, 1980). Management strategies therefore depend on the rate of recovery of communities and on the pattern of recovery (i.e. species changes following control) (Wade, 1990). Long term management will be enhanced when either: recovery is slow, or when the replacement community is less of a nuisance than the community originally controlled.

This study was implemented to investigate two aspects of the mechanical harvesting programme in North Island hydro-power lakes: (1) the rates of aquatic weed regrowth following mechanical harvesting to estimate cutting frequency for weedbed control; (2) species changes in the native and introduced aquatic plant communities following mechanical harvesting.

Table 1. Comparison of physical and chemical properties of Lake Aratiatia and Lake Ohakuri over the study period. Winter–summer surface water temperature ranges, attenuation coefficient for PAR (K_o), nutrient concentration ranges and current velocity ranges at the experimental sites in the lakes.

	Aratiatia	Ohakuri
Lake Area (km²)	0.34	7.2
Residence time (days)	0.3	5–10
Mean depth (m)	4–5	19.6
Inflowing water	fast flows of clear Lake Taupo water	slow flows of algal-rich water
Dominant weed species	*Lagarosiphon major* and *Ceratophyllum demersum*	*Egeria densa* and *Ceratophyllum demersum*
Temperature (°C)	12–25	10–23
K_0	0.2–0.25	0.51–0.72
DRP mg m^{-3}	2–9	2–20
NH_4-N mg m^{-3}	6–67	5–150
Velocity m s^{-1}	0.01–0.3	< 0.01

Table 2. Summary of regrowth following harvesting on two occasions.

	Aratiatia	Ohakuri
1st harvest		
Season	winter	summer
RGR (m m^{-1} $week^{-1}$)	0.025	0.079
50% surface time time (months)	6–7	5.5
2nd harvest		
Season	summer	winter
RGR (m m^{-1} $week^{-1}$)	0.06	0.046
50% surface time time (months)	9	7

Methods

Study sites

Between 1988 and 1991 experiments were conducted in two contrasting hydro-lakes, Lake Aratiatia and Lake Ohakuri. These are the first of eight hydro-lakes on the Waikato River in the central North Island. The lakes differed markedly in size, residence time, and clarity (Table 1). In Lake Aratiatia, a small shallow lake, the weed beds which were harvested grew to the surface from a depth of 4 metres and ca. 100 metres out from the lake edge. Lake Ohakuri, in contrast, had steeper shores and the nuisance weeds grew up from a maximum depth of 5 metres, but this was only 50 metres out from the shore. The dominant weed species in Lake Aratiatia were *Lagarosiphon major* and *Ceratophyllum demersum*. Both these species were present in Lake Ohakuri as well as *Egeria densa*. Low growing native *Nitella* spp. and *Potamageton* spp. were also present at the study sites in both lakes.

Environmental properties

Photosynthetically Active Radiation (PAR) was measured in both lakes using a Licor underwater scalar irradiance probe. Samples for inorganic nutrient analysis were collected from the littoral zone and analysed using a Technicon autoanalyser II system, and temperature and dissolved oxygen were measure using a YSI Model 54A temperature and oxygen meter. In Lake Aratiatia current velocity along a littoral zone transect was measured 1 m above the lake bed using a Teledyne-Gurley PYGMY current meter operated by a SCUBA diver. In Lake Ohakuri velocities were below the sensitivity level of the meter (0.01 m s^{-1}) at the study site.

Weed harvesting procedure

Harvesting began on the inshore part of the weedbed from where the harvester cut strips parallel to the shore, slowly moving outwards. The cut weed was transferred by conveyor on to the banks and removed where possible. The cutting depth was to a maximum of 2.5 m and in shallow water the blades often hit the sediments. To minimise the risk of transfer of undesirable weed species, the harvester was cleaned prior to moving to a different lake by removal of all visible weed fragments and then spraying with diquat.

Field sampling studies

Lake Aratiatia

Aquatic weed regrowth prior to and following harvesting, was monitored in Lake Aratiatia at two to four week intervals from February 1988 until September 1990. Over this period the weedbeds were subjected to mechanical harvesting three times. Two permanent transects (Transect D and Transect P), 100 m long, were laid perpendicular to the shore in areas of weedbeds that were to be harvested. A third transect (Transect C),

70 m long was laid in an area which was unharvested for the duration of the study to act as a control.

Sampling was carried out using SCUBA. To minimise disturbance of the vegetation divers moved only along the downstream side of each transect rope and assessed the vegetation on the upstream side. At each sampling time the following information was recorded at two metre intervals along the transects; presence/absence, plant height and % cover of each species in a 1 m^2 quadrat pushed into the weedbed. Biomass measurements were made adjacent to the transects using a 0.25 m^2 ring quadrat with a large net attached to trap plant material. Vegetation was separated into species, dried and weighed. From changes in plant height, species specific relative growth rates (RGR) were calculated as: $\ln (L_t/L_o)/t$; where L = shoot height (m) and t = time (weeks).

Lake Ohakuri

The study area in Lake Ohakuri was harvested in August 1988 and again in April 1989. Regrowth of the harvested area was monitored at approximately monthly intervals until 11 months after the second harvest. Vegetation regrowth was assessed using SCUBA in the same manner as Lake Aratiatia from eight permanent transects, 10 m long, which were laid parallel to the shore at 0.5 m depth intervals from 1.0 m to 4.5 m.

Results

Environmental conditions

The two lakes had a similar temperature range and high levels of inorganic plant nutrients were found in both lakes (Table 1) which occur below urban and several geothermal and agricultural inputs. Although nutrient levels were similar, the high flushing rate in Lake Aratiatia (Table 1) prevented the levels of phytoplankton growth which contributed to the high PAR attenuation coefficient in Lake Ohakuri. The most marked differences between lakes were in water velocity and in attenuation of PAR (Table 1).

In Lake Aratiatia, current velocities along a transect showed marked differences before and after harvesting. Prior to harvest the highest velocity (0.25 m s^{-1}) was along the inner and outer (deep) margin of the weed bed but this was considerably reduced after harvesting as the current was able to spread across the width of the lake.

Regrowth following harvesting

Patterns of upward growth

The weed harvester had an effective cutting depth of 2.5 m but the efficacy depended on weather and weedbed conditions. Consequently when the transects were analysed after harvesting the weedbeds often appeared patchy at depths above 2.1 m (Figure 1). Markedly different patterns of regrowth were observed in the two lakes. In the harvested area in Lake Aratiatia the weed bed remaining was often patchy in appearance and subsequent regrowth was highly variable. In some areas there was virtually no regrowth; such as the area 50 m from shore on the harvested transect D when compared with the area 75 m from shore (Figure 1a). Regrowth involved a variety of species which varied with time in relative importance. The height of *L. major* was reduced markedly with cutting. The height of *Nitella* increased with time in the deeper section of the harvested transects with a corresponding increase in % cover.

In Lake Ohakuri, the appearance of the weedbed immediately after harvesting was the same as in Lake Aratiatia. However regrowth in this lake was not patchy. The change in overall weedbed height increased at a relatively uniform rate at all depths (Figure 1b).

Changes in species composition

In Lake Aratiatia there was a dramatic change in species dominance following harvesting. Total biomass on the transect declined from 462 to 230 g m^{-2} (dry weight). This was almost entirely attributed to a decline in *L. major*. The native *Nitella* spp. and filamentous algae showed marked increases in their contribution to total biomass from ca. 2% to ca. 10% of the total. The increase in filamentous algae was particularly marked, to the extent that dense blankets of *Cladophora* developed over the cut weedbeds in the high light, high nutrient environment. These decreased later as the beds of *L. major* and *C. demersum* grew upwards. *Nitella* increased in % cover over the same period from <5% to 80% (Figures 2a and b).

In Lake Ohakuri a shallow water (1–2 m depth) *C. demersum* dominated community occurred immediately after harvest (Figure 2c). This was probably due to the removal of rooted species leaving loose fragments of *C. demersum*. As time progressed this community changed to one dominated by *E. canaden-*

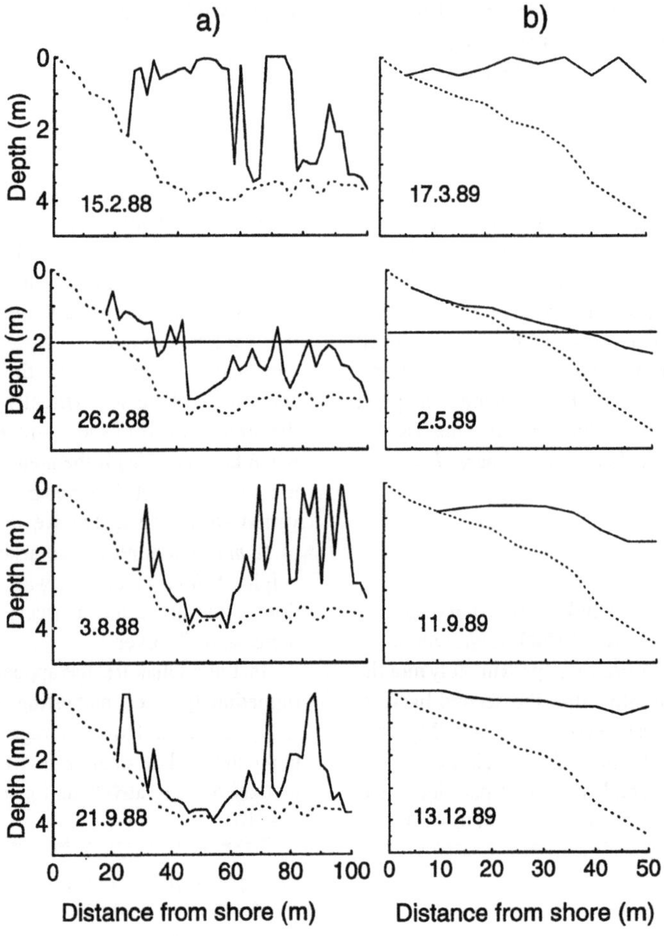

Figure 1. Patterns of aquatic weed regrowth (solid lines) prior to harvesting, 2 days after harvesting and at two intervals after harvesting in (a) Lake Aratiatia and (b) Lake Ohakuri. Horizontal lines are shown on profiles measured 2 days after harvesting and indicate harvester cutting depth. Dotted lines indicate the lake bed.

sis, *Egeria densa* and *Potamogeton crispus*. In deep water (2.5–4.5 m), the % cover of *C. demersum* and *E. densa* remained high over the entire period (Figure 2d). % cover of *E. canadensis* increased but there was no significant change in *P. crispus* or the native *Nitella*. Thus the extent of community changes following harvesting in Lake Ohakuri depended on water depth with least change occurring in deeper water at 2.5 to 4 m. Growth rates ranged from 0.025 to 0.06 week^{-1} in Lake Aratiatia and from 0.046 to 0.079 week^{-1} in Lake Ohakuri. Rates were temperature dependent in both lakes as illustrated by the summer and winter values (Table 2). In winter, regrowth rates were slower in Lake Aratiatia than for Lake Ohakuri but there is little difference between summer growth rates.

Of interest to the lake managers was the time taken for the weedbeds to form surface growths. In Lake Ohakuri 50% of the weedbed reached the surface after approximately 6 months after both harvests (summer and winter). In Lake Aratiatia the weedbed was reduced markedly by the second harvest so that little material remained. In spite of a high regrowth rate it took 9 months for 50% surface growths to form.

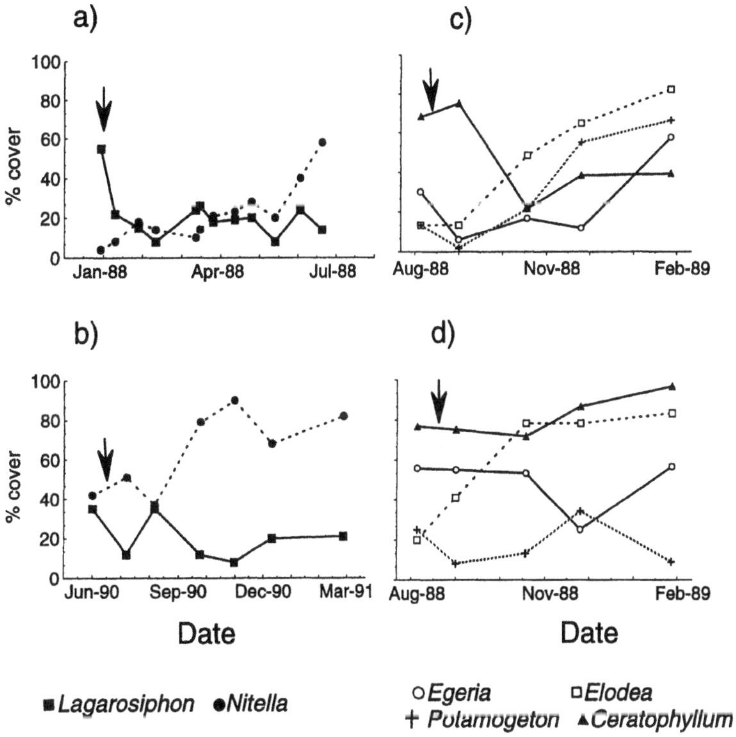

Figure 2. Changes in % cover of dominant species in (a) Lake Aratiatia, 1988 (b) Lake Aratiatia, 1990–91, (c) Lake Ohakuri, 1–2 m depth zone and (d) Lake Ohakuri 2.5–4.5 m depth zone. Arrows indicate harvest.

Discussion

The two lakes showed markedly different responses in aquatic plant regrowth following harvesting. In small low residence time Lake Aratiatia regrowth was patchy, and large areas showed little or no regrowth. In this lake, where *L. major* did not regrow, native species flourished. By contrast in Lake Ohakuri regrowth was regular, not patchy, rapid, and native plants were not able to take advantage of the increased potential habitat following cutting. Changes in species composition were manifested by the introduced weed species and occurred only in shallow (<2 m) water.

The reasons for the patchy regrowth following harvesting in L. Aratiatia are almost certainly due to water currents. Areas where little or no regrowth occurred were all where velocity regularly exceeded 0.15 m s^{-1}. A dynamic cause-effect relationship exists with vegetation and water currents (Dawson & Robinson, 1985). Vegetation in a watercourse can cause major variations in the resistance to flow and while the interaction between waterflow and vegetation is complex (Pitlo & Dawson, 1990) a relationship can be drawn between high current velocities in Lake Aratiatia and reduced weedbed height. From the modelled relationship of macrophyte biovolume as a function of velocity proposed by Henriques (1987), a velocity increase from <0.1 m s^{-1} (Lake Ohakuri and 80–90 m section of harvested transect D in Lake Aratiatia) to 0.25 m s^{-1} (30–50 m section in Figure 1a) would translate to a 50% decrease in total macrophyte volume. A further reason for the patchy regrowth may have been the blanketing effect of the *Cladophora* which developed on the cut weed beds in L. Aratiatia. The low attenuation coefficient for PAR in L. Aratiatia meant that 40% of subsurface light reached a depth of 4 m. The slower height increase of *Lagarosiphon major* in the clear waters of Lake Aratiatia after harvesting, particularly in areas where currents were rapid, meant that the low growing *Nitella* had an opportunity to increase in % cover in a high light environment. *Chara* and *Nitella* species have been noted elsewhere as revegetating lakes where open areas in the littoral zone have been created because of changes to the vegetation due to management practices (Engel & Nichols, 1984; Wade, 1990). A proportional increase in filamentous algae such as recorded for Lake Aratiatia on harvested transects was also noted after harvesting of milfoil beds in Lake Mendota (Nichols et al., 1992).

The strong currents in Lake Aratiatia meant that *C. demersum* was unlikely to grow to the surface in the absence of *L. major* which provided a stable rooted support. In contrast, in Lake Ohakuri, where the water flow was negligible and a base of Hydrocharitaceans remained below the harvester cutting depth, % cover of *C. demersum* remained high in the 2.5 to 4.5 m depth zone.

Wade (1990) identified three patterns of re-establishment of vegetation following harvesting: (1) dominance by species not present prior to management; (2) dominance by species which were dominant prior to management; (3) dominance by species which were present but not dominant prior to management. Patterns 2 and 3 were both exhibited by the lakes in this study. In Lake Ohakuri, pattern 2 was evident in deeper water where the harvester could not reach the lake bed and pattern 3 occurred in shallow water which was maximally disturbed by the harvester. In Lake Aratiatia a mosaic of patterns 2 and 3 occurred where *Nitella* flourished among areas that were not rapidly influenced by regrowing *L. major*.

It was concluded from this study that in the high velocity, clear water of Lake Aratiatia a single cut per year was effective in reducing the overall height of the weedbed. Harvesting of the dense beds of Hydrocharitaceae was beneficial to the reestablishment of *Nitella* communities, particularly where water currents limited regrowth of the *L. major*. In Lake Ohakuri twice yearly harvesting was necessary because of the reduced time taken for the weed beds to reach the surface and the regular upward growth of these beds. The high % cover of *E. densa* and *C. demersum* remaining after harvesting in Lake Ohakuri meant that there was little chance of a change in species composition.

Acknowledgments

Dr Ian Johnstone is thanked for many discussions on the project which was funded by ECNZ (Ltd) (North Island Hydro-Group). Mark James assisted with field work.

References

Dawson, F. H. & W. N. Robinson, 1985. Submerged macrophytes and the hydraulic roughness of a lowland chalkstream. Verh. int. Verein. Limnol. 22: 1944–8.

Engel. S. & S. A. Nichols, 1984. Lake sediment alteration for macrophyte control. J. Aquatic Plant Mgmt 22: 38–41.

Henriques, P. R., 1987. Aquatic macrophytes. In P. R Henriques (ed.), Aquatic biology and hydro-electric power development in New Zealand, Oxford University Press, Oxford, 280 pp.

Howard-Williams, C., 1993. Processes of aquatic weed invasions: The New Zealand example. J. Aquat. Plant Mgmt, 31: 17–23.

Johnstone, I. M., 1981. Management strategies for aquatic weeds in hydro-lakes, In The Waters of the Waikato Vol 1. University of Waikato, Hamilton, New Zealand: 35–38.

Nichols, S. A., R. C. Lathrop & S. R. Carpenter, 1992. Long-term vegetation trends: A history. In J. F. Kitchell (ed.), Food web management, A case study of Lake Mendota, Academic Press, NY, 553 pp.

Pitlo, R. H. & F. H. Dawson, 1990. Flow-resistance of aquatic weeds In A. H. Pieterse & K. J. Murphy (eds), Aquatic weeds, The ecology and management of nuisance aquatic vegetation. Oxford University Press, Oxford. 593 pp.

Wade, P. M., 1990. Physical control of aquatic weeds In A. H. Pieterse & K. J. Murphy (eds), Aquatic weeds, The ecology and management of nuisance aquatic vegetation, Oxford University Press, Oxford, 593 pp.

Wade, P. M. & R. W. Edwards, 1980. The effect of channel maintenance on the aquatic macrophytes of the drainage channels of the Monmouthshire levels, South Wales 1840–1976. Aquat. Bot. 8: 307–322.

Hydrobiologia **340**: 235–239, 1996.
J. M. Caffrey, P. R. F. Barrett, K. J. Murphy & P. M. Wade (eds), Management and Ecology of Freshwater Plants.

Hydrilla control with split treatments of fluridone in Lake Harris, Florida

Alison M. Fox, William T. Haller & Donn. G. Shilling
Center for Aquatic Plants and Department of Agronomy, University of Florida, 7922 NW 71st Street, Gainesville, FL 32653, USA

Key words: residue sampling, herbicide dissipation

Abstract

After several unsuccessful management efforts, a split treatment of fluridone was applied to the 6700 ha Lake Harris in March and June 1987, at a rate of 3.4 kg ha^{-1} (680 and 340 kg fluridone, respectively) to two 3 m deep, hydrilla-infested bays. Fluridone concentrations in the water were sampled following the June treatment. Average fluridone concentrations were 2.1 μg l^{-1} prior to this second application, and a maximum concentration of 30.2 μg l^{-1} was detected in the treated area on the day following application. Fluridone residues dissipated out of the plot quickly due to dilution but concentrations declined lake-wide more slowly, following a logarithmic model, with an estimated fluridone half-life of 97 days. Control of hydrilla in Lake Harris resulted from the long exposure (over 25 weeks due to the split application) to fluridone concentrations of 2 μg l^{-1}, well below the maximum labelled rate of 150 μg l^{-1}.

Introduction

Lake Harris, Florida, USA (28 ° 47′N; 81 ° 50′W) is a eutrophic lake with an area of 6700 ha and an average water depth of 3.7 m. The major inflow into Lake Harris is the Palatlakaha River and the outflow is the Dead River, which connects Lake Harris with other large waterbodies in the Oklawaha chain of lakes. Depending on stream discharge and wind direction, water flow in the Dead River may occur in either direction (into or out of Lake Harris). Submersed plant growth in Lake Harris has historically been limited to depths of less than 2 m by low light penetration, and in these shallow areas wind and wave action has further reduced macrophyte establishment.

Hydrilla (*Hydrilla verticillata* (L.f.) Royle) was first discovered in Lake Harris in 1981, approximately 20 years after this exotic species was introduced into Florida. Reflecting its relentless expansion throughout the state, hydrilla coverage increased in Lake Harris during the following six years, despite attempts to control it with various herbicides.

By 1987, fluridone had been used to control large areas of hydrilla successfully in many Florida lakes, often achieving several hectares of weed control for each hectare treated. However, fluridone applications to small infestations within large lakes, as in Lake Harris, were less effective than whole-lake treatments, and research efforts were developed to understand and improve the efficiency of such fluridone applications.

For example, a hydrilla treatment in the Venetian Gardens Bay of Lake Harris (Figure 1) using 470 kg active ingredient (a.i.) fluridone at a cost of $200 000 was unsuccessful in 1986. In a concerted effort to control this infestation, a fluridone treatment and residue monitoring program was devised in 1987. The objective of the residue sampling program was to monitor herbicide movement out of the treated plots and to determine fluridone persistence in the lake water.

Materials and methods

Herbicide application

In March 1987, 61 ha of hydrilla in the Venetian Gardens Bay of Lake Harris (Figure 1) were treated by helicopter with 136 kg a.i. of fluridone AS (aqueous

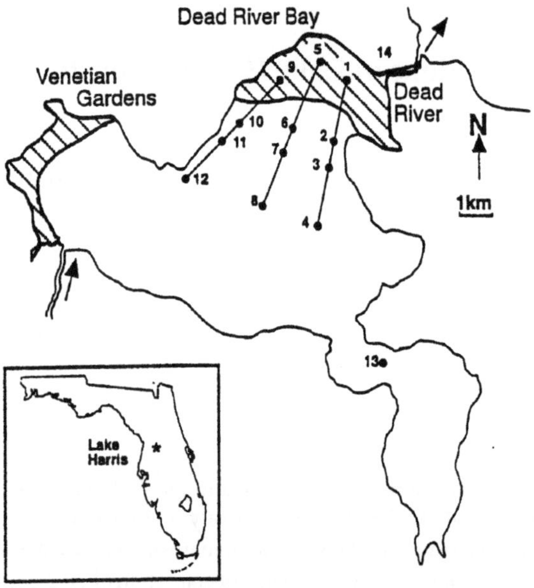

Figure 1. Map of Lake Harris indicating hydrilla infestations (hatched areas) in June 1987, and fluridone residue sampling stations (●) 1 to 14. Inset shows location of Lake Harris in Florida.

suspension), and this treatment was repeated in April. Helicopters were also used to apply 408 kg a.i. fluridone AS to 121 ha of hydrilla in Dead River Bay in March 1987. It was intended to repeat the Dead River Bay treatment in April, but because heavy rainfall resulted in high water flows out of the Dead River (Figure 1), re-treatment of this bay was postponed until 4 June, 1987. At that time 100 ha were treated with 340 kg a.i. fluridone, which was applied to 10 plots of 10 ha each, distributed throughout the hydrilla infestation. The theoretical fluridone concentration within the treated plots would have been 111 $\mu g\ l^{-1}$.

Fluridone residue sampling

Prior to the June 1987 fluridone treatment, 14 water sampling stations were established throughout Lake Harris (Figure 1). Three sampling stations (1, 5 and 9, Tier A) were within the hydrilla infestation in Dead River Bay (but not within treatment plots) and these formed the ends of three transects that radiated away from the treated area. Stations 2, 6 and 10 (Tier B) were 0.8 km from the edge of the hydrilla infestation along each transect, with stations 3, 7 and 11 (Tier C) a further 0.8 km beyond. The ends of the transects (stations 4, 8 and 12, Tier D) were in the open lake, being 3.2 km from the edge of the hydrilla beds. Station 13 was by the highway bridge over Little Lake Harris (Figure 1), over 7 km from the treatment area, and station 14 was approximately 1 km downstream in the Dead River towards Lake Eustis.

Water was collected from 0.5 m below the water surface and from 0.5 m above the lake bottom using a Van Dorn water sampler, and was frozen in 1 l brown polyethylene bottles. These samples were collected the day prior to the June 4 treatment, and again 1, 4, 7, 14, 21, 27, 42, 56, 98 and 174 days after treatment (d.a.t.).

Fluridone residues were analyzed by HPLC following the methods of Fox et al. (1991) with a limit of quantification of 0.05 $\mu g\ l^{-1}$. These data were statistically analyzed, using SAS Institute software to perform analyses of variance (ANOVA) and linear regressions. A natural logarithm transformation was used in all ANOVA to normalize the data. A topographical mapping program, Surfer (Version 4, Golden Software, Inc., Colorado, USA) was used to plot isopleths of fluridone residue concentrations in Dead River Bay and along the transects. An inverse distance gridding method with a weighting of 2 and a normal search was selected, and a spline interpolation was applied to smooth the resulting contour map.

Results

Fluridone concentrations remaining in the lake water from the March and April treatments, collected from stations 1 to 12 on 3 June 1987, averaged 2.15 $\mu g\ l^{-1}$, despite an estimated inflow of 10% of the lake volume due to rainfall in April and May. There were no significant differences between surface and bottom samples at this time because Florida lakes are not strongly stratified. Neither factors of sample depth nor transect showed significant interactions with time in ANOVA of fluridone concentrations from stations 1 to 12. Fluridone was shown to dissipate from this part of the lake following a logarithmic model ($r^2 = 0.99$; Figure 2). This model predicts that fluridone concentrations would be reduced to 2.0, 1.0 and 0 $\mu g\ l^{-1}$ by 89, 275 and 847 d.a.t., respectively. A fluridone half-life of 97 days was estimated from these data but the exponential dilution model accounted for a smaller proportion of the data variability ($r^2 = 0.82$).

An ANOVA comparing fluridone dissipation from stations on different tiers (A to D) along the transects showed a significant interaction between time and distance from the treatment area, with different tiers fit-

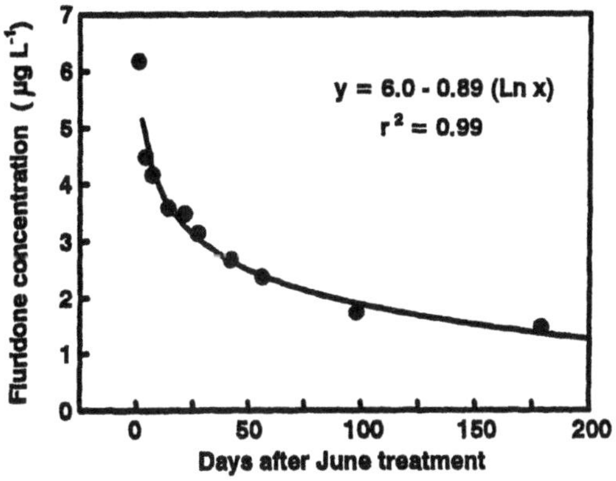

Figure 2. Fluridone concentrations in water sampled between 1 and 174 days after the 4 June 1987 treatment in Dead River Bay on Lake Harris.

ting different dissipation models. However, much of this variation resulted from the data collected 1 and 4 d.a.t. when coefficients of variation (CV) of the average lake concentrations were 120% and 61%, respectively. Average concentrations at all subsequent sampling times had CVs of less than 40%, with most being less than 30%. An ANOVA omitting the first two sampling times and comparing tiers with number of weeks after treatment showed no significant differences in dissipation rates between the tiers.

The isopleths of fluridone concentrations around the transects (Figure 3) indicate that fluridone rapidly dissipated from the treated areas into the open lake. Within 1 day of treatment, average fluridone concentrations 0.8 km from the hydrilla infestation (Tier B) were 4.6 $\mu g\ l^{-1}$, double the pre-treatment value. Average concentrations reached 6.6 $\mu g\ l^{-1}$ at a distance of 3.8 km from the hydrilla (Tier D) by 4 d.a.t.

The maximum fluridone concentration 1 d.a.t. was 30.2 $\mu g\ l^{-1}$ at the surface of station 9, contrasting with the surface concentration at station 1 of 3.0 $\mu g\ l^{-1}$. Fluridone concentrations were generally lower in the eastern transect (stations 1 to 4) compared to the others because of internal currents in the lake combined with outflow from the east side of the bay though the Dead River (station 14). Surface concentrations at station 14 were 10.3, 14.6, 7.3 and 3.2 $\mu g\ l^{-1}$ on 1, 4, 7 and 14 d.a.t., respectively, indicating some loss of herbicide from the lake.

By 14 d.a.t., fluridone concentrations were evenly distributed throughout the transects with an average value of 3.6 $\mu g\ l^{-1}$ and CV of 20%. A maximum fluridone concentration of 2.25 $\mu g\ l^{-1}$ at station 13 (over 7 km from the treatment area) attained 14 d.a.t. was 0.5 $\mu g\ l^{-1}$ higher than the pre-treatment sample.

Figure 3. Fluridone concentration isopleths at 1 $\mu g\ l^{-1}$ intervals around the sampling transects in the Dead River Bay in Lake Harris, 1, 4 and 7 days after the 4 June 1987 treatment, (see Figure 1 for sampling station positions on transects). Arrow indicates location of Dead River outflow.

Hydrilla was observed to be dying and decaying in Dead River Bay within three months of the June treatment. Biennial vegetation surveys reported by the

Table 1. Annual fluridone treatments for hydrilla control in Lake Harris with theoretical concentrations within the treated plots and within the whole lake (6,700 ha), assuming average water depths of 3 m and 3.7 m respectively.

Year	Total applications (kg)	Total applications (ha)	Theoretical concentration ($\mu g\ l^{-1}$) Treated plots	Theoretical concentration ($\mu g\ l^{-1}$) Whole lake
1985	228	86	88	0.9
1986	467	127	121	1.9
1987 Mar/Apr	680	243	93	2.8
June	340	101	111	1.4
1987 Annual total	1020	344	98	4.2

Department of Natural Resources indicate less than 0.5 ha of hydrilla has been found in recent surveys. The lack of hydrilla regrowth was probably due to the marginal light penetration in the lake, which prior to 1987 had caused the slow establishment of the hydrilla infestation. The eutrophic states of many of the lakes in this region are such that little submersed vegetation occurs.

Discussion

Despite the rapid dissipation of fluridone out of the treatment area and into the open lake, hydrilla was controlled throughout Lake Harris in 1987. The documentation of fluridone concentrations was critical in explaining why this treatment was effective, when apparently similar methods using less herbicides had failed in previous years. Application rates within the treatment plots were no higher than in previous years, with initial theoretical concentrations of 111 $\mu g\ l^{-1}$ in 1987, compared with 121 $\mu g\ l^{-1}$ in 1986 (Table 1).

The important feature of the 1987 treatment was not the high initial dose of fluridone, but the long exposure of hydrilla to a lower concentration, resulting from dilution of residues throughout the lake, slow residue degradation, and the three month split between treatments. Pre-treatment samples indicated that the hydrilla had already been exposed to at least 2 $\mu g\ l^{-1}$ of fluridone for three months. This resulted from the dilution of the March/April treatments throughout the whole lake to a theoretical concentration of 2.8 $\mu g\ l^{-1}$ (Table 1). Although the June treatment alone would have diluted to only 1.4 $\mu g\ l^{-1}$ of fluridone in the whole lake, it supplemented the existing fluridone concentrations (Table 1), resulting in a total exposure of hydrilla to at least 2 $\mu g\ l^{-1}$ for 180 days (March to June + 89 days predicted from Figure 2) or at least 1 $\mu g\ l^{-1}$ for 365 days (90 + 275).

In enclosed ponds where dilution was not significant, fluridone degradation typically followed a logarithmic model, but often reaching 0 $\mu g\ l^{-1}$ more quickly than found in this study (West et al., 1983; Langeland & Warner, 1986). Fluridone degradation occurs principally by photolysis (Mossler et al., 1989) and it is likely that low light penetration in Lake Harris, may account for the relatively slow rate of fluridone dissipation from the whole lake.

This study was one of the first to document the susceptibility of hydrilla when exposed to low concentrations of fluridone for long periods of time, and has lead to the development of split applications extending over several weeks or months. This strategy has been progressively refined and has proved highly successful, even in flowing water where daily applications to maintain concentrations of 10 to 15 $\mu g\ l^{-1}$ over periods of 10 weeks have been extremely effective (Haller et al., 1990; Fox et al., 1994). Laboratory studies of concentration/exposure time relationships have supported these field applications (Netherland et al., 1993), and recent data indicate that 1 $\mu g\ l^{-1}$ is potentially the minimum concentration at which fluridone is effective on hydrilla under controlled conditions (M. D. Netherland, unpublished data).

This could explain why the treatments in 1985 were not effective, being insufficient to raise the whole lake concentration to above 1 $\mu g\ l^{-1}$ (Table 1). Although the 1986 treatments could have resulted in at least this concentration throughout the lake, the duration of exposure was probably insufficient as the treatments were not split but were all applied in September.

The realization from this study that fluridone applications to hydrilla infested bays cannot always be considered in isolation from the whole waterbody, explained the unpredictability of spot treatments in many of Florida's other large, wind-swept lakes. In such cases, the application strategy must be either to treat the whole waterbody, or to make frequent split applications to the target area in order to maintain elevated local fluridone concentrations for a sufficient period of time. Dye studies to estimate rates of plot dilution have been valuable in determining optimal split application strategies in many waterbodies (Fox et al., 1991).

Variables, such as plant age, growth rate, light penetration, and factors that influence herbicide degradation, can potentially influence the threshold of concentration/exposure time relations below which fluridone treatments are ineffective on hydrilla. The lowest threshold of phytotoxicity to hydrilla apparently occurs somewhere between 1 and 15 $\mu g\, l^{-1}$ fluridone, depending upon plant and environmental factors. Although hydrilla was controlled in Lake Harris with 2 to 3 $\mu g\, l^{-1}$ fluridone, these treatment rates have not been successful in other lakes.

Aquatic plant managers and researchers are attempting to approach the minimum concentration/exposure time threshold in the field, cautiously trying to save money (e.g., each 1 $\mu g\, l^{-1}$ applied to Lake Harris costs approximately $130 000) but also trying to ensure weed control. This study in Lake Harris was an important first step in realizing the practicality of this valuable hydrilla management strategy.

Acknowledgments

This project was funded in part by the Lake County Water Authority and the U.S. Department of Agriculture and IFAS Center for Aquatic Plants cooperative agreement No. ARS 58-43YK-9-0001. The field assistance of Margaret Glenn and Sue Newman and the laboratory assistance of James Dickson, Sean Ragland, Richard Napier and Eric Milgran are gratefully acknowledged. Published with the approval of the Florida Agricultural Experiment Station as Journal Series No. R-04533.

References

Fox, A. M., W. T. Haller & D. G. Shilling, 1991. Correlation of fluridone and dye concentrations in water following concurrent application. Pestic. Sci. 31: 25–36.

Fox, A. M., W. T. Haller & D. G. Shilling, 1994. Use of fluridone for hydrilla management in the Withlacoochee River, Florida. J. aquat. Plant Mgmt 32: 47–55.

Haller, W. T., A. M. Fox & D. G. Shilling, 1990. Hydrilla control program in the upper St Johns River, Florida, USA. Proc. Eur. Weed Res. Soc. Symp. Aquat. Weeds 8: 111–116.

Langeland, K. A. & J. P. Warner, 1986. Persistence of diquat, endothall, and fluridone in ponds. J. aquat. Plant Mgmt 24: 43–46.

Mossler, M. A., D. G. Shilling & W. T. Haller, 1989. Photolytic degradation of fluridone. J. aquat. Plant Mgmt 27: 69–73.

Netherland, M. D., K. D. Getsinger & E. G. Turner, 1993. Herbicide concentration/exposure time requirements for Eurasian watermilfoil and hydrilla. Proc. 27th Annu. Meet., Aquat. Plant Control Res. Prog. Misc. Paper A-93-2. USAE Waterways Experiment Station, Vicksburg, MS: 146–155.

West, S. D., R. O. Burger, G. M. Poole & D. H. Mowery, 1983. Bioconcentration and field dissipation of the aquatic herbicide fluridone and its degradation products in aquatic environments. J. Agric. Food Chem. 31: 579–585.

Hydrobiologia **340**: 241–245, 1996.
J. M. Caffrey, P. R. F. Barrett, K. J. Murphy & P. M. Wade (eds), Management and Ecology of Freshwater Plants.

Crassula helmsii: attempts at elimination using herbicides

F. H. Dawson
NERC Institute of Freshwater Ecology, River Laboratory, East Stoke, Wareham, Dorset BH20 6BB, UK

Key words: Aquatic plant control, *Crassula helmsii*, elimination attempts, herbicides

Abstract

The high resistance to control by herbicides of stands of the aggressively-invasive water plant *Crassula helmsii* (Australian swamp stone crop or New Zealand Pygmy Weed, also sold as *Tillaea recurva*) was shown during a series of tank and field trials aimed at:

firstly, selecting the most appropriate UK-approved herbicide showed that diquat, either directly or in alginate form was effective on submerged plants particularly at low biomass, whereas for emergent stands, although glyphosate was initially selected as effective, diquat was subsequently recommended;

secondly, the efficacy of the herbicides selected under a range of conditions of biomass, season of application and, particularly, field conditions showed that whilst low biomasses could be controlled and the plant could probably be eliminated, elevated or multiple applications would be necessary at the very high biomasses (up to 45 kg fresh weight per m^2) achieved by this plant, unless the bulk of the biomass could be physically reduced prior to herbicide application; further trials were considered necessary to meet legal current constraints.

Introduction

The aggressively-invasive water plant *Crassula helmsii* (T. Kirk) Cockayne which continues to invade ponds and lakes in Britain and to outcompete native flora, has plant stands which show a high resistance to control by herbicides (Dawson, 1988, 1989). Other methods have been attempted such as grass carp and shade material, but these have proved either difficult to implement over the range of conditions, ineffective, aesthetically displeasing or prone to damage (Dawson & Warman, 1987). Herbicides were therefore considered essential to any control strategy and particularly in nature reserves where the absence of control may often lead to rapid suppression or loss of the rarer flora (Cooke, 1986; Dawson, 1994).

The plant has an amphibious habit with growth forms adapted to a wide range of habitats from drying soils surrounding temporary pools to submerged growing from depths of 3 m in Britain. Maximum biomasses were found in emergent stands at the margins of nutrient-rich static waters (Dawson, 1994).

The objectives of this study were to:

(1) determine, in tank and field trials, which were the most effective herbicides to control submerged and marginal/emergent stands of *C. helmsii*, and to

(2) extend recommendations for the use of the selected herbicides to eliminate this plant from nature reserves by amending existing guidelines.

Methods and materials

The methods, experimental design and work programme included:

(a) tank-trials on the control of both emergent and submerged stands of *C. helmsii* using proprietary UK-approved aquatic herbicides; and

(b) field trials over a representative range of water levels.

Herbicide selection trials

Tank trials

Plant material of *C. helmsii* was collected as turves of 0.25 by 0.25 m in area, from shallow (emergent form) and deeper (submerged form) water from Mockbeggar Pond, adjacent to the New Forest, Hampshire, UK, in September 1988 and grown as emergent (7) and submerged (7) stands respectively in large domestic water tanks (0.6 × 1.0 × 0.55 m deep, 300 l) in an almost unshaded area protected from the wind, away from normal access and without direct drainage. Turves were transported sealed in suitably-sized thick polythene bags and placed in pairs in each of the tanks. Turves from emergent stands were supported on wooden boards fitted 0.25 m from the top of the tanks to create shallow-water but to avoid excessive shading from the tanks sides; turves from submerged stands, were placed upon the bottom of the tanks. Tanks were filled with calcareous tap water to near the top.

Water levels were maintained by addition of tap water as necessary; plant nutrients were monitored and replenished to maintain good plant growth, typically c. 10 ml of a commercial general fertilizer (8-4-4 NPK) every ten days.

Plant material for further trials was collected in January and July 1989. Subsequent trials were also undertaken in 1993–1994, following the culture from 1991 of sufficient 'sediment-free' plant material from that material previously used as 'controls' in 1989, on stands typical of natural 'high' biomass stands (30–50 kg m^{-2} fresh weight).

Field trials

Three types of site were selected to include the habitat range of this plant, firstly, on the basis of the uniformity and extent of plant stand to allow for adjacent single plots for each herbicide separated by control areas and secondly, on relevant permission being available:

(i) stands drying in summer with the plant growing as a short turf of 0.02–0.1 m in height around the margins of a shallow gravel lake, (2 m × 2 m);
(ii) emergent stands of 0.3–0.6 m in height growing at the margins of a lake in water depths from about 0.2 m depth up to the moist margins (10 m × 10 m); and
(iii) submerged stands at water depths of about 1 m in a lake (10 m × 10 m).

Herbicide application

In the initial 'low' biomass herbicide-selection trials (13–16 kg m^{-2} fresh weight, 7% dry weight) in October 1988, July 1989 and July 1990, single doses of herbicides (emergent – asulam, 2,4-D amine, dalapon & glyphosate, submerged – dichlobenil, diquat & terbutryne, were applied in an appropriate manner and at the maximum 'permitted' dose rates to field sites and to the tanks after 'good growth' had re-established following transfer (minimum 6 weeks). Subsequent trials in March 1993, May 1993 and May 1994, on plant material cultured in tanks to 'higher' biomasses (up to 50 kg m^{-2}), were treated in an attempt to achieve full control with a series of elevated dose rates up to × 50 with both glyphosate, with and without adjuvant (Mixture B) to optimise efficacy, and diquat as supplied, diluted to highest water volumes (× 50) and in the alginate formulation. Dose rates were calculated according to individual tank volumes for submerged stands or surface areas of tanks for emergent because of tank flexibility.

Assessment

After the initial trial, the plant material in each tank was weighed prior to and at intervals after treatment by sliding the plant mat onto a large pre-weighed wire frame, draining for five minutes before weighing using a large spring balance (0–50 kg) and returning it to its tank. At the conclusion of a trial (5–12 weeks), plant material was removed for assessment of living material, fresh and dry weight (105 °C); dried samples were sorted, by hand, to remove any stones, etc. originating from field sites and the weights corrected.

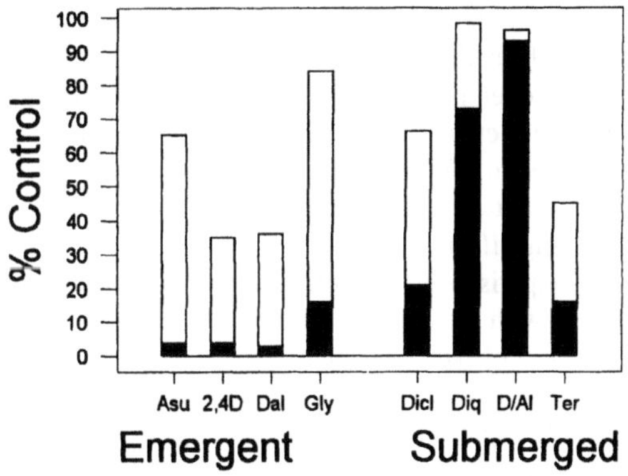

Figure 1. Comparison of the reduction in plant biomass, as percentage control in 'low biomass' trials (13–16 kg m^{-2}) with differing UK-approved herbicides during autumn (lower lines) and spring or summer (upper lines) at approved concentrations of herbicides as specified on the product label. Key: Asu = Asulam; 2,4D = 2,4 diamine; Dal = Dalapon; Gly = Glyphosate; Dicl = Dichlobenil; Diq = Diquat; D/Al = Diquat alginate; Ter = Terbutryne.

Results

Tank trials

The environment of the tanks. The water temperature varied daily and seasonally but no heat stress was observed in any plants; some plant stems and leaves were however killed when crushed in ice during winter when temperatures fell to −7 °C. Changes in water chemistry were never excessive compared to local ponds. Nutrient additions were never enough to cause more than minor algal growth.

'Low' biomass trials 1988–89

Submerged plants

In tank trials at biomasses of 3–5 kg fresh weight, diquat and diquat-alginate were both rapid and effective in killing the majority of the plant and reducing the biomass to 0.1 kg fresh weight within a few weeks (Figure 1). Dichlobenil and terbutryne were considerably slower and only partially reduced biomasses to 1–2 kg, as might be expected by their mode of action and the time of application in the growing season. Material treated with dichlobenil was brittle, easily fragmented and had poor roots.

Emergent plants

At biomasses of 8–10 kg, glyphosate was the most effective reducing biomass to 1.6 kg by December 1988, although little lasting control was found in the repeat trials of August 1989 and at higher biomasses in July 1990, apart from direct scorching showed rapidly regenerated; this was also observed at the field sites. However the biomass in several of the treated tanks was lower than the control tanks, particularly with glyphosate, indicating some degree of reduction in growth rates and therefore some control (1–2 kg).

Correction of the July 1989 results for growth of the plants compared to the controls showed that whereas glyphosate was only partially effective, diquat was very effective with reductions to 0.5–0.1 kg followed by dichlobenil (to ~1 kg) and terbutryne (to 0.7 kg).

Field trials

Field trials on the short drying turves were inconclusive with much scorching but reductions in biomass appeared to show a similar pattern to the tank trials. Plant stands were however almost entirely killed in the deeper water trials, but differing degrees of control between herbicides were not observed as material may have drifted into or out of the areas under wind action and water movements may have occurred creating a herbicide 'cocktail' despite the use of buffer zones.

High biomasses tank trials

Trials at 'higher' biomasses (30–45 kg m^{-2} fresh weight) in May 1993 and July 1994 in which glyphosate and diquat were applied at higher than the normal application rates, resulted in more reliable reductions in biomass particularly at the × 5 and × 10 concentrations (Figure 2a & b). Diquat was effective when tested at higher concentrations but even when tested at a single dose rate of × 50, plant material was not fully killed at the highest biomasses of 45 kg m^{-2} fresh weight. However a disproportionally higher quantity of dead material remained compared to low biomass trials, some of which remained viable in the waste pit even after several months under black polythene.

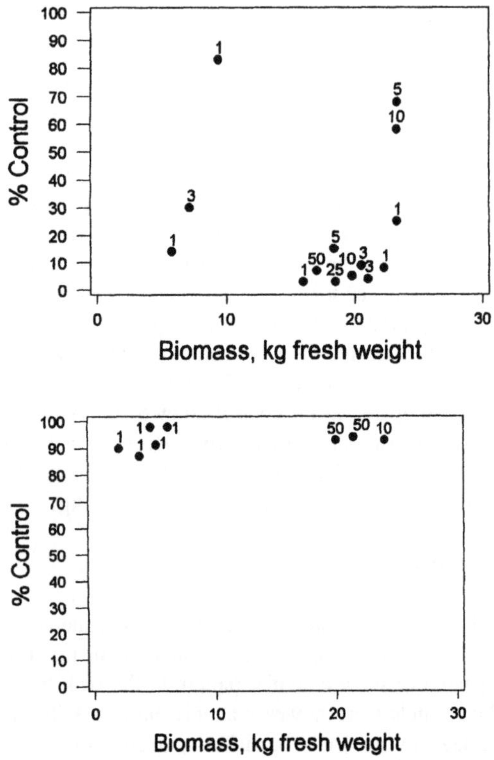

Figure 2. Comparison of the reduction in plant material at differing biomass with (a) Glyphosate and (b) Diquat and diquat alginate at different concentrations of herbicide. The concentrations, × 1 to × 50, are indicated by the numbers by each symbol.

Discussion, conclusion and recommendations

The results are discussed in relation to the objectives of this study:

1. determining from tank and field trials the most effective and acceptable herbicide to control both submerged and marginal/emergent stands of *C. helmsii*;

2. extending recommendations for the use of herbicides to eliminate this plant from nature reserves, and amending existing guidelines.

The 'comparison of the effectiveness of different aquatic herbicides' in tank trials showed conclusive results for tank trials but not for field trials particularly on emergent stands. In initial trials, both untreated emergent and underwater stands showed that the degree of control varied with the type of herbicide but that low biomass underwater stands (up to 10 kg m^{-2}) were best controlled by diquat and diquat-alginate. Emergent stands were however only partially susceptible to one herbicide glyphosate at normal concentrations despite the addition of adjuvant.

Hydrogen peroxide (100 volume, 100 ml m^{-2}), a potentially environmentally-acceptable chemical was also tested in field and tank trials and was quite effective initially but its effect was limited to a direct kill or scorching effect.

Comparison of the normal requirements of weed control with those of elimination illustrate two problems both of which through 'legal' constraint lead to limitation in the effectiveness of control:

(i) the typical biomass of a maturing emergent stand of *Crassula helmsii* is considerably higher during the approved application period (50 kg m^{-2}) fresh weight, than for normal aquatic 'weeds'; this results in an unusually low ratio of herbicide to 'weed' biomass,

(ii) elimination of this plant is desired not merely the firm control as required for most other aquatic sites.

These problems are exemplified by the field trials on both drying turves and emergent stands, in which a single application at the approved dose limited biomass but produced little observable effect; underwater stands, typically lower in biomass, were far more successful controlled if the problem of die-back to shoot tips and release of fragments of differing viability for recolonisation, is neglected. (Fine wire-mesh enclosures were previously recommended to contain these fragments).

Dose rate and timing have been investigated in the tank trials on both glyphosate and diquat in anticipation of 'off-label' trials at higher concentrations but even ten times the normal maximum dose was insufficient to give more than two-thirds control in a single application. If eradication of this plant from nature reserves is to be achieved, multiple applications are indicated and these should be fully tested in further tank and field trials.

The second objective, that of extending the existing recommendations for use of herbicides to eliminate this plant from nature reserves by synthesizing the results of herbicide tank and field trials with the variety of conditions and range of sites to prevent the further continuing spread of *Crassula helmsii*, indicated:

1. different guidelines are required for sites with differing degrees of dominance, e.g. 'low' versus 'high' biomass areas or sites, and

2. noting that some techniques were ineffective or counter-productive.

Thus for example

(i) manual or mechanical normally results in the increase in the number of viable fragments within the wet area with a consequential loss of local flora and also an increase in the risk of spread to other sites, e.g. on operators boots, equipment or machinery.

(ii) herbicides should be applied as effectively as possible on the first application because reduced or subsequent applications only causes the loss of other more susceptible species effectively enhancing the dominance of *C. helmsii.* However alternatives available including the application of herbicide in winter when native species are less common may cause adverse less effect and the natural seed bank can help to regenerate the local flora.

Recommendations for further work on the control of *Crassula helmsii* include:

1. Improving the control of emergent stands by
   (i) using glyphosate in an off-label way for the UK,
   (ii) optimizing the concentration of surfactant or 'adjuvant' additive,
   (iii) undertaking new user trials with glyphosate, to determine appropriate concentrations or
   (iv) extending the recommended use of diquat to emergent plants as well as submerged.

Further research to assess the herbicidal properties and by-products of simple chemical compounds for use in nature reserves (or by the use of other herbicides) could be undertaken. The use of hydrogen peroxide in these trials resulted in some direct control but this was not exploited; further investigation of 'peroxygen' generating compounds, in particular, is suggested.

2. Trials on (i) multiple doses of herbicide diquat at intervals of 2–4 weeks and (ii) in a concentration appropriate to the biomass present, as part of a considered elimination strategy for this plant.

3. Continuation of the study of the primary and secondary dispersion of this plant, both by man and other mechanisms to reduce the invasion of important and rare natural site, and to reduce the invasion of large site in which control is likely to be impractical eg Cumbrian lakes and Scottish locks.

Acknowledgments

Thanks are due to: ICI Agrochemicals for giving the diquat alginate ('Midstream') and diquat ('Reglone'); Monsanto Agricultural Company for glyphosate ('Roundup') for use in these trials; Drs P. Boon, M. Palmer and M. Gibson formerly of The Nature Conservancy Council, UK (now Scottish Natural Heritage or English Nature); Paul Henville for his technical assistance; S. Shin & Graham for their manual assistance and maintenance around the tanks.

References

Cooke, A. S. (ed.), 1986. The use of herbicides on nature reserves. Focus on Nature Conservation Series 14. Nature Conservancy Council, Peterborough, 80 pp.

Dawson, F. H., 1988. The alien aquatic Crassula helmsii continues to expand its distribution in Britain. BSBI News 49: 43.

Dawson, F. H., 1989. Some attempts at the control of the alien aquatic Crassula helmsii (T. Kirk) Cockayne. BSBI News 51: 46.

Dawson, F. H. & E. A. Warman, 1987. Crassula helmsii; is it an aggressive alien aquatic plant in Britain? Environmental Conservation 42: 247–272.

Dawson, F. H., 1994. The spread of *Crassula helmsii* (Kirk) Cockayne in Britain. In L. de Waal, L. Child, P. M. Wade & J. H. Brock (eds). J. Wiley & Sons, Chichester: Ecoloy & Management of Invasive Riverside Plants: 1–14.

Hydrobiologia **340**: 247–251, 1996.
J. M. Caffrey, P. R. F. Barrett, K. J. Murphy & P. M. Wade (eds), Management and Ecology of Freshwater Plants.

Hydrilla tuber formation in response to single and sequential bensulfuron methyl exposures at different times

K. A. Langeland
University of Florida, Institute of Food and Agricultural Sciences, Department of Agronomy, Center for Aquatic Plants, 7922 NW 71st Street, Gainesville, Florida, USA

Key words: tubers, inhibition, herbicide, photoperiod, herbicide registration

Abstract

Hydrilla (*Hydrilla verticillata* (L.f.) Royle) grown in outdoor tanks was exposed to bensulfuron methyl concentrations of 25, 50, or 100 ppb on June 16, August 20 or October 15; 50 ppb June 16 and August 20, or 25 ppb on June 16, July 21, August 20, and October 15, 1990, with a 35-day contact time. Hydrilla was also exposed to the compound on August 9, 1991 at concentrations of 10, 20, 30, 40, or 50 ppb. In 1990, the August 20 exposure resulted in the greatest inhibition of tuber production for a single application. Exposure in June caused hydrilla to produce at least twice as many tubers as unexposed plants by April 10, 1991. Exposure in October arrested tuber production, which had already begun. Exposure in June and August delayed tuber formation until after February 9, 1991. Exposure in June, July, August, and October inhibited tuber formation for the entire growing season. Hydrilla treated with all concentrations of bensulfuron methyl on August 9, 1991 produced tubers only sporadically through March 16, 1992, compared to unexposed hydrilla, which produced an average of 48 tubers/531 sq cm by January 4, 1992. With the onset of warmer weather after March 16, tubers produced by unexposed hydrilla more than doubled, and comparable numbers of new tubers were produced by plants that were exposed to 10 or 20 ppb. Tuber production was inhibited for the entire growing season by exposure to 50 ppb on August 9, 1991. In spite of the promise that bensulfuron methyl showed for use in aquatic plant management, the Experimental Use Permit was not renewed in 1992 and efforts to register the compound were discontinued.

Introduction

Sulfonyl urea compounds were recognized as potential herbicides and growth regulators in 1966, and by the mid 1970s had become 'one of the most exciting breakthroughs in the field of herbicide research in several decades' (Beyer et al., 1988). The sulfonyl urea compound, bensulfuron methyl, is used for weed control in rice production and studies have indicated potential usefor aquatic plant management (Anderson & Dechoretz, 1988; Haller et al., 1992; Langeland & Laroche, 1992; Van & Vandiver, 1992, Bowmer et al., 1992, Getsinger et al., 1994; Langeland, 1993). It was registered for experimental use as an aquatic herbicide or growth regulator in 1989.

Hydrilla (*Hydrilla verticillata* (L.f.) Royle), a submersed plant, which is a serious weed in many parts of the world, produces large numbers of vegetative propagules, commonly called tubers, in the hydrosoil (Bowes et al., 1979; Sutton & Portier, 1985; Sutton et al., 1992). Hydrilla tubers (monoecious) can persist for up to four years under experimental conditions (Van & Steward 1990). Following successful spring-early summer fluridone treatments in the Upper St. Johns and Withlacoochee Rivers, hydrilla regrowth occurred almost exclusively from tubers and turions, which demonstrates their tolerance to application of aquatic herbicides (Haller et al., 1992). Because of the large numbers, longevity, and tolerance to aquatic herbicides, depletion of hydrilla tubers is a key factor in hydrilla regrowth following control measures.

Bensulfuron methyl has been shown to inhibit hydrilla tuber formation (Anderson, 1988; Van & Vandiver, 1992; Haller et al., 1992, Langeland & Laroche,

1992; Van & Vandiver, 1994). Hydrilla (dioecious) produces tubers as a short day photoperiodic response when day lengths become 13 hours or less, or September through April in North Florida (Van et al., 1978). Therefore, timing of bensulfuron methyl application with respect to initiation of tubers should be important in the effectiveness of bensulfuron methyl to inhibit tuber formation.

The objectives of the studies reported in this paper were to determine influence of timing and rate of bensulfuron methyl application on hydrilla growth and tuber production. The paper will also discuss, briefly, reasons why this potentially valuable aquatic plant management tool was not registered for aquatic use.

Materials and methods

Experiment 1

On May 10, 1990, apical cuttings, 20 cm in length, were collected from dioecious (female) hydrilla plants in Newnans Lake, Alachua Co., Florida. Nine cuttings were planted in 26 × 30 cm plastic pans filled with a depth of 20 cm of a soil mix consisting of equal amounts by volume of Metro-mix 200 ® and top soil overlain with sand. Ten grams of a slow release fertilizer (18-6-12) was added to each pan. Ten pans were placed in each of 36 tanks which contained 900 l of water.

Bensulfuron methyl was added to each tank to result in 25, 50, or 100 ppb on June 16, August 20 or October 15; 50 ppb June 16 and August 20, or 25 ppb on June 16, July 21, August 20, and October 15, 1990. Three replications each of these treatments and a control were randomly assigned to each of the 36 tanks. Water was replaced in the tanks once every 35 days and coordinated with herbicide applications so that each application resulted in this contact time.

One pan from each tank was harvested in mid-June and monthly thereafter through December and in February, March, and May, 1991. All soil was rinsed away from the plant material over a 0.25-cm mesh screen. Tubers were collected from the screen or detached from the plants and counted. Plant tissue, excluding tubers, was dried to constant weight in a forced-air desiccating oven and weighed.

Experiment 2

Nine 20-cm apical hydrilla cuttings were planted in each of 162 plastic buckets (531 sq cm), which contained 20 cm of Metro-mix 200 ® and 10 g of 18-6-12 slow release fertilizer overlain with a layer of course builders sand. Nine buckets were placed in each of eighteen concrete tanks, which contained 900 l of well water. On August 9, 1991, 6 weeks after planting the cuttings, three tanks each were treated with sufficient bensulfuron methyl to result in nominal concentrations of 10, 20, 30, 40, or 50 ppb and three tanks were left untreated for checks. Water was not replaced in the tanks for the duration of the experiment. One bucket per tank was removed per month for 9 months to determine plant tissue dry wt and tuber production, as described for Experiment 1.

Results and discussion

Experiment 1

Untreated hydrilla increased in dry wt through December, after which no additional increases were observed ($y = -4.5 + 0.8x - 0.008x^2$, where y is tissue dry wt and x is days after planting, $r^2 = 0.65$,data not presented). Chlorosis and bud abortion of bensulfuron methyl treated plants was observed, and plants treated with 25 ppb in June, July, August, and October were only 23 cm tall compared to untreated plants that were 45 cm tall (LSD = 7 cm, data not presented) at the end of the experiment. However, a difference in growth rate expressed as plant dry wt (excluding tubers) was not observed between treated and untreated plants (comparison of slope coefficients by Student's $t, p<0.05$).

The first observation of earnest tuber formation by untreated hydrilla was observed on September 7, 1990, which indicates that tuber initiation occurred between this date and the August 18, 1990 sampling (Table 1). This is consistent with the observation of Van et al. (1978) that hydrilla tuber formation occurs when day length is 13 hours or less because this photoperiod occurs during the first of September at the latitude of Gainesville, FL (29 °40′). The maximum cumulative number of tubers produced by untreated hydrilla, during the experimental period, was observed on December 26, which suggests that additional tuber formation did not occur after this date.

Hydrilla treated with 25, 50, or 100 ppb bensulfuron methyl in June produced at least as many tubers

Table 1. Cumulative tubers produced by untreated hydrilla (female) and after exposure to various concentrations of bensulfuron methyl at various times. Values are average numbers of tubers per 970 sq cm in three replications and numbers in parenthesis are standard errors of means.

Date of application	Untreated	June 16			August 20			October 15			Jun 16 & Aug 20	Jun 16, Jul 21, Aug 20, & Oct 15
Bensulfuron methyl concentration (ppb)	–	25	50	100	25	50	100	25	50	100	50	25
Date												
6/9/90	0 (0)	1 (1)	1 (1)	1 (1)	0 (0)	1 (1)	1 (1)	0 (0)	1 (0)	1 (0)	1 (1)	0 (0)
7/19/90	2 (0)	0 (0)	0 (0)	0 (0)	1 (1)	3 (2)	2 (1)	2 (1)	1 (1)	1 (1)	1 (1)	0 (0)
8/18/90	1 (0)	0 (0)	1 (1)	2 (2)	2 (0)	1 (0)	1 (1)	3 (0)	3 (2)	3 (1)	5 (5)	0 (0)
9/7/90	6 (2)	30 (4)	10 (4)	8 (6)	3 (1)	2 (1)	2 (1)	16 (5)	19 (3)	14 (8)	5 (3)	1 (0)
10/7/90	17 (2)	28 (4)	25 (3)	17 (6)	1 (0)	1 (1)	1 (1)	28 (5)	21 (1)	13 (3)	0 (0)	1 (1)
11/6/90	42 (5)	64 (12)	65 (5)	51 (4)	1 (1)	1 (1)	0 (0)	25 (10)	18 (4)	32 (2)	1 (1)	1 (1)
12/26/90	57 (9)	81 (12)	102 (9)	60 (1)	12 (2)	2 (1)	2 (0)	16 (3)	19 (3)	32 (8)	1 (0)	2 (2)
2/9/91	48 (7)	65 (18)	104 (0)	54 (14)	12 (6)	5 (2)	1 (1)	24 (3)	30 (8)	24 (8)	2 (1)	1 (1)
4/10/91	50 (11)	100 (14)	124 (10)	127 (29)	55 (9)	27 (6)	8 (4)	37 (12)	29 (3)	20 (5)	42 (1)	0 (0)
5/22/91	56 (3)	78 (12)	110 (25)	76 (6)	54 (2)	28 (9)	27 (9)	8 (8)	23 (12)	16 (2)	43 (12)	1 (0)

Table 2. Hydrilla tuber production after treatment with various concentrations of bensulfuron methyl on August 9, 1991. Values are average numbers of tubers per 531 sq cm in three replications and numbers in parentheses are standard errors of means.

Date	Untreated	Bensulfuron methyl concentration (ppb) 10	20	30	40	50
9/18/91	2 (1)	0 (0)	0 (0)	0 (0)	0 (0)	0 (0)
10/23/91	13 (1)	0 (0)	0 (0)	0 (0)	0 (0)	0 (0)
11/29/91	26 (2)	2 (2)	0 (0)	2 (2)	0 (0)	0 (0)
1/4/92	48 (5)	3 (1)	1 (1)	1 (1)	2 (2)	0 (0)
2/13/92	45 (19)	3 (2)	1 (1)	1 (1)	2 (2)	0 (0)
3/16/92	32 (20)	0 (0)	0 (0)	0 (0)	0 (0)	0 (0)
4/19/92	82 (5)	34 (6)	24 (7)	12 (4)	3 (1)	2 (1)
5/25/92	81 (11)	59 (14)	58 (24)	48 (23)	25 (13)	2 (1)
6/30/92	73 (7)	49 (13)	55 (13)	39 (17)	11 (2)	10 (6)

as untreated plants through February 9, 1991 and by April 10, 1991 these plants produced at least two-fold as many tubers as untreated plants (Table 1). This suggests that exposure of hydrilla to bensulfuron methyl prior to tuber initiation may cause increased tuber production after initiation occurs. An explanation for this is beyond the scope of this paper and deserves additional study.

Bensulfuron methyl applied in August delayed tuber formation at all application rates (Table 1). Twenty five ppb delayed tuber formation until November 6, 1990 through December 26, 1990. However, by April 10, 1991 these plants had produced as many tubers as the untreated plants. Fifty ppb applied in August delayed tuber formation until December 26, 1990 through February 9, 1991, and these plants produced only about half as many tubers as untreated plants by May 22, 1991. One hundred ppb applied in August reduced tuber production to a greater extent than 50 ppb applied in August through March 4, 1991. However, by May 5, 1991 plants treated with either 50 or 100 ppb in August produced the same number of tubers.

Hydrilla exposed to 25, 50, or 100 ppb bensulfuron methyl in October did not produce substantially more tubers following exposure, which suggests that tuber initiation was arrested even when exposure occurred after initiation (Table 1). This inhibition resulted in lower maximum tuber production compared to untreated plants.

Tuber formation was delayed through February 9, 1991 by application of 50 ppb bensulfuron methyl in June and August (Table 1). However, based on the lack of inhibition by applications in June alone, the June application is not necessary. By April 10, 1991 these plants had produced a similar number of tubers as untreated plants, but had produced more than those treated with a single application of 50 ppb in August. This suggests that the stimulatory effect of the June application may have antagonized the potential effect of the August application.

Applications of 25 ppb bensulfuron methyl in June, July, August, and October inhibited tuber production throughout the growing season. However, based on the lack of inhibitory effect of the June applications in the present study, the June application is probably not necessary.

Results of Experiment 1 suggest that the optimum time to apply bensulfuron methyl to inhibit tuber formation is August at the latitude of Gainesville, Florida (USA). This is just prior to tuber formation.

Experiment 2

Experiment 2 was conducted to verify results of Experiment 1 on application of bensulfuron methyl just prior to tuber initiation, and to determine the effects of lower concentrations than used in experiment one.

Untreated hydrilla grew during Experiment 2 at a rate defined by the equation, $y = 27 + .0008x^2$, where y is whole plant biomass expressed as g dry wt and x is days after planting. As in Experiment 1, vegetative growth of hydrilla was not different when exposed to any of the bensulfuron methyl concentrations tested as compared to untreated plants (comparison of slope coefficients by Student's $t, p<0.05$).

Tuber initiation was delayed by all concentrations of bensulfuron methyl tested and for longer periods of time at the higher rates (Table 2). Untreated hydrilla produced an average of 48 tubers by January 4, 1992, after which tuber production slowed, probably because of low water temperature, until after March 16, 1992. In contrast, hydrilla treated with all concentrations of bensulfuron methyl produced tubers only sporadically through March 16, 1992 and essentially no tubers were produced by plants treated with 50 ppb.

With the onset of warmer weather after March 16, tuber numbers produced by unexposed hydrilla more than doubled, and comparable numbers of new tubers were produced by plants that were exposed to 10 or 20 ppb. Tuber production by plants treated with 30, 40, or 50 ppb bensulfuron continued to be lower than untreated plants through April 19, 1992. By May 25, 1992, plants treated with 30 and 40 ppb had apparently lost inhibition, while tuber production by plants treated with 50 ppb was inhibited through June 30.

Conclusions

Bensulfuron methyl has potential for use in managing hydrilla, especially for inhibition of tuber production. Based on the data of this study, a single application of 50 ppb bensulfuron methyl just prior to tuber initiation, or perhaps sequential applications of only 10 ppb (or lower) just prior to tuber initiation and just prior to warming of water in Spring, would be sufficient to prevent tuber formation. However, the experimental use permit for bensulfuron methyl was not renewed by the manufacturer in 1992, and efforts to register the compound for use in aquatic sites were discontinued. The apparent reason for this was the potential for liability associated with contamination of irrigation water when it was learned that bensulfuron methyl is relatively persistant (half-lives up to 77 days) in deep lakes and ponds (Anderson, 1992; Langeland, 1994) compared to shallow ponds (Langeland & Laroche, 1994) and rice paddies (Anderson, 1992). This scenario is indicative of the difficulty of registering compounds for use in aquatic sites.

Acknowledgments

This material is based upon research supported in part by IFAS/ARS Cooperative Agreement No. 58-43YX-9-001 and in part by E. I. du Pont de Nemours & Co., Inc. Patty Mikell assisted with manuscript preparation.

This article was published with the approval of the Florida Agricultural Experiment Station as J. Series No. R-04555. Any opinions, findings, conclusions, or recommendations expressed in this publication are those of the authors and do not necessarily reflect the view of the U.S. Department of Agriculture. Mention of trade names is not intended to recommend the use of one product over another.

References

Anderson, L. W. J., 1988. Growth regulator activity of bensulfuron methyl in aquatic plants. In J. E. Kaufman & H. E. Westerdahl (eds), Chemical Vegetation Management. Plant Growth Regulator Society of America, San Antonio, Texas: 127–145.

Anderson, L. W. J., 1992. Dissipation of bensulfuron methyl in aquatic sites. USDA/ARS Annual Report of Aquatic Weed Control Investigations, USDA/ARS Aquatic Weed Control Research Laboratory, Davis, CA: 12–13.

Anderson, L. W. J. & N. Dechoretz, 1988. Bensulfuron methyl: A new aquatic herbicide. In Proceedings, 22nd Annual Meeting, Aquatic Plant Control Research Program, 16–19 November 1987, Portland Oregon. 1988. Environmental Laboratory, US Army Engineers Waterways Experiment Station, Vicksburg, MS: 225–235.

Beyer, E. M., M. J. Duffey, J. V. Hay & D. D. Schluter, 1988. Sulfonylureas. In P. C. Kearney & D. D. Kaufman (eds), Herbicides: Chemistry, Degradation and Mode of Action, V. 3. Marcel Dekker Inc. NY 1988: 117–190.

Bowes, G., A. Scott Holaday & W. T. Haller, 1979. Seasonal variation in the biomass, tuber density, and photosynthetic metabolism of hydrilla in three Florida, lakes. J. Aquat. Plant Mgmt 17: 61–65.

Bowmer, K. H., G. McCorkelle & G. R. Sainty, 1992. Potential use of bensulfuron methyl for sediment application in irrigation systems in Australia. J. Aquat. Plant Mgmt 30: 44–47.

Getsinger, K. D., G. O. Dick, R. M. Crouch & L. S. Nelson, 1994. Mesocosm evaluation of bensulfuron methyl activity on Eurasian watermilfoil, vallisneria, and American pondweed. J. Aquat. Plant Mgmt 32: 1–6.

Haller, W. T., A. M. Fox & C. A. Hanlon, 1992. Inhibition of hydrilla tuber formation by bensulfuron. J. Aquat. Plant Mgmt 30: 48–49.

Langeland, K. A., 1993. Hydrilla response to Marinerþ applied to lakes. J. Aquat. Plant Mgmt 31: 175–178.

Langeland, K. A., 1994. Bensulfuron methyl residues in Florida lakes. J. Aquat. Plant Mgmt 32: 80–81.

Langeland, K. A. & F. B. Laroche, 1992. Hydrilla growth and tuber production in response to bensulfuron methyl concentration and exposure time. J. Aquat. Plant Mgmt 30: 53–58.

Langeland, K. A. & F. B. Laroche, 1994. Persistence of bensulfuron methyl and control of hydrilla in shallow ponds. J. Aquat. Plant Mgmt 32: 12–15.

Sutton, D. L. & K. M. Portier, 1985. Density of tubers and turions of hydrilla in South Florida. J. Aquat. Plant Mgmt 32: 64–67.

Sutton, D. L., T. K. Van & K. M. Portier, 1992. Growth of dioecious and monoecious hydrilla from single tubers. J. Aquat. Plant Mgmt 30: 15–20.

Van, T. K., W. T. Haller & L. A. Garrard, 1978. The effect of day length and temperature on hydrilla growth and tuber production. J. Aquat. Plant Mgmt 16: 57–59.

Van, T. K. & K. K. Steward, 1990. Longevity of monoecious hydrilla propagules. J. Aquat. Plant Mgmt 28: 74–76.

Van, T. K. & V. V. Vandiver, 1992. Response of monoecious and dioecious hydrilla to bensulfuron methyl. J. Aquat. Plant Mgmt 30: 41–44.

Van, T. K. & V. V. Vandiver, 1994. Response of hydrilla to various concentrations and exposures of bensulfuron methyl. J. Aquat. Plant Mgmt 32: 7–11.

Hydrobiologia **340:** 253–257, 1996.
J. M. Caffrey, P. R. F. Barrett, K. J. Murphy & P. M. Wade (eds), Management and Ecology of Freshwater Plants.

Glyphosate as a management tool in carp fisheries

A. Krüger, G. Okoniewska[1], Z. Pochitonow, Z. Król[2] & R. P. Garnett[3]
[1]*Instytut Rybactwa Sródlądowego, ul. Glowna 48, 05–500 Piaseczno-Zabieniec, Poland*
[2]*Monsanto Europe S.A., Branch Office, ul. Stawki 2, 00–193 Warszawa, Poland*
[3]*Monsanto plc, P.O. Box 53, Lane End Road, High Wycombe, Bucks, HP12 4HL, UK*

Key words: glyphosate, ecotoxicology, fisheries, *Cyprinus carpio*, weed control

Abstract

Glyphosate was evaluated for use as a novel management tool to improve the efficiency of intensive carp (*Cyprinus carpio*) production in Poland. The survival and growth of the carp fry was greatest in ponds in which natural vegetation had been treated with glyphosate prior to flooding, which favoured the natural development of food organisms. The yield was greater than merely flooding the vegetation or the alternative technique of maintaining a bare fallow prior to flooding. Using glyphosate as part of the pond management programme proved cost effective and had no deleterious effect on the carp fry or their food organisms.

Introduction

The isopropylamine salt of glyphosate (N-(phosphonomethyl)glycine) has become a standard treatment for controlling emergent and floating water weeds. In Europe, regulatory approval for aquatic use of Roundup (360 g a.i./l glyphosate) was first gained in 1977 in the United Kingdom, followed by Ireland, Poland, France and Belgium. The success of glyphosate as an aquatic herbicide is due to a unique combination of highly effective weed control and favourable safety and environmental characteristics (Seddon, 1981).

It is readily adsorbed by soil, where it is degraded by soil micro-organisms (Sprankle et al., 1975). In aquatic environments, glyphosate dissipates rapidly due to dilution, adsorption and microbial degradation (Brønstad & Friestad, 1985; Goldsborough & Beck, 1989). Once adsorbed by the sediment it is not easily removed even by flowing water (Comes et al., 1976). Glyphosate is of low toxicity to aquatic organisms (Tooby, 1985; Chen et al., 1989; Feng et al., 1990; Garnett et al., 1992).

This paper presents results from trials in which glyphosate was used as part of a novel management technique in commercial carp (*Cyprinus carpio*) fisheries. The programme aimed to improve intensive carp production in Poland by optimising the natural development of food organisms. In the standard production system, ponds are flooded in early spring. Typically, newly hatched carp fry are introduced into the 'first transition ponds' in late spring and feed on organisms which have developed naturally. After a few weeks, when they weigh about 2 g, the fry are transferred to the 'second transition ponds' which are flooded in spring or early summer. In October, when they reach 30–60 g, the fry are moved to their final ponds where they remain for one or two years until they are ready for sale, at a weight of 800–1500 g. The first and second transition ponds are about 0.5 m and 1 m deep respectively.

Materials and methods

Experiments were undertaken in 1992 and 1993 at the Zabieniec Experimental Fishery Station, near Warsaw. The techniques were evaluated in three replicates, each comprising a pond of 0.2 ha. The details of the techniques and the management of the fisheries are summarised in Tables 1 and 2. Ponds maintained as bare fallow were mechanically cultivated at regular intervals to prevent the natural vegetation from establishing before flooding. Where the triticale mulch was evaluated, spring triticale was sown into the bare fallow and allowed to grow prior to flooding the pond.

In 1992 the first transition ponds were fed at the time of flooding with liquid cattle manure, at a rate of 15 000 l ha^{-1}. The following year, no cattle manure was applied so that the effect of using green manure could be studied. In both years the fry introduced to the second transition ponds originated from first transition ponds in which natural vegetation had been treated with glyphosate. The second transition ponds were fed liberally with supplementary carp food.

Throughout the season the abundance of the three main groups of food organisms (*Rotatoria, Copepoda* and *Cladocera*) was assessed. The hydro-chemical characteristics of the waters are reported in Krüger (1993) and Krüger & Okoniewska (1994). Carps were weighed at the end of the periods spent in the first and second transition ponds.

The weather in 1992 was unusually dry and hot. The average air temperature from June to September was 20.7 °C compared to 18.7 °C in 1993, while the number of hours of sun during the same period was 27% lower in 1993 (Biultetynu Agrometeorologicznego IMiGW). Due to the drought conditions in 1992, the first transition ponds were at 70–80% and the second transition ponds at 50% of their normal levels. In 1992 the average water temperature in the first transition ponds was 21.7 °C (18.7 °C in 1993) and in the second transition ponds 20.5 °C (17.6 °C in 1993).

Results in 1992

In the first transition ponds the highest biomass of food organisms was achieved using glyphosate to destroy natural vegetation before flooding (Table 3). The yield of carp fry was 10% greater than from the bare fallow method and 25% higher than from flooding untreated vegetation. The survival rate of the carp in the bare fallow was low due to competition from a high population of a phyllopod, *Branchipus schaefferi,* favoured by the warm conditions of 1992. Competition was less severe where the vegetation remained in the ponds (with or without glyphosate), either because of the protective effect of the vegetation or a 3 to 5 days delay in appearance of the phyllopod.

Survival of carp fry in the second transition ponds was greatest where glyphosate had been sprayed on natural vegetation seven days before flooding the ponds (963 kg ha^{-1}) and in the bare fallow technique (973 kg ha^{-1}). The yield from these techniques was over three times greater than from the traditional technique of flooding the ponds in late March (Table 3), where the survival of the fry was poor. The slow development of food organisms in the traditionally managed ponds provided insufficient food at the critical early stages of carp development.

Results in 1993

Food organisms were scarcer, and the biomass of carp was lower, than in 1992 (Table 4) due to the lower water temperatures. In the first transition ponds the highest yield (373 kg ha^{-1}) was achieved by treating natural vegetation with glyphosate. This was 1.4 times greater than that of the conventional bare fallow. The triticale mulch technique yielded 318–340 kg ha^{-1}. The ponds in which natural vegetation was flooded yielded very poorly (163 kg ha^{-1}) due to slow development of the natural food sources.

The highest yield of carp in the second transition ponds was given by natural vegetation treated with glyphosate. The survival rate was the same as that in ponds using the triticale mulch technique, but the growth rate was superior because food organisms were more abundant (particularly *Copepoda*). One of the second transition ponds, a replicate in which triticale was treated with glyphosate before flooding, gave an unexplained extremely low survival and yield of carp.

Discussion

New technology can increase the potential yield from carp fisheries compared to traditional management techniques. The use of glyphosate on vegetation prior to flooding allowed the rapid development of natural food organisms for the carp fry. The overall quality of the surface water in the ponds, particularly the second transition ponds, was clearly more favourable under the moderately intensive techniques using glyphosate than the least intensive method of flooding natural vegetation (Krüger, 1993). An equivalent increase in the biomass of periphyton communities after treating water with glyphosate was noted by Austin et al. (1991).

If vegetation is first treated with glyphosate there is no need to flood the ponds in early spring to boost the abundance of food organisms. The problem of unreliable water supplies is reduced. It further eliminates

Table 1. Management techniques evaluated

	Technique	Condition of pond	Date of spraying or cultivation	Date of flooding
	1992: first transition ponds			
1	bare fallow until flooding	bare soil	final cultivation 17 May	24 May
2	natural vegetation (control)	disused 3 years	none	24 May
3	glyphosate (1800 g a.i./ha)1 on natural vegetation	disused 3 years	sprayed 21 May	24 May
	1992:second transition ponds			
1	bare fallow until flooding	bare soil	final cultivation 9 June	16 June
2	glyphosate (2% solution)2 on natural vegetation	disused 3 years	sprayed 9 June	16 June
3	glyphosate (2% solution)2 on emergent weeds in flooded ponds	disused 3 years	sprayed 9 June	end March
4	natural vegetation until flooding	disused 3 years	none	end March
	1993: first transition ponds			
1	bare fallow until flooding	bare soil	final cultivation 8 June	9 June
2	spring triticale mulch sown in fallow	green mulch	none	9 June
3	glyphosate (1800 g a.i./ha)1 on spring triticale mulch sown in fallow	green mulch	sprayed 5 June	9 June
4	glyphosate (1800 g a.i./ha)1 on natural vegetation	disused 3 years	sprayed 5 June	9 June
5	natural vegetation	disused 3 years	none	9 June
	1993: second transition ponds			
1	spring triticale mulch sown in fallow	green mulch	none	5 July
2	glyphosate (1800 g a.i./ha)1 on spring triticale mulch sown in fallow	green mulch	sprayed 2 July	5 July
3	glyphosate (1800 g a.i./ha)1 on natural vegetation	disused 3 years	sprayed 2 July	5 July

[1] Glyphosate applied as the standard 360 g a.i./l formulation using:
[1] boom sprayer at 300 l/ha total spray volume
[2] knapsack sprayer
[3] *Triticum × Secale*

Table 2. Fish management. The carp fry were introduced and caught at different times in each year, and were fed using different regimes.

Ponds	Feeding	Fry introduced	No. fry introduced	Fry harvest
1992				
1st transitional	manure 24 May	1 June	500,000/ha	6–8 June
2nd transitional	artificial from 20 July	7 July	25,000/ha	5–9 October
1993				
1st transitional	none[1]	16 June	500,000	17 July
2nd transitional	artificial from 6 August	22 July	10,000	13 Sept

[1] One pond (replicate) in each treatment was fed artificially.

the need for regular mechanical cultivation, which is difficult due to the difficulty of access for machinery to the bottom of the ponds. The yield of carp is increased, which justifies the new technique financially. In 1992 the cost of Roundup was about 200 000 zloty per litre. The net cost of 5 l ha^{-1} would be one million zloty.

Table 3. Results 1992. The biomass of food organisms, and the yield of carp were measured for each of the management techniques.

	Food organisms biomass mg/litre			Yield of carp			
	Rotatoria	*Copepoda*	*Cladocera*	% survival	no. fish per ha	av. weight g/fish	catch kg /ha*
First transition ponds		**stocking rate 500,000/ha**					
1 bare fallow	1.40	1.25	5.03	21.0	105095	5.0	519
2 natural vegetation	0.50	1.90	1.85	34.5	176075	2.6	457
3 natural vegetation + glyphosate	1.87	3.82	2.41	38.5	191450	3.0	569
Second transition ponds		**stocking rate 25,000/ha**					
1 bare fallow	4.25	39.16	2.67	83.8	20962	43.3	908
2 glyphosate pre summer flooding	0.67	38.43	2.10	81.9	20485	43.8	898
3 glyphosate after spring flooding	2.18	51.43	1.32	82.4	20620	36.7	758
4 untreated after spring flooding	10.27	37.77	1.70	24.9	6247	38.8	243

* Catch is not an exact multiple of survival and weight due to rounding off.

Table 4. Results 1993. The biomass of food organisms and the yield of carp were measured for each of the management techniques.

	Food organisms biomass mg/litre			Yield of carp			
	Rotatoria	*Copepoda*	*Cladocera*	% survival	no. fish per ha	av. weight g/fish	catch kg /ha*
First transition ponds		**stocking rate 500,000/ha**					
1 bare fallow	1.07	0.003	0.03	73.2	366250	0.71	267
2 spring triticale	0.46	0.003	0.11	84.7	423750	0.81	340
3 spring triticale + glyphosate	1.07	0.002	0.00	60.0	300000	1.06	318
4 natural vegetation + glyphosate	0.99	0.045	0.45	71.5	357500	1.05	373
5 natural vegetation	1.85	1.119	0.04	34.5	172500	0.95	163
Second transition ponds		**stocking rate 10,000/ha**					
1 spring triticale	0.215	7.40	2.13	70.5	7055	46	322
2 spring triticale + glyphosate	0.06	17.06	3.22	32.8	3280	55	181
3 natural vegetation + glyphosate	0.10	14.81	4.51	70.6	7061	53	375

* Catch is not an exact multiple of survival and weight due to rounding off.

For summer fry from the first transition ponds, at a price of 100 zloty per fish, an increase in production of 10 000 fry per hectare is needed to meet the cost of the herbicide. This was easily achieved in both years of the experiments. The sales value of carp was 25 000 zloty per kilogramme in 1992, so an increased final yield of 40 kg ha^{-1} would be needed to compensate for the cost of herbicide. This was achieved in the production of the second transition ponds in 1992 and 1993.

The use of a green mulch of spring sown triticale (*Triticum* × *Secale*) was evaluated in 1993. The results were promising, although the yield of carp fry was not as great as in the ponds in which natural vegetation was treated with glyphosate. The catch was greater when the triticale was flooded directly than if it had been sprayed with glyphosate. This may be caused by the relatively rapid desiccation of the young triticale by glyphosate, compared to the slower rotting of the untreated triticale after flooding, which would have been less beneficial for the development of food organisms. The more established and varied natural vegetation in the other ponds was less rapidly affected by the herbicide.

When the ponds are flooded 3 to 7 days after treatment with glyphosate, the majority of the herbicide is expected to remain undegraded. Much of this will have been inactivated by adsorption on to sediment or soil (Tooby, 1985). The results of these studies clearly show that any remaining glyphosate was not deleterious to the development of the invertebrate food organisms or the carp fry themselves. This supports previous field work in which fish were unaffected by exposure to glyphosate (Tooby, 1985; Mitchell et al., 1987).

Recently, Monsanto has identified a superior surfactant technology to enhance the safety characteristics of its glyphosate formulation (Clemence & Merritt, 1989). It performs equally to standard glyphosate, but is non-irritant to the skin and eyes, and shows very high margins of safety to aquatic life. For example, the LC_{50} for carp is greater than 895 mg l^{-1} of glyphosate, the highest concentration tested, compared only to 5.8 mg l^{-1} for the standard formulation used in the experiments reported in this paper. In future, glyphosate will be used with even more confidence in fisheries and other environmentally sensitive situations.

References

Austin, A. P., G. E. Harris & W. P. Lucey, 1991. Impact of an organophosphate herbicide (glyphosate) on periphyton communities developed in experimental streams. Bull. Envir. Contam. Toxicol. 47: 29–35.

Brønstad, J. O. & H. O. Friestad, 1985. Behaviour of glyphosate in the aquatic environment. In E. Grossbard & D. Atkinson, The Herbicide Glyphosate. Butterworths, London: 200–205.

Chen, Y., H. Chaing, L. Wu & Y. Wang, 1989. Residues of glyphosate in an aquatic environment after controlling water hyacinth *Eichhornia crassipes*. Weed Res, Japan 34: 117–122.

Clemence, T. G. A. & C. R. Merritt, 1989. New glyphosate formulations set new standards of operator and environmental safety. British Crop Protection Conference – Weeds 1993: 359–364.

Comes, R. D., K. F. Bruns & A. D. Kelley, 1976. Residues and persistence of glyphosate from irrigation water. J. Agric. Food Chem. 25: 517.

Feng, J. C., D. G. Thompson & P. E. Reynolds, 1990. Fate of glyphosate in a Canadian forest watershed 1. Aquatic residues and off target deposit assessment. J. Agric. Food Chem 38: 1110–1118.

Garnett, R. P., G. Hirons, C. Evans & D. O'Connor, 1992. The control of *Spartina* (cord-grass) using glyphosate. Aspects of Applied Biology 29: Vegetation management in forestry, amenity and conservation areas: 359–364.

Goldsborough, L. G. & A. E. Beck, 1989. Rapid dissemination of glyphosate in small forest ponds. Arch. Envir. Contam. Toxicol. 14: 537–544.

Krüger, A., 1993. Improvement of the technology of rearing summer and autumn carp fry using a herbicide roundup. Monsanto Agricultural Company, Komunikaty Rybackie I/1993: 1–6.

Krüger, A. & G. Okoniewska, 1994. Experiments on the use of Roundup for green manure management in production of summer and autumn carp fry. Komunikaty Rybackie 1/1994: 1–6.

Mitchell, D. G., P. M. Chapman, T. J. Long, 1987. Seawater challenge testing of coho salmon smolts following exposure to Roundup herbicide. Environ. Toxicol. Chem. 6: 875–878.

Seddon, J. C., 1981. The control of aquatic weeds with the isopropylamine salt of N phosphono methyl glycine. In Association of Applied Biologists Symposium: Aquatic Weeds and their Control: 141–148.

Sprankle, P., W. F. Meggitt & D. Penner, 1975. Adsorption, mobility and microbial degradation of glyphosate in soil. Weed Sci. 23: 229–234.

Tooby, T., 1985. Fate and biological consequences of glyphosate in the aquatic environment. In E. Grossbard & D. Atkinson (eds), The Herbicide Glyphosate.Butterworths, London, 206–217.

Hydrobiologia **340:** 259–263, 1996.
J. M. Caffrey, P. R. F. Barrett, K. J. Murphy & P. M. Wade (eds), Management and Ecology of Freshwater Plants.

Glyphosate in fisheries management

J. M. Caffrey
Central Fisheries Board, Mobhi Boreen, Glasnevin, Dublin 9, Ireland

Key words: glyphosate, weed managment, reed control, fisheries, *Schoenoplectus lacustris*

Abstract

Glyphosate is the active ingredient of the broad-spectrum, translocated herbicide 'Roundup'. Glyphosate is cleared for safe use in or near watercourses, being rated virtually non-toxic by the World Health Organisation. Trials in and alongside Irish fishery watercourses first commenced in 1989 and are continuing to date. The aim of this work is to evaluate the product's efficacy in clearing nuisance 'reed' species in recreational fisheries. The longevity of control and impact on the habitat and its fauna is also investigated. Trials in canal fisheries have demonstrated the capacity of glyphosate to remove obstructive stands of reeds (mainly *Schoenoplectus lacustris*, *Glyceria maxima*, *Phragmites australis*, *Sparganium erectum* and *Typha latifolia*), so creating reed-free areas and swims for anglers. These swims remained open for three years following a single application. In 1992 a trial over a 3 km length of the River Boyne, a renowned salmonid fishery, was undertaken. The results clearly demonstrated the ability of glyphosate to provide long-term control of dense (354 shoots m^{-2}) *Schoenoplectus* infestations in a large watercourse. In the year following, less than one shoot per m^2 was present in the channel. In 1994 a small increase in density (7.6 shoots m^{-2} was recorded, so enabling unobstructed angling in a stretch of river that had been virtually unfishable for years. Trout (*Salmo trutta* L.) and salmon (*Salmo salar* L.) also used the newly exposed gravels for spawning in the winter of 1993, thereby improving fish recruitment and production in the fishery.

Introduction

Dense instream stands of emergent monocotyledonous aquatic plant species (here, collectively termed 'reeds') pose serious problems for the beneficial use of affected lakes, rivers, canals, irrigation channels and drainage ditches (Haslam, 1987; Caffrey, 1990; Murphy et al., 1990). In Ireland, the increasing incidence of obstructive 'reed' stands in amenity watercourses has caused widespread concern because of their adverse impact on amenity exploitation and on the consequent loss of local and national tourist revenue. In particular, abundant 'reed' growth in prime angling waters throughout the country has presented difficulties for fishery managers, both in terms of reduced angler exploitation and in significantly reduced spawning recruitment and fish productivity in the affected fisheries (Caffrey, 1991).

Clearing nuisance vegetation by traditional manual or mechanical methods is both labour intensive and costly, whilst providing only limited and short-term control (Caffrey, 1993). For this reason, preliminary trials using glyphosate were undertaken. Results presented by researchers in Europe and the U.S. (e.g. Evans, 1978; Tooby, 1981; Bronstad & Friestad, 1985; Goldsborough & Beck 1989; Smith et al., 1993) demonstrated the efficacy of this herbicide in controlling a broad spectrum of emergent and floating-leaved aquatic plant species, whilst also highlighting its favourable toxicological and environmental profile.

To evaluate the efficacy of glyphosate in Ireland, small-scale preliminary trials were initiated on the Grand Canal near Ferbane (Grid Ref. N 11 24) in August 1989. Plots (c. 10 m^2) which supported monodominant and mixed stands of *Schoenoplectus lacustris* L., *Phragmites australis* (Cav.) Trin. ex Steud., *Sparganium erectum* L., *Typha latifolia* L. and *Glyceria maxima* (Hartman) Holm. were treated at the recommended rate of 1.8 kg ai ha^{-1} (5 l ha^{-1} formulated product in a water volume of 200 l ha^{-1}). Assessments conducted in the following summers showed a virtual

100% control in most treated plots (Caffrey & Monahan, 1992, 1993). The success of these trials led to the establishment in 1992 of a large-scale trial on a densely infested section of the River Boyne, a highly prized salmon and brown trout fishery.

The objective was to evaluate the efficiency with which glyphose cleared the target vegetation and to determine the longevity of effective control following a single treatment. Implications for the management of fishery watercourses are discussed.

Table 1. Estimated percentage cover and mean shoot density ($N = 30$) of *Schoenoplectus lacustris* prior to (1992) and after (1993, 1994) glyphosate treatment.

	% cover	No. shoots m^{-2}
1992	75–100	354.4 (142)
1993	1	0.9 (2.8)
1994	5	7.6 (21.7)

Study area

The study site was located immediately upstream of Navan (Grid Ref. N 87 68) on the River Boyne. The Boyne is one of Ireland's larger catchments, draining an area of 2500 km^2; limestone is the dominant rock type in the catchment. A major arterial drainage programme in this catchment was commenced in 1969 and continued until 1985. The effect of this, in combination with siltation from peat harvesting operations upstream of Navan and of the effects of eutrophication in the river may have collectively contributed to the proliferation of reed species in the main channel over past years. A 3 km section of the River Boyne was selected for study. The river in this section was c. 35 m wide, with a mean water depth of 1.2 m and a mean summer flow velocity of 30c s^{-1}. The substrate comprised deep deposits (to a maximum of 1.3 m) of coarse sand and silt, overlying coarse gravel shoals. The bank were relatively high (1–2 m) and, in places, densely tree-lined. Prior to the expansion of the reed populations in the river, this section was intensively fished for brown trout and salmon.

Materials and methods

The dominant plant species over the trial site was *Schoenoplectus lacustris*, although stands of *Sparganium erectum* were locally abundant. The emergent foliage present in the designated section of river was sprayed at 5 l ha^{-1} formulated glyphosate in August 1992. No additional surfactant was required. At this time of year water levels are low, so making the reed beds accessible by wading. In addition, most of the plants have flowered. This is considered to be the optimum time for effective glyphosate application (Evans, 1978). The herbicide was applied using conventional knapsack sprayers fitted with nozzles which enabled a swath width of 1.5 m to be treated. In order to effectively treat all the foliage at the trial site, it was necessary to subdivide the area into manageable stretches which could be sprayed in one pass by three operatives working side-by-side. Thus, contiguous plots measuring c. 100 × 12 m were marked and individually treated until the operation was completed.

Prior to treatment, the shoot density per unit area was recorded from 30 randomly selected quadrats, each of 1 m^2. Results were similarly assessed during the summers of 1993 and 1994, and details relating to the plant community and physical characteristics of the river bed were also recorded. Facts relating to angler utilisation and angler satisfaction were obtained from members of the Navan and District Angling Association.

Results

For a number of years prior to 1992 long sections of the River Boyne were totally overgrown with *Schoenoplectus lacustris*, predominantly in its emergent form. In parts of the study channel this plant presented c. 100% cover, with no water being visible beneath the tall shoots. Overall cover within the 3 km section was c. 75%. This abundant plant cover is reflected in the mean shoot density figures recorded immediately pre-treatment in 1992 (Table 1).

Following spraying, in late-September 1992, while the shoots were discoloured, none had wilted or decayed. This reflects the slow-acting nature of glyphosate, as observed in similar trials conducted at the Aquatic Weed Research Unit (P. Barrett & J. Newman, pers. comm.).

A site assessment in August 1993 revealed a very different physiognomy to that present in previous years. On this occasion, the channel was practically devoid of emergent shoots and only a few isolated stands of *S. lacustris* were present. Percentage cover decreased from a mean value of c. 75% in 1992 to c. 1% in the year following treatment. Shoot density on this occasion was at 0.9 m^{-2}. A mosaic of submerged, trailing stands of *S. lacustris* now occupied the river bed. In total, this submerged vegetation cover occupied c. 5% bottom cover. Where these stands grew, relatively deep accumulations of sand and silt were present. Elsewhere, the erosive winter floods had scoured much of the fine substrate that had previously been consolidated within and beneath the emergent reed stands. Clean, coarse gravel was visible where the sand and silt had been removed. In places, deeply anchored and intertwined rhizome masses were visible. The winter floods had not been sufficiently erosive to dislodge these dead rhizomes and they continued to accumulate and consolidate deep islands of fine substrate.

Two years after treatment, the percentage cover of emergent shoots within the 3 km trial section had increased from c. 1% in 1993 to c. 5% in 1994. This compares with a cover of c. 75% in 1992. Shoot density increased to 7.6 ± 21.7 m^{-2}, although long sections of channel supported no emergent shoots. On this occasion the percentage bottom cover with submerged *S. lacustris* vegetation increased from c. 5% to c. 15%. Some of the stands were dense and trailed just beneath the water surface. Deep silt shoals were present beneath this vegetation. Where the flow velocity was slow (< 20c s^{-1}) small numbers of emergent shoots were present within these stands.

In areas where gravels had been exposed, small stands of *Ranunculus penicillatus* (Dumort) Babs. and *Potamogeton crispus* L. were recorded. In slightly less erosive areas, stands of *Callitriche stagnalis* agg. and *Zannichellia palustris* L. were present. Trailing filaments of *Cladophora glomerata* L. (Kutz.) clung to the exposed gravels and to the submerged foliage present in the channel. Deep cushions of *Fontinalis antipyretica* Hedw. and *Rhynchostegium riparioides* (Hedw.) Rich. covered exposed rocks in areas of brisk flow.

Discussion

The fact that a single application of glyphosate can effectively control a wide range of emergent 'reeds' in aquatic situations is well documented (e.g. Seddon, 1981; Evans, 1982; Barrett, 1976, 1985; Bovey, 1985; Barrett & Gibson, 1990; Caffrey, 1993) and is supported by results from the present study. Less information, however, is available on the longevity of control. Results from the River Boyne trial show that a high level of control can be achieved over at least two growing seasons following a single correctly timed application of glyphosate. This is supported by results from other trials on *S. lacustris* and on other emergent species in Irish canals where effective control for three years was achieved following a single August application (Caffrey & Monahan, 1993). Based on the findings from the 1994 assessment on the River Boyne, it is probable that an increase in both shoot cover and density will be recorded in 1995, although this is unlikely to adversely affect the beneficial use of the watercourse.

To maintain the channel in its current relatively reed-free condition, it would be advisable to re-treat any emergent shoots during the summer of 1995. This should be a relatively simple operation in the absence of the dense obstructive stands that previously occupied the river. Interim spraying of emergent shoots every two to three years should keep the vegetation under control and ensure optimum exploitation of the river as an amenity resource.

The application technique adopted for this trial was cumbersome and labour intensive. It was, however, necessary to ensure that all the vegetation was treated over the same period of application. Under different circumstances this operation might more efficiently be conducted over two seasons. In the first year operatives fitted with chest waders and knapsack sprayers could treat 4–5 m wide strips along the length of the channel. The following season these strips should be clear of vegetation, so enabling boats fitted with spray booms to travel unobstructed whilst spraying the remaining foliage on either side.

Although glyphosate is translocated, only those plants on which the spray actually lands will be killed, so enabling partial or localised treatment in aquatic situaitons. This enables the water or fishery manager to tailor the control operation to meet the specific requirements of any particular watercourse under his control. In this manner, marginal 'reed' stands may be left untreated and maintained as refugia for fish and wildlife. The untreated vegetation may also stabilise banksides or, ultimately, provide secondary banks. These banks may then reduce the river cross-sectional area, so speeding flow in the main channel. Glyphosate may also be used for the benefit of anglers to provide vantage points for fishing key pool or swim areas. Likewise, islands of vegetation may be left at specific locations in the channel to guide water flow along a predetermined course or to create a meander in an otherwise straight section of river. Thus, judicious weed treatment may serve to rehabilitate drained or otherwise impacted rivers.

Since the late seventies long, 'reed'-infested sections of the River Boyne have been unfishable for most of the angling season. Under these conditions, relatively unobstructed angling is only possible from February to mid-May, by which time shoots emerge and interfere with the free movement of anglers' baits and lures. As the growing season advances, the 'reed' stands can become so dense that the passage of fish becomes impeded. Furthermore, even after the dead shoots are washed from the river in late-Autumn, deep deposits of sand and silt, bound by masses of well anchored rhizomes, continue to cover original salmon spawning gravels. The inaccessibility of these spawning beds to both salmon and trout represents a significant loss of recruitment to the River Boyne fishery and seriously interferes with the overall productivity of this important, national angling resource (O'Reilly, 1991).

The scouring action of major floods during the winter and spring following herbicide treatment not only removed the dead shoots from the channel but, in so doing, exposed large areas of gravel. With little reed regrowth in the treated channel in 1993 and 1994, the river was fishable throughout the whole angling season (February to September). Members of the Navan and District Angling Association and the Game Angling Adviser to the Central Fisheries Board reported quality brown trout in this channel, of a standard not witnessed for a decade and a half. Another beneficial aspect of the glyphosate treatment was that, in the winter of 1993–94, brown trout and salmon were observed to spawn in the newly exposed gravels. From a fisheries management viewpoint, this is an important and very favourable result. Efforts should now be focused on maintaining the gravels in a clean, silt-free condition whilst also working to remove deeply anchored rhizome masses in other sections, and so expose larger areas of spawning habitat.

Concern that filamentous algae might proliferate in the treated section following the removal of *S. lacustris* was unfounded and, while small carpets of *Cladophora glomerata* were recorded in 1993 and 1994, abundant growth did not develop.

Whilst glyphosate is recognised as having a low toxicity to man and a minimum harmful effect on wildlife (Atkinson, 1985), the surfactant in the standard formulation used for the present trial is regarded as being more toxic to aquatic organisms and more hazardous to operators than the active ingredient itself (Clemence & Merritt, 1993). Recent advances in surfactant technology have resulted in the production of two new glyphosate formulations which have improved operator and environmental safety characteristics. That the new formulations are no longer classified as 'hazardous', either to operator safety or to aquatic life, represents a large improvement in safety margins over the standard product (Garnett, pers. comm.).

Acknowledgments

Thanks are due to the Navan and District Anglers Association who part funded the trial and who provided angling information in relation to the fishery. Ms C. Monahan and other C.F.B. staff are also gratefully acknowledged for their considerable field assistance. Thanks are also due to Dr P. Fitzmaurice for his editorial comments.

References

Atkinson, D., 1985. Toxicological properties of glyphosate – a summary. In E. Grossbard & D. Atkinson (eds), The Herbicide Glyphosate. Butterworths, London: 127–133.

Barrett, P. R. F., 1976. The effect of dalapon and glypohsate on *Glyceria maxima*. Proc. Brit. Crop Prot. Conf. - Weeds: 79–82.

Barrett, P. R. F., 1985. The efficacy of glyphosate in controlling aquatic weeds. In E. Grosslard & D. Atkinson (eds), The Herbicide Glyphosate. Butterworths, London.

Barrett, P. R. F. & M. T. Gibson, 1990. The control of some emergent weeds with glyphosate. Proc. EWRS 8th Symp. on Aquat. Weeds: 21–25.

Bovey, R. W., 1985. Efficacy of glyphosate in non-crop situations. In E. Grossband & D. Atkinson (eds), The Herbicide Glyphosate. Butterworths, London: 435–448.

Bronstad, J. O. & H. O. Friestad, 1985. Behaviour of glyphosate in the aquatic environment. In E. Grossbard & D. Atkinson (eds), The Herbicide Glyphosate. Butterworth, London.

Caffrey, J. M., 1990. The classification, ecology and dynamics of aquatic plant communities in some Irish rivers. Ph.D. Thesis, University College, Dublin, 254 pp.

Caffrey, J. M., 1991. Aquatic plant management in Irish rivers. In M. Steer (ed.), Irish Rivers: Biology and Management, Royal Irish Academy, Dublin: 85–98.

Caffrey, J. M., 1993. Aquatic plant management in relation to Irish recreational fisheries development. J. aquat. Plant Mgmt 31: 162–168.

Caffrey, J. M. & C. Monahan, 1992. Aquatic plant management in Irish canals. Annual Report, 1991. Office of Public Works commissioned report, Central Fisheries Board, Dublin, 69 pp.

Caffrey, J. M. & C. Monahan, 1993. Aquatic plant management in Irish canals. Annual Report, 1992. Office of Public Works commissioned report, Central Fisheries Board, Dublin, 53 pp.

Clemence, T. G. A. & C. R. Merritt, 1993. New Roundup herbicide formulations set new standards of operator and environmental safety. Brighton Crop Protection Conf. - Weeds: 1337–1340.

Evans, D. M., 1978. Aquatic weed control with the isopropylamine salt of N-phosphonomethyl glycine. Proc. EWRS 5th Internat. Symp. on Aquatic Weeds: 171–178.

Evans, D. M., 1982. *Phragmites* control with glyphosate through selective equipment. Proc. EWRS 6th Symp. on Aquat. Weeds: 209–211.

Goldsborough, L. G. & A. E. Beck, 1989. Rapid dissemination of glyphosate in small forest ponds. Arch. of Envir. Contam. and Tox. 14: 537–544.

Haslam, S. M., 1987. River plants of Western Europe. Cambridge Univeristy Press, Cambridge, 512 pp.

Murphy, K. J., T. O. Robson, M. Arsenovic & W. van der Zweerde, 1990. Aquatic weed problems and management in Europe. In A. Pieterse & K. J. Murphy (eds), Aquatic Weeds, the Ecology and Management of Nuisance Aquatic Vegetation. Oxford University Press, Oxford: 295–317.

O'Reilly, P., 1991. Trout and salmon rivers of Ireland. Merlin Unwin Books, London, 327 pp.

Seddon, J. C., 1981. The control of aquatic weeds with the isopropylamine salt of N-phosphonomethyl glycine. Aquat. Weeds and their Contr., Association of Applied Biologists: 141–148.

Smith, B. E., D. G. Shilling, W. J. Haller & G. E. MacDonald, 1993. Factors influencing the efficacy of glyphosate on Torpedograss (*Panicum repens* L.). J. aquat. Plant. Mgmt 31 : 199–202.

Tooby, T. E., 1981. Predicting the direct toxic effects of aquatic herbicides to non-target organisms. Conf. on Aquat. Weeds and their Contr., Association of Applied Biologists: 265–274.

Hydrobiologia **340**: 265–269, 1996.
J. M. Caffrey, P. R. F. Barrett, K. J. Murphy & P. M. Wade (eds), Management and Ecology of Freshwater Plants.

The use of herbicides for weed control in flooded rice in North Italy

A. C. Sparacino, S. Bocchi, R. Ferro, N. Riva & F. Tano
Istituto di Agronomia, via Celoria 2, 20133 Milano, Italy

Key words: flooded rice, weed control

Abstract

The use of chemicals to control weeds in flooded rice in the Po Valley, Italy is restricted by limited knowledge of potentially useful herbicides. Different strategies of chemical weed control, including different application timings of several types of herbicides are compared based on effectiveness and selectivity. *Echinochloa* spp., the most common weed was, in the experimental conditions, better controlled with early applications (pre-sowing and pre-emergence) of new herbicides which were effective substitutes for molinate, the use of which is no longer allowed in some Italian areas. *Alisma plantago-aquatica, Bolboschoenus maritimus* and *Schoenoplectus macronatus* were controlled by different types of herbicides applied post-emergence. Cinosulfuron (sulfonylurea) showed the highest effectiveness and selectivity in controlling the more common *Hetherantera* spp., even when applied in very low amounts.

Introduction

In Italian flooded rice fields the main weed species are *Echinochloa* spp. (*Gramineae*) and *Alisma plantago-aquatica* (Alismataceae), *Bolboschoenus maritimus* and *Schoenoplectus mucronatus* (Cyperaceae) (Ferrero & Sparacino, 1988). *Heteranthera reniformis* and *H. limosa* (Pontederaceae) are spreading within the area, together with red rice (Sparacino & Sgattoni, 1993). Large amounts of herbicides are used to control these weeds: the quantities employed being around three times those used in other crops. Ninety per cent of the total amount of herbicide (around 10 kg a.i. ha^{-1} of active chemical) consists of graminicides (Bassi, 1993).

Since 1987 problems have arisen due to restrictions on the use of molinate and bentazon in the rice growing areas of Piedmont and Lombardy, following the detection of high quantities in the ground water. The consumption of these products has decreased and herbicides with higher concentrations of active ingredients have become widely used. At present an average of two treatments per crop are carried out, one being a two-product mixture. Propanil, molinate, oxadiazon and bensulfuron-methyl were used on about 45 per cent of the rice area in 1991. Pre-sowing treatments are increasing compared to some years ago and post-emergence treatments in May, when rice has only 3–4 leaves, are anticipated. There is concern about the use of mixtures with dimepiperate, thiocarbazil, thiobencarb and molinate to control *E. crus-galli* (barnyard grass) in the 1–2 leaf stage rice and even more so about bensulfuron-methyl or pirazzosifen and triclopyr (with MCPA and propanil) used for the control of Cyperaceae, Alismataceae, Butomaceae Pontederaceae.

This study was aimed at controlling weeds efficiently when competition with rice is high. It involved testing a diverse range of herbicides (Table 1) in an

Table 1. Herbicide and dosage of active substance used at S. Alessio - 1990 at different growth stages

Herbicide	Dosage ha^{-1}	Growth Stage
Quinclorac / Bentazon	0.6 l / 2 l	
Molinate / Bentazon	4 l / 2 l	
Oxadiazon + Dimepiperate	0.5 l + 2.4 l	3 days. pre-sowing
Pirazzosifen	2.7 l	4 days post-sowing
Cinosulfuron / Molinate	80 g / 4 l	3–4 rice leaves
Bensulfuron-methyl + molinate	60 g + 4 l	1.5 - 2 rice leaves

Table 2. Herbicides and dosage of active substance used at Rosate - 1991 at different growth stages

Herbicide	Dosage ha^{-1}	Growth Stage
Molinate Cinosulfuron	4 l 80 g	3–4 rice leaves
Cinosulfuron Quinclorac	80 g 0.75 l	3–4 rice leaves
Molinate Metosulam	30 ml	1–2 rice leaves
Molinate Metosulam	60 ml	1–2 rice leaves
Molinate Bentazone	4 l 2 l	37 days after seeding
Oxadiazon + Dimepiperate	0.5 l + 2.4 l	4 days before sowing
Molinate Bensulfuron - methyl	4 l 60 g	1.5–2 rice leaves
Pirazzosifen	2.7 l	4 days before sowing

Table 3. Herbicides and dosage of active substance used at Rosate - 1992 at different growth stages

Herbicide	Dosage ha^{-1}	Growth Stage
Quinclorac /Bentazon	0.6 l / 2 l	Beginning of tillering
Molinate /Bentazon	4 l/2 l	"
Pretilachlor + Fenchlorim	1.2 + 0.6	"
Pirazzosifen	2.7 l	3 days before sowing
Cinosulfuron / Molinate	80 g / 4 l	3–4 rice leaves
Bensulfuron-methyl / Molinate	60 g / 4 l	1.5–2 rice leaves
Metosulam 30/ Molinate	30 ml / 4 l	3–4 Cyperaceae leaves
Metosulam 60 / Molinate	60 ml / 4 l	3–4 Cyperaceae leaves

attempt to avoid both accumulation of synthetic chemicals which pose environmental risks and the rise of floristic associations which are particularly difficult to control. Other significant goals were to decrease herbicide dose rates and to reduce the cost of weed control treatments.

Materials and methods

In 1990, at S. Alessio (Pavia), the effectiveness of herbicides used for the control of *Echinochloa* spp., *Heteranthera reniformis, Alisma plantago-aquatica, Bolboschoenus maritimus* and *Schoenoplectus mucronatus* was tested (Table 1). At the end of March plots (100 m^2) were divided by putting plastic sheets (ondolux) as partitions into mechanically cut furrows. A randomised block design with four replications was adopted.

The rainfall in early April caused some stagnation in the plots, promoting early emergence of *B. mar-*

Table 4. Percentage weed control at S. Alessio in 1990

	E. crus-galli			*H. reniformis*			*B. maritimus*			*S. mucr.*		*A. plantago-aquatica*		Grain
Days after application	15	30	45	15	30	45	15	30	45	30	45	30	45	yield ($t\ ha^{-1}$)
Quinclorac/ Bentazon	39 ab	39 b	76 b	-	-	-	100 c	100 c	100 c	100 c	100 b	100 b	100 b	5.04 cd
Molinate /Bentazon	100 b	100 c	100 b	-	-	-	100 c	100 c	100 c	100 c	100 b	100 b	100 b	6.38 ab
Oxadiazon + Dimepiperate	100 b	100 c	100 b	99 c	91 b	98 cd	-	-	-	-	-	-	-	5.74 cd
Pirazzosifen	100 b	93 bc	80 b	91 bc	84 b	83 b	23 a	67 b	70 b	100 c	100 b	100 b	100 b	5.94 bc
Cinosulfuron /Molinate	-	-	-	81 bc	89 b	100d	0 a	78 b	91 c	100 c	100 b	100 b	100 b	5.54 cd
Bensulfuron-methyl / Molinate	100 b	100 b	100 b	70 b	73 b	76 b	68 b	89 c	88 c	100 c	100 b	100 b	100 b	7.29 a
		plants m^{-2}		weed cover %		plants m^{-2}			plants	weed cover %				
Control	3.3	3.6	5.8	3.4	11.3	38.7	32	75	78	4.4	5.1	2.5	5.0	4.27 d

Values followed by different letters are significantly different ($P < 0.05$)

itimus which, even when controlled by a pre-sowing treatment, might have interfered with subsequent treatments. In 1991 and 1992 the trials continued at Rosate (Milano) with the addition of metosulam for controlling *Heteranthera* spp., *B. maritimus* and *S. mucronatus* (Tables 2 and 3). To avoid the early weed emergence observed in 1990, the plastic partition walls of the plots were placed in muddy soil just before sowing. In 1991 and 1992 a portable pump with three TK nozzles was used. The experimental design and plot size were the same as in 1990. In all three years, observations on weed density, percentage weed control and percentage soil cover were made for each weed species at 15, 30 and 45 days after herbicide application. Transformed data were statistically analysed using analysis of variance and significantly different means compared using the Duncan test.

Results and discussion

S. Alessio (Pavia, 1990)

The early spread of *B. maritimus* before rice sowing exerted some control on *Echinochloa crus-galli*, *Schoenoplectus mucronatus* and *Alisma plantago-aquatica*; whereas *Heterantera reniformis* increased during the season. *B. maritimus* also partly decreased the effectiveness of the treatments against the other weeds. Complete control of *Echinochloa crus-galli* during the period studied (from maximum tillering to stem elongation) was achieved with the mixture of oxadiazone + dimepiperate but only partial control using mixtures with molinate and with quinclorac/bentazon and pirazzosifen (Table 4). *Heteranthera reniformis* was controlled by cinosulfuron (100 per cent of control 45 days after treatment (DAT)) and by oxadiazon + dimepiperate (more than 98 per cent). Pirazzosifen and the mixture of pretilachlor also gave a good control of *H. reniformis* but bensulfuron was less effective. The density of *B. maritimus* ranged from 32 to 78 plants m^{-2} during the period studied. Bensulfuron provided the quickest and best control of this weed (68 per cent of control at 15 DAT and 89 percent at 45 DAT) (Table 4). Cinosulfuron and pirazzosifen had no clear effect on *B. maritimus* until one month after treatment and were then less effective than bensulfuron methyl. However, 45 DAT cinosulfuron and bensulfuron both gave about 90 percent control. *S. mucronatus* and *A. plantago-aquatica* infestations were completely controlled by all the chemicals, cinosulfuron, bensulfuron and pirazzosifen being active at the earliest observation time. Molinate and bentazon completely controlled all the *Echinochloa* spp. and all the dicotyledons weeds. Grain yield was significantly improved by reducing competition through weed control. The best yield (7.28 t ha^{-1}) was obtained with bensulfuron-methyl + molinate, which controlled all the dicotyledons weeds well. Good results were also obtained with molinate and bentazon, whereas all other chemicals gave less satisfactory control.

Rosate (Milano) 1991

The main infestation in these plots was of *Echinochloa crus-galli* and *H. reniformis* with a presence of *B. maritimus* and *S. mucronatus* (Table 5). Pirazzosifen applied four days before sowing controlled *E. crus-galli* most effectively (around 100%). The lower effectiveness of molinate and metosulam applied at the 1–2 leaf stage of *E. crus-galli* (barnyard grass) could be due to low temperatures during the period of treatment. The oxadiazon + dimepiperate mixture controlled *Echinochloa* spp. well during rice tillering but less during the stem elongation phase.

The ground cover by *H. reniformis* in untreated plots increased from 55 per cent (June 21) to 92 per cent (July 21). As in the previous year cinosulfuron and the oxadiazon + dimepiperate mixture controlled *H. reniformis* best, as in the previous year and good results were obtained also with pirazzosifen and pretilachlor + fenclorim, although their effectiveness reduced with time. Metosulam did not exert good control at any dose.

B. maritimus, which was not widespread, was controlled reasonably well by all the herbicides, confirming the good results for bensulfuron of the previous year at S. Alessio. By July 4 cinosulfuron and pirazzosifen had reduced the infestation of *B. maritimus*. Metosulam gave similar results to those obtained with the other herbicides.

Cinosulfuron, pirazzossifen, metosulam and bentazon gave the best control of *S. mucronatus*. Improvement of grain yield reached 236 per cent with pirazzosifen treatment, while the oxadiazon + dimepiperate mixture increased yield by 201 per cent. Pretilachlor + fenclorin did not completely control *Echinochloa* spp., *H. reniformis* (especially during the last phases) and *B. maritimus*, but still increased grain yield by 92 per cent. It is concluded that the effectiveness of control of the predominant species

Table 5. Percentage weed control at Rossate in 1991.

	E. crus-galli			*H. reniformis*			*B. marit.*		*S. mucr.*	Yield
Days after application	15	30	45	15	30	45	30	45	45	(t ha^{-1})
Quinclorac / Cinosulfuron	56 b	74 b	77 b	90 cd	90 cd	100 d	95 b	95 b	100	7.05 bc
Molinate / Bentazon	74 bc	78 b	62 bc	-	-	-	100 b	100 b	100	7.03 bc
Oxadiazon + Dimepiperate	98 c	97 bc	83 cd	96 cd	93 cd	91 d	-	-	-	8.18 ab
Pirazzosifen	100 c	100 c	100 d	99 d	98 d	89 d	89 b	97 c	100 b	9.15 a
Cinosulfuron / Molinate	74 bc	78 b	62 bc	93 cd	100 d	96 d	91 b	99 c	100 b	6.94 bc
Bensulfuron-methyl/ Molinate	80 b	80 b	70 bc	85 bcd	72 bc	30 b	96 b	92 c	94 b	5.74 cd
Metosulam 30/ Molinate	bc	78 b	65 bc	77 bc	57 b	11 ab	89 b	79 ab	100 b	6.24 cd
Metosulam 60 / Molinate	70 bc	70 b	70 bc	77 b	52 b	13 ab	94 b	88 b	98 b	4.97 d
		plants m^{-2}		weed cover %		plants m^{-2}	plants m^{-2}			
Control	25.3	33.8	57.5	55	85	92.3	15.8	19	6.75	2.72 e

Values followed by different letters are significantly different ($P < 0.05$).

(e.g. *E. crus*-galli or *H. reniformis*) is linked to yield increases.

Rosate (1992)

The infestation of *Echinochloa crus-galli* during 1991 was greater than in previous years reaching 47.5 per cent ground cover in the untreated plots within 15 days and 70 per cent 30 days later (Table 6). Of the products used, molinate gave the best results, 97 per cent control of *Echinochloa* spp. during the whole period. Pirazzossifen and pretilachlor + fenclorim gave fair weed control during the first phase but their effectiveness decreased later. Quinclorac was not sufficiently effective.

The infestation of *Heteranthera reniformis* was also higher than in the previous two years. Cinosulfuron gave control between 94 and 100 per cent during the experiment. The other chemicals, especially metosulam, were not sufficiently effective. Cinosulfuron also controlled *B. maritimus* well by 45 DAT when ground cover was decreased by 92.2 per cent. *B. maritimus* was also controlled by bensulfuron methyl, pirazossifen and the double dose of metosulam. All these herbicides exercised good control from the beginning. *A. plantago-aquatica* was controlled early by all the herbicides studied. In 1992 the infestation was so high that rice yield in the untreated plot was only 0.79 t ha^{-1}, and every type of control increased production. Nevertheless, the yields obtained with some herbicides such as pirazzosifen (3.54 t ha^{-1}) quinclorac (2.88 t ha^{-1}) and prelilaclor + fenclorim (3.27 t ha^{-1}) were quite low, probably because of their low efficacy in controlling *Echinochloa* spp. (quinclorac) or *H. reniformis* (pretilachlor + fenchlorim and pirazzossifen). The highest yield and greatest weed control was obtained with cinosulfuron (9.31 t ha^{-1}) and good yields were produced by bensulfuron + molinate (6.34 t ha^{-1}) and metosulam (c. 6 t ha^{-1}) which did not completely control *H. reniformis* and *B. maritimus*.

Conclusions

The weed flora in the experimental rice fields used in 1990–92 was typical, consisting of *E. crus-galli, B. maritimus, S. mucronatus, A. plantago-aquatica* and *H. reniformis,* the former species being the most important weed reducing grain production in Italy. Molinate, traditionally used to control it, can now be integrated with or replaced by lower dosage herbicides to avoid problems caused by its use in some environments. Among the chemicals effective against *Echinochloa* are oxadiazone + dimepiperate and pirazzosifen: the first mixture also providing effective control for *H. reniformis*; while the pirazzosifen had a wide spectrum of action. Pretilachlor + fenclorim and quinclorac were not always able to control *E. crus-galli*: the first showed a limited residual action and the latter had little effect during early weed growth.

In addition to the traditional bentazon some other products, with less impact on the environment, provided satisfactory control of Cyperaceae and Alismataceae weeds. The best control of Cyperaceae, Alismataceae and Pontederaceae was obtained with cinosulfuron, with bensulfuron and pirazzosifen also being effective. Metosulam showed good control of *A. plantago* and Cyperaceae but its effect on *H. reniformis* was discontinuous and short. General analysis of the results shows that the best yields were more regularly obtained with pirazzossifen, oxadiazon + dimepiperate, bensulfuron + molinate, cinosulfuron + molinate and the tra-

Table 6. Percentage Weed control at Rosate in 1992

Days after application	*E. crus-galli* 15	30	45	*H. reniformis* 15	30	45	*B. maritimus* 15	30	45	*A. plantago-aquatica* 15	30	45	Grain Yield (t ha^{-1})
Quinclorac / Bentazon	-	44 b	27 b	-	-	-	90 c	95 c	95	100 b	100 b	100 b	2.88 b
Molinate / Bentazon	99 c	98 d	97 d	-	-	-	95 c	95 c	100	100 b	100 b	100 b	4.75 b
Pretilachlor + Fenchlorim	84 b	78 c	68 c	53 b	70 b	48 b	-	-	-	96 b	97 b	100 b	3.27 b
Pirazzosifen	81 b	80 c	63 c	1.5 a	0.a	0.a	52 ab	59 b	65 b	99 b	99 b	99 b	3.54 b
Cinosulfuron / Molinate	95 c	95 d	95 d	94 c	100 c	100 c	73 b	82 b	92 c	79 b	100 b	100 b	9.31 d
Bensulfuron-methyl / Molinate	95 c	97 d	96 d	52 b	62 b	53 b	65 b	71 b	77 bc	76 b	98 b	100 b	6.34 c
Metosulam 30/Molinate	95 c	100 d	100 d	19 a	5.a	8.a	31 ab	43 ab	72 bc	100 b	100 b	100 b	5.57 c
Metosulam 60/ Molinate	95 c	100 d	100 d	13 a	5.a	0.a	59 b	72 b	79 c	100 b	100 b	100 b	6.27 c
		plants m^{-2}		weed cover %		plants m^{-2}		weed cover %					
Control	47.5	57.5	70	85	100	100	32.5	37.5	43.8	18.8	24	27	0.79 a

Values followed by different letters are significantly different ($P < 0.05$).

ditional mixture, molinate + bentazon. The possibility to use different molecules and systems for better control of the complex weed flora in flooded rice fields could enable farmers to maintain high yield levels without causing problems due to selection of tolerant weeds or unacceptable environmental impact as in the past when few herbicides were used. Crop rotation, when possible, might also reduce pollution risks. It is important to improve knowledge of the biology of weeds to better define the period of highest competition between rice and weeds, in relation to rice growth rate and weather condition, when control effectiveness should be maximum.

References

Bassi, A., 1993. Evoluzione del consumo di diserbanti. L'Inf.Agr. 7: 27–34.

Ferrero, A. & A. C. Sparacino, 1988. Lotta alle infestanti delle colture erbacee in Lombardia. Atti Conv. Laureato in Sc. Agr. Cremona, 16 Sett. 33–43.

Sparacino, A. C. & P. Sgattoni 1993. Le erbe infestanti il riso. L'Inf.Agr. 7, 1993: 37–55.

Hydrobiologia **340:** 271–275, 1996.
J. M. Caffrey, P. R. F. Barrett, K. J. Murphy & P. M. Wade (eds), Management and Ecology of Freshwater Plants.

The interaction between *Cyprinus carpio* L. and *Potamogeton pectinatus* L. under aquarium conditions

N. S. Sidorkewicj, A. C. López Cazorla & O. A. Fernández
CERZOS and Departments of Biology and Agronomy. Universidad Nacional del Sur, 8000-Bahía Blanca, Argentina

Key words: Cyprinus carpio, Potamogeton pectinatus, turbidity, aquaria

Abstract

Plants and seedlings of *Potamogeton pectinatus* were obtained from tubers grown under laboratory conditions. Four plants (mean total length: 14.3 m) and two seedlings (mean height: 10.9 cm) were placed in each of twenty 100 l aquaria illuminated with fluorescent lighting. A 5 cm-thick layer of muddy sediment was then put in each aquarium together with two size-matched fish (mean size classes: 6.8, 14.1 and 23.0 g) of the species *Cyprinus carpio*. After four weeks, the total length of the plants in the control and small fish aquaria had increased by 71% and 3% respectively, whereas plant total length in the aquaria with medium and large fish had diminished by 33% and 76%, respectively. Few seedlings survived in the presence of the fish. The reduction in plant growth was associated with an increase in water turbidity in all treatments as a result of the benthic feeding habit of *C. carpio*, and of direct herbivory action in the case of medium- and large-sized fish.

Introduction

Cyprinus carpio L. (common carp) has only recently been introduced into the drainage channels (Cazzaniga, 1981) of an irrigation area in Argentina known as the Valle Inferior del Río Colorado (VIRC) (62 ° 37′W, 39 ° 23′S), where approximately 90 000 ha of cultivated land are under irrigation. This species is considered to have been introduced accidentally into some of the VIRC's drainage channels through the transfer of water from the Colorado River, where it is abundant. The first indication that carp may be responsible for environmental changes in the water medium was reported by Sabbatini (1989), when he observed that the water in several experimental vegetation sites within the drainage channels of the area had become turbid owing to suspended matter. A simultaneous reduction was observed in the growth of *Potamogeton pectinatus* L., a commonly occurring weed which causes major problems in the channels.

The objective of the present work was to test under laboratory conditions the hypothesis that the benthic feeding habit of *C. carpio*, by stirring up silt and organic particles among the loose sediment on the bottom, is responsible for the turbidity effect observed in the drainage channels. It is postulated that the turbidity of the water, plus the deposition of suspended material on leaf surfaces, limits *P. pectinatus* growth by increasing shade-stress and leads to the deterioration and even disappearance of the vegetation. An additional factor restricting plant growth could be the direct herbivore action of the fish. It is sought to gain a deeper understanding of the interaction in this particular irrigation system between the common carp and its environment, including the vegetation, with a view to incorporating a more rational management of this species into a long-term integrated weed management programme (Fernández, 1982; Fernández et al., 1987a, b). It is thereby hoped to place constraints on the growth of target weed populations.

Materials and methods

Two seedlings and four plants of *P. pectinatus*, obtained from tubers grown under laboratory conditions in

200 ml plastic pots containing soil, were placed into each of twenty 100 l glass aquaria, on a 5 cm-thick bed of mud collected from a channel bottom in the CORFO channel system of southern Argentina. Fluorescent lighting (40 W) was supplied during 12 h per day.

The total foliage length (TL) of each plant was measured by the Modified Line Intersect Method (Tennant, 1975), commonly used for the estimation of root length. Seedlings were measured in terms of their height (Table 1).

After allowing the water to clear completely, one-year-old common carp were measured, weighed, and added to the aquaria in size-matched pairs in accordance with the three size classes shown in Table 1. The experimental design thus consisted of a control and three treatments based on fish size, with five replicates. A battery of pumps provided subsuperficial aeration of the water to prevent any air stirring effect. No food was provided during the experiment.

Every week for four weeks the following environmental parameters were measured: turbidity with an AV 311 Nefelometric Turbidimeter (NTU units), conductivity (mS cm^{-1}), temperature (C), pH, and oxygen (mg l^{-1}). On completion of the experiment the plants were again measured, and fish were measured and weighed.

Data on the difference in total length of the plants ($\Delta TL = TL_f - TL_i$), based on the mean value per aquarium, were analyzed. Comparison among consecutive mean values was carried out by means of an *a priori*, one-tailed t test, following the Levene test for homoscedasticity. A regression analysis was carried out to determine the relationship between initial carp weight and the difference in plant total length. In view of the high level of dispersion and the erratic behaviour of the turbidity data, it was sought to introduce a centralizing parameter which would not be sensitive to extreme values. To this end the median and semi-interquartile range of each treatment for each week of the experiment were used.

Results

The changes in the median value of water turbidity and the semi-interquartile range can be observed in Figure 1. The initial and final values for the control and the fish treatments are shown in Table 1. Removal of the fish after four weeks resulted in the clearing of the water; keeping them in the system, on the other hand, stabilized the turbidity values at 35–40 NTU, as measured in additional aquaria kept under the same laboratory conditions as described for the experiments, and which retained fish for several months.

Electrical conductivity, temperature, pH and oxygen content remained almost constant throughout the experimental period; initial and final values are shown in Table 1. Final total length of the plants and the percentage of variation are shown in Table 1. The Levene test did not detect heteroscedasticity in the data of ΔTL ($P = 0.13$). The one-tailed t test detected differences among control-T1 ($P<0.01$), T1-T2 ($P<0.05$) and T2-T3 ($P<0.05$) pairs. The regression analysis carried out between ΔTL and the initial weight of the fish did not show any deviation from the linear adjustment ($P>0.20$); the regression coefficient was significant ($P<0.01$), and it was found that 79% of the variability in the dependent variable could be accounted for by the independent variable ($Y = 8.43 - 0.88X$).

Few seedlings survived in the small fish treatment and in the remaining treatments they disappeared altogether (Table 1). No plant remains were observed floating on the surface.

Weight and length of the fish remained practically constant for the duration of the experiment (Table 1).

Discussion

The precise effect of the presence of *C. carpio* on the water environment and on submerged vegetation is generally agreed to be negative. Some authors suggest that the carp's habit of feeding on the bottom of the channel serves to increase the turbidity of the water (Cahn, 1929; Anderson, 1950; McCrimmon, 1968; Breukelaar et al., 1994; Cline et al., 1994), indirectly causing the disappearance of the vegetation by reducing the rate of photosynthesis (Robel, 1961; Westlake, 1971; Barko et al., 1986; Engel & Nichols, 1994). Others note no significant increase in turbidity, probably because of particular soil characteristics or climatic conditions which do not favour the normal behaviour of the species; they do however report the uprooting of vegetation (Crivelli, 1983; Fletcher et al., 1985). The issue of the existence of herbivory is more controversial. Some authors maintain that the fish feed on soft-leaved vegetation in the absence of animal-based feed (for instance invertebrate benthic species), as reviewed by Alikuhni (1966), Sarig (1966) and Kantrud (1990) and stated by King & Hunt (1967), Fletcher et al. (*loc. cit.*) and Chapman & Fernando (1994). According to Sibbing (1988: 161) 'the carp is very limited in process-

Table 1. Initial and final values of physical and biological parameters.

	Control (mean ± s.d.)	T1 (mean ± s.d.)	T2 (mean ± s.d.)	T3 (mean ± s.d.)
Turbidity (NTU)$^{(*)}$				
Initial	7	10	8	13
Final	1	42	33	60
Conductivity (mS cm^{-1})				
Initial	2.1 ± 0.4	1.6 ± 0.4	1.7 ± 0.3	1.9 ± 0.4
Final	2.1 ± 0.5	1.9 ± 0.5	1.7 ± 0.2	2.0 ± 0.3
Temperature (C)				
Initial	15.3 ± 0.3	15.4 ± 0.3	15.3 ± 0.2	15.4 ± 0.2
Final	17.1 ± 1.4	17.5 ± 0.3	17.1 ± 0.5	17.8 ± 0.4
pH				
Initial	7.8 ± 0.2	7.8 ± 0.1	8.0 ± 0.2	7.9 ± 0.2
Final	8.0 ± 0.2	8.0 ± 0.1	7.9 ± 0.2	7.9 ± 0.2
O_2 (mg l^{-1})				
Initial	7.9 ± 0.3	7.8 ± 0.6	7.5 ± 0.5	7.6 ± 0.5
Final	7.2 ± 0.8	7.3 ± 0.6	7.1 ± 0.7	6.4 ± 0.9
Fish weight (g)				
Initial		6.8 ± 1.5	14.1 ± 1.4	23.0 ± 3.3
Final		6.7 ± 1.4	13.1 ± 1.4	21.6 ± 3.5
Fish total length (cm)				
Initial		8.0 ± 0.7	10.2 ± 0.3	11.8 ± 0.7
Final		8.3 ± 0.7	10.2 ± 0.4	12.0 ± 0.6
Plant total length (m)				
Initial	14.0 ± 2.6	14.2 ± 2.8	15.1 ± 3.7	13.9 ± 2.4
Final	24.2 ± 9.0	14.8 ± 6.9	10.1 ± 6.9	3.3 ± 2.7
$\%V = \frac{TL_f - TL_i}{TL_i} 100$	(+) 71 %	(+) 3 %	(−) 33 %	(−) 76 %
Seedling height (cm)				
Initial	11.7 ± 8.2	9.4 ± 6.4	9.9 ± 7.8	12.5 ± 9.5
Final	27.2 ± 7.4	3.2 ± 3.6	—	—

$^{(*)}$ Median value

ing long and struggling prey ... as well as vegetal matter, due to the lack of oral teeth and the specialized morphology of its pharyngeal chewing apparatus' so that 'the reported herbivorism illustrates its opportunism in feeding behaviour'. Nevertheless, Crivelli (1981, 1983) claims not to have found any green matter in dissected guts of 32.8–51.2 cm-long animals, and categorically dismisses the possibility that the species eats submerged macrophytes.

In the present study it was observed that the fish cause a significant increase in water turbidity owing to the suspension of particles on the water column. The rapid increase in turbidity during the first two weeks is considered to be the combined result of two circumstances: (a) the rapid movements of the fish early on in the experimental period until they became accustomed to their new environment, (b) the initial intensive dredging behaviour of the fish in their search for prey (arthropods, snails, etc.), which were easily detectable on the benthic substratum of the aquaria. Later on, when the availability of prey appeared to be exhausted, the feeding activity on the bottom acquired less significance. After four weeks the water continued to show a significant degree of turbidity, a state which appeared to take on a permanent character in the presence of fish.

Plant growth and development were seriously affected in all fish treatments, confirming that the larger the fish the greater the damage caused to the plants, but without an uprooting effect. The vegetal physiognomy varied substantially; the smaller fish appeared to affect plant development indirectly through a shade-stress effect, as indicated by the development of an upper canopy and the chlorotic, weak morphology of

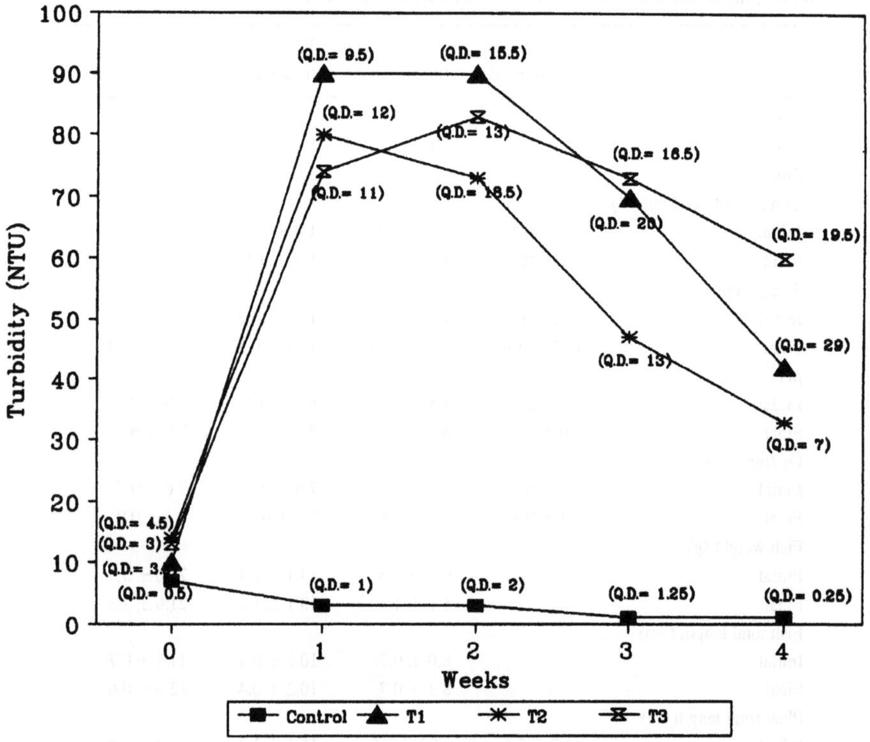

Figure 1. Median value and semi-interquartile range (Q.D.) of water turbidity for the control and the fish treatments, at each week of the experiment.

the remaining basal plant tissues. In the two larger fish treatments the damage was higher owing to probable direct herbivory, most of the plants apparently having been heavily browsed. All effects were conspicuous toward the third week of the experiment.

The seedlings suffered even greater damage than the plants, since most disappeared altogether. Unlike in the case of the plants, the side branches of seedlings placed in the aquaria with small fish showed signs of browsing, probably because of the more delicate structure of the young tissues.

Some of the experimental results of this paper agree with observations in the field. They support the hypothesis that the permanent turbidity of the water recorded in some of the drainage channels of the CORFO system over the last few years and the disappearance and deterioration of the vegetation could be a direct consequence of the presence of the carp. Further research is required to lay the foundations for the adequate management of this fish in combination with channel maintenance and other weed control techniques on a long-term basis.

Acknowledgments

The authors thank G. C. Möckel for her laboratory assistance. Financial support from the Consejo Nacional de Investigaciones Científicas y Técnicas of Argentina, Grant 1332-00-88, and that of the European Communities STD 3 Programme Contract N° TS3*-CT92-0125, is greatly appreciated. Thanks are also due to R. Camina and N. Winzer for their valuable comments and suggestions.

References

Alikuhni, K. H., 1966. Synopsis of biological data on *Common carp* (Linnaeus), 1758 (Asia and the Far East). FAO Fish. Synop. 31.

Anderson, J. H., 1950. Some aquatic vegetation changes following fish removal. J. Wildl. Mgmt 14: 206–209.

Barko, J. W., M. S. Adams & N. L. Clesceri, 1986. Environmental factors and their consideration in the management of submersed aquatic vegetation: a review. J. Aquat. Plant Mgmt 24: 1–10.

Breukelaar, A. W., E. H. R. R. Lammens, J. G. P. Klein Breteler & I. Tátrai, 1994. Effects of benthivorous bream (*Abramis brama*) and carp (*Cyprinus carpio*) on sediment resuspension and concentrations of nutrients and chlorophyll *a*. Freshw. Biol. 32: 113–121.

Cahn, A. R., 1929. The effect of carp on a small lake: the carp as dominant. Ecology 10: 371–374.

Cazzaniga, N. J., 1981. Caracterización químico-faunística de canales de drenaje del Valle Inferior del Río Colorado. Ecosur 8: 24–46.

Chapman, G. & C. H. Fernando, 1994. The diets and related aspects of feeding of Nile tilapia (*Oreochromis niloticus* L.) and com mon carp (*Cyprinus carpio* L.) in lowland rice fields in northeast Thailand. Aquaculture 123: 281–307.

Cline, J. M., T. L. East & S. T. Threlkeld, 1994. Fish interactions with the sediment-water interface. Hydrobiologia 275/276: 301–311.

Crivelli, A. J., 1981. The biology of the common carp, *Cyprinus carpio* L. in the Camargue, Southern France. J. Fish Biol. 18: 271–290.

Crivelli, A. J., 1983. The destruction of aquatic vegetation by carp. A comparison between Southern France and the United States. Hydrobiologia 106: 37–41.

Engel, S. & S. A. Nichols, 1994. Aquatic macrophyte growth in a turbid windswept lake. J. Freshwat. Ecol. 9: 97–109.

Fernández, O. A., 1982. Manejo integrado de malezas. Pl. Dan. 5: 69–79.

Fernández, O. A., J. H. Irigoyen, M. R. Sabbatini & R. E. Brevedan, 1987a. Aquatic plant management in drainage channels of Southern Argentina. J. Aquat. Plant Mgmt 25: 65–67.

Fernández, O. A., J. H. Irigoyen, M. R. Sabbatini & O. Svachka, 1987b. Recomendaciones para el control de *Potamogeton striatus* y *Chara contraria* en distritos de riego. Malezas 15: 5–44.

Fletcher, A. R., A. K. Morison & D. J. Hume, 1985. Effects of carp, *Cyprinus carpio* L., on communities of aquatic vegetation and turbidity of waterbodies in the Lower Goulburn River Basin. Aust. J. Mar. Freshwat. Res. 36: 311–327.

Kantrud, H. A., 1990. Sago pondweed (*Potamogeton pectinatus* L.): a literature review. U.S. Fish and Wildlife Service, Resour. Publ. 176, Washington, D.C., 89 pp.

King, D. R. & G. S. Hunt, 1967. Effect of carp on vegetation in a Lake Erie marsh. J. Wildl. Mgmt 31: 181–188.

McCrimmon, H. R., 1968. Carp in Canada. Bull. Fish. Res. Bd Can.: 165.

Robel, R. J., 1961. Water depth and turbidity in relation to growth of sago pondweed. J. Wildl. Mgmt 25: 436–438.

Sabbatini, M. R., 1989. Biología y manejo de *Chara contraria* A. Braun ex Kutz. Mg Thesis. Univ. Nac. del Sur, Arg. 102 pp.

Sarig, S., 1966. Synopsis of biological data on *Common carp* (Linnaeus), 1758 (Near East and Europe). FAO Fish. Synop. 31.

Sibbing, F. A., 1988. Specializations and limitations in the utilization of food resources by the carp, *Cyprinus carpio*: a study of oral food processing. Envir. Biol. Fish. 22: 161–178.

Tennant, D., 1975. A test of a modified Line Intersect Method of estimating root length. J. Ecol. 63: 995–1001.

Westlake, D. F., 1971. Water plants and the aqueous environment. Biol. Hum. Aff. 36: 10.

Hydrobiologia **340**: 277–284, 1996.
J. M. Caffrey, P. R. F. Barrett, K. J. Murphy & P. M. Wade (eds), Management and Ecology of Freshwater Plants.

Long-term effects of sheep grazing on giant hogweed (*Heracleum mantegazzianum*)

Ulla Vogt Andersen[1]* & Birgitte Calov[2]
[1]*Royal Veterinary and Agricultural University, Botanical Section, Rolighedsvej 23, DK-1958 Frederiksberg C, Denmark (*Address for correspondence: COWI Consulting Engineers and Planners, Parallelvej 15, DK-2800 Lyngby, Denmark)*
[2]*Danish Forest and Landscape Research Institute, Hoersholm Kongevej 11, DK-2970 Hoersholm, Denmark*

Key words: Heracleum mantegazzianum, seed bank, germination, grazing, meadow vegetation, biodiversity

Abstract

A meadow dominated by a mature, flowering stand of Giant Hogweed (*Heracleum mantegazzianum*) was grazed by sheep in the years 1987–1993. After two years Giant Hogweed cover was much reduced and a typical meadow vegetation was established. By 1993 Giant Hogweed was completely eliminated though total species diversity was much reduced. Soil sampled from the grazed area developed no Giant Hogweed seedlings in a germination test and contained no viable seeds of the species. In contrast soil from adjacent stands produced numerous seedlings with a peak emergence from samples taken after the winter. Seeds collected in October showed a viability in tetrazolium test of 88%. Germination averaged 22% after storage at room temperature and 25% following three weeks treatment at −18 °C. It was concluded that the persistence of Giant Hogweed seeds in Danish meadow soils is less than seven years.

Introduction

Giant Hogweed, *Heracleum mantegazzianum* Somm. & Lev., has spread vigorously throughout Denmark since the 1960s due to its invasive nature. It is most successful in semi-natural habitats and is not restricted to riparian sites. The success of the species is related to its competitive abilities, its large seed production and the well developed dispersal strategies (Pyšek & Prach, 1994). *H. mantegazzianum* is considered as an undesired invasive weed, and many studies are concerned with the control of the species. Both the use of herbicides such as glyphosate (Williamson & Forbes, 1982; Rubow, 1990; Caffrey, 1994) and sheep grazing (Andersen, 1994) have been found to be efficient control methods.

Each mature individual can produce more than 50 000 seeds (Tiley & Philp, 1994). The fruit is adapted to both water and wind dispersal, and it is believed that the seeds can retain their viability in soil for at least 15 years (Lundström, 1989). This would imply the formation of a huge persistent seed bank though there is no research evidence to support this hypothesis.

The aims of the present study were to investigate the long-term effects of sheep grazing on Giant Hogweed, and to determine if a period of seven years of grazing was sufficient to prevent the reestablishment from the seed bank. The enormous seed production indicates a wasteful reproduction strategy (Salisbury, 1975), and therefore attention was focused on the germinability of freshly collected seeds. According to Muller (1978) the seeds of *Heracleum* species usually germinate in the spring, indicating that seed dormancy must be broken by a cold period. Both untreated and frozen seeds were therefore used in the experiments.

Study area

The study was carried out in a 1.7 ha open, mesotrophic meadow situated adjacent to Lake Furesø in NE Zealand, Denmark. Previous to grazing in 1987, the meadow was dominated by 4 m high *H. mantegazzianum*. The number and frequency of other species were very low. The meadow had not been grazed for the previous 25 years and no fertilizers or pesticides had been applied. Stocking rate in the trial was 5 sheep per ha 1987–1989 and increased to 10 sheep per ha in 1990–1993. The area was grazed each year from May to October.

In addition to the two grazed areas (East and West), two ungrazed control sites (East and West) situated outside the grazed area were selected for soil seed bank studies (Figure 1). The Reference area East was covered by a dense stand of *H. mantegazzianum* which had flowered and fruited annually for at least 10 years. The Giant Hogweed in the Reference area West, near to a summer cottage, had only flowered and fruited for two years, having been cut down regularly with a lawn mower until the cottage was abandoned in 1992.

Materials and methods

Surveys of the vegetation were carried out in 1989 and 1993 in the west and the east ends of the meadow. The floristic composition was recorded using the Raunkiær circle method (Raunkiær, 1909). Ten circles (0.1 m^2) were placed at random, and all plants rooted inside the circles were recorded and the results expressed as percentage frequency. Cover percentage was measured at both grazed sites using the Hult-Sernander-DuRietz scale of cover classes (Malmer, 1974) which were then converted to percentage values as follows: 1=2%, 2=9%, 3=18%, 4=36%, 5=72% (Hansen & Jensen, 1972). The vegetation surveys were carried out twice per growing season to minimize seasonal fluctuations.

The cover percentages of all species were recorded in three quadrats at each of the four sites on 18 April 1994. In the ungrazed reference areas the numbers of vegetatively resprouted *H. mantegazzianum* plants were counted and the number of seedlings of the species estimated within the quadrats.

Soil samples (2 × 1 litre) from the grazed areas were collected on 5 August 1993 at a depth of 0–10 cm and transferred to shallow polystyrene trays. The trays were kept in greenhouse at 18 °C and irrigated regularly for three months. All emerging seedlings were recorded. Soil sampling was repeated on 12 October 1993 and 3 March 1994, when samples from the reference areas were also included.

Seeds of *H. mantegazzianum* were collected from the reference areas on 12 October 1993 and dried and stored at room temperature. Half of the samples were frozen at −18 °C for three weeks. Germination capacity was tested using a standard laboratory germination test, following the International Seed Testing Association, ISTA (1993). Eighteen hundred seeds from each of the frozen and non-frozen samples were kept at 20–30 °C in light on a Jacobsen's germination apparatus for 28 days. Counts for speed of germination were made after 10 days.

Four hundred of each of the frozen and non-frozen seeds were stained with tetrazolium (Moore, 1972). The seeds were cut longitudinally through the seed coat and soaked in tapwater for half an hour, followed by soaking in 1% solution of Triphenyltetrazolium chloride at 22 °C for 4 hours. After treatment all seeds were examined under a dissecting microscope. Seeds showing a red colour, that did not obviously originate from fungi or other living tissue on the seed coat, were regarded as viable. The data were analyzed statistically using a t-test.

Plant nomenclature follows Hansen (1981) except for Giant Hogweed: *Heracleum mantegazzianum* Somm. & Lev.

Results

Vegetation development

A comparison of the survey results in May 1989 and May 1993 showed that frequency of *H. mantegazzianum* was reduced from 70–0% in the grazed area West and cover from 7% to 0% during this period. In the grazed area East the respective values were 80–0% and 19–0%. *H. mantegazzianum* had completely disappeared from the grazed area except for three seedlings recorded in April 1994. There were decrease in the total number of species in both grazed sites from 1989 to 1993 (Figure 2).

Figure 3 shows a comparison among the four sites of average cover of *H. mantegazzianum*, other species, total number of species and species diversity (measured as average number of species per m^2 on 18 April 1994). Both average species diversity and total number of species were highest at the grazed sites after control of *H. mantegazzianum*, whereas at the ungrazed sites this species accounted for nearly 50% of the total vegetation cover, and the numbers of other species present were very low.

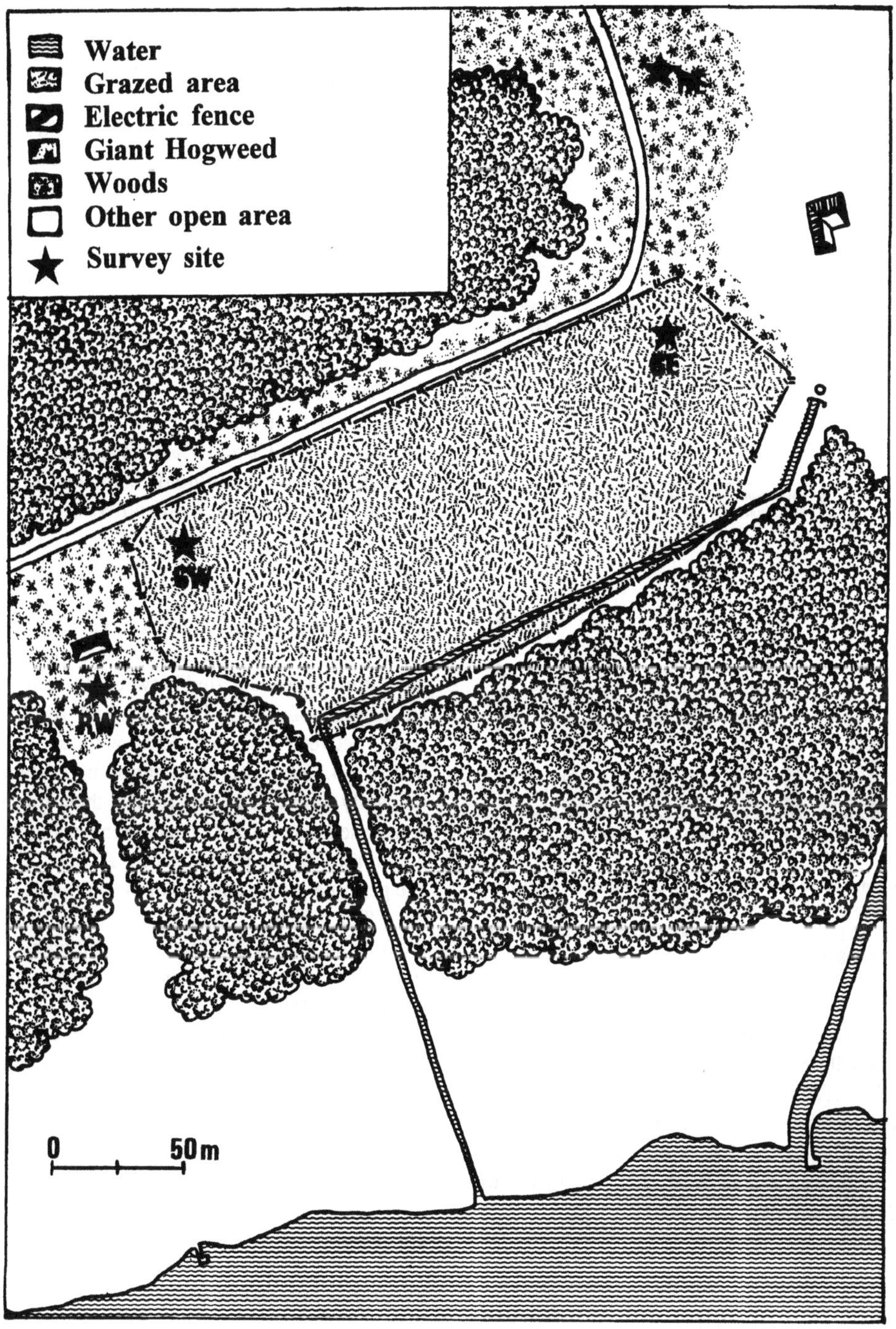

Figure 1. Map of the study area showing the location of the survey sites. RE = Reference area East, RW = Reference area West, GE = Grazed area East, GW = Grazed area West.

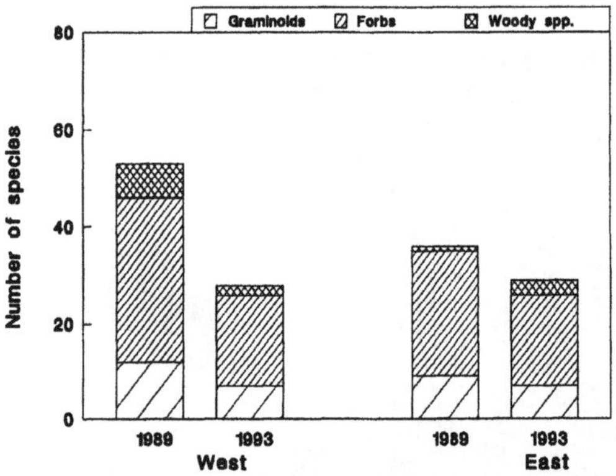

Figure 2. Number of graminoids, forbs, and woody species in the grazed sites East and West. Comparison between the years 1989 and 1993.

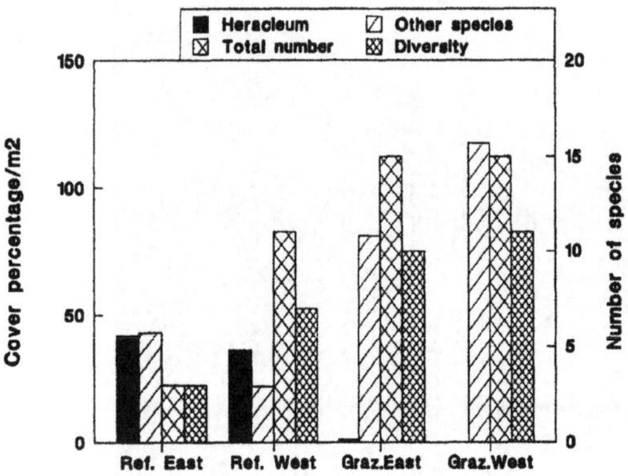

Figure 3. Average cover percentage of *Heracleum mantegazzianum* and other species (axis left), and total number of species and number of species per m^2 (= diversity) (axis right). Comparison between the grazed and the ungrazed areas.

Significant differences in numbers of vegetative shoots and seedlings of *H. mantegazzianum* were found between the reference areas East and West. There were more vegetative plants in reference area East (21) than in West (7) ($t_{veg}=9.23^{**}$), but there were more seedlings in reference area West (225) compared to East (192) ($t_{seed}=3.59^{*}$).

Seed bank

Figure 4 shows examples of the trays used for germination. *H. mantegazzianum* emergence was recorded in the soil sampled from both the reference areas but not in any of the samples from the grazed areas. Emergence of *H. mantegazzianum* showed seasonal variation. When soil samples were collected in October only 2 seedlings and 1 vegetative shoot appeared compared to 73 seedlings and 9 vegetative in the March–May surveys. There were differences in germination of all species between the sampling dates (Table 1). In general the highest number of species and individuals emerged from the 3rd sample date compared to the 1st and the 2nd. Figure 5 shows the numbers of *H. mantegazzianum* and other species which emerged in the 3rd analysis.

Reference area East showed the lowest species diversity of the four areas examined with a total of only 8 species germinating from the soil samples compared to 20, 23 and 21 respectively from the other three sites. Species numbers in each analysis are given in Table 1. In soil from the reference areas East and West only 13% and 9% of the germinated species respectively were graminoids compared to 30% and 26% in the grazed areas East and West.

Germination

The laboratory germination test gave the following results: At day 10, the germination percentage for the untreated seeds varied between 3% and 12% with an average of 6.39% ± 2.75%, and for the frozen seeds between 2% and 15% with an average of 8.39% ± 4.26%. The germination percentage is significantly higher for the frozen than for the non-frozen seeds ($t=3.20^{**}$) indicating a higher germination rate for the treated seeds. At day 28, the average germination percentage of the untreated seeds was 22.11% ± 4.25%, compared to 24.67% ± 8.16% for the frozen seeds, again showing a significant difference between the final germination percentages measured at day 28 ($t=3.08^{**}$).

In the tetrazolium test an average of 87.6% ± 2.4% of the *H. mantegazzianum* seeds were determined as viable. 2.6% ± 1.4% of the seeds were non-viable and 9.8 ± 3.3% were empty. A *t*-test showed no significant differences between frozen and non-frozen seeds.

Discussion

Control and species diversity

Seven years after start of the grazing regime *H. mantegazzianum* had been eliminated. In the early spring of 1994 a few seedlings of the species were observed before the start of grazing season. Seeds of *H. man-*

Fig. 4 (boven) HYDR 3473 JC

Figure 4. Germination trays (3rd analysis) from Reference area West (above) and Grazed area East (below). Photographed 9 June 1994.

Table 1. Results from the soil seed bank analyses. 1st analysis: August to November 1993, 2nd analysis: October 1993 to February 1994, 3rd analysis: March to May 1994. Number of emerged individuals of each species and the total number of species are listed for the four areas examined for each of the three analyses. The species are listed alphabetically with graminoids first followed by forbs and woody species. Means of two replicates at each analysis are given. - : No emerging individuals.

	Grazed East			Grazed West			Ref. East		Ref. West	
Species/Analysis	1st	2nd	3rd	1st	2nd	3rd	2nd	3rd	2nd	3rd
Agrostis stolonifera	-	5	15	-	-	2	-	-	-	-
Carex spp.	2	-	-	1	-	-	-	-	-	-
Juncus bufonius	-	-	1	-	-	3	-	-	-	-
Juncus effusus	>85	-	90	13	-	64	-	-	-	1
Hordeum vulgare	-	-	5	-	-	-	-	-	-	-
Poa annua	-	-	-	5	-	-	-	-	-	-
Poa pratensis	28	34	158	10	17	82	3	10	9	16
Aegopodium podagraria	-	-	-	-	-	-	-	-	-	2
Angelica sylvestris	-	-	-	-	-	-	-	-	-	1
Cardamine amara	-	1	-	-	2	-	-	-	-	2
Cerastium fontanum ssp.*triviale*	3	1	3	2	2	4	-	-	-	1
Chrysosplenium alternifolium	-	-	-	-	-	-	-	-	-	2
Cirsium arvense	2	-	6	2	2	1	-	-	1	-
Cirsium oleraceum	-	-	-	-	-	-	-	-	-	4
Epilobium roseum	-	-	-	-	-	-	13	14	-	2
Epilobium spp.	-	-	-	2	-	-	-	-	-	-
Galeopsis speciosa	1	-	-	-	-	-	-	-	-	1
Glechoma hederacea	-	-	-	-	-	-	-	-	-	2
Gnaphalium uliginosum	6	3	9	-	-	-	-	-	-	1
Heracleum mantegazzianum	-	-	-	-	-	-	2	59	1	23
Hypericum tetrapterum	-	-	-	2	-	-	-	-	-	-
Plantago major	1	1	2	-	-	1	1	-	-	1
Polygonum laphatifolium ssp. *pallidum*	-	-	-	-	-	-	-	-	1	1
Ranunculus repens	17	6	22	13	5	2	-	-	3	1
Sagina procumbens	3	4	23	-	5	18	-	-	3	-
Scrophularia nodosa	-	-	-	3	-	3	-	-	-	-
Senecio vulgaris	-	-	-	-	1	-	-	-	-	-
Stellaria media	-	-	-	1	-	-	-	-	-	-
Stellaria nemorum ssp.*glochidisperma*	-	-	-	-	-	1	-	-	-	-
Urtica dioica	99	11	26	76	14	25	32	12	2	3
Veronica beccabunga	20	-	14	1	-	11	1	-	-	-
Veronica filiformis	-	1	-	-	-	2	1	-	-	1
Acer campestre	-	-	1	-	-	-	-	-	-	-
Crataegus monogyna	-	-	17	-	-	6	-	17	-	7
Prunus padus	-	-	1	-	-	2	-	-	-	-
Number of species	**12**	**10**	**16**	**13**	**8**	**16**	**7**	**5**	**7**	**19**

tegazzianum could easily have been dispersed by water from adjacent areas during the previous very wet winter.

H. mantegazzianum is generally undesirable as its presence can lead to a great loss of local biodiversity in a few years (Pyšek, 1991, 1994). Control of *H. mantegazzianum* using herbicides or other traditional methods can cause unwanted changes in the plant community, and involve risks of contamination of nearby waters (Dodd et al., 1994; Sampson, 1994). Chemical control also creates open sites with a potential for the establishment of other invasive species, such as *Petasites hybridus* (Caffrey, 1994). By contrast grazing is an environmentally safe control method. Sheep grazing sustains a dense, short vegetation of forbs and grasses (Andersen, 1994).

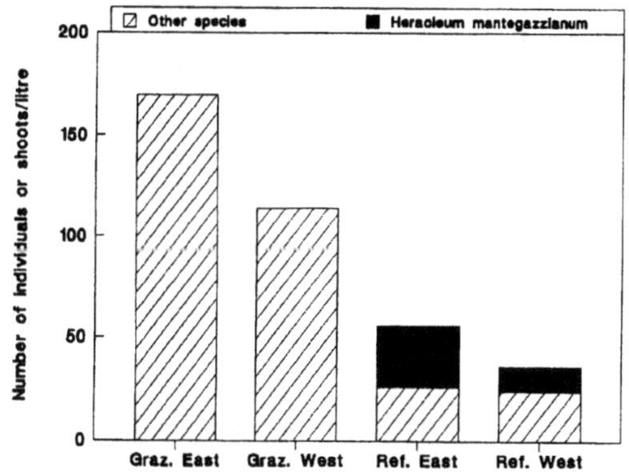

Figure 5. Number of individuals emerging per litre soil in 3rd seed bank analysis. Comparison of *Heracleum mantegazzianum* and other species.

For a complete eradication of Giant Hogweed a high grazing pressure is needed. Grazing by 10 sheep per ha was found to change the vegetation towards a less species rich community dominated by grazing tolerant species, such as *Poa pratensis*, *Ranunculus repens* and *Juncus effusus*. Conservationists and local authorities must decide whether they require total eradication of *H. mantegazzianum* or just sufficient control to prevent flowering and fruiting. We believe the latter is preferable, in which a grazing pressure of 5 sheep per ha is used to control the species and prevent it from flowering, without reducing the number of other species present in the meadow vegetation.

Soil seed bank

Our results confirm that the soil seed bank of *H. mantegazzianum* becomes eliminated when the species is not allowed to set seed for seven years. The 15 year period suggested by Lundström (1989) thus appears to be an overestimate.

In soil sampled from areas where *H. mantegazzianum* had been allowed to flower and set seed, numerous seedlings appeared. The results from Reference area West indicated that a seed bank of *H. mantegazzianum* could be established in the soil after seed set for just one or two years. In the area where the species had never been controlled (Reference area East) vegetative regrowth appeared to be more important than germination from seeds. Species diversity is much lower in this area compared to the grazed areas and Reference area West, where control had been practised except for the last two years. This indicates that the depressing effect of *H. mantegazzianum* on species diversity requires more than two years. However, the number of individuals/shoots of other species per litre soil (Figure 5) is much smaller from the reference areas than from the grazed areas. This could partly arise from an allelopathic effect of *H. mantegazzianum* on other germinating species. According to Rice (1974) the furanocoumarins produced by *Heracleum* species possess such properties.

Germination of Heracleum mantegazzianum

In the soil seed bank analyses (Table 1) *Heracleum mantegazzianum* seeds germinated practically only in the 3rd (March–May) analysis. Many species of *Apiaceae* and woodland herbs are spring-germinating (Thompson & Grime, 1979) due to a chilling requirement which imposes winter dormancy and delays germination until favourable conditions in spring (Grime et al., 1981). The fact that no seedlings of *H. mantegazzianum* appeared in the 1st analysis, when soil was sampled in August shortly after seed dispersal, indicates that dormancy is of the innate type (Harper, 1977). This type of dormancy is found in the related species *Heracleum sphondylium,* where the embryo is immature and several months of further development is needed after seed dispersal (Harper, 1977). Maturation of seeds and breaking of dormancy in *H. mantegazzianum* may also require a minimum cold period in humid soil during winter. Our results from the laboratory germination test support this assumption as the frost treatment resulted in a higher speed of germination and a higher total germination.

In the tetrazolium test on average 88% of the seeds were determined as viable. The tetrazolium test is not very easy to evaluate as there are several factors that can cause overestimation of the number of viable seeds (Moore, 1985). Seeds with weak tissue, that would not be able to germinate may be counted as viable. But the test has the advantage that dormant seeds are also counted as viable (Ellis et al., 1985). This enables the estimation of number of viable *H. mantegazzianum* seeds in the wild without fully understanding the mechanisms of dormancy.

Compared to germination in the wild, the laboratory germination were probably underestimated as dormancy was not broken prior to the test. The addition of chilling as a pre-treatment (ISTA, 1993) would probably result in a higher germination percentage. The tetrazolium test, however, has most likely overestimated the viability. Therefore the germination potential is probably between 23% and 88% in natural environments.

Acknowledgments

The authors wish to express their gratitude to the employees at The Plant Directorate, Department for Seed Analysis, to The Royal Veterinary and Agricultural University, Section for Horticulture, for providing greenhouse facilities and to Vinnie Deichmann for technical help and provision of laboratory equipment.

References

Andersen, U. V., 1994. Sheep grazing as a method of controlling *Heracleum mantegazzianum*. In de Waal, L. C., L. E. Child, P. M. Wade & J. H. Brock (eds), Ecology and management of invasive riverside plants. John Wiley, London: 77–91.

Caffrey, J. M., 1994. Spread and management of *Heracleum mantegazzianum* (Giant Hogweed) along Irish river corridors. In de Waal, L. C., L. E. Child, P. M. Wade & J. H. Brock (eds), Ecology and management of invasive riverside plants. John Wiley, London: 67–76.

Dodd, F. S., L. C. de Waal, P. M. Wade & G. E. D Tiley, 1994. Control and management of Giant Hogweed (*Heracleum mantegazzianum*). In de Waal, L. C., L. E. Child, P. M. Wade & J. H. Brock (eds), Ecology and management of invasive riverside plants. John Wiley, London: 111–126.

Ellis, R. H., T. D. Hong & E. H. Roberts, 1985. Handbook of seed technology for genebanks: No. 3, Volume II, Compendium of Specific Germination Information and Test Recommendations. International Board for Plant Genetic Resources, Rome: 211–667.

Grime, J. P., G. Mason, A. V. Curtis, J. Rodman, S. R. Band, M. A. G. Mowforth, A. M. Neal & S. Shaw, 1981. A comparative study of germination characteristics in a local flora. J. Ecol. 69: 1017–1059.

Hansen, K. (ed.), 1981. Dansk feltflora. Gyldendal, Copenhagen, 758 pp.

Hansen, K. & J. Jensen, 1972. Vegetation on roadsides in Denmark. Qualitative and quantitative composition. Dansk Botanisk Arkiv. 28: 1–143.

Harper, J. L., 1977. Population Biology of Plants. Academic Press, London, 892 pp.

International Seed Testing Association (ISTA) 1993. International rules for seed testing. Seed Sci. & Technol. 21, Supplement.

Lundström, H., 1989. New experiences of the fight against the giant hogweed *Heracleum mantegazzianum*. 30th Swedish Crop Protection Conference. Reports Swedish University of Agricultural Sciences. 2: 51–58.

Malmer, N., 1974. Scandinavian approach to vegetation science. Meddelanden från Avdelingen för Ekologisk Botanik, Lunds Universitet 2: 30 pp.

Moore, R. P., 1972. Tetrazolium staining for assessing seed quality. In Heydecker, W. Seed Ecology. Butterworths, London: 347–366.

Moore, R. P., 1985. Handbook on tetrazolium testing. The International Seed Testing Association, Zürich, 99 pp.

Muller, F. M., 1978. Seedlings of the North-western European lowland. Junk & Pudoc, Wageningen, 654 pp.

Pyšek, P., 1991. *Heracleum mantegazzianum* in the Czech Republic, dynamics of spreading from the historical perspective. Folia Geobot. Phytotax. Praha. 26: 439–454.

Pyšek, P., 1994. Ecological aspects of invasion by *Heracleum mantegazzianum* in the Czech Republic. In de Waal, L. C., L. E. Child, P. M. Wade & J. H. Brock (eds), Ecology and management of invasive riverside plants. John Wiley, London: 45–55.

Pyšek, P. & K. Prach. 1994. How important are rivers for supporting plant invasions? In de Waal, L C., L. E. Child, P. M. Wade & J. H. Brock (eds), Ecology and management of invasive riverside plants. John Wiley, London: 18–26.

Raunkiær, C., 1909. Formationsundersøgelse og formationsstatistik. Bot. Tidsskr. 30: 20–132.

Rice, E. L., 1974. Allelopathy. Academic Press, London, 353 pp.

Rubow, T., 1990. Giant Hogweed. Importance and control. Dansk Planteværns Konference. 7: 201–209.

Salisbury, E., 1975. The survival value of modes of dispersal. Proc. R. Soc. Lond. B. 188: 183–188.

Sampson, C., 1994. Cost and impact of current control methods used against *Heracleum mantegazzianum* (Giant Hogweed) and the case for investigating a biological control programme. In de Waal, L. C., L. E. Child, P. M. Wade & J. H. Brock (eds), Ecology and management of invasive riverside plants. John Wiley, London: 55–65.

Tiley, G. E. D. & B. Philp, 1994. *Heracleum mantegazzianum* (Giant Hogweed) and its control in Scotland. In de Waal, L. C., L. E. Child, P. M. Wade & J. H. Brock (eds), Ecology and management of invasive riverside plants. John Wiley, London: 101–109.

Thompson, K. & J. P. Grime, 1979. Seasonal variation in the seed banks of herbaceous species in ten contrasting habitats. J. Ecol. 67: 893–921.

Williamson, J. A. & J. C. Forbes, 1982. Giant Hogweed (*Heracleum mantegazzianum*). Its spread and control with glyphosat in amenity areas. Proceedings British Crop Protection Conference – Weeds: 967–972.

Hydrobiologia **340**: 285–290, 1996.
J. M. Caffrey, P. R. F. Barrett, K. J. Murphy & P. M. Wade (eds), Management and Ecology of Freshwater Plants.

Effects of grazing by fish and waterfowl on the biomass and species composition of submerged macrophytes

Ellen Van Donk[1] & Adrie Otte[2]
[1]*Dept. of Water Quality Management and Aquatic Ecology, Agricultural University, P.O. Box 8080, 6700 DD Wageningen, The Netherlands*
[2]*AquaSense, P.O. Box 41125, 1009 EC Amsterdam, The Netherlands*

Key words: herbivory, biomanipulation, waterplants, coots, rudd, Lake Zwemlust

Abstract

Biomanipulation improved water transparency of Lake Zwemlust (The Netherlands) drastically. Before biomanipulation no submerged vegetation was present in the lake, but in summer 1987, directly after the measure, submerged macrophyte stands developed following a clear-water phase caused by high zooplankton grazing in spring. During the summers of 1988 and 1989 *Elodea nuttallii* was the most dominant species and reached a high biomass, but in the summers of 1990 and 1991 *Ceratophyllum demersum* became dominant. The total macrophyte biomass decreased in 1990 and 1991. In 1992 and 1993 *C. demersum* and *E. nuttallii* were nearly absent and *Potamogeton berchtholdii* became the dominant species, declining to very low abundance during late summer. Successively algal blooms appeared in autumn of those years reaching chlorophyll-*a* concentrations between 60–130 $\mu g\, l^{-1}$. However, in experimental cages placed on the lake bottom, serving as exclosures for larger fish and birds, *E. nuttallii* still reached a high abundance during 1992 and 1993. Herbivory by coots (*Fulica atra*) in autumn/winter, and by rudd (*Scardinius erythrophthalmus*) in summer, most probably caused the decrease in total abundance of macrophytes and the shift in species composition.

Introduction

Submerged vegetation is of great importance to the functioning of shallow lakes, affecting both abiotic and biotic processes (Carpenter & Lodge, 1986). Macrophytes play a key role in several feed-back mechanisms that tend to keep the water clear at relatively high nutrient loadings (Moss, 1990), e.g. by reducing nutrient levels in the water (Van Donk et al., 1993), providing a refuge for herbivorous zooplankton (Timms & Moss, 1984), and preventing resuspension of the sediment by wind and benthivorous fish (Meyer et al., 1990).

With increase of lake trophy, the area occupied by submerged macrophytes may decrease. The disappearance of submerged macrophytes from eutrophic waters has been attributed mainly to poor light availability due to shading by epiphyton and phytoplankton (Phillips et al., 1978). Recently, it has been suggested that top-down effects, like herbivory by vertebrate and invertebrate grazers, are also important in structuring and lowering the macrophyte biomass (e.g. Lodge, 1991). Often an increase in herbivorous birds has been observed when submerged macrophyte communities were restored (Hanson & Butler, 1990; Hargeby et al., 1994; Lauridsen et al., 1994a; Schutten et al., 1994; Van Donk et al., 1994).

Disappearance of submerged macrophytes in shallow eutrophic lakes can lead to an alternative stable turbid state dominated by phytoplankton (Scheffer, 1990). Restoration by means of nutrient reduction seems often to be retarded or even prevented by strong buffering mechanisms (Moss, 1990; Scheffer et al., 1993). Return of aquatic vegetation is crucial for a successful restoration of such lakes (Moss, 1990). Reduction of the planktivorous fish stock ('biomanipulation') has been recently applied successfully to several turbid shallow lakes to induce a switch from the phytoplankton-dominated state to a clear-water state with submerged macrophytes (e.g. Jeppesen et al., 1990; Lauridsen et al., 1994b; Meijer et al., 1994; Hanson & Butler, 1994). Some of these lakes tended to return abruptly to the turbid state during the summer months, some five years after restoration (Meyer et al., 1994). These changes started with a decline of submerged macrophytes (Van Donk et al., 1993; Meijer et al., 1994).

In this paper we present the results of a study in Lake Zwemlust analysing the factors causing changes in submerged macrophyte biomass and species composition after restoration by biomanipulation. Especially the impact of herbivory by waterfowl and fish is discussed.

Study area

Lake Zwemlust is a small water body (1.5 ha), with a mean depth of 1.5 m and a maximum depth of 2.5 m. It is located in Nieuwersluis in the Province of Utrecht, The Netherlands. The water quality in the lake had deteriorated until the biomanipulation measures in 1987, due to nutrient-rich seepage water from the polluted River Vecht running about 50 m from the lake. Besides precipitation, seepage water is the main source of the lake's water input. Prior to biomanipulation, the lake was highly turbid, especially in summer (Secchi-depth, 0.3 m), primarily because of high biomass of the cyanobacterium *Microcystis aeruginosa*. The recurrent and persistent blooms of algae led to deterioration of the light climate and a complete disappearance of submerged macrovegetation. Only small strips of emerged and floating plants (*Phragmites australis* and *Nuphar lutea*) were present in the littoral zone of the lake (Van Donk et al., 1989). Consequently, an alteration in the structure of the fish community occurred: the piscivore pike (*Esox lucius*) vanished and the planktivore bream (*Abramis brama*) became dominant.

In March 1987 biomanipulation was carried out as a stand-alone restoration measure, because reduction of the external nutrient loading was not feasible within a reasonable time-span. The total mass of fish was removed, ca. 1500 kg, including about 75% bream (length 10–15 cm). After the lake was refilled by seepage, in ca. 3 days, it was restocked with juvenile fish: 1600 O^+ pike fingerlings (4 cm), and 140 specimen of rudd (*Scardinius erythrophthalmus*), fork length 9–13 cm. The offsprings of rudd were meant to serve as food for pike. The biomanipulation measures are discussed at some length by Van Donk et al. (1989, 1990). The effects of the experimental biomanipulation on the lake's ecosystem have been studied for more than seven years.

Material and methods

Field observations

Biomass and composition of submerged macrophytes in the lake were estimated according to Ozimek et al. (1990) during six successive years starting August 1987. Fish biomass and composition were determined yearly in October using the mark-recapture method (Ricker, 1975). In the period September–February, when more than 20 coots (*Fulica atra*) are present in and around the lake, the number of coots grazing on the macrophytes in the lake was counted fortnightly. Gut contents of rudd were analysed according to Prejs & Jackowska (1978). The techniques of sampling and monitoring lake's limnology are further outlined in Van Donk et al. (1989).

Estimation herbivory

We estimated consumption of macrophytes by rudd, according to Prejs (1984); Prejs considers 0^+ rudd as planktivorous, whereas 1^+ and $>1^+$ rudd feed mainly on macrophytes. Consumption of macrophytes by rudd is based on the average daily consumption of ca. 8–10 mg DW macrophyte per day per gram fish and multiplied by total biomass of rudd 1^+ and $>1^+$ and number of days of intense grazing on the plants. Besides, only the periods when water temperature is above 16 °C have to be considered as the periods of high feeding. Further, we assumpted that the biomass of rudd was constant over the feeding periods and the same as in October. This assumption may give an overestimation of the grazing by rudd.

Total consumption of macrophytes by coots was assessed from the number of 'birds days' (average number of birds d^{-1} × number of days) and the daily consumption per coot. The mean daily intake of ca. 45 g DW plant per coot, as measured by Hurter (1979), was used for calculation of total consumption.

Exclosures

To evaluate grazing effects by fish and birds on macrophyte composition, six cages made of an iron frame with dimensions of 4 m (length) × 1.5 m (width) × 0.6 m (height) and covered by wire-netting (1 × 1 cm mesh width), were used. The cages were

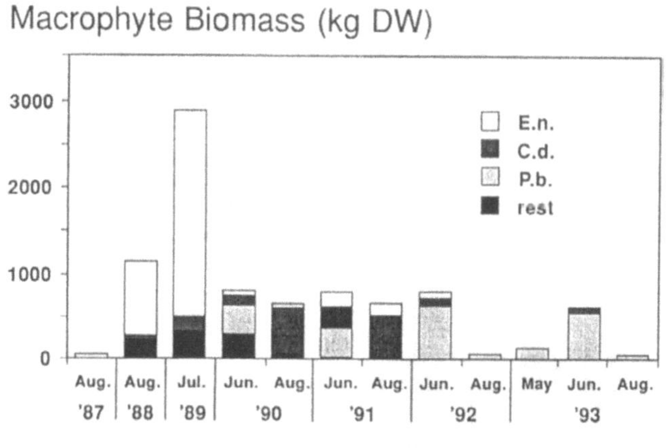

Figure 1. Contribution of the different species of submerged macrophytes (kg Dw) to their total biomass in Lake Zwemlust after biomanipulation.

Table 1. Biomass of rudd > 1^+ (kg/ha) and coverage (%) of submerged macrophytes (E.n. = *Elodea nuttallii*; P.b. = *Potamogeton berchtoldii*; C.d. = *Ceratophyllum demersum*) outside (lake) and inside the exclosures at the start of the experiment in May 1992.

	Rudd (> 1+)	Coverage %		
	kg/ha	E.n.	P.b.	C.d.
Lake	297	30	30	5
Exclosures				
1	0		as in the lake	
2	413		" "	
3	363		" "	
4	750		" "	
5	1263		" "	
6	1575		" "	

placed on the lake bottom at a depth of 2.0 m in May 1992, containing an identical macrophyte composition as in the lake. An increasing biomass of rudd was introduced in these cages (0–1575 kg ha^{-1}) (Table 1), while grazing by birds was excluded. During the experiment that lasted up to July 1993 the cages were inspected monthly by divers.

Results

Macrophyte composition and development in the lake (1987–1993)

Total biomass of submerged macrophytes and contribution to it of the different species differed enormously in lake Zwemlust during the years following biomanipulation (Figure 1). In summer 1988 macrophytes occupied ca. 70% of the lake bottom (total biomass ca. 90 g DW m^{-2}) and in summer of 1989 almost 100% (total biomass ca. 200 g DW m^{-2}), with *Elodea nuttallii* dominating. However, in summers of 1990 and 1991 total biomass of the macrophytes decreased (total biomass ca. 60 g DW m^{-2}), *Ceratophyllum demersum* being the dominant species. In 1992 and 1993 *C. demersum* and *E. nuttallii* were nearly absent and *Potamogeton berchtholdii* became the dominant species in spring (biomass ca. 45 g DW m^{-2}), declining to very low abundance during late summer.

Rudd

The development of biomass of rudd >1^+ is given in Figure 2. In 1988 and 1989 biomass of rudd >1^+ was quite low, but increased in the following years to 297 kg ha^{-1} in 1991, declining to 200 kg ha^{-1} in 1993.

Coots

From 1989 onwards coots invaded Lake Zwemlust extensively during autumn and winter. In 1989/1990, when the lake was dominated by *Elodea*, coots were present in high numbers (ca. 150) from Sept.–Feb. In 1990/1991 and 1991/1992 a maximum was observed at the beginning of December. But the numbers declined in the following months when submerged macrophytes became scarce. In Sept.–Feb. 1992/1993, i.e. after the collapse of *Potamogeton* in August 1992, number of coots was maximally 30 (recorded in October 1992) (Figure 3).

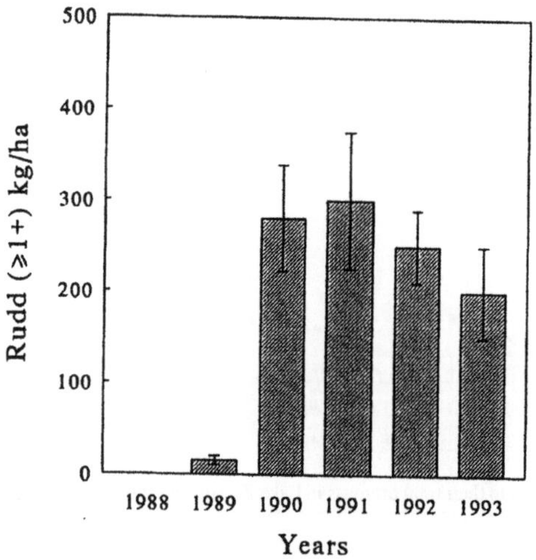

Figure 2. Biomass of rudd (>1$^+$) in Lake Zwemlust after stocking with rudd in March 1987. The 95% confidence intervals are indicated.

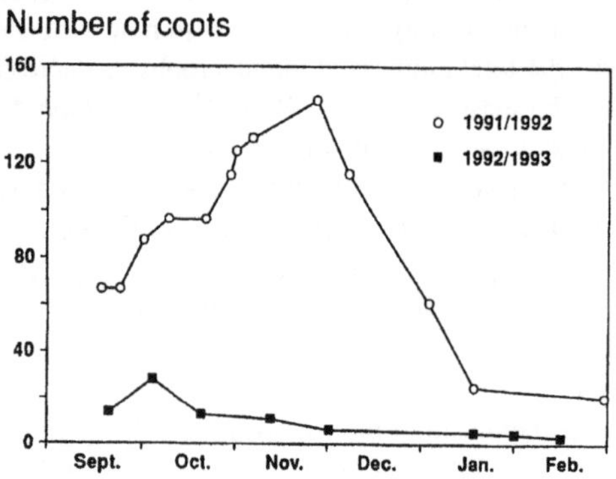

Figure 3. The mean number of coots grazing on submerged macrophytes in Lake Zwemlust during the winter 1991/1992 and 1992/1993.

Herbivory by fish and birds (1989–1993)

In guts of 0$^+$ rudd, the contribution of macrophytes was low, less than 10% weight of total food. However, in guts of 1$^+$ and >1$^+$ rudd macrophytes constituted >85% of food weight. From 1989 onwards the total consumption of macrophytes by 1$^+$ and >1$^+$ rudd increased to ca. 360 kg DW in 1991 (ca. 40% of maximum macrophyte biomass in 1991) and decreased to 200 kg DW in 1992 and 170 kg DW in 1993 (Table 2).

The highest consumption by coots (1200 kg DW) was found in Sept.–Feb. 1989/1990 (ca. 40% of maximum macrophyte biomass in 1989) (Table 2). In 1990/1991 and 1991/1992 the total consumption by coots decreased but its relative proportion to the maximum macrophyte biomass increased to 70–80% of the preceding years. The consumption by coots in 1992/1993 was the lowest (ca. 7% of the maximum macrophyte biomass of 1992). At the start of autumn 1992 nearly all submerged macrophytes had disappeared and only a low number of coots foraged on the lake during the following winter.

Table 2. Estimates of herbivory (kg DW macrophytes) by rudd (June–Sept.) and coots (Sept.–Feb.) in Lake Zwemlust for the years 1988–1993.

	Consumption (kg DW macrophytes)		
Period (June–Sept.)	By rudd	Period (Sept.–Feb.)	By coots
1989	0	1988/1989	0
1990	330	1989/1990	1200
1991	360	1990/1991	800
1992	200	1991/1992	600
1993	170	1992/1993	40

Exclosures

The percentage vegetated area occupied by the different macrophyte species in and outside the cages, observed at the end of the experiment (July 1993), are given in Figure 4. Outside the cages, *P. berchtholdii* was the dominant species (ca. 90%), while *E. nuttallii* and *C. demersum* were scarce (resp. 1 and 5%). The macrophyte composition of cage 1 differed considerably from the other 5 cages (nos. 2–6) and the lake. In cage 1, excluding herbivory by birds and fish, *E. nuttallii* dominate 100%. The percentages macrophyte cover between cages 2–6 (excluding herbivory by birds, but including fish) did not differ, although total macrophyte biomass varied. In these cages ca. 30% was occupied by *P. berchtholdii* and 10% each by *E. nuttallii* and *C. demersum*. Unlike in Cages 2 and 3, the plants *P. berchtholdii* and *E. nuttallii* did not reach the top of the cages in 4, 5 and 6. Especially young apical leaves of *E. nuttallii* were grazed.

Discussion

Grazing pressure by rudd is unevenly distributed among macrophyte species. Prejs & Jackowska (1978) found for rudd a strong preference for *Elodea* and a low preference for *Ceratophyllum*. In laboratory experiments with macrophytes from Lake Zwemlust rudd fed selectively on *E. nuttallii* followed by *P. berchtholdii*,

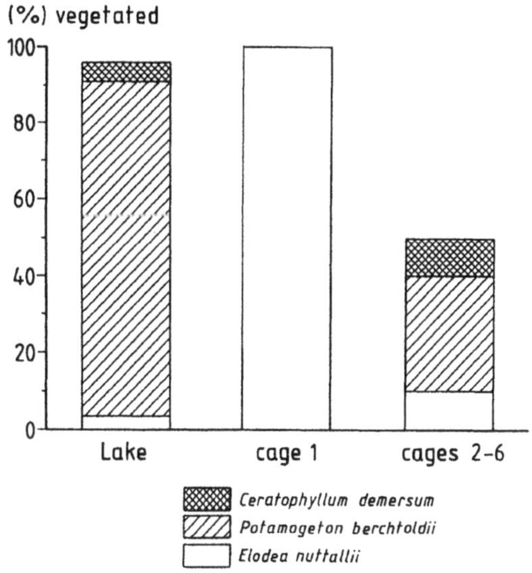

Figure 4. Coverage percentages of the different macrophyte species in the lake and in the exclosures as measured at the end of the experiment in July 1993 (see Table 1).

but did not graze on *C. demersum*, which calcareous structure is apparently much less edible. Rudd, grazing only during the growing season on the macrophytes, prefer young shoots. Prejs (1984) stated that this grazing behaviour may sometimes even stimulate the production of the macrophytes. The difference in macrophyte composition between exclosure 1 (no rudd) and exclosures 2–6 (with rudd) confirmed the results of the laboratory experiments, that grazing by rudd may result in a shift in dominance from *E. nuttallii* (cage 1) to expansion of less edible species (cages 2–6) as *P. berchtholdii* and *C. demersum* (Figure 4).

Kiørboe (1980) stated that grazing by coots has only a minimal effect on macrophyte growth because grazing often takes place outside the growing season of the plants. Coots, however, pull out whole plants and may influence the macrophyte composition and succession by removing especially plants still present during autumn and winter. Contrary to many other submerged macrophytes, *Elodea* is rather unaffected by cold water in late autumn and winter (Wallsten, 1980). It also does not form overwintering structures as in case of *Ceratophyllum* (ca. 10 cm long dormant buds) and *Potamogeton* (turions or tubers). *Potamogeton* starts to form these structures already during the early summer (Sastroutomo, 1981) and decreasing light conditions, e.g. due to epiphyton growth, may even accelerate the formation (Van Vierssen et al., 1994). During winter these tubers or turions lay in or on the sediment, not accessible for coots, while the ca. 10 cm long dormant buds of *Ceratophyllum* are only available when *Elodea* is not dominant. After grazing-induced losses of *Elodea* by coots in Lake Zwemlust during autumn and winter 1989/1990, other macrophyte species like *Ceratophyllum* and *Potamogeton* were able to occupy the whole available area in the subsequent spring period. In winter 1990/1991 and 1991/1992 the coots started to graze on *Ceratophyllum*. Thus, grazing by coots in Lake Zwemlust on *Elodea* and *Ceratophyllum* may eventually result in a dominance of *Potamogeton*. This is confirmed by the exclosure experiments. The macrophyte composition in the lake was dominated by *Potamogeton*, while in cages 2–6, with herbivory by rudd but not by coots, also the other species were relatively abundant (Figure 4). The collapse of *P. berchtholdii* in the lake already in August of 1992 and 1993 may be explained by both grazing of rudd and the formation of turions.

Both fish and bird grazing on macrophytes may affect the internal balance among autotrophic components by reducing the biomass of macrophytes, thereby reducing their competition with algae for nutrients (Lodge, 1991). Furthermore, since some macrophytes species incorporate nutrients from the sediment, these nutrients may be remobilized to the water after the macrophytes eaten by fish and birds are egested, giving phytoplankton access to a supplementary nutrient source (Hansson et al., 1987). Thus, in Lake Zwemlust both fish and bird herbivory reduced the total macrophyte biomass making nutrients more available for phytoplankton growth.

In 1992 and 1993 the amount of filamentous green algae increased and phytoplankton blooms occurred again, chlorophyll-*a* concentrations reaching 60–130 $\mu g\ l^{-1}$. This occurred despite that the external nutrient load to the lake did not differ from previous years and the amount of zooplankton (e.g. cladocerans) even increased (Gulati, 1995). In 1992 and 1993 *P. berchtoldii* was the dominant macrophyte and due to its natural life cycle and herbivory by rudd, nutrients for phytoplankton growth became available already late summer when temperature and light conditions were still suitable for phytoplankton growth.

Fish manipulation in Lake Zwemlust switched the lake from a turbid state (dominating by phytoplankton) to a clear water state (dominated by macrophytes). After five years, however, fish and bird grazing on macrophytes affected the internal balance among autotrophic components by changing composition and lowering the standing crop of the macrophytes, thereby reducing their competition with algae for nutrients. To study in more detail the effects of selective herbivory by waterfowl and fish on longterm recovery of Lake Zwemlust, supplementary experiments in the laboratory and *in situ* exclosures are planned.

Acknowledgments

We thank Drs M. Laterveer-de Beer and Dr J. T. Meulemans for the inspections of the cages by diving, the OVB (Organisation for Improvement of Inland Fisheries) for determining the biomass of rudd and Dr R. D. Gulati for a critical reading of the manuscript. The project was partly financed by the Province of Utrecht.

References

Carpenter, S. H. & D. M. Lodge, 1986. Effects of submersed macrophytes on ecosystem processes. Aquat. Bot. 26: 341–370.

Gulati, R. D., 1995. Food chain manipulation as a tool in the management of small lakes in the Netherlands: the Zwemlust example. In Biomanipulation in lakes and reservoirs management. Bernardi R. & G. Guissani (eds). International Lake Environmental Commitee (ILEC) Vol. 7: 147–163.

Hanson, M. A. & M. G. Butler, 1994. Responses to food web manipulation in a shallow waterfowl lake. Hydrobiologia 279/280: 457–466.

Hansson, L.-A., L. Johansson & L. Persson, 1987. Effects of fish grazing on nutrient release and succession of primary producers. Limnol. Oceanogr. 32: 723–729.

Hargeby, A., G. Anderssen, I. Blindow & S. Johansson, 1994. Trophic web structure in a shallow eutrophic lake during a dominance shift from phytoplankton to submerged macrophytes. Hydrobiologia 279/280: 83–90.

Hurter, H., 1979. Nahrungsökologie des Blässhuhn (*Fulica atra*) an den Uberwinterungsgewässern in nördlichen Alpenvorland. Der Ornithologische Beobachter 76: 257–288.

Jeppesen, E., J. P. Jensen, P. Kristensen, M. Sondergaard, E. Mortensen, O. Sortkjaer & K. Olrik, 1990. Fish manipulation as a lake restoration tool in shallow, eutrophic, temporate lakes 2: threshold levels, long-term stability and conclusions. Hydrobiologia 200/201: 219–227.

Kiørboe, T., 1980. Distribution and production of submerged macrophytes in Tripper Ground, and the impact of waterfowl grazing. J. Appl. Ecol. 17: 675–687.

Lauridsen, T. L., E. Jeppesen & F. Østergaard Andersen, 1994a. Colonization and succession of submerged macrophytes in shallow fish manipulated Lake Vaeng: impact of sediment composition and waterfowl grazing. Aquat. Bot. 46: 1–15.

Lauridsen, T. L., E. Jeppesen & M. Søndergaard, 1994b. Colonization and succession of submerged macrophytes in shallow Lake Vaeng during the first five years following fish manipulation. Hydrobiologia 275/276: 233–242.

Lodge, D. M., 1991. Herbivory on freshwater macrophytes. Aquat. Bot. 41: 195–224.

Meijer, M-L., M. W. De Haan, A. W. Breukelaar & H. Buitenveld, 1990. Is reduction of the benthivorous fish an important cause of high transparency following biomanipulation in shallow lakes? Hydrobiologia 200/201: 303–317.

Meijer, M-L, E. Jeppesen, E. Van Donk, B. Moss, M. Scheffer, E. Lammens, E. Van Nes, B. A. Faafeng, J. P. Jensen, 1994. Long-term responses to fish-stock reduction in small shallow lakes: Interpretation of five year results of four biomanipulation cases in the Netherlands and Denmark. Hydrobiologia 275/276: 457–467.

Moss, B., 1990. Engineering and biological approaches to the restoration from eutrophication of shallow lakes in which aquatic plant communities are important components. Hydrobiologia 200/201: 367–379.

Ozimek, T., E. Van Donk & R. D. Gulati, 1990. Can macrophytes be useful in biomanipulation of lakes? The lake Zwemlust example. Hydrobiologia 200/201: 399–409.

Phillips, G. L., D. Eminson, B. Moss, 1978. A mechanism to account for macrophyte decline in progressively eutrophicated freshwaters. Aquat. Bot. 4: 103–126.

Prejs, A., 1984. Herbivory by temperate freshwater fishes and its consequences. Environmental Biology of Fishes 10: 281–296.

Prejs, A. & H. Jackowska, 1978. Lake macrophytes as the food of roach (*Rutilus rutilus* L.) and rudd (*Scardinius erythrophthalamus* L.) I. Species composition and dominance relations in the lake and the food. Ekol. pol. 26: 429–438.

Ricker, W. E., 1975. Computation and interpretation of biological statistics of fish populations. Bull. Fish. Res. Bd Can. 191: 382 pp.

Sastroutoma, S. S., 1981. Turion formation, dormancy and germination of curly pondweed, *Potamogeton crispus* L.. Aquatic Bot. 10: 161–173.

Scheffer, M., 1990. Multiplicity of stable states in freshwater systems. Hydrobiologia 200/201: 475–487.

Scheffer, M., S. H. Hosper, M.-L. Meijer, B. Moss & E. Jeppesen, 1993. Alternative equilibria in shallow lakes. Trends in ecology and evolution 8: 275–279.

Schutten, J., A. van der Velden & H. Smit, 1994. Submerged macrophytes in the recently freshened lake system Volkerak-Zoom (The Netherlands), 1987–1991. Hydrobiologia 275/276: 207–218.

Timms, R. M. & B. Moss, 1984. Prevention of growth of potentially dense phytoplankton populations by zooplankton grazing in the presence of zooplanktivorous fish in a shallow wetland ecosystem. Limnol. Oceanogr. 29: 472–486.

Van Donk, E., R. D. Gulati & M. P. Grimm, 1989. Food-web manipulation in Lake Zwemlust: positive and negative effects during the first two years. Hydrobiol. Bull. 23: 19–34.

Van Donk, E., M. P. Grimm, R. D. Gulati & J. P. G. Klein Breteler, 1990. Whole-lake food-web manipulation as a means to study community interactions in a small ecosystem. Hydrobiologia 200/201: 275–291.

Van Donk, E., R. D. Gulati, A. Iedema & J. T. Meulemans, 1993. Macrophyte-related shifts in the nitrogen and phosphorus contents of the different trophic levels in a biomanipulated shallow lake. Hydrobiologia 251: 19–26.

Van Donk, E., E. De Deckere, J. P. G. Klein Breteler & J. T. Meulemans (1994). Herbivory by waterfowl and fish on macrophyten a biomanipulated lake: effects on long term recovery. Verh. int. Ver. Limnol. 25: 2139–2143.

Van Vierssen, W., M. J. M. Hootsmans & J. E. Vermaat, 1994. Lake Veluwe, a macrophyte-dominated system under eutrophication stress. Geobotany 21. Kluwer Academic Publishers, 373 pp.

Wallsten, M., 1980. Effects of the growth of *Elodea canadensis* Michx. in a shallow lake (Lake Tämnaren, Sweden). Dev. Hydrobiol. 3: 139–146.

Hydrobiologia **340**: 291–294, 1996.
J. M. Caffrey, P. R. F. Barrett, K. J. Murphy & P. M. Wade (eds), Management and Ecology of Freshwater Plants.

Biological control of the wetlands weed purple loosestrife (*Lythrum salicaria*) in the Pacific northwestern United States

G. L. Piper
Department of Entomology, Washington State University, Pullman, WA 99164-6382, USA

Key words: purple loosestrife, *Lythrum salicaria*, wetlands, biological control, insects, weeds

Abstract

Purple loosestrife (*Lythrum salicaria*) is an Eurasian perennial hydrophyte that has become naturalized in wetlands and in and along waterways throughout temperate North America. The ecological integrity of such areas is threatened by rapidly forming monotypic infestations that displace valued flora and diminish critical fish and wildlife habitat. The inability of physical, cultural, and chemical methods to provide adequate control of the weed has led to the development of an insect-based biological control program. The first field releases of the bud and leaf feeding beetles, *Galerucella calmariensis* and *G. pusilla*, and a root-mining weevil, *Hylobius transversovittatus*, were made in the United States and Canada in 1992. A total of 4740 *Galerucella* spp. adults were released in central Washington during 1992 and 1993 at eight sites and 471 *H. transversovittatus* egg inoculations were made in 1993 at three locations. Establishment of both *Galerucella* spp. was confirmed and *Hylobius* colonization was achieved.

Introduction

Purple loosestrife, *Lythrum salicaria* L. (Lythraceae), is an emergent, herbaceous perennial hydrophyte of Eurasian origin. Thompson et al. (1987) and Mal et al. (1992) have chronicled the plant's introduction into and subsequent spread throughout temperate North America. Its naturalized North American distribution includes southern Canada and the northern tier of the United States where it infests alluvial floodplains, wetlands, wet pastures, stream and river margins, pond and lake shores, irrigation canals and wasteways, and roadside ditches (Stuckey, 1980). Purple loosestrife has become a particularly troublesome weed in the Pacific Northwestern states of Idaho, Oregon, and Washington during the last several decades.

Flowering begins in late June and continues until September. A mature, 0.5 to 3.0 m tall, multi-stemmed plant may produce between two and three million seeds. Propagule dispersal is largely by drift in moving water; long distance transport occurs when seeds become embedded in mud adhering to wildlife, livestock, humans, and vehicles. Purple loosestrife also spreads vegetatively by resprouting from detached stems and rootstock fragments (Thompson et al., 1987). Extensive monotypic stands can develop in just a few years and infestations are unusually long-lived, often persisting for more than 20 years (Thompson et al., 1987).

Purple loosestrife is an aggressive weed that rapidly displaces wildlife-supporting indigenous plant species in wetland communities while offering no substitute value to animal occupants. Reduced floral diversity results in the elimination of natural foods and cover essential for many birds and fur-bearing mammals. Infestations also diminish other wildlife-related recreation opportunities such as hunting, fishing and bird watching, decrease storage capacities of impounded waterbodies, increase siltation, and clog irrigation waterways.

Although various methods can be used to control small populations of purple loosestrife, all are costly and require continued long-term application and, in some situations, may be environmentally undesirable (Thompson et al., 1987). The most ecologically prudent, cost-efficient, and enduring *L. salicaria* management technique may be biological control (Malecki et al., 1993). Biological control involves the intentional utilization of host-specific herbivores to reduce the density and distribution of a noxious plant to the point where it becomes a nondisruptive component of the occupied ecosystem.

Surveys for purple loosestrife herbivores, conducted by United States Department of Agriculture (USDA) and Commonwealth Institute of Biological Control (CIBC) [now the International Institute of Biological Control (IIBC)] scientists between 1979 and 1981 in northern and central Europe, revealed the occurrence of 120 phytophagous insect associates (Batra et al., 1986; Blossey & Schroeder, 1986). Subsequent field and laboratory investigations indicated that six species demonstrated excellent potential as plant suppressants. In 1992, three highly host-specific insects were approved for release in North America by both the USDA and Agriculture Canada (Malecki et al., 1993). Included were the bud and leaf destroying beetles *Galerucella calmariensis* (L.) and *G. pusilla* (Duftschmidt) (Coleoptera: Chrysomelidae), and the root-infesting weevil *Hylobius transversovittatus* Goeze (Coleoptera: Curculionidae).

Galerucella spp. adults appear in May and chew holes in young leaves (Blossey & Schroeder, 1991). From mid-May to mid-July, females lay up to 500 eggs in small batches on stems, leaves, and in leaf axils. Upon hatching, early stage larvae feed on leaf and flower buds; older larvae skeletonize leaves. Feeding by the larvae reduces shoot growth and often prevents flower and seed formation (Blossey & Schroeder, 1991). At high beetle densities, both larval and adult feeding may result in complete defoliation and subsequent plant death (Blossey & Schroeder, 1991). Mature larvae pupate in the soil and first generation adults appear in early summer. Adults feed until September before seeking overwintering sites in the soil.

During early May, overwintered *H. transversovittatus* adults emerge and feed on purple loosestrife leaves. Egg deposition soon commences and may continue into August, with each female producing up to 300 eggs (Blossey & Schroeder, 1991). Approximately 70% of the eggs are laid in the soil close to the root; the remainder are inserted into stems near the soil surface. Larvae enter the root and consume vascular and carbohydrate storage tissues over a one to two year period. Attacked plants are stunted and produce fewer stems and consequently less seed. Most well-established rootstocks must be successively infested for several years before they succumb to the effects of the beetle (Blossey & Schroeder, 1991). Pupation occurs within the mined root.

Table 1. Status of purple loosestrife natural enemy field releases in Washington in 1992 and 1993.

Colonization		Beetles colonized		Status[b]
Site	Date(s)	Species[a]	No.	
WDFW-G1	7/29/92	Gc & Gp	840	E
WDFW-G2	7/29/92	Gc & Gp	840	E
WDFW-G3	7/20/93	Gc & Gp	500	E
WDFW-G4	7/20/93	Gc & Gp	500	E
WDFW-G5	8/25/93	Gc & Gp	200	E
WDFW-G6	8/25/93	Gc & Gp	200	E
WDFW-G7	8/25/93	Gc & Gp	260	E
USBR-G1	7/20/93;	Gc & Gp	600	
	8/25/93	Gc & Gp	800	E
WDFW-H1	7/ 9/93	Ht	213	C
WDFW-H2	7/16/93	Ht	223	C
WDFW-H3	8/25/93	Ht	35	C

aGc = *Galerucella calmariensis*; Gp = *Galerucella pusilla*; Ht = *Hylobius transversovittatus*.
bE = establishment; C = colonization.

This paper presents information on the status of the first field releases of both *Galerucella* spp. and *H. transversovittatus* against purple loosestrife in Washington.

Study site description

All insect releases were made at the South Columbia Basin Wildlife Area (Desert and Potholes Units), located in central Washington approximately 30 km south of the city of Ephrata. Two irrigation water drainages, the Winchester and Frenchman Hills wasteways, traverse the land in a southeasterly direction. Dense monocultures of purple loosestrife dominate these waterways and adjacent wetlands areas. Site selection was based upon criteria recommended by Hight & Drea (1991) for optimum bioagent survival and dispersal.

Materials and methods

Galerucella *spp.*

Galerucella calmariensis and *G. pusilla* larvae, field-collected during June 1992 and 1993 near Gelnhausen in southcentral Germany, were reared to adulthood in an overseas insectary by IIBC entomologists. Adults were then airfreighted to and processed through federally approved U.S. quarantine facilities prior to transshipment to Washington. Beetles were field released within 24 h of receipt. Uncaged releases of both species were made by gently tapping the insects directly onto plants within a 4 m^2 area from ventilated paperboard cartons used for their temporary confinement. Individual release sites were clearly marked and mapped to facilitate subsequent relocation.

Post-release recovery surveys, conducted during May and July in 1993 and 1994, involved the visual examination of plants for *Galerucella* eggs, adults or adult feeding injury. Searches began at the release point and radiated outward along compass-based transects to assess insect occurrence and spread.

Hylobius transversovittatus

Forty-eight insectary-reared adult weevils, derived from colonies collected in Germany and Finland (Blossey, 1993), were provided by IIBC entomologists in 1993. The quarantine-processed beetles were confined to 62 × 62 × 62-cm saran screen cages (24 adults/cage; 1:1 sex ratio), provisioned with cut *L. salicaria* stems for feeding and oviposition (Blossey, 1993). Rearings were maintained at 298K, 60–70% RH, and an 18:6 h (light:dark) photoperiod from mid-June to mid-August. Eggs were excised from the stems weekly and retained for field use. Colonization of *Hylobius* was effected by inserting eggs into holes punched into stems or rootstocks and then sealing the punctures with modeling clay. Inoculated, uncaged plants were marked and mapped for future evaluation.

A follow-up survey to ascertain weevil status was undertaken in June 1994. Since the nocturnal adults are rarely observed, the discovery of foliage injured by their feeding was used to gauge colonization success.

Results and discussion

Galerucella spp.

Details of releases at each of the sites and the current status of establishment are summarized in Table 1. Recovery surveys in 1993 and 1994 confirmed the presence of overwintered (1992 and 1993) and F1 generation (1993 and 1994) adults of both beetle species. Establishment was achieved at all field locations irrespective of the initial number of beetles released per site. However, *Galerucella* spp. population development was more accelerated at three sites where either 840 or 1400 adults had been introduced. The survivorship rate of overwintering adults at the two 1992 release sites was estimated at 3–5% but improved to 8–10% at these sites in 1994. The survival rate of overwintered F1 adults varied from 3–8% at the six 1993 release locations. The factors responsible for the high adult mortality observed are undetermined. Beetle dispersal from initial release points at the 1992 and 1993 sites was omnidirectional, but detectable spread did not exceed 50 m at any location during the first year after release. It is possible, however, that some adults may have moved greater distances and were overlooked during the surveys.

Hylobius transversovittatus

Manual insertion of eggs into purple loosestrife stems and roots was an effective method of attaining colonization of the weevil at all three 1993 release sites (Table 1). Plants with irregularly scalloped leaf margins, a feature associated with adult *Hylobius* feeding, were detected at each site. The number of injured plants observed per location within a 10 m diameter circle surrounding the central release point varied from 10–30%. Establishment can be confirmed if 1994 F1 adults successfully overwinter and produce feeding injury and oviposit in 1995.

It is premature to speculate on the impact the three introduced natural enemies will have on *L. salicaria* in Washington. However, Blossey & Schroeder (1991) have predicted that a 90% reduction in purple loosestrife infestation density in North America will eventually be realized upon the successful establishment of these and possibly several other yet to be introduced biological control agents.

Acknowledgments

Thanks are extended to B. Blossey, S. Hight, R. Malecki, G. Petersen, and D. Schroeder for facilitating natural enemy acquisitions, to R. D. Kent and C. J. Perry, Washington Department of Fish and Wildlife, and C. Conley, U.S. Bureau of Reclamation, for technical assistance and permission to make releases on lands under their jurisdictions, and to S. A. Voss for her invaluable assistance with laboratory and field studies. The financial support of the Washington State Department of Agriculture and Washington Department of Fish and Wildlife is gratefully acknowledged.

References

Batra, S. W. T., D. Schroeder, P. E. Boldt & W. Mendl, 1986. Insects associated with purple loosestrife (*Lythrum salicaria* L.) in Europe. Proc. ent. Soc. Wash. 88: 748–759.

Blossey, B., 1993. Herbivory below ground and biological weed control: life history of a root-boring weevil on purple loosestrife. Oecologia 94: 380–387.

Blossey, B. & D. Schroeder, 1986. Final report. A survey of arthropods and fungi associated with *Lythrum salicaria* in selected areas in northern Europe. European Station, Commonw. Inst. Biol. Contr., Delemont, Switzerland: 38 pp.

Blossey, B. & D. Schroeder, 1991. Study and screening of potential biological control agents of purple loosestrife (*Lythrum salicaria* L.). Final report. European Station, Int. Inst. Biol. Contr., Delemont, Switzerland. 41 pp.

Hight, S. D. & J. J. Drea, Jr., 1991. Prospects for a classical biological control project against purple loosestrife (*Lythrum salicaria* L.). Nat. Areas J. 11: 151–157.

Mal, T. K., J. Lovett-Doust, L. Lovett-Doust & G. A. Mulligan, 1992. The biology of Canadian weeds. 100. *Lythrum salicaria*. Can. J. Plant Sci. 72: 1305–1330.

Malecki, R. A., B. Blossey, S. D. Hight, D. Schroeder, L. Kok & J. R. Coulson, 1993. Biological control of purple loosestrife. BioScience 43: 680–686.

Stuckey, R. L., 1980. Distributional history of *Lythrum salicaria* (purple loosestrife) in North America. Bartonia 47: 3–20.

Thompson, D. Q., R. L. Stuckey & E. B. Thompson, 1987. Spread, impact, and control of purple loosestrife (*Lythrum salicaria*) in North American wetlands. Fish and Wildlife No. 2. U.S. Dept. Interior, Fish and Wildlife Serv., Washington, D.C. 55 pp.

Hydrobiologia **340:** 295–300, 1996.
J. M. Caffrey, P. R. F. Barrett, K. J. Murphy & P. M. Wade (eds), Management and Ecology of Freshwater Plants.

Filamentous freshwater macroalgae in South Africa – a literature review and perspective on the development and control of weed problems

Margaret Anne Joska & John J. Bolton
Botany Department, University of Cape Town, Rondebosch, 7700, South Africa

Key words: South Africa, filamentous macroalgae, copper sulphate

Abstract

Studies on freshwater filamentous algae have not been undertaken in South Africa for some thirty years. Early investigations were mainly of a taxonomic nature and ecological information is virtually non-existent. However, in the recent years the spread of urban settlement and increasing demand for both industrial and domestic water have highlighted the problems facing current water supplies. Irrigation systems in particular have suffered increasing interferences from filamentous algal blooms, mainly *Cladophora glomerata*. As nutrient loads have increased in impoundments and rivers, the presence of this alga has become more obvious, causing decreased water flow in canals and an escalation in costs associated with its control. Copper sulphate and predosing with commercial sulphuric acid to reduce pH are now the commonest control method in most of the irrigation systems. A synopsis of current conditions is presented and proposed avenues of research are discussed.

Introduction

The earliest descriptions of filamentous freshwater algae in the southern African literature (here defined as the area south of the Zambezi River and the northern border of Namibia) are probably those by Harvey (1838) and Kützing (1849). Braun (1868), Reinsch (1878) and West & West (1897) published accounts of collections of Charophyta, Cyanophyta and Chlorophyta from 'southern Africa', the 'Cape of Good Hope' and Namibia, respectively. Marloth (1913) mentions a few species of Chlorophyta that had been found at the Cape. Yamanouchi (1913) described a new species of *Hydrodictyon, H. africanum* (Chlorophyta), which he had cultivated from some soil samples from the Cape Flats, sent to the University of Chicago (Pocock, 1960b). A number of papers pertaining to southern African species were published by West & West (1897) and West (1912). In 1912, West gave an account of species collected by the Percy Sladen Memorial Expedition from Angola, Namibia and the north-eastern Cape Province of South Africa. A major set of publications was the 'Contributions to our Knowledge of Freshwater Algae of Africa', published in a series of 12 reports from 1914 to 1937 (Fritsch, 1914; Fritsch, 1918; Fritsch & Stephens, 1921; Fritsch & Rich, 1924; Hodgetts, 1926; Fritsch & Rich, 1929a; Fritsch & Rich, 1929b; Nygaard, 1932; Rich, 1932, 1935, 1936; Fritsch & Rich, 1937). All these reports were based on preserved specimens sent to Professor F. E. Fritsch of the Botany Department at the University of London, initially by the then head of the Department of Botany at the University of Cape Town, Professor H. H. W. Pearson. (Exceptions were those specimens reported on by Nygaard (1932), which had been sent by Professor C. E Moss of the Department of Botany at the University of the Witwatersrand to Prof. C. H. Ostenfeld at the University of Copenhagen.) In the early 1920's, Miss E. L. Stephens, also of the Department of Botany in Cape Town, had initiated and maintained a large collection of preserved specimens now held in the Bolus Herbarium (BOL) at the University of Cape Town. Isolated papers such as those by Printz (1920) (sub aerial algae) and Huber-Pestalozzi (1930) (freshwater algae from Knysna) were published on samples sent from South Africa. Investigation of some southern African Charophyta were published by Groves (1925) and Groves & Stephens (1926, 1933).

Miss M. A. Pocock of Rhodes University, Grahamstown published two papers on *Hydrodictyon* (Pocock, 1937, 1960b) and one on *Haematococcus* (Pocock, 1960a). Her unpublished manuscript entitled 'Freshwater Algae in Southern Africa' (Pocock, 1966) lists most genera and many species of freshwater algae that had been thus far been recorded, and comprises brief descriptions with comments on occurrence and ecology. This manuscript is held in the Bolus Herbarium Library in the Botany Department at the University of Cape Town.

Claassen (1961), Welsh (e.g. 1961, 1965) and Archibald (1966, 1967) contributed to our knowledge of the Cyanophyta in the Transvaal and eastern Cape, these publications being primarily taxonomic in nature. Major taxonomic publications by Gauthier-Lièvre (1964a, 1964b, 1965) dealt, in part, with southern African species of Conjugales (Chlorophyta). Shillinglaw (1980) and Truter (1987) published identification guides to algae in some South African impoundments, but *Cladophora glomerata* (Linneaus) Kützing (Chlorophyta), *Anabaena* sp., *Microcystis* sp., *Spirulina* sp. and *Oscillatoria* sp. (Cyanophyta) are the only species listed which can commonly produce visible macroscopic growths.

Thus, prior to the 1930's, algal publications relating to southern Africa were almost entirely of a taxonomic nature, but increasing population and mining activity in the Transvaal with the consequent pressure on available water supplies led to some isolated investigations of a more ecological nature (Hutchinson et al., 1932; Schuurman, 1932; Weintroub, 1953). An exception to this is the publication of Harrison et al. (1960), who investigated the purifying effects of a marsh lying adjacent to the polluted Klip and Klipspruit Streams, near Johannesburg. Also Hancock (1973) identified a number of filamentous algae in an ecological survey of mineral- and acid-polluted streams on the Witwatersrand. Reports on rivers and estuaries in southern African made scant or no reference to filamentous macroalgae (e.g. Hart & Allanson, 1984; O'Keeffe, 1986; Ferrar, 1989). The period between the 1940s and 1960s saw very few publications on southern African freshwater macroalgae. This dearth of information, especially in ecological research, can be ascribed to the fact that few major growths of 'pest' species, usually caused by high nutrient loading, had occurred. Further, the few phycologists who were active in southern Africa were mainly engaged in marine research and those research scientists involved in water management had little or no botanical training. This lack of indigenous freshwater macroalgal research has been noted in Africa as a whole (John, 1986) and in Australia (Entwisle, 1989) and New Zealand (Biggs & Price, 1987; Quinn, 1991).

Table 1. Water usage in South Africa (from Rawhani, 1991)

Demand sector	Year 1980	1990	2000
Municipal & Industrial	2 547	3 729	5 263
Mining & Power Station	709	916	1 322
Irrigation	8 504	9 695	10 974
Stock & Conservation	440	470	503
Estuaries & Lakes	2 768	2 767	2 767
Forestry & Other	1 323	1 466	1 609
Total	16 291	19 043	22 438

Table 2. Copper and sulphuric acid dosage of irrigation canals in 1993. (Figures supplied by the Dept. of Water Affairs & Forestry, Pretoria)

Irrigation scheme	Canal	Length (km)	H_2SO_4 kg/annum	Cu^{2+} kg/annum
Hartbeespoort	West	55	2500	59
(Transvaal)	East	60	2500	59
Oranje-Riet (O.F.S.)	Kalkfontein	132	12500	292
Olifants River (W. Cape)	Right	160	150	117
Total			17650	527

A project 'Preliminary Investigation into Algal Weeds in Inland Waters', funded by the Water Research Commission in Pretoria, was initiated in 1992 as a result of complaints sent to the Department of Water Affairs and Forestry in Pretoria by some irrigation boards in the south-western Cape. Nutrient eutrophication with concomitant noxious growth of *Cladophora glomerata* in irrigation canals and other water impoundments in the Transvaal and Orange Free State was reported by Toerien (1975). This paper reviews the current status of filamentous algal weed problems in this area and also in Cape river systems where excessive algal growths have appeared more recently.

Transvaal and Orange Free State

It is in these two provinces that water demands are greatest. Irrigation, mining, power stations, munici-

Table 3. Results of questionnaire sent to irrigation boards and engineering firms in the Cape, detailing perceived problems with macroalgae [data expressed as a percentage of number of respondents (n = 82)].

Water use	Perceived problem			
	No	Yes	Don't know	Total usage
Irrigation	36.2	24.1	0.0	60.3
Recreation	6.0	4.8	0.0	10.8
Drinking & industrial	14.5	13.3	1.1	28.9
Total	56.7	42.2	1.1	

pal and industrial requirements used 75% of available water in 1990 in South Africa (Rawhani, 1991, Table I). In the early 1970's, growths of *C. glomerata* in the irrigation canals below the Hartbeespoort Dam had caused disruptions to water flow and associated problems. These growths occurred during spring and summer, from September to January. Initially they were removed manually at a cost of approximately R40 000 per annum (in 1990) (approximately US$ 12 000), but chemical control, using copper sulphate, was found to be more efficient (Bruwer, 1991). Whitton (1990) and Whitton et al. (1989) discuss the sensitivity of *Cladophora glomerata* to copper sulphate.

The Transvaal and Orange Free State receive their rainfall mainly in the summer months, from October to February. However, these provinces frequently suffer from drought cycles and water stored in a comprehensive dam system is used to maintain water flow in irrigation canals. Water supplied from these dams is often turbid during periods of normal rainfall, but turbidity declines during periods of drought. It was noted that *C. glomerata* growth dropped markedly in the Rand Water Board's Zuikerbosch Canal, where water is supplied from the Vaal Dam, when turbidity levels dropped below 40 NTU (M. Steynberg & A. J. Pieterse, pers.comm.). Rand Water currently supplies 3500 ml of potable water per day to agricultural, industrial, mining and municipal users in these provinces. The Lesotho Highlands Water Project (LHWP), currently under construction will, when completed, supply almost double this amount of water per day to South Africa. This water from the LHWP will be less alkaline (softer) than water presently supplied from dams in the Transvaal and Orange Free State (Rand Water Board, 1993). It is suspected that this, probably clear, water supply will promote a greater *Cladophora* problem (M. Steynberg, pers.comm.). Rivers in the Transvaal and Orange Free State have natural pH ranges of 6–8 and electroconductivity (EC) levels seldom >1000 μS cm^{-1} (Dallas & Day, 1993). However, large shanty settlements, agricultural, industrial and mining pollution have led to increased pH and EC levels being recorded in rivers, which can therefore no longer be considered 'natural'. Almost all excess water originally abstracted for irrigation purposes, is returned to that river system closest to the irrigation end-point. Thus, any chemicals used to control aquatic growth in water impoundments and irrigation systems have the possibility of affecting the downstream river systems.

Individual *C. glomerata* growths may reach 20 m in length in irrigation canals during peak summer growth periods (Bruwer, 1991). Attempts at various methods of control of this growth were reported by Bruwer (1980, 1991), and Bruwer et al. (1980). In the Hartbeespoort (Transvaal) and Oranje-Riet (Orange Free State) irrigation canal systems, copper sulphate in conjunction with a predosage of sulphuric acid has been used to control algal growth. The predosing with sulphuric acid serves to reduce the pH of the canal water to between 5.5 and 6 which allows the copper sulphate to remain in its most effectively phytotoxic ionic form (Cu^{2+}). Dose calculation and methods of application are outlined in Du Plessis 1992a and 1992b. Amounts of sulphuric acid and total copper applied to specific sections of three canal systems are shown in Table II (figures supplied by the Dept. of Water Affairs). This chemical dosage of these canals, especially in the Transvaal and Orange Free State, is carried out with precision, with pH measurements being recorded at each dosage site.

Cape river systems, especially the Breede River

In the Western and Northern Cape provinces, the rivers arise from the coastal escarpment and are subjected to a regular seasonal (winter) flushing. pH values, especially in the headwaters, are often low (4–4.5) due to the specific 'fynbos' vegetation (Dallas & Day, 1993). EC values are dependent on the prevailing geological profile. Thus, waters draining areas with Table Mountain Sandstone will have low EC (<150 μS cm^{-1}) whilst lower river reaches which run over Malmesbury Shales have considerably higher EC (>1000 μS cm^{-1}) (Dallas & Day, 1993). Irrigation, degradation of natural riverine vegetation and sewage effluent have led to increased salinization in many South African rivers

(Du Plessis & Van Veelen, 1991) and we found EC levels >1000 μS cm^{-1} at Bonnievale, a mid point on the Breede River. *Enteromorpha* cf. *flexuosa* (Wulfen ex Roth) J. G. Agardh and *Compsopogon coeruleus* (Balbis) Montagne, were recorded at this site (Joska & Bolton, 1994). Before initiating the sampling programme, we sent a questionnaire to all Irrigation Boards, Municipalities and engineering firms concerned with water related projects. We received a 42% response to this questionnaire. The majority of respondents used water for irrigation and over 40% of these stated that they had had algal problems (Table III). As a result of this response, we were able to ascertain that major algal problems occurred in the Breede River canal systems. This system comprises four separate canals, La Chasseur, Robertson, Sanddrift and Angora.

During October and November 1992 a bloom of *Oedogonium capillare* Kützing occurred in the La Chasseur canal. Subsequent information received from the Cape Town City Council Scientific Services Branch indicated that a bloom of this same alga also occurs annually in the Theewaterskloof Dam, one of the main reservoirs supplying greater Cape Town (W. Harding pers.comm.). Apart from this algal bloom, algal cover in the canal system was never >30%.

Oedogonium spp., *Cladophora fracta* (Müll. ex Vahl) Kütz, and *Spirogyra* spp. were the most common species in the canal systems. Control of algal growth is the duty of the water bailiff responsible for each canal. The water bailiffs take no pH measurements and use a simple method of copper sulphate application from November to March (late spring to the end of summer). A 50 kg bag of copper sulphate is suspended from a spar which lies across the canal and a small hole is cut in the base of the bag which lies just below the water level. Use of these bag systems effect a slow dispersal of copper sulphate crystals in the water at various points in the canal system. During the one year period of our sampling, the bailiffs used a total 2500 kg of copper sulphate in the entire system (Mr A. Baard pers.comm.) This method of algal control, very different from that used in the Transvaal and Orange Free State, appears to suppress algal blooms. In July 1994, during a continuation of our monitoring of this canal system, we found that another bloom of *Oedogonium* had occurred in the same canal. Since copper sulphate was not being applied at this time, it can be inferred that the copper sulphate dosing by this method is effective and/or that the *O. capillare* bloom is seasonal with a possible light/temperature growth 'trigger'. At no time during our sampling period in the Breede River canal systems did we find that turbidity levels would preclude algal growth. In the Olifants River canal system, the problem algae were two species of *Nitella*. Here, the growths were controlled with copper sulphate and predosage of sulphuric acid.

Conclusion

Macroalgae are significant problem organisms in South african irrigation canals and potable water systems, especially *C. glomerata* in the Transvaal and Orange Free State. Biological knowledge of macroalgae especially in South Africa in these systems is minimal. Presently, copper sulphate is successfully used to control these growths (sometimes combined with sulphuric acid to lower the pH). However, long term effects of these control methods have not been studied and research to investigate the causal factors, seasonality of recruitment and growth of the algae, and efficiency of control methods is presently being undertaken.

Acknowledgments

We would like to thank the Water Research Commission for their financial support and the University of Cape Town for laboratory and administrative facilities.

References

Archibald, C. G. M., 1966. Some new and interesting cyanophytes from the Kowie river system in the eastern Cape Province (South Africa). Nova Hedwigia 12: 429–538.

Archibald, C. G. M., 1967. New and interesting cyanophytes from the Kowie river system in the eastern Cape Province (South Africa). II. Nova Hedwigia 13: 389–395.

Biggs, B. J. F. & G. M. Price, 1987. A survey of filamentous algal proliferations in New Zealand rivers. N.Z. J. Mar. & Freshwat. Res. 21: 175–191

Braun, A., 1868. Die Characeen Africas. Monatsb. d. Königl. Akad. Berlin: 782–872.

Bruwer, C. A., 1980. Aksieplan vir die ontwerp van 'n beheerstrategie teen die oormatige groei van *Cladophora* in besproeiingskanale. Report BC:001 for the Department of Water Affairs, Forestry and Environment, Pretoria, 7 pp.

Bruwer, C., 1991. Chemical control of benthic algae in Hartebeespoort irrigation canals. (Brochure) Department of Water Affairs, Private Bag X313, Pretoria.

Bruwer, C., D. J. Krynauw & J. S. De Wet, 1980. Chemiese, fisiese en biologiese veranderinge met betrekking tot kanaallengte onderkant Hartbeespoortdam soos bepaal in die oos-kanaal op 8 September, 1980. Report BC 004, Department of Water Affairs, Forestry and Nature Conservation. Pretoria, 9pp.

Claasen, M. I., 1961. A contribution to our knowledge of the freshwater algae of the Transvaal province. Bothalia 7: 559–666.
Dallas, H. F. & J. A. Day, 1993. The effect of water quality variables on riverine ecosystems: A review. Report No. TT 61/93. Water Research Commission, Pretoria, 240 pp.
Du Plessis, B. J., 1992a. 'n Ligginspesifieke handleiding vir die chemiese dosering van probleemalge in besproeiingkanaalstelsels: Hartbeespoort- Staatswaterskema. Report No. N A210/09/DIQ/0192, Department of Water Affairs and Forestry, Pretoria, 35 pp.
Du Plessis, B. J., 1992b. 'n Liggingspesifieke handleiding vir die chemiese dosering van probleemalge in besproeiingskanaalstelsels: Oranje-Rietrivier-Staatswaterskema. Report No. N C510/00/DIQ/0592, Department of Water Affairs and Forestry, Pretoria, 43 pp.
Du Plessis, H. M. & M. Van Veelen, 1991. Water quality: salinization and eutrophication time series and trends in South Africa. S. Afr. J. Sci. 87: 11–16.
Entwisle, T. J., 1989. Macroalgae in the Yarra River Basin: Flora and distribution. Proc. r. Soc. Victoria 101: 1–76.
Ferrar, A. A., 1989. Ecological flow requirements for South African rivers. S. A. National Scientific Programmes Report No. 162. C.S.I.R., Pretoria, 118 pp.
Fritsch, F. E., 1914. Contributions to our knowledge of the freshwater algae of Africa. 1. Some freshwater algae from Madagascar. Annales Biolog. lacustre 7: 40–59.
Fritsch, F. E., 1918. Contributions to our knowledge of the freshwater algae of Africa. 2: A first report on the freshwater algae of the S.A. Museum. Ann. S.A. Mus. 9: 433–611.
Fritsch, F. E. & F. Rich, 1924. Contributions to our knowledge of the freshwater algae of Africa.4. Freshwater and subaerieal algae from Natal. Trans. r. Soc. S.A. 11: 298–398.
Fritsch, F. E. & F. Rich, 1929a. Contributions to our knowledge of the freshwater algae of Africa. 7. Freshwater algae (exclusive of diatoms) from Griqualand West. Trans. r. Soc. S.A. 18: 1–92.
Fritsch, F. E. & F. Rich, 1929b. Contributions to our knowledge of the freshwater algae of Africa. 8. Bacillariales (diatoms) from Griqualand West. Trans. r. Soc. S.A. 18: 93–123.
Fritsch, F. E. & F. Rich, 1937. Contributions to our knowledge of the freshwater algae of Africa. 13. Algae from the Belfast Pan, Transvaal. Trans. r. Soc. S.A. 25: 153–228.
Fritsch, F. E. & E. L. Stephens, 1921. Contributions to our knowledge of the freshwater algae of Africa. 3. Freshwater algae (exclusive of diatoms), mainly from the Transkei territories, Cape Colony. Trans. r. Soc. S.A. 9: 1–72.
Gauthier-Lievre, L. 1964a. Oedogoniacees Africaines. Nova Hedwigia 7: 153–255.
Gauthier-Lievre, L., 1964b. Oedogoniacees Africaines. Nova Hedwigia 7: 257–568.
Gauthier-Lievre, L., 1965. Zygnemacees Africaines. Nova Hedwigia Beiheft No. 20, Cramer, Weinheim.
Groves, J., 1925. *Chara macropogon* Braun in South Africa. J. Bot. 63: 183.
Groves, J. & E. L. Stephens, 1926. New and noteworthy South African Charophyta. 1. Trans. r. Soc. S.A. 13: 145–157.
Groves, J. & E. L. Stephens, 1933. New and noteworthy South African Charophyta. II. Trans. r. Soc. S.A. 21: 271–289.
Hancock, F. D., 1973. Algal ecology of a stream polluted through gold mining in the Witwatersrand. Hydrobiologia 43: 189–229.
Harrison, A. D., P. Keller & D. Dimovic, 1960. Ecological studies on Olifantsvlei, near Johannesburg. Hydrobiologia 15: 243–284.
Hart, R. C. & B. R. Allanson, 1984. Limnological criteria for management of water quality in the southern hemisphere. S.A. National Scientific Programmes Report No. 93 C.S.I.R. Pretoria.
Harvey, W. H., 1838. Genera of South African plants. A. S. Robertson. 21 Heerengracht, Cape Town, 427 pp.
Hodgetts, 1926. Contributions to our knowledge of the freshwater algae of Africa. 6: Some freshwater algae from Stellenbosch. Trans. r. Soc. S.A. 13: 49–103.
Huber-Pestalozzi, G., 1930. Algen aus dem Knysnawalde in Sudafrika. Z. Bot. 23: 443–480.
Hutchinson, G. E., G. E. Pickford & J. F. M. Schuurman, 1932. A contribution to the hydrobiology of pans and other inland waters of South Africa. Arch. Hydrobiol. 24: 1–154.
John, D. M., 1986. The inland waters of tropical West Africa. Advances in Limnology 23: 1–244.
Joska, M. A. P. & J. J. Bolton, 1994. Preliminary investigation into algal weeds in inland waters 1992–1993. Report No. 426/1/94. Water Research Commission, Pretoria, 66 pp.
Kützing, F. T., 1849. Species algarum. Brockhaus, Leipzig.
Marloth, I., 1913. The flora of South Africa. Vol. 1. London, 264 pp.
Nygaard, G., 1932. Contributions to our knowledge of the freshwater algae of Africa. 9. Freshwater algae and phytoplankton from the Transvaal. Trans. r. Soc. S.A. 20: 101–148.
O'Keeffe, J. H., 1986. Ecological research on South African rivers – a preliminary synthesis. S.A. National Scientific Programmes Report No. 121 F.R.D., Pretoria, 121 pp.
Pocock, M. A., 1937. *Hydrodictyon* in South Africa. With notes on the known species of *Hydrodictyon*. Trans. r. Soc. S.A. 24: 263–280.
Pocock, M. A., 1960a. *Haematococcus* in Southern Africa. Trans. r. Soc. S.A. 36: 5–55.
Pocock, M. A., 1960b. *Hydrodictyon*: A comparative biological study. J. S. Afr. Bot. 26: 167–319.
Pocock, M. A., 1966. Freshwater algae in southern Africa. Unpublished Manuscript, 231 pp.
Printz, A., 1920. Subaerial algae from South Africa. Det. Kgl. Norske Videnskabers Selskabs Skrifter 1: 3–41.
Quinn, J. M., 1991. Guidelines for the control of undesirable biological growths in water. Consultancy Report No. 6213/2 Ministry for the Environment, Wellington, New Zealand.
Rand Water Board 1993. Water – the essence of life : Water – ons lewensbron. Rand Water Board, P.O. Box 1127, Johannesburg 2000, 16 pp. (Brochure)
Rawhani, S., 1991. To irrigate or not to irrigate: the South African dilemma. In Extended abstracts of the Southern African Irrigation Symposium, Durban 1991: 19–20.
Reinsch, P., 1878. Freshwater algae from the Cape of Good Hope. J. Linn. Soc. Bot. 16: 232–248.
Rich, F., 1932. Contributions to our knowledge of the freshwater algae of Africa. 10. Phytoplankton from S. African pans and vleis. Trans. r. Soc. S.A. 20: 149–188.
Rich, F., 1935. Contributions to our knowledge of the freshwater algae of Africa. 11. Algae from a pan in Southern Rhodesia. Trans. r. Soc. S.A. 23: 107–160.
Rich, F., 1936. Contributions to our knowledge of the freshwater algae of Africa. 12. Some diatoms from the Victoria Falls. Trans. r. Soc. S.A. 24: 207–220.
Schuurman, J. F. M., 1932. A seasonal study of the microflora and microfauna of Florida Lake, Johannesburg, Transvaal. Trans. r. Soc. S.A. 20: 230–386.
Shillinglaw, S. N., 1980. Common algae found in South African impoundments. Report No. TR 106. Department of Water Affairs, Forestry & Environmental Conservation, Pretoria, 34 pp.
Toerien, D. F., 1975. A review of eutrophication and guidelines for its control in South Africa. C.S.I.R. Special Report. Wat. 48. Pretoria. 110 pp.

Truter, E., 1987. An aid to the identification of the dominant and commonly occurring genera of algae observed in some South African impoundments. Report No. TR 135. Department of Water Affairs, Pretoria, 101 pp.

Weintroub, D., 1953. A preliminary account of the aquatic and sub-aquatic vegetation and flora of the Witwatersrand. J. Ecology 21: 43–57.

Welsh, H., 1961. Two new cyanophytes from the Transvaal. Nova Hedwigia, 3: 3.

Welsh, H., 1965. A Contribution of Our Knowledge of the Blue-Green Algae of South West Africa and Bechuanaland. Nova Hedwigia 9: 1–4.

West, G. S., 1912. Percy Sladen memorial expedition in South-West Africa, 1908–1911. Report No. 14. Fresh-water algae. Ann. S.A. Mus. 9: 61–90.

West, W. & G. S. West, 1897. Welwitsch's African freshwater algae. J. Bot. 35: 113–122.

Whitton, B. A., 1970. Biology of *Cladophora* in freshwaters. Water Res. 4: 457–476.

Whitton, B. A., I. G. Burrows & M. G. Kelly, 1989. Use of *Cladophora glomerata* to monitor heavy metals in rivers. J. Appl. Phycol. 1: 292–299.

Yamanouchi, S., 1913. *Hydrodictyon africanum*, a new species. Bot. Gazette 55: 74–79.

Hydrobiologia **340:** 301–305, 1996.
J. M. Caffrey, P. R. F. Barrett, K. J. Murphy & P. M. Wade (eds), Management and Ecology of Freshwater Plants.

Towards understanding the nature of algal inhibitors from barley straw

Irene Ridge & J. M. Pillinger*
*Departments of Biology and *Earth Sciences, The Open University, Milton Keynes, MK7 6AA, UK*

Key words: algal control, barley straw, lignin, oxidised polyphenolics, brown-rotted wood, white-rotted wood.

Abstract

The algal inhibitors released from barley straw decomposing in water and providing the basis for its use in algal control could be either of microbial origin or derived from straw components. We report here that unrotted straw releases algal inhibitors if finely chopped or autoclaved, providing further support for the view that straw, and not microbial colonists, is the primary source of inhibitors. Further support is also provided for the suggestion that inhibitors are or derive from oxidised lignin. Comparisons of lignin-enriched wood (brown-rotted) with lignin-depleted wood (white-rotted) from various deciduous trees show high antialgal activity of the former and little or no activity of the latter. Preliminary studies have shown that solubilised lignin is present in the liquor from rotted barley straw and brown-rotted wood. Since, however, the antialgal effects of deciduous leaf litter appear to depend initially on release of tannins and given that alkaline, oxidising conditions are usually essential for antialgal activity, it is proposed that oxidised polyphenolics, derived from lignin or tannins, are a source of algal inhibitors from plant litter.

Introduction

Barley straw decomposing in water releases algal inhibitors (Welch et al., 1990; Gibson et al., 1990) and is now well established as an effective means of controlling nuisance algae (Ridge & Barrett, 1992; Newman & Barrett, 1993). In order to optimize the technique, predict the effects on freshwater communities and assess more carefully the likely side-effects, it is necessary (1) to characterise the inhibitor(s) and understand (2) how it is produced and (3) acts on algae. Here we address the first two points and present further evidence that: (a) the role of micro-organisms relates primarily to the release from straw and not the synthesis of algal inhibitors; (b) inhibition is associated closely with the oxidation and solubilization of lignin.

Barley straw typically becomes antialgal after 1–3 months in well-aerated water and remains active against a wide range of green algae and cyanobacteria for at least 6 months at dose rates in the range 3–50 g m^{-3}. Active components are unstable in water (Welch et al., 1990). It must be pointed out, however, that the very low dose rates of straw active in field situations do not always inhibit test algae (*Chlorella vulgaris* Beijerinck and *Microcystis aeruginosa* Kutz emend Elenkin 1924) in standard, two to four-day laboratory bioassays. We commonly find that consistent inhibition in bioassays requires dose rates of straw in the range 0.5–1.0 kg m^{-3}; Jelbart (1993) was unable to inhibit *M. aeruginosa* in bioassays using liquor from rotting straw at an effective rate of 0.1 kg m^{-3}. Newman and Barrett (1993), however, obtained inhibition of *M. aeruginosa* in bioassays in the presence of liquor from the decomposition tank and rotting straw at 2.57 g m^{-3}, which is close to the minimum rate effective in field conditions.

Microbial decomposition is generally considered to be necessary for straw to become antialgal. Welch et al. (1990) found that unrotted straw was not antialgal; and Gibson et al. (1990) reported that when rotted straw was autoclaved to stop microbial action, the straw no longer prevented growth of *C. vulgaris, Cladophora glomerata* (L.) Kuetz or *Selenastrum capricornutum* Printz 1913 an observation which we have confirmed using the cyanobacterium *M. aeruginosa* as the test organism (unpublished data). Newman & Barrett

(1993) suggested that the main requirements for straw to be active are (i) maintenance of aerobic conditions and (ii) an active and diverse microflora. The apparent requirement for some degree of rotting has thus led to the assumption amongst many workers that the inhibitory compounds are of microbial origin. We have studied the mycoflora of rotting straw and concluded that, on balance, the widely varying fungal flora which is associated with different batches of decomposing straw in different water bodies is unlikely to account totally for the prolonged antialgal activity of the straw. It is quite possible that some fungi on straw produce antialgal substances and we did indeed isolate two fungi, albeit uncommon and unusual species, which inhibited *C. vulgaris*, although not *M. aeruginosa* (Pillinger et al., 1992). To our knowledge no one has investigated the production of algal antibiotics in water by either bacteria or actinomycetes. Our working hypothesis, however, is that algal inhibitors associated with rotting barley straw do not arise primarily by microbial synthesis.

If the above hypothesis is correct, the most likely explanation of the need for an active microflora is that it acts to liberate some inhibitory component(s) from straw and/or enhance the activity of such components. Any procedure which mimics microbial decomposition might then be expected to release inhibitors and we present evidence here that non-decomposed straw can, after certain physical or chemical treatments, cause algal inhibition.

Regarding the chemical nature of inhibitors, only phenolics among the relatively few components of barley straw have a known potential to inhibit algae (Dedonder & van Sumere, 1971) and only covalently-bound wall components could persist for the long period over which straw remains active. Simple phenolics (0.6% w/w) have been ruled out as likely sources of inhibitors because they are active only at millimolar concentrations and present in insufficient amounts to explain the prolonged antialgal effects (Newman & Barrett, 1993; Pillinger et al., 1994). By a process of elimination, therefore, lignin (15% w/w) emerged as the most likely source of algal inhibitors from barley straw. Chemical analysis, using a technique applicable to the detection of macromolecular material, pyrolysis gas chromatography-mass spectrometry (pyGCMS), indicated the presence of lignin-derived material in the aqueous phase of barley straw rotting in tanks of water (Pillinger, 1993; Pillinger et al., 1993). Lignin, which is generally regarded as highly insoluble, can therefore be solubilised from barley straw in conditions where the straw is antialgal. Evidence that solubilised lignin has antialgal acitivity derives from work with lignin-enriched wood samples which result from the action of brown-rot fungi. Brown-rotted wood inhibits algal growth (Ridge et al., 1995) and pyGCMS analysis showed that a crude preparation of soluble lignin from the steep liquor of this wood contained lignin derivatives (Pillinger et al., 1993 and unpublished). After redissolving the lignin preparations, they showed approximately the same level of antialgal activity as the original wood liquor from which they were obtained. We present further evidence here that solubilised lignin is closely involved in the generation of algal inhibitors and argue that the procedures used to release inhibitors from unrotted straw are those which would bring about lignin solubilisation.

Methods

Barley straw showing antialgal activity was obtained by incubating in tap water for at least four weeks at 18–25 °C with vigorous aeration (Pillinger et al., 1994). Details of the laboratory bioassay used to demonstrate antialgal activity using cultures of either *C. vulgaris* (CCAP 211/12) or *M. aeruginosa* (CCAP 1450/6) are described in Pillinger et al., 1994. The standard algal culture medium modified after Jaworski (Pillinger, 1993) included Hepes buffer (20 mM) to maintain a constant pH of 8.2 and bioassays were of three or four days duration. Algal growth was quantified by chlorophyll *a* extraction and results for different treatments (five replicates) are expressed as a percentage of the control (10 replicates). Controls were always in log phase growth with an absorbance value at 665 nm for extracted chlorophyll *a* of between 0.4 and 0.6 arbitrary units (equivalent to 1000–1550 mg l^{-1}) at the end of the bioassay; this equated to a cell count ml^{-1} in the range $2\text{–}3 \times 10^6$.

In experiments using fresh (unrotted) straw, samples were chopped coarsely (5 cm) or finely (to pass through a 1.4 mm sieve) and either added directly to sterile algal medium immediately prior to bioassay or autoclaved in the medium (121 °C, 15 min) at a rate of 10 g straw/l and bioassayed after cooling (16–18 h). In experiments using rotting wood, samples were collected in the field, dried, crushed in a mortar and pestle to a powdery consistency and passed through a 1 mm sieve before use. Weighed samples were autoclaved in algal medium (121 °C, 15 min) and left to cool for 16–18 h before carrying out bioassays. Species

Table 1. Inhibition of *C. vulgaris* by fresh, unrotted barley straw after physical disruption or autoclaving. Results (average of 5 replicates) are expressed as % of control growth; ***, significantly different from control ($p<0.001$, *t*-test).

Straw autoclaved	% Control growth	
	5 cm straw	< 1.4 mm straw
–	102	30***
+	35***	21***

used were crack willow (*Salix fragilis*), elm (*Ulmus procera*), birch (*Betula* sp.), oak (*Quercus robur*) and sycamore (*Acer pseudoplatanus*) collected from various locations in the UK.

Results

Antialgal activity of unrotted straw

Fresh, coarsley chopped barley straw (approximately 5 cm lengths) is inactive in algal bioassays. However, if the straw is chopped more finely (< 2 mm) activity is observed against *C. vulgaris* and, if autoclaved in the algal medium, coarsely chopped straw also shows activity (Table 1). Preliminary experiments using *M. aeruginosa* indicate that it is considerably more sensitive than *C. vulgaris* to the inhibitors from unrotted straw: finely chopped straw produced 90% inhibition of growth at dose rates of 7 g/l without autoclaving and 4 g/l after autoclaving in the algal medium (*cf* data for 1.4 mm straw in Table 1).

Comparisons of brown- and white-rotted wood

Brown-rotted wood from a range of deciduous tree species strongly inhibited the growth of *C. vulgaris* at dose rates between 1 and 10 g/l (Figure 1 and Ridge et al., 1995). By contrast, lignin-depleted wood resulting from attack by white-rot fungi shows little or no activity against *C. vulgaris* at similar dose rates (Figure 1). Only white-rotted willow at the highest dose rate (4 mg l^{-1}) produced algal inhibition. Brown- and white-rotted wood from the same species (elm and birch) showed this difference so that type of rotting rather than type of wood correlates with antialgal activity.

Discussion

Making the assumption that inhibitors from fresh barley straw are chemically similar to those released during microbial decomposition, the data in Table 1 provide further evidence that components of straw rather than microbial products are the source of algal inhibition. This conclusion is strengthened by the preliminary finding that alkaline pretreatment of straw effectively mimics the early stages of microbial decomposition and eliminates the lag period before straw becomes antialgal: straw treated with concentrated sodium hydroxide solution, washed and then placed in an aerated decomposition tank became antialgal within 4 days (unpublished data). Alkaline hydrolysis thus provides a model for at least some aspects of microbial action which lead to the formation of active, rotted straw. Both autoclaving and alkali treatment have been reported to effect the abiotic liberation of lignin from straw (Vered et al., 1981; Moss et al., 1990). We have also shown using pyGCMS analysis that lignin-derived material is present in the aqueous phase of unrotted barley straw autoclaved in bioassay medium (Pillinger & Gilmour, unpublished). Since soft-rot bacteria and fungi, a physiological group frequently represented by isolates from straw rotting in water (Pillinger, 1993), can facilitate lignin release, there is a clear similarity between the conditions, biotic or abiotic, which lead to antialgal activity and those which cause solubilisation of lignin.

The comparison of brown- and white-rotted wood (Figure 1) provides additional evidence that solubilised lignin is associated with antialgal activity. Using pyGCMS, we have analysed in more detail the differentially rotted samples of elm and made the further suggestion that the chemical structure, in addition to overall amount, of lignin may be significant in conferring algal toxicity (Pillinger et al., 1995). A critical factor appears to be the oxidation potential or concentration of readily oxidised groups present, which links to our earlier suggestion (Pillinger et al., 1994) that phenolic hydroxyl groups will undergo abiotic oxidation during or after lignin solubilisation (given the alkaline, oxidising conditions necessary for antialgal activity). Oxidation may facililtate lignin solubilization and/or enhance toxicity of the solubilised material (Pillinger et al., 1994).

Our tentative conclusion, therefore, is that algal inhibition by barley straw (and brown-rotted wood) is associated with the solubilization and oxidation of lignin. The size range of active lignin fragments and

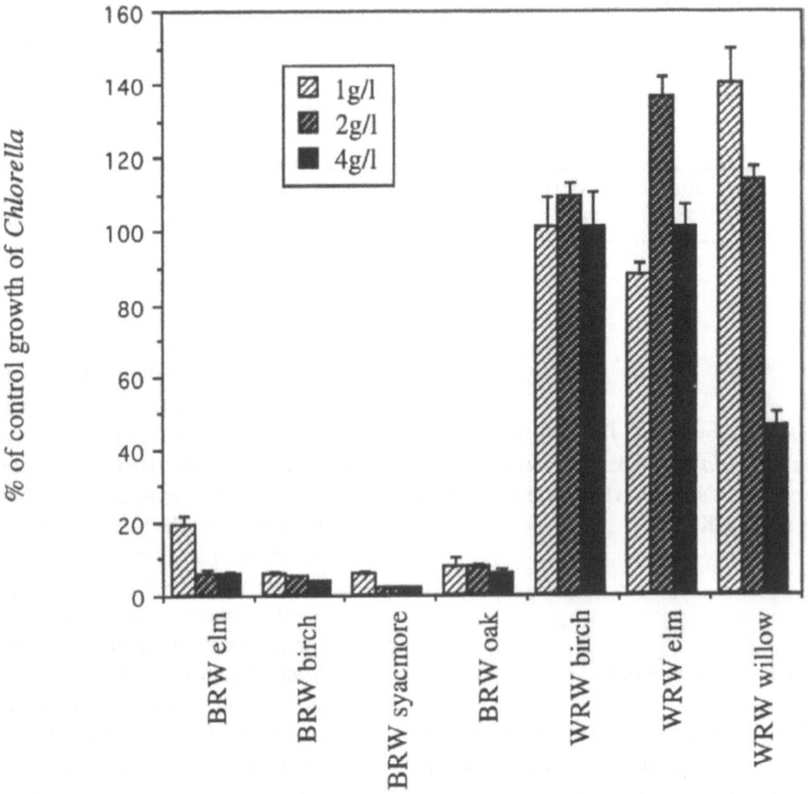

Figure 1. Effect on the growth of *C. vulgaris* of brown-rotted and white- rotted wood samples (BRW and WRW) from deciduous trees. Error bars show SE of the normalised means (n = 5).

possible synergistic effects of low molecular weight lignin-derived phenolics have not yet been investigated. However, deciduous leaf litter is also antialgal (Ridge et al., 1995) and inhibition here appears to depend initially on oxidised tannins. The general source of algal inhibitors can thus be defined more accurately as oxidised polyphenolics. These naturally occurring substances enter water bodies constantly through litter inputs and their oxidation leads ultimately to the formation of humic substances. We can only speculate as to why barley straw is such a potent source of algal inhibitors. The explanation may lie in the structure of its lignin, which could influence both solubilisation and ease of oxidation: barley straw decomposes more slowly than wheat or linseed straw, neither of which shows prolonged antialgal activity (J. Newman, pers. comm.), suggesting that gradual release of lignin over a long period is a prerequisite for sustained high activity. An urgent need now is to identify more precisely the structural features of polyphenolics that influence their ease of oxidation and potency in algal inhibition.

Acknowledgements

We thank Dr Iain Gilmour for collaboration involving pyGCMS studies of lignin and Dr J. R. Newman and Mr P. R. F. Barrett for helpful discussions. Technical help from Mrs Tina Wardhaugh and Mr John Walters is gratefully acknowledged.

References

Dedonder, A. & C. F. van Sumere, 1971 The effect of phenolics and related compounds on the growth and respiration of *Chlorella vulgaris*. Z. Pflanzenphysiol. Bd 65: 70–80.

Gibson, M. T., I. M. Welch, P. R. F. Barrett & I. Ridge, 1990 Barley straw as an inhibitor of algal growth II: laboratory studies. J. appl. Phycol. 2: 241–48.

Jelbart, J. 1993 A laboratory investigation of the effect of rotting barley straw on the cyanobacterium *Microcystis aeruginosa*. Water 20: 31–33.

Moss, A. R., D. I. Givens & J. M. Everington, 1990 The effect of sodium hydroxide treatment on the chemical composition, digestibility and digestible energy content of wheat, barley and oat straws. Animal Feed Science Technol. 29: 73–87.

Newman, J. R. & P. R. F. Barrett, 1993 Control of *Microcystis aeruginosa* by decomposing barley straw. J. Aquat. Plant Mgmt 31: 203–206.

Pillinger, J. M., 1993 Algal control by barley straw: an interdisciplinary study. PhD Thesis, The Open University, Milton Keynes, UK

Pillinger, J. M, J. A. Cooper & I. Ridge, 1994 Role of phenolic compounds in the antialgal activity of barley straw. J. Chem. Ecol. 20: 1557–1569.

Pillinger, J. M, I. Gilmour & I. Ridge, 1993 Control of algal growth by lignocellulosic material. In J. C. Duarte, M. C. Ferreira & P. Ander (eds), Lignin biodegradation and transformation: Biotechnological Applications. Abstr.. FEMS Symp., Lisbon, Portugal, April 1993: 57–58.

Pillinger, J. M, I. Gilmour & I. Ridge, 1995. Comparison of the antialgal acitivity of brown-rotted and white-rotted wood and *in situ* analysis of lignin. J. Chem. Ecol. in press.

Pillinger, J. M., J. A. Cooper, I. Ridge & P. R. F. Barrett, 1992 Barley straw as an inhibitor of algal growth III: the role of fungal decomposition. J. Appl. Phycol. 4: 353–355.

Ridge, I. & P. R. F. Barrett, 1992 Algal control with barley straw. In Vegetation management in forestry, amenity and conservation areas. Aspects of applied Biology 29, Association of Applied Biologists: 457–62.

Ridge, I., Pillinger, J. M. & J. Walters, 1995. Alleviating the problems of excessive algal growth. In D. M. Harper & A. J. D. Ferguson, (eds), The Ecological basis for river management. Wiley Publishers, Chichester, England: 211–18.

Vered, Y., O. Milstein, H.M. Flowers & P. Gressel, 1981. Biodegradation of wheat straw lignocarbohydrate complexes (LCC) 1. Dynamics of liberation of hot aqueous LCCs from wheat straw and partial characterisation of the products. Eur. J. appl. Microbiol. Biotechnol. 12: 183–188.

Welch, I. M., P. R. F. Barrett, M. T. Gibson & I. Ridge, 1990 Barley straw as an inhibitor of algal growth I: studies in the Chesterfield Canal. J. Appl. Phycol. 2: 231–39.

Hydrobiologia **340:** 307–311, 1996.
J. M. Caffrey, P. R. F. Barrett, K. J. Murphy & P. M. Wade (eds), Management and Ecology of Freshwater Plants.

The control of diatom and cyanobacterial blooms in reservoirs using barley straw

[1]P. R. F. Barrett, [2]J. C. Curnow, [3]J. W. Littlejohn
[1] *8 Sunderland Avenue, Oxford OX2 8DX, England*
[2] *Environmental Medicine, Grampian Health Board, Aberdeen, Scotland*
[3] *Grampian Regional Council, Depatment of Water Services, Aberdeen, Scotland*

Key words: Barley straw, algal control, potable supply reservoirs, diatoms, cyanobacteria

Abstract

A potable supply reservoir, with a long history of diatom blooms in spring and cyanobacterial blooms in summer, was treated with barley straw in March 1993 with subsequent additions in December 1993 and June 1994. Within two months of the initial treatment, algal numbers started to fall compared with previous years and have remained consistently lower throughout 1993 and 1994. Cyanobacteria have not bloomed and cell numbers remained low. Chemical analysis of the water showed locally elevated concentrations of geosmin close to the straw on one occasion but the overall concentration of this and a range of other organic molecules remained within acceptable limits and at concentrations similar to those found in other untreated reservoirs in the region. Observed and potential advantages to public health and potable supply management resulting from the use of barley straw are discussed.

Introduction

The production of high quality water from potable supply reservoirs can be adversely affected by algal blooms. Diatom blooms regularly occur in spring causing taint and odour to the water and blocking filtration systems. Blooms of cyanobacteria occur mainly during the summer months also causing taint and odour problems but may also produce toxins. These blooms are not new but problems caused by these potentially toxic algae escalated dramatically in Britain in 1989 during a period of high temperatures and low rainfall. This was recorded in a report by the National Rivers Authority (NRA, 1990). Subsequently, there appears to have been a sustained increase in the numbers of toxic blooms. These were originally thought to have been caused by the hot dry conditions prevailing in the summers of 1990 and 1991 but problems continued to occur during the subsequent summers of 1992 and 1993 which were cooler and had higher rainfall. The NRA Report (NRA, 1990) listed a number of control measures which were thought to control cyanobacteria. Although some of these measures have produced beneficial levels of control, none of them has been completely successful so far in eliminating problem growths of cyanobacteria or diatoms. Many of these control measures are expensive to operate and only achieve an effect in a limited range of water bodies.

A novel technique for the control of at least some species of algae, mentioned only briefly in the NRA Report (1990), involves the introduction of barley straw into water. The controlling properties of barley straw when rotting in water were first reported by Welch et al. (1990). Subsequent research by a number of authors, for example Barrett and Newman (1993), has shown the susceptibility of a wide range of algal species to the effects of barley straw. Both green algae and cyanobacteria are inhibited by an unknown substance or substances, released presumably during the decomposition of the straw. These authors presented data from laboratory and field trials which showed that effective algal control could be achieved with as little as 2.5 g straw m^{-3} of water. The publication of this early research led to worldwide interest and the Aquatic Weeds Research Unit (AWRU) of Long Ashton Research Station has received reports of successful control of algae from many countries including Australia, Canada, Ireland, South Africa, Sweden and the USA, as well as many regions of the United Kingdom. These reports show that the technique has wide potential and is not limited to British conditions. A detailed description of the research which led to the development of this technique is given by Barrett (1994).

The observation that cyanobacteria appear to be particularly susceptible to the active agent released by barley straw (Newman & Barrett, 1993) offers a possible method of algal control in potable supply reservoirs, particularly where other methods have failed or are inappropriate. Apart from the relatively low cost of barley straw, it appears to have a number of advantages over most other forms of control. Firstly, it has been shown to be selective to algae (Newman et al., 1994) so that suppression of algal blooms may allow recolonisation by vascular plants which would subsequently compete with, and so further reduce, algal growth. Secondly, a single application of barley straw can inhibit the growth of algae for 6–8 months (Ridge & Barrett, 1992) so that only one or two applications each year may be necessary. Thirdly, straw applied in a form suitable for algal control also provides a good habitat for some species of invertebrate animals which can increase in numbers and so benefit fish and wildfowl (Street, 1978). Fourthly, the long-lasting 'slow release' effect of the straw allows treatment before bloom formation and inhibition of growth throughout the summer.

The data from trials carried out by the AWRU and from reports received from other treatments with straw showed that many species of green algae as well as cyanobacteria are susceptible to the effects of barley straw. However, no trials had been carried out specifically on the control of diatoms although some chance observations during other trials suggested that these algae are either resistant or only slightly susceptible to the effects of straw.

Pillinger (1993) proposed that the anti-algal factor in straw might be associated with oxidised phenolic compounds some of which are highly anti-algal and suggested that the active fraction in barley straw was of a lignin origin. Although the anti-algal factor has not been identified, Barrett (1994) proposed the hypothesis that the lignin fraction of the straw might be oxidised under aerobic conditions, which are known to be essential to the production of the anti-algal factor, to quinones and, subsequently, to humic acids. The action of sunlight on humic acids in water catalyses the production of singlet oxygen or hydrogen peroxide (Cooper & Zika, 1983). Hydrogen peroxide has been shown to inhibit the growth of *Microcystis aeruginosa* at concentrations as low as 2 mg l^{-1} (Barrett & Newman, 1992). Cooper and Ziba, (1983) showed that considerably higher levels of peroxide could occur in waters containing naturally occurring levels of humic acids.

Table 1. Mean Monthly Algal Counts in Reservoir 1 Cells per ml

Year	1991	1992	1993	1994
January	N/C	10,000	10,000	400
February	13,000	18,000	17,500	6,200
March	21,500	28,000	22,800*	7,800
April	57,400	38,000	29,000	8,700
May	67,500	25,200	14,500	3,400
June	45,000	16,500	6,800	106*
July	N/C	17,700	3,000	57
August	2,000	10,500	4,500	440
September	N/C	OS	1,500	
October	4,000	OS	1,000	
November	10,000	4,000	1,000	
December	2,000	7,000	1,000*	

* Straw introduced after the sampling dates in March and December 1993 and June 1994. OS Dense blooms of Anabaena caused reservoir to be taken temporarily out of service and cell counts were suspended. N/C Not counted. Samples for algal enumeration were fixed using Lugol's Iodine solution, concentrated by sedimentation and counted under the microscope in a Sedgewick- Rafter chamber. For Anabaena spp, each filament length of 22 cells (about one loop) was taken as equivalent to one unit.

Peroxides are known to have bactericidal properties and, as straw had already been shown to control cyanobacteria following exposure for several weeks, it was thought possible that the same control of some of the potentially pathogenic organisms found in water might also be achieved. However, there was concern that rotting barley straw might release chemicals into the water which could cause taint and odour or other problems complicating the treatment process.

In order to test the ability of barley straw to control cyanobacteria and pathogenic micro-organisms, and to determine if any undesirable chemicals were released into the water from the straw, a trial was established in a water body with a known history of algal blooms and where routine analysis of chemical and bacterial contaminants was available. This water body suffered not only from a regular summer bloom of cyanobacteria but also from spring blooms of diatoms and thus offered the opportunity to test the effects of straw on

Table 2. Dissolved organic compounds found in the Reservoir following treatment with straw.

Substances	Concentration (ng l-1) in water					
	Above straw		Close to straw		At exit point	
	1993	1994	1993	1994	1993	1994
Geosmin	3.3	<0.1	7.0	1.0	4.3	1.2
2 Methylisoborneol	<1.0	<1.0	<1.0	<1.0	1.0	<1.0
Phenol	<10.0	<10.0	<10.0	<10.0	<10.0	<10.0
para-Cresol	<10.0	<10.0	91.0	<10.0	17.0	<10.0
4-Ethylphenol	<10.0	<10.0	<10.0	<10.0	17.7	<10.0

diatom blooms. The trial was set up in the spring of 1993 in the Grampian Region of Scotland as a collaborative project between the Grampian Health Board, Grampian Water Services and the AWRU. The data presented below are the results from 1993 and the early part of 1994 of an ongoing study which is projected to continue for at least two more years.

Methods and materials

The principal site was an impounding reservoir with a surface area of 25 000 m^2 and a normal capacity of 250 000 m^3 although a recent survey suggests that roughly 25% of the volume is now silt. It is fed by two streams which drain surrounding land used for mixed farming. The associated treatment works, which includes coagulation, direct pressure filtration and chlorination, supplies 3.5 megalitres of water per day to a nearby town. Normally, the reservoir is subject to two annual blooms of algae (Table 1), one of the diatom *Asterionella* sp, generally lasting from March to June and a second of, usually, the cyanobacteria *Anabaena* sp occurring later in the summer and autumn. This autumn bloom is prone to raft formation which severely disrupts the treatment process and, in two recent years when the blooms became toxic, an alternative water source has had to be found for customers.

In March 1993, a total of seven tonnes of barley straw was introduced into the reservoir at three sites situated at each stream inlet and around the draw-off tower. The straw was contained in marine trawler nets supported by floats. This quantity was equivalent to a uniform dose of 38 g of straw m^{-3} throughout the reservoir. In the following December, more barley straw was added at regular spacings throughout the reservoir giving an additional dose equivalent to 6.5 g straw m^{-3}. This straw was also contained in fishing net but packed more loosely to improve water movement and aeration. In June 1994, an additional quantity of straw equivalent to 7.6 g straw m^{-3} was added but, in this instance, it was packed into tubular plastic netting using a tree wrapping device. At this time, the remains of the original seven tonnes of straw was removed.

Water samples were taken and analysed throughout the year to establish algal type, cell numbers and chlorophyll-*a* concentration. In addition, a general survey of dissolved organic compounds was made using gas chromatography/mass spectrometric analysis (GC-MS) at three points representing inlet water, water adjacent to the straw and water just prior to draw-off.

Results

Cell counts

The addition of straw to the reservoir coincided with a fully established bloom of the diatom *Asterionella* sp. As expected, the straw had little immediate effect and the bloom ran its course and ended in mid-June. Thereafter, the pattern of algal growth changed from that experienced in previous years and the water remained clear for the remainder of 1993 with the bottom of the reservoir clearly visible for several metres from the bank, which was unusual. Algal counts supported the visual observations. Cell numbers at the time of treatment were in excess of 20 000 cells ml^{-1} with chlorophyll-*a* concentrations at 100 $\mu g\ l^{-1}$ but then fell progressively. There was a bloom of Tabellaria sp in late May but this was short lived and by early July counts had fallen to 3000 cells ml^{-1} with a chlorophyll-*a* concentration of about 10 $\mu g\ l^{-1}$. These levels were sustained with minor variations throughout the year and have remained at significantly lower lev-

Table 3. Concentrations of compounds released into distilled water over 24 hours.

Substance	Concentration (ng l-1) in water
Geosmin	11.4
2-Methylisoborneol	<1.0
Phenol	370.5
para-cresol	694.1
4-Ethylphenol	549.8

Table 4. Mean monthly geosmin concentration (ng l^{-1}) in selected reservoirs in Grampian Region, 1993.

	Reservoir				
Month	1*	2	3	4	5
April	–	–	<1.0	–	–
May	4.3	<1.0	<1.0	4.0	1.7
June	5.6	3.2	3.6	4.5	<1.0
July	3.6	2.0	3.3	3.3	<1.0
August	1.3	2.0	3.7	1.0	–
September	1.7	<1.0	–	0.8	<1.0
October	<1.0	<1.0	–	<1.0	<1.0
November	<1.0	<1.0	–	<1.0	–

* Straw treated

els compared with previous years to the present time (Table 1).

Chemical analysis

Screening for over 60 organic compounds, including a range of herbicides, is routinely carried out either by purge and trap or by extraction technique as part of the water quality monitoring programme. Chemical analysis of water from the reservoir identified a number of organic compounds, including para-cresol and 4-ethylphenol which may have been associated with the introduction of the straw but they were at very low concentrations (1–100 ng l^{-1}). A taste-producing compound, geosmin (Trans–1,10-dimethyl-trans-9-Decalol) was detected but exit concentrations were commensurate with those found in other reservoirs not containing straw. On one occasion, 2-Methylisoborneol, another known taste-producing compound, was also detected (Table 2). Other than these, the compounds monitored have either not been detected, or detected at limit of detection values. Further work to expand the range of compounds monitored is currently being planned.

The potential for these compounds to be released from straw was demonstrated by taking some straw from one of the nets in the reservoir and soaking it in a beaker of distilled water for 24 hours. The results obtained from analysis of the filtrate (Table 3) indicate relatively high concentrations of phenolic compounds and geosmin.

Geosmin is one of the taste producing chemicals commonly associated with lowland sources and is generally attributed to biological activity in these waters. Analysis for this chemical is carried out in many of the reservoirs in the Grampian Region. Table 4 shows the concentrations of geosmin in five of these reservoirs during 1993. Reservoir 1, which was treated with straw, shows a relatively high concentration of geosmin during May, June and July after the straw had been added but it also had a high algal density in April and May which could have been responsible for this level.

Discussion

The reservoir used in this trial has a long history of cyanobacterial blooms with well recorded observations of algal types and cell counts. In attempting to control these blooms with barley straw, it was necessary to establish that no adverse effects could be demonstrated, particularly with respect to taste-producing compounds or other polluting substances released from the straw. While locally high concentrations of geosmin were found in water near the straw after the initial treatment in 1993, they had fallen to levels considered normal at the draw-off point. In subsequent treatments (autumn, 1993 and spring, 1994), the straw was applied in a more dispersed form and in smaller individual quantities. No elevated local concentrations of geosmin were detected around these smaller quantities and the concentration at the draw-off point remained well within acceptable limits. During the 17 month period of the trial, the level of all those chemicals tested under routine analysis has remained well within acceptable limits for a potable supply reservoir and there have not been any customer complaints attributable to the straw.

Although there appears to have been a marked reduction in algal populations in the reservoir over the two summers since the straw was first applied, the absence of a legitimate control means that no definite conclusions can yet be drawn. However, a comparison with algal levels in the two previous seasons suggests that the straw has greatly reduced the growth of cyanobacteria. In addition, there appears to have been an inhibition of diatom growth. Whether this is due to a direct effect of the anti-algal chemical or to some indirect change in water quality is unclear. Nevertheless, the control of diatom levels in the water in the spring of 1994, produced a marked reduction in the algal-produced odour which, in the past, was very noticeable at the treatment works.

The range of organic chemicals monitored after the addition of straw was necessarily limited to those routinely analysed by the water industry. It is possible that other unusual chemicals were produced but not identified by the extraction process. Nevertheless, the absence of any detected adverse effects from organic chemicals released into the water and the apparent improvement in water quality have been sufficiently encouraging to enable other reservoirs in the Region to be considered for treatment with straw. Apart from the benefits of reducing the growth of potentially toxic cyanobacteria, the reduction in cell numbers of any species or group of algae reduces the processing costs and improves the efficiency of the treatment process enabling other pathogens to be removed more effectively. Further results from the continuation of this, and other projected trials in the Grampian Region, will be published later.

Acknowledgments

The authors wish to acknowledge with thanks the considerable help given by the following: The Water Treatment Staff at Banff and Buchan Division; Water Services Laboratory, Turrif; Department of Microbiology, Aberdeen Royal Hospitals NHS Trust; Lorraine Ryan, Jason Nicol and Diane McGregor.

References

Barrett, P. R. F. & J. R. Newman, 1992. Annual Progress Report. Aquatic Weeds Research Unit, Broadmoor Lane, Sonning, Reading, U.K.

Barrett, P. R. F. & J. R. Newman, 1993. The control of algae with barley straw. PIRA Conference Proceedings: Straw - a valuable raw material. Paper 41.

Barrett, P. R. F., 1994. Field and laboratory experiments on the effects of barley straw on algae. Proceedings, BCPC Monograph 59, Comparing field and glasshouse pesticide performance II. 191–200.

Cooper, W. J. & R. G. Zika, 1983. Photochemical formation of hydrogen peroxide in surface and ground water exposed to sunlight. Science 220, 711–712.

NRA, 1990. Toxic blue-green algae. National Rivers Authority Report, Water Quality Series No 2.

Newman, J. R. & P. R. F. Barrett, 1993. Control of Microcystis aeruginosa by decomposing barley straw. J. aquat. Plant Mgmt 31: 203–206.

Newman, J. R., P. R. F. Barrett & T. G. Cave, 1994. The use of straw to control algae in drainage ditches – an ecological survey. Conference Proceedings, 'Nature Conservation in Drainage Habitats' Nottingham University, 1993.

Pillinger, J. M., 1993. Algal control by barley straw. Ph D Thesis, Department of Biology, The Open University, Milton Keynes. U.K.

Ridge, I. & P. R. F. Barrett, 1992. Algal control with barley straw. Aspects of Applied Biology 29, 457–462.

Street, M., 1978. Research on the improvement of gravel pits for waterfowl by adding straw. Game Conservancy Annual Review, 10, 56–61.

Welch, I. M., P. R. F. Barrett, M. T. Gibson & I. Ridge, 1990. Barley straw as an inhibitor of algal growth 1: studies in the Chesterfield Canal. J. Appl. Phycol. 2, 231–239.

Hydrobiologia **340**: 313–316, 1996.
J. M. Caffrey, P. R. F. Barrett, K. J. Murphy & P. M. Wade (eds), Management and Ecology of Freshwater Plants.

Multiple use of aquatic green biomass for food/feed protein concentrate, bioenergy and microbial fermentation products

V. N. Pandey & A.K. Srivastava
Experimental Botany Laboratory, Department of Botany, University of Gorakhpur–273009 (U.P.), India

Key words: Green biomass, aquatic weeds

Abstract

Fractionation of aquatic green biomass of three water weeds for multiple use through yield of protein concentrate, fibrous residue and whey (deproteinised juice) has been studied. The potential of protein concentrate for use as food/feed supplement, that of fibrous residue as ensilaged fodder/substrate for mushroom growth/production of bioenergy and of whey as substrate for microbial fermentation has been studied and is discussed.

Introduction

In North Eastern Uttar Pradesh region of India, there are many reservoirs and low lying areas where water persists throughout the year. These water bodies harbour abundant aquatic vegetation, which is of little use at present. Since removal of water weeds involves time and cost, it has been suggested that their use may be more advantageous than destruction (Mitchell, 1974). Utilization of aquatic plants as an additional source of food has been suggested (Bates & Hentages, 1976) and several workers have reported the promising potential of aquatic vegetation as raw material for leaf protein extraction (Boyd, 1968; Pandey & Srivastava, 1989, 1991a, 1991b). Such leaf protein concentrates (LPC) have high nutritive value and can be used for food/feed supplementation (Pirie, 1987). Fractionation of fresh, green biomass for its multiple use through yields of food/feed protein concentrate, and employment of by products, *viz.* fibrous residue and whey (deproteinised juice) for production of bioenergy (biogas) and microbial fermentation products has been suggested (Pirie, 1987).

The present study was undertaken to ascertain the use of three aquatic weeds, viz. *Eleocharis dulcis, Monochoria hastata* and *Veronica anagallis-aquatica* for yield, composition and nutritive value of their LPC, and employment of deproteinised juice (whey) as substrate for microbial fermentation as well as of the fibrous residue as substrate for mushroom and biogas production (Figure 1).

Materials and methods

Extraction and analysis of LPC

Fresh plants were collected randomly from dense stands of species growing naturally in water-logged sites near Gorakhpur. For each cut, fresh and dry biomass yields were recorded. Dry weight was determined after drying the sample at 80°C until a constant weight was achieved. For extracting LPC, 200g fresh leaf sample was homogenized and squeezed through muslin cloth to obtain the leaf extract, which was heated to 80°C for 20 min. The coagulated protein was separated by centrifugation and washed with water prior to vacuum drying (60°C).

Nitrogen content was determined by the micro method of Doneen (1932). Total protein content was calculated by multiplying the protein N value by 6.25. Lipid content was determined by Soxhlet extraction. Ash content was estimated by heating the samples to 54°C for 4h in a muffle furnace. Starch and crude fibre contents of LPC were determined according to Snell & Snell (1953) and AOAC (1980), respectively. Total soluble sugar was estimated by determining reducing and non-reducing sugars according to Snell & Snell (1951).

β-carotene content was measured following Arkcoll & Holden (1973). Percent extractabes of LPC, total N and protein N were calculated according to Pandey & Srivastava (1991 a).

The *in vitro* enzymatic digestibility of protein in LPC was assayed with pepsin followed by trypsin according to Saunders et al. (1973) and the results are expressed as percent protein in LPC digested on dry weight basis. Protein efficiency ratio was determined according to AOAC (1980). For determination of amino acid composition, 25 mg of defatted LPC was hydrolysed with 6N HCL in a sealed tube for 24h, at 110°C. After hydrolysis HCI was removed *in vacuo* and the residue was analysed in an amino acid analyzer (LKB–4101).

Analysis of whey

The whey sample, after autoclaving, was analysed for soluble carbohydrate by anthrone colorimetric method of Snell & Snell (1953) and total nitrogen by microkjeldahl method of Doneen (1932). For analysis of mircobial load of whey, the solidified whey sample (0.5 g agar powder/25 ml whey) was exposed to air for 6 hours, followed by incubation, as well as counting the colonies that appeared at 37°C for 24 h, for the growth of bacterial colonies and at 28°C for 48 h, for fungal colonies. Sodium propionate (0.5%) and benzyl penicillin (500 units/ml) were added to molten agar to check the bacterial colonies. In order to obtain actinomycetes colonies, 0.5% sodium propionate and 150μg/l actidione (cyclohaximide) were added to the molten agar to control the bacteria and fungi. Plates were incubated at 28°C for 7 days.

Cultivation of yeasts in whey

Sacchromyces cerviseae (AS–7) was grown in whey using a rotary shaker (100–120 rpm) at 28°C for 48 h. Growth of yeast was measured by counting the number of viable cells/ml of the whey sample after incubation for 48 h. The number of viable cells per ml of the original inoculum (20 × 10^3 cells/ml) was taken as control. Dry weight of yeast was measured after incubation.

Production of citric acid in Eleocharis *whey, fortified with sugar*

Spore suspension (5 ml) of *Aspergillus niger(AN–4)* was inoculated (inoculum size 22 × 10^6/ml) in whey (95ml, pH 6.0) fortified with sucrose (total sugar 10%).

Table 1. Yield and composition of green biomass, extractables of Nitrogen and LPC**

Parameters	*Eleocharis dulcis*	*Monochoria hastata*	*Veronica anagillis-aquatica*
Fresh biomass yeild (kg ha^{-1})	39830.00	39850.00	27035.0
Dry biomass yield (kg ha^{-1})	3855.00	3805.4	2838.6
Ash content in plants (%)*	11.68	7.3	11.4
Crude Protein in plants (%)*	13.75	29.3	29.3
Dry LPC yield (kg ha^{-1})	610.50	955.1	380.8
LPC extractability (%)	27.65	24.5	18.4
Protein N extractability (%)	64.85	62.6ww	26.4

* Dry weight basis. ** Leaf Protein Concentration.

Table 2. Composition and nutritive value of LPC

Parameters	*Eleocharis dulcis*	*Monochoria hastata*	*Veronica anagillis-aquatica*
Protein*	64.34	71.13	69.96
Lipids*	6.30	4.34	13.13
Crude fibre*	6.37	1.50	2.20
Starch*	12.70	8.60	9.80
Ash*	11.68	8.34	7.82
Total soluble sugars*	7.95	7.40	6.90
B-Carotene ($\mu g\ g^{-1}$)	465.00	460.00	480.00
Vitamin A (IU)**	776.55	768.20	801.60
Pepsin-trypsin digestibility*	67.87	61.70	61.87
Protein Efficiency Ratio	1.72	1.70	1.70

* Per cent dry weight basis.
* Conversion factor.
β-carotene 1 μg=1.67 IU of Vitamin A.

Incubation was done for 3,6,9, and 12 days, separately on a rotary shaker, at 28 ± 1°C. After incubation, the amount of sugar consumed was estimated colorimetrically (Morris, 1978). Titrable acidity was determined with standard sodium hydroxide solution. Citric acid produced was measured quantitatively according to Marrier & Boulet (1958).

Oyster mushroom *(Pleurotus ostreatus)* was grown on fibrous residue of *Eleocharis dulcis* alone or mixed with paddy straw (1:1). Production of biogas was studied from fibrous residue of *E!eocharis dulcis* alone or mixed with cowdung (1:1 and 1:3, separately).

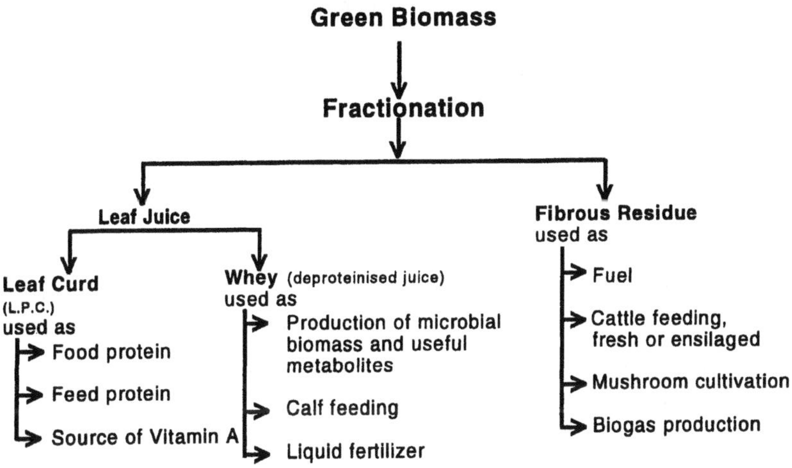

Figure 1. Fractionation of green biomass

Table 3. Composition and microbial load of whey* samples and growth of yeast in whey

Parameters	Source of whey/values per 100 ml of whey		
	Eleocharis dulcis	*Monochoria hastata*	*Veronica anagillis-aquatica*
Ash**	2.0	1.00	1.55
Dry weight**	4.5	3.40	3.80
Total glucose equivalent** (anthrone +ve material)	1.00	0.80	0.80
Total N**	0.25	0.17	0.22
Lipid**	0.50	0.30	0.20
Bacteria****	60	43	40
Actinomycetes***	16	12	12
Fungi***	40	16	32
Propagation of yeast in whey****	80×10^{10}	70×10^{10}	75×10^{10}
Biomass of yeast in whey*****	0.52	0.46	0.52

* Deproteinized leaf juice.
* % dry weight basis.
** Number of colonies/25 ml whey.
*** *Saccharomyces cereviseae* (inoculum size 2×10^3 cell/ml; incubated at 28°C, 48h).
**** *S. cereviseae* (g/100 ml dry wt. of yeast, incubated at 28°C, 48h).

Table 4. Citric acid production by *Aspergillus niger* in *Eleocharis* whey fortified with sugar

Incubation period ** (days)	Sugar consumer (g/100 ml)	Titrable acidity normality	Citric adic yield (g/100 ml)	Mycelial dry weight (g/100 ml)
3	4.50	0.226	0.90	0.595
6	7.70	0.400	1.95	0.870
9	9.80	0.600	2.80	1.251
12	9.90	0.550	0.55	1.408

* Total sugar in whey before fermentation 10g/100 ml.
* At 28±1 °C.

Results and discussion

Yields of biomass, LPC and percent extractables of total N and protein N of three aquatic weeds (Table 1) were promising as compared with other aquatic weeds (Boyd, 1968). Table 2 shows the composition and nutritive value of LPC of the three aquatic weeds. These protein concentrates have good protein content (64.34% to 71.93% protein on dry weight basis) and high *in vitro* digestibility (61.70% to 67.87%). Amino acid analyses of LPCs show that these contain all the essential amino acids. The lysine content in the LPCs (4.19–5.5 g/100 g protein) appears adequate from nutritional viewpoint. Methionine and cysteine contents are, however, appreciably lower, probably due to partial or complete destruction of sulphar-containing amino acids during hydrolysis prior to amino acid analysis. Analysis of whey reveals that whey of *E. dulcis*, *M. hastata* and *V. anagallis-aquatica* contain sufficient nutrients to support the growth of bacteria, actinomycetes and fungi, including yeasts (Table 3). The rich microbial load of these whey samples offers the hope that these may be used as industrial fermentation media for conversion into useful products. Use of whey has also been suggested earlier as substrate for producing microbal metabolites and biomass (Chandra et al., 1984) and growing yeast as single cell protein (Pardez & Gagagro, 1973). The production of protein as yeasts offers the best hope for additional food/feed protein. Whey supplemented with sugar, has also been recorded suitable for production of citric acid by *Aspergillus niger* (Table 4).

Studies on the utilization of fibrous residue of *Eleocharis dulcis* for cultivation of oyster mushroom *(Pleurotus ostreatus)* show that it yields 880 g mushroom/bag (50 × 20cm), as compared to the lesser yield on paddy straw (850 g/bag). Studies on the production of bioenergy as biogas from fibrous residue of *E. dulcis* show that it has good potential for production of biogas which was the best when cowdung and *Eleocharis* fibre were mixed in 3:1 ratio.

This study and also the earlier ones (Pirie, 1987; Telex & Graham, 1983) confirm that aquatic weeds such as *E. dulcis, M. hastata* and *V. anagallis aquatica* can be put to practical use by fractionation of their green biomass as Figure 1.

Acknowledgement

Authors are thankful to Prof. P.C. Misra, Head, Botany Department, University of Gorakhpur for facilities, V.N. Pandey thanks the D.S.T. and C.S.I.R., New Delhi, for financial assistance.

References

AOAC, 1980. Official methods of analysis A.O.A.C., Washington, DC, 1018 pp.

Arkcoll, D. B. & M. Holden, 1973. Changes in chloroplast pigments during the preparation of leaf protein. J. Sci. Food Agric. 24: 12–17.

Bates, R. P. & J. D. Hentages, 1976. Aquatic weeds – eradicate or cultivate? Econ. Bot., 30: 39–50.

Boyd, C. E. 1968. Freshwater plants: a potential source of protein. Econ. Bot. 22: 393–396.

Chandra S., S. Chakraborti & S. Matai, 1984. 'Whey' as the medium for production of microbial metabolites and biomass. Progress in Leaf Protein Research (N. Singh). Today and Tomorrow's Printers and Publishers, New Delhi, 377–389.

Doneen, L. D., 1932. A micro method for nitrogen in plant material. Plant Physiol. 7: 717.

Marrier, J. R. & M. Boulet, 1958. Direct determination of citric acid in milk with an improved pyridine acetic anhydride method. J. Dairy Sci. 41: 1685.

Mitchell, D. S. (ed.), 1974. Aquatic vegetation and its use and control. UNESCO, Paris.

Morris, D. L., 1978. Science. 107: 254.

Pandey, V. N. & A. K. Srivastava, 1989. *Veronica anagallis aquatica* L. A potential source of leaf protein. Aquat. Bot. 34: 385–388.

Pandey V. N. & A. K. Srivastava, 1991a. Yield and quality of leaf protein concentrates from *Monochoria hastata* (L) Solms. Aquat. Bot. 295–299.

Pandey V. N. & A. K. Srivastava, 1991b. Yield and quality of leaf protein concentrates from *Eleocharis dulcis* (Burm.f.) Hensch. Aquat. Bot. 41: 369–374.

Pardez-Lopez, O. & E. Gagagro, 1973. Experientia 29 :1233.

Pirie, N. W. 1987. Leaf protein and its by-products in human and animal nutrition. Cambridge University Press, London, 209 pp.

Saunders, R. M., M. A. Connor, A. N. Booth, E. M. Bickoff & Kohler, 1973. Measurements of digestibility of alfaalfa protein concentrates by *in-vivo* and *in-vitro* methods. J. Nutr. 103: 530–535.

Snell, F. D. & C. T. Snell, 1953. Colorimetric methods of analysis, D.Van Nostrand, New York, p215. 3: 606.

Telek, L. & H. D. Graham (eds), 1983. Leaf protein concentrates. AVI Publishing Co., West Port, CT, USA.

Hydrobiologia **340**: 317–321, 1996.
J. M. Caffrey, P. R. F. Barrett, K. J. Murphy & P. M. Wade (eds), Management and Ecology of Freshwater Plants.

Morphology and nutritional value of *Aponogeton undulatus* Roxb. growing in deeply flooded areas in Bangladesh

Q. R. Islam
Northeast Regional Water Management Project (Flood Action Plan 6) Gulshan, PO Box 6096, Dhaka 1212, Bangladesh

Key words: low-lying basin, floating plant, rootstock, food

Abstract

Aponogeton undulatus Roxb. grows in abundance throughout the deeply flooded land in the Northeast Region of Bangladesh. The plant has the ability to elongate and tolerate submergence after deep flooding. The elongation ability is expressed chiefly by the elongation of petiole. The rootstock of the plant is an important food item for the low-income people in this deeply flooded area. Rice is the main crop in this area providing the major source of energy and protein. However, the crop is often damaged by floods. The nutrient composition of rootstock of *Aponogeton undulatus* Roxb. shows that it can provide an adequate supply of carbohydrate, protein and some minerals. This food can be exceptionally useful as a nutrient supplement in many areas where purchasing power is limited because of low incomes. Four species of *Aponogeton* are found in Bangladesh. The aquatic environment prevailing in the deeply flooded area in Bangladesh has great potential in terms of propagation of these species. If this potential can be realized it would improve the nutrition of the people and maintain biodiversity and traditional ecosystems. Further study on the chemical composition of rootstock, and ecological, morphological and physiological characteristics of all the *Aponogeton* species would be of great value.

Introduction

The deeply flooded lands in the Northeast Region of Bangladesh support large natural plant communities. A number of submerged and rooted floating plants which grow profusely are used as food. Many people with low incomes depend on these food sources during times of scarcity. Among them, the rootstock of *Aponogeton* is the most important and dominant food item.

The demand for the rootstock increases in the years when there is a poor harvest of rice. In average years rice is the dominant calorie source and supplies more than one-half of daily available protein. In the deeply flooded area rice is cultivated only in the dry season and is often damaged by flash floods when the crop is at maturing stage. The rootstock of *Aponogeton*, therefore, provides some food security during the times of stress and scarcity in low-lying areas. There has been little research into the growth habits of the plant and nutritional value of the rootstock. The present study examines (1) the possibilities of *Aponogeton* improving the nutritional status of the people and (2) the significance of the plant in the management of natural resources in deeply flooded area of Bangladesh.

Description of sites where the plants grow

Aponogeton grows in abundance throughout the bowl-shaped low-lying areas, known as *Sylhet* basin, in the Northeast Region of Bangladesh. The basin comprises an area of 6000 square kilometres (4.2% of Bangladesh), between latitude 24°27′ and 24°44′ North, and longitude 90°52′ and 91°40′ East. It is bounded to the north, south, and east by the hill ranges and highland of India, and to the west by the Old Brahmaputra River catchment. The basin is traversed by several rivers and extends over Netrokona, Kishoreganj, Sunamganj, Habiganj and Sylhet districts. It is characterized by the presence of numerous flooded depressions, known as haors, which are situated between rivers. These haors vary in size from as little as few hectares to many thousands of hectares. In many cases two or more neighbouring haors link up to form larger water bodies. There are 47 major haors in the basin (Scott & Rashid, 1992; Khan et al., 1994). The whole basin is subject to early floods and a rapid rise in flood-levels. The numerous rivers rising in the hills provide an abundant supply of water to the plains and cause extensive flooding during the monsoon season (June to September). Virtually all of the basin is flooded to a depth of 1 to 5 m (Brammer et al., 1988). The flooding restricts crop production in the wet season. Deepwater or floating rice is grown over a small area where flood waters drain early. Flooding is mainly by clear water, but silt laden water affects the northern part of the basin when sudden floods occur. Flood-waters drain rapidly from the ridges after the rainy season ends in September, but haor centres can retain some standing water for most or all of the dry season.

Historically, much of the basin was forested with water tolerant trees, and these provided a well-protected sanctuary for various species of fish, and a wealth of biodiversity. About 500 local rice varieties, more than two dozen submerged and rooted floating plant species, and 120 fish species have been identified in the basin (Karim, 1993; Bernacsek et al., 1993). Currently the haors support major subsistence and commercial fisheries, the seasonally flooded plains support a major rice-growing industry, and the abundant aquatic vegetation provides a source of food, fuel and fertilizer for local people.

Materials and methods

The angiosperm family of *Aponogetonacea* is a perennial water plant. Four species have been identified in Bangladesh (Khan & Halim, 1987; Mamun, 1989). The species are: *Aponogeton undulatus* Roxb., *Aponogeton appendiculatus* Bruggen, *Aponogeton echinatus* Roxb., and *Aponogeton natans* (L.) Engl. & Krause. These species are found in the flooded lands from early in the monsoon season (April) to the post-monsoon period (October). The rootstock of all these species is edible. Among the species, *Aponogeton undulatus* Roxb. was selected for the present study. This species grows profusely on the lands which flood to a depth of 1 to 2 m in the monsoon season. The depth of water decreased to about 0 to 50 cm at the end of the growth period.

Table 1. Morphological characteristics of *Aponogeton undulatus* Roxb.

Item	Range	Average±S.D.
Number of submerged leaves	11–15	13.40±1.62
Number of floating leaves	3–7	5.20±1.60
Total number of leaves	16–22	18.60±1.96
Total length of plant	83.0–113.0	99.8±10.79
Length of rootstock (cm)	1.83–3.81	2.57±0.53
Thickness of rootstock (cm)	0.67–1.34	1.02±0.21
Weight of rootstock (g)	1.33–4.40	2.21±0.95

S.D.: standard deviation

The rootstock of *Aponogeton undulatus* Roxb. was collected from a flooded depression, known as Shanir haor, in November 1992. The material was stored at room temperature (23° to 31°C). Before storing, the length and thickness of ten rootstocks was measured using a slide calliper. Ten well-developed rootstocks at 42.8% moisture content were weighed on a precision balance to obtain the average weight of each rootstock. The moisture content was determined according to the method described by Pearson, 1970. The method involves the measurement of weight lost due to the evaporation of water. Samples were supplied to the Institute of Nutrition and Food Science, University of Dhaka and the nutritional value was determined.

The rootstocks started germination in the first week of April. Five germinated rootstocks were sown in a pot to examine the elongation ability of the plant. The pot contained 4 kg of clay soil. It was watered every other day. When the plants reached second to third leaf emerging stage, the pot was transferred to a water tank and the plants were submerged with water. The water level was raised by increment of 5 cm at 3 to 4 day intervals until flowering when the total depth was 1.3 m. The plant was later confirmed as *Aponogeton undulatus* Roxb. at the National Herbarium, Dhaka. Throughout the growing period, the total number of leaves, elongation of petiole and leaf, submergence tolerance, flowering, and vegetative reproduction of the plant were recorded. Average temperature during the study ranged from 20° to 31°C.

Ten households in the deeply flooded area were interviewed on the contribution of rootstock in the diet of people, use of the rootstock in the preparation of food and preservation methods. Marketing of the rootstock were recorded from visits to markets. Data on the nutrient composition of the rootstock were taken from the M. Sc. thesis on 'A Study on the Nutritional

Value of *Aponogeton undulatus*'- submitted by Tania Jesmin, Institution of Nutrition and Food Science, to the University of Dhaka in June 1994.

Results

The plant is a tuberous herb with submerged and floating leaves. As shown in Table 1, the average number of submerged leaves was 13.40 ± 1.62. The average number of floating leaves formed at the flowering stage was 5.20 ± 1.60. The submerged leaves were transparent, and the floating leaves were leathery. Leaf margins were undulating. The average number of total leaves produced by a plant was 18.60 ± 1.96.

The plant exhibited the capacity to elongate after submergence. Plant elongation involved the petiole and leaf blade with a gradual rise of water level. However, petiole elongation was the most important because leaf elongation was limited. Petioles of the floating leaves elongated faster than those of submerged leaves. They exhibited an initial increase of 2.08 ± 1.02 cm d^{-1}. The elongation rate in the plants varied from 0.7 to 3.6 cm d^{-1}. The average ratio of petiole to the leaf blade was 0.13 ± 0.02 at the submergence time. The ratio varied from 0.11 to 0.15. It was 5.23 ± 0.35 for floating leaves and 1.12 ± 0.06 for submerged leaves at the flowering time. The ratio varied from 4.68 to 5.71 for floating leaves and 1.04 to 1.18 for submerged leaves. By the elongation of petiole plants were found to grow 99.8 ± 10.79 cm long.

This plant showed an extra-floral vegetative reproduction capacity with the formation of buds. Four to six weeks after rootstock germination, it produced a shoot which formed 2–3 buds in the water. The buds subsequently developed into young plants with several tiny leaves. These young plants were later dispersed from the shoot presumably by wave action, and after floating on the water for a period became established as independent plants. Subsequently, the plant produced 1–2 flowering spikes. Fruit had a short terminal and curved beak. Seeds had a smooth testa. Generally, each plant produces one rootstock which serves for vegetative reproduction. The rootstocks are irregular, round shaped, and covered with flimsy roots. The colour is straw brown and the skin is thin. Data on the size and weight of the rootstock are provided in Table 1. People start to glean the rootstock from fields at the beginning of dry season (October/November) when the monsoon flood water recedes. Mainly, women and children of low-income families glean until fields are ploughed for transplantation of rice in December/January. The rootstock is sold in the market. The price is a little below that of rice. Wealthy people never glean rootstock, but they sometimes exchange rice for it. The use of the rootstock is extended until the harvest of rice in June. Both rice and rootstock are preserved by drying in the homestead.

The rootstock is neutral in taste and has little flavour. People have evolved a number of ways to serve this rootstock. It is mostly eaten boiled, often in combination with vegetables or fish, or mixing with rice. Rootstock is also simmered with vegetables and spices. The dried rootstock is coarsely ground and combined to make *chapattis*, the bread usually eaten in the morning. The starchy flour is also used to prepare cakes, mainly by wealthy people. The dried rootstock is sometimes served fried. During the rice growing period in the dry season, when people with low incomes in the low-lying area wait for the next rice harvest, the rootstock is served at all three meals. It was estimated from the households interviews that the rootstock contributes 25 to 60% of the calories and 30% of the protein in the diet of low-income people during this time.

Discussion

Several specialized characters that enable *Aponogeton undulatus* Roxb. to adapt to deeply flooded lands are evident in the study. Among these, the extra-floral vegetative reproduction capacity with the development of young plants in the water is the most important morphological features of the plant. It seems that the young plants can increase plant density and rootstock yield. Further experiments are needed to test the assertion. An understanding of the environment and the characterization of the growing areas will also assist in elucidating the factors controlling plant growth.

Considering the nutritional value of the rootstock, the *Aponogeton* can play an important role in enhancing food supplies in Bangladesh. The nutritional value of rootstock appears to be superior to those of root and tuber crops (Table 2). The rootstock contains starch (soluble carbohydrate) as the principal component. The second highest component is protein. The fat content is relatively low. This low fat content demonstrates that the rootstock has a low energy density (224 kcal/100 g). The crude fibre concentration is also low in the rootstock. This, however, may explain the high total digestibility of protein. Eggum (1979) suggests that dietary crude fiber has a negative

Table 2. Nutritional value of rootstock of *Aponogeton undulatus* Roxb. in comparison with commercial root and tuber foods and rice.

Botanical name/ Common name	Moisture (g/ 100 g)	CHO (g/ 100 g)	Protein (g/ 100 g)	Fat (g/ 100 g)	Fiber (g/ 100 g)	Vit C (mg/ 100 g)	Ca (mg/ 100 g)	Zinc (mg/ 100 g)	Mg (mg/ 100 g)	K (mg/ 100 g)	Fe (mg/ 100 g)
A. undulatus[a]	42.8	46.0	8.3	0.7	0.7	0.3	37.2	2.4	110.3	955.8	18.2
Root and tuber food[b]											
Potato	74.7	22.6	1.6	0.6	0.4	10.0	11.0	0.2	22.0	249.0	0.7
Sweet potato	68.5	28.2	1.2	0.3	0.8	2.0	20.0	–	31.0	373.0	0.8
Wetland Taro[c]	70.0	24.4	3.0	0.8	1.0	6.0	40.0	–	–	464.0	1.7
Rice[d]	13.3	79.0	6.4	0.4	0.2	–	9.0	–	–	91.9	4.0

[a] Source: Jesmin, 1994.
[b] Source: Damtin-Hill et al., 1988; Pennington & Church, 1980; and Ali et al., 1992.
[c] *Colocasia esculenta* L.
[d] Source: Damtin-Hill *loc. cit.*, Eggum *loc. cit.* and Ali *loc. cit.*

effect on digestibility of both protein and energy in rice. Since protein quality depends on the total amino acid composition, more study is required to investigate the biological value of the protein. The rootstock can be an important source of calcium, potassium, iron, and zinc. One hundred grammes of rootstock has been reported to contain 33.7 mg of phytic acid (inositol hexaphosphoric acid), which impedes mineral absorption by forming insoluble compounds with the minerals (Jesmin, 1994). The effect of phytic acid can be overcome by adding calcium carbonate to the food (Eggum *loc. cit.*).

The dominant nutritional problem in rice-eating Bangladesh is an inadequate quantity of food, which stems from the linked issues of rice supplies, rice prices, and low consumer purchasing power. Rice provides 69% of calories and 51% of protein to the people (Salam et al., 1991). The rootstock compares favourably with rice in protein content. Calorie-protein malnutrition in Bangladesh reflects low consumption levels because people are poor. The recommended daily protein intake is 56 g for men and 44 g for women at approximate calorie levels of 2700 and 2000, respectively. This recommendation represents a diet that contain 8 to 9% of the calories as protein. The rootstock of *Aponogeton undulatus* Roxb. exceeds this protein-to-calorie concentration. Therefore, the rootstock of the plant can be the dominant calorie source in the diets of many low-income people. Moreover, increased utilization of *Aponogeton* will directly enhance protein supplies.

About one-third of total area in Bangladesh, including 26 800 square kilometres of cultivated land, is deeply flooded between May and October (MPO, 1987). Propagation of the plant in this flooded area would be especially beneficial to low-income people and to those living in low-lying areas where market foods are limited both in availability and variety. The plant can be used for commercial production. There is also a great potential for marketing this food in the domestic urban market and in the international market as wild food. Further study is required on the protein quality and presence of other nutrients in the rootstock, and into the mechanism involved in the elongation ability and vegetative reproduction capacity.

Acknowledgments

The author is particularly grateful to Dr Salar Khan, Honorary Advisor to the Bangladesh National Herbarium, and Ex-Professor of the Department of Botany, University of Dhaka, and Harry King, Team Leader, Northeast Regional Project who reviewed the entire manuscript. Appreciation is expressed to Dr A. T. M. A. Rahim of the Institute of Nutrition and Food Science, University of Dhaka for his interest in the study of nutrient composition of rootstock. The author is deeply indebted to Ms Sara Camblin Breault, First Secretary, Canadian High Commission, for granting CIDA (Canadian International Development Agency) funds for this study.

References

Ali, K. M. S., M. A. Malek, K. Jahan & K. Salamatullah (eds), 1992. Nutritional value of local foods (in Bengali). Institute of Nutrition and Food Science, Dhaka University, Bangladesh, 35 pp.

Bernacsek, G. M., S. Nandi, N. C. Paul & T. Akbar, 1993. Fisheries specialist study. Northeast Regional Water Management

Project, Canadian International Development Agency, Dhaka, Bangladesh, 291 pp.

Brammer, H., J. Antoine, A. H. Kassam & H. T. van Velthuizen, 1988. Land resources appraisal of Bangladesh for agricultural development. Report 2: Agroecological Regions of Bangladesh. UNDP/FAO, 570 pp.

Damtin-Hill, I., N. Hassan, R. Karim & M. R. Duthie, 1988. Tables of nutrient composition of Bangladesh foods. Hellen Keller International, Dhaka, Bangladesh, 24 pp.

Eggum, B. O., 1979. The nutritional value of rice in comparison with other cereals. In proceedings of the workshop on chemical aspects of rice grain quality. International Rice Research Institute, Manila, Philippines: 91–111.

Islam, Q. R., M. Huq, F. Siddique & J. Fenton, 1993. Special study: Agriculture in the Northeast Region. Northeast Regional Water Management Project, Canadian International Development Agency, Dhaka, Bangladesh, 110 pp.

Jesmin, T., 1994. A Study on the Nutritional Value of *Aponogeton undulatus*. M. Sc. thesis. Institution of Nutrition and Food Science, University of Dhaka, Bangladesh, 83 pp.

Karim, A., 1993. Plant diversity and their conservation in freshwater wetlands. In A. Nishat, Z. Hussain, M. K. Roy & A. Karim (eds), Freshwater wetlands in Bangladesh: Issues and approaches for management. IUCN, the World Conservation, Gland, Switzerland: 75–104.

Khan, M. S., E. Haq, S. Huq, A. A. Rahman, S. M. A. Rashid & H. Ahmed (eds), 1994. Wetlands of Bangladesh. Bangladesh Centre for Advanced Studies, Dhaka, Bangladesh, 90 pp.

Khan, M. S. & M. Halim, 1987. Aquatic angiosperms of Bangladesh. Bangladesh National Herbarium, Bangladesh Agricultural Research Council, 120 pp.

Mamun, A. A., 1989. Agro-ecologocal studies of weed in Bangladesh. Japan International Cooperation Agency, Bangladesh, 142 pp.

Master Plan Organization, 1987. Technical Report No. 14: Agricultural Production System. Ministry of Irrigation, Water Development and Flood Control, UNDP and World Bank, Dhaka, Bangladesh, 57 pp.

Pearson, D., 1970. The chemical analysis of foods. Churchill, J & A. London: 4–5.

Pennington, J. A. & H. N. Church, 1980. Food values of portions commonly used, 13th ed. Harper and Row, New York, 1980.

Salam, A., K. Kamal, S. Hossain, T. Islam, A. R. Islam, D. K. Bhadra, H. Rashid & T. U. Ahmed (eds), 1991. Report on the household expenditure survey 1988–89. Bangladesh Bureau of Statistics, Dhaka, Bangladesh, 201 pp.

Scott, D. & S. M. A. Rashid, 1992. Wetland assesment and ornithology main surveys. Northeast Regional Water Management Project, Canadian International Development Agency, Dhaka, Bangladesh, 55 pp.

Hydrobiologia **340**: 323–331, 1996.
J. M. Caffrey, P. R. F. Barrett, K. J. Murphy & P. M. Wade (eds), Management and Ecology of Freshwater Plants.

Constructed wetlands for waste water treatment: the use of laterite in the bed medium in phosphorus and heavy metal removal

R. B. Wood & C. F. McAtamney
Freshwater Laboratory, University of Ulster, Traad Point, Ballyronan, Co. Derry, BT45 6LR, Northern Ireland

Key words: Constructed wetlands, laterite, phosphorus removal, heavy metal removal

Abstract

In Northern Ireland, phosphorus enrichment of lakes due to agriculture is a significant problem. Heavy metal exports from landfill sites, often located on water-logged land, are also of concern. Locally available laterite, a low grade bauxite which is rich in iron and aluminium, is used in acid solution with subsequent precipitation to remove phosphorus and heavy metals at several sewage treatment works. Constructed wetlands offer an attractive alternative to conventional waste water treatment in certain circumstances but removal of phosphorus is strongly dependent on the bed medium. Calcium-, iron- and aluminium-rich solid media are recommended. A brief introduction to the use and cost-effectiveness of constructed wetlands (CWs) in treating a range of effluents is given. This study, using both laboratory tests and pilot-scale constructed wetlands, reports the effectiveness of granular laterite in removing phosphorus and heavy metals from landfill leachate. Initial laboratory studies have shown that laterite is capable of 99% removal of phosphorus from solution. A pilot-scale experimental CW containing laterite achieved 96% removal of phosphorus. This removal is much greater than that reported in other systems. Initial removals of aluminium and iron by pilot-scale CWs have been up to 85% and 98% respectively. Percolating columns of laterite reduced Cd, Cr and Pb to undetectable concentrations. Possible application of this low cost, low technology, visually unobtrusive yet efficient system to rural areas with dispersed point sources of pollution is discussed.

Introduction

Constructed wetlands

There is considerable evidence of the effectiveness of constructed wetland systems (CWs) in treating a range of water borne effluents. This range covers both different effluent types (sewage, agriculture, fish farming, mine spoil, oil industry, paper/pulp production and landfill) and varying loadings. Operational systems known to us range from 18 ha (British Steel) down to a few m^2 serving individual houses; in concentration, BODs of around 300 mg l^{-1} (in one instance up to 1300 mg l^{-1} (Vymazal, 1993)), phenols, (O.2 mg l^{-1}), total chromium (0.1 mg l^{-1}) and faecal coliforms (ca. 2×105 CFU ml^{-1}) have been treated, often achieving 90 to 99.9% removal efficiency. Other examples are reported in Moshiri (1993) and Wood & McAtamney (1994). Treatment of nitrogen-rich effluents can be very effective, both in decomposing organic nitrogen and in oxidising the principal product, ammonia, to nitrate (beneficial to fisheries if less so to drinking water suppliers) and also in de-nitrifying up to 70% of the Total N input (Vymazal, 1993).

The whole 'wetland' is effectively a complex filter, in which physical, chemical, macrophyte and microbial components integrate into an effective system. Table 1 outlines the principal processes in this complex filter. Unplanted filter beds do not perform the wide range of functions found in well-established CWs and the principal contribution of macrophytes is in transporting oxygen to their water-logged rhizomes and roots. Leakage of oxygen into the rhizosphere creates local niches for aerobic microbes in the surrounding bed and some oxygen can be used in decomposition (Brix, 1993). Away from the rhizosphere, anaerobic conditions develop supporting facultative and obligate anaerobes and processes such as denitrification.

Table 1. Outline of effluent treatment processes involved in constructed wetlands.

Process	Role or function
Physical	
Filtration	As the water passes through the plant detritus and substrata , colloidal and other particles are filtered. Rhizomes open up the soil to provide a hydraulic pathway through the media.
Sedimentation	Flocculation and gravitational settling of solids.
Adsorption	Adsorption of colloidal solids by interparticular molecular forces.
Chemical	
Precipitation	Formation and precipitation of insoluble compounds such as CaPO4 . The addition of Al, Fe, or Ca will precipitate phosphorus.
Decomposition	Oxidation, reduction and irradiation of less stable compounds.
Adsorption	Adsorption of heavy metals by the soil.
Biological	
Bacterial Metabolism	Bacterial decomposition of organics and nitrification- denitrification. (The principal mechanism for B.O.D. and N removal)
Plant Growth	Temporary seasonal up-take of nutrients and trace metals by plant. (May be permanently removed by harvesting.) However uptake of nutrients negligible compared to those lost by microbial transformations. Water uptake and transpiration.
Residence Time	Natural decay of pathogens and other organisms.

This matrix of mixed redox conditions also has significant impact on mobilisation and solubility of components such as iron (Fe), manganese (Mn) and phosphate (PO_4). Several workers (e.g. Lawson, 1985; Armstrong et al., 1990) have shown that a well-established Phragmites bed may deliver up to 15 g oxygen m^{-2} d^{-1} into the bed. The generalised 'design rule' of allowing up to 5 m^2 bed for a 40 g d^{-1} BOD load (one person equivalent) provides modest spare capacity.

Macrophytes may also take up phosphorus (P) or heavy metals to varying degrees (Denny, 1980; Shierup & Larsen, 1981; Howard-Williams, 1989; Delgado et al., 1993) which may remain immobilised for the life of the bed. The growth and decay of rhizomes was thought to help maintain hydraulic pathways through the bed, but the modern preference for coarse gravel, not fine silts, makes this of less concern. With regard to the removal of phosphate, bed materials rich in calcium (Ca), iron (Fe) and/or aluminium (Al) are recommended (Cooper, 1990). This recognises the role of chemical adsorptions relative to uptake by the vegetation or redox mediated processes alone.

Given the widespread occurrence of phosphorus controlled eutrophication of lakes, we are seeking to enhance the use of CWs for efficient P-removal. In addition, domestic waste in Northern Ireland has for decades been land-filled, usually to poor quality, waterlogged soils, from which leachate, rich in heavy metals and ammonia ($N\text{-}NH_4$) seeps, contaminating possibly for long periods. New legislation should reduce this practice but integrated CWs may have a role in longer term amelioration. The ready growth of Phragmites and Phalaris in Ireland, its unobtrusive contribution to landscape, and the local availability of laterite has led us to explore the possibilities of combining them in a CW system. A further advantage is the widely recognised low cost involved (Table 2).

Laterite

In Ireland laterites are intercalated in the Tertiary basaltic lava flows in County Antrim. Laterite is a hydrated mixture of Al, Fe and Titanium (Ti) derived from the decomposition of alumino-silicate bauxite rocks and the subsequent loss of alkalis, lime, magnesia and silica. Eyles (1952) reports an advanced lateritised sample containing 6.9% SiO_2, 39.8% Al_2O_3, 26.7% Fe_2O_3 and 3.6% TiO_2.

The mechanisms of phosphate adsorption are complex but in broad terms phosphate ions are chemically adsorbed onto surfaces of hydrous oxides of Fe and Al by ligand exchange. Laterite, dissolved in acid, dosed to effluent and precipitated again at pH >6.0 was found effective in removing organic matter including natural peat colour, P, and heavy metals from effluents. This led to its use as DIFCO, now known as Ferric Aluminium Sulphate, for decolourising drinking water and as tertiary treatment to reduce P-loading to Lough Neagh (Gray, 1985).

Table 2. Some relative costs of effluent treatment using constructed wetlands (CW)

Company	Site and Scale	Costs (per annum) (£/K)	Method	Comments and Reference
Severn Trent Water plc Municipal sewage	103 m^3 d^{-1}	35	C.W.	Preferred option: 50 by 1995 + 200 by 1997; low maintenance [1]
Amoco Oil	3.7 × 103 m^3 d^{-1}	1 4–12 (= U$3 M)	C.W. alternatives	Plus 3 EPA and other awards [2]
Agricultural run-off to Long Lake Maine, USA	78 m^3 s^{-1}	1 8.6 6.6	C.W. Diversion Change Agriculture	[3]
Solid waste leachate Escambia Co Florida	1000T d^{-1} solids + seepage		C.W.	Savings $1 Mp.a. operating $5M in construction $1 M p.a. in escrow { Supplementary paper with 4}
Septage NH_3 treatment Arroyo Hondo New Mexico	1.3 m^3 d^{-1}	1 5–8	C.W. + ponds standard	[5]

Costs are actual for Severn Trent Water plc; others are relative.

Key to references
[1] Green and Upton (1993); [2] Lienard et al. (1990); [3] Higgins, Rock, Bouchard & Wrengrezynek (1993); [4] Martin et al. (1991); [5] Ogden (1993).

Methods

Before incorporating laterite into a mixed CW bed the capacity of laterite to adsorb P and heavy metals under laboratory conditions, both from pure solutions and from leachate (spiked as necessary) was determined with sorbent concentrations, particle size (surface area) and contact time as variables.

Analytical methods

Soluble Reactive Phosphorus (SRP), which is a measure of phosphate in its soluble form was analysed by a spectrophotometric method devised by Murphy & Riley (1958 & 1962). The samples were first filtered through prewashed Whatman GF/C filter papers. Molybdate antimony and ascorbic acid were added to the samples and the concentration of the resulting blue coloured complex was measured spectrophotometrically at 882 nm.

Total iron was determined by a spectroscopic method (HMSO 1978a) using 2,4-tripyridyl-1,3,5,-triazine (TPTZ). Total aluminium was determined by a spectroscopic method (HMSO 1978b) using formaldoxime. Total manganese was determined by a spectroscopic method (HMSO 1980) using pyrocatechol violet. Samples were preserved prior to analysis by adding 2.0 ml HCl (5M) per 100 ml of sample. Cd, Cr and Pb were determined by graphite furnace AA spectrophotometry using a Perkin Elmer 4100 with Zeeman correction. An average of two readings were taken. Samples were preserved prior to analysis by adding 1 ml of concentrated nitric acid per 100 ml of sample.

Laboratory experiments

Batch sorption experiments
10 g of laterite with a particle diameter of 2.00–3.35 mm (BS 410 test sieves) was shaken on an orbital shaker at 65 r.p.m with 100 ml of $P\text{-}PO_4$ solution. Residual concentrations of $P\text{-}PO_4$ in solution were determined over a period of 7 days. A series of experiments was conducted covering initial concentrations (Co) of 5, 10, 20, 30, 40 and 50 mg $P\text{-}PO_4$ l^{-1}.

Percolation columns
Mixed solutions of heavy metals Pb, Cr, Ni, Ag, Sb (all 100 $\mu g\ l^{-1}$), Al, Se, Fe, Mn (all 400 $\mu g\ l^{-1}$) and Cd, Zn, Ba (all 5 $\mu g\ l^{-1}$) were percolated at 10 ml min^{-1} through a 13 mm diameter × 25 mm column packed with crushed, washed laterite of nominal 6 mm diameter. Up to 10 passages through the column were made. Only the Cd, Cr, Pb, Fe, Mn and Al results are presented in this paper.

Pilot scale CWs

The design of the pilot scale experimental CW is shown in Figure 1. The beds (5 m long × 1 m wide × 0.5 m deep) had been planted for almost two years with rhizomes of Phragmites australis and Phalaris arundinacea at 25 cm intervals in a medium of crushed (6 mm diameter) granite gravel with one unplanted control in order to distinguish plant from bed action. Two 0.5 m long panels in each planted bed were filled with crushed laterite or granite gravel. The control unplanted bed was completely gravel. Raw leachate was obtained from the settlement pond of the Letterloan Road landfill site at Macosquin, Co. Londonderry. This site covers some 14.09 ha and is now nearly full after eleven years of use. The leachate was run through the beds at either 4 or 8-day retention times. The difference in performance between the bed types was determined using ANOVA (analysis of variance) and interpretation was based on significant differences at 95% confidence limits.

Results

Batch sorption experiments

Figure 2 shows the kinetics of phosphate sorption on laterite. Approximately 60% of P from an initial concentration of 50 mg l^{-1} P was removed in about 24 hours. At the more environmentally likely concentrations of 5 to 10 mg l^{-1} P, 80–90% removal was achieved.

Percolation columns

Figure 3 shows the effect on Cd, Cr, Pb, Fe, Al and Mn concentrations of percolating solutions, chosen to simulate leachate from aged wastes, through laterite packed columns. Cd, Cr and Pb were undetectable after 10 runs. Fe, Al and Mn were respectively 84.3%, 57% and 35.6% removed. Other results, not reported here, show that removal of metals is not significantly different under mildly acid or the circumneutral conditions characteristic of our leachate. In these laboratory scale percolations the columns were hydraulically loaded up to 290 times greater than in the the 5 m long pilot-scale CW described above. Ongoing work on flow rates, surface area of laterite granules, leachate/laterite ratio and the phosphate and heavy metal sorption capacity of laterite will hopefully optimize CW design for improved long-term performance.

Constructed wetland beds

The performance of the 5 m CWs in removing phosphorus from landfill leachate is shown in Figure 4. S.R.P. is usually undetectable in the raw leachate we have studied and under these conditions some phosphorus originating in the beds was leaked. Overall though, concentrations are not high and the bed with laterite leaked much less than the other beds (Figure 4a). Spiking the influent leachate with S.R.P. to approximately 1.5 mg l^{-1} in a second trial showed that all beds reduced P in the effluent, with the laterite bed achieving in excess of 95% removal (Figure 4b). Statistical analysis (ANOVA) showed a significant difference between the three treatments. Comparison of the laterite planted bed to the gravel planted bed indicated that laterite was significantly more efficient at

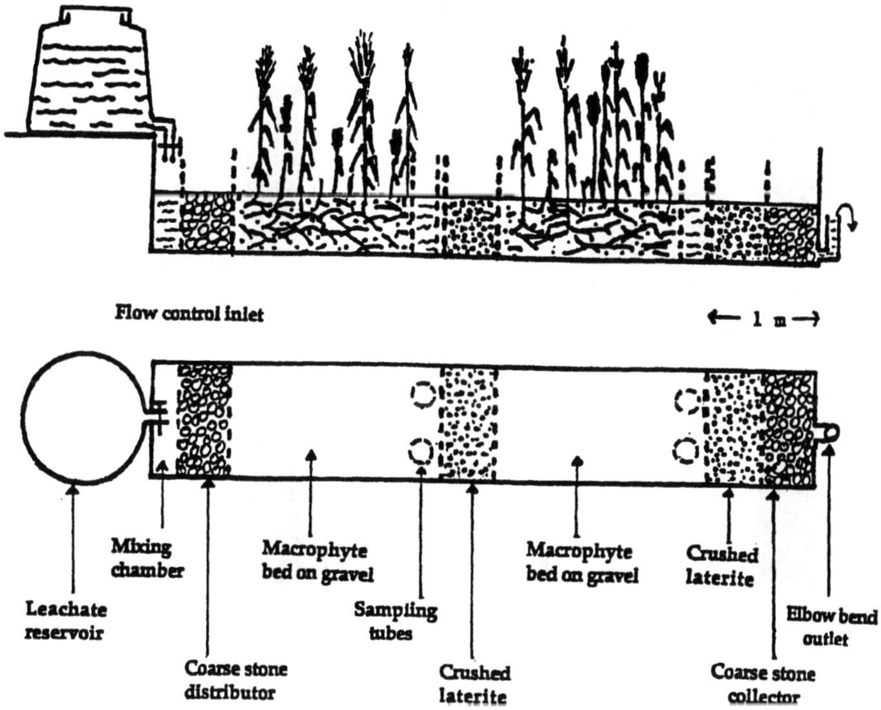

Figure 1. Section and plan through pilot-scale Constructed Wetland System at the University of Ulster's Freshwater Laboratory, Ballyronan, Co. Londonderry.

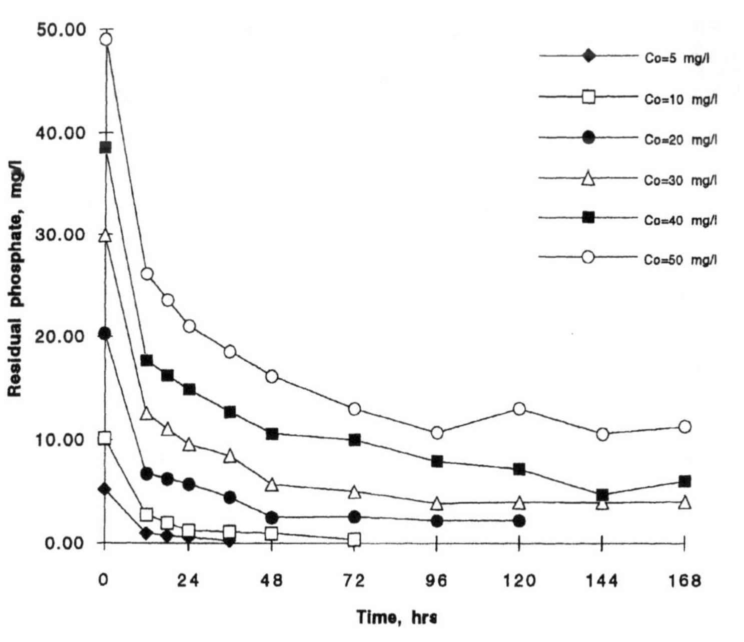

Figure 2. Kinetics of phosphate sorption on laterite. (2.00-3.35 mm). Conditions: 10 g laterite, 100 ml phosphate solution, 65 rpm. (Co is initial phosphate concentration)

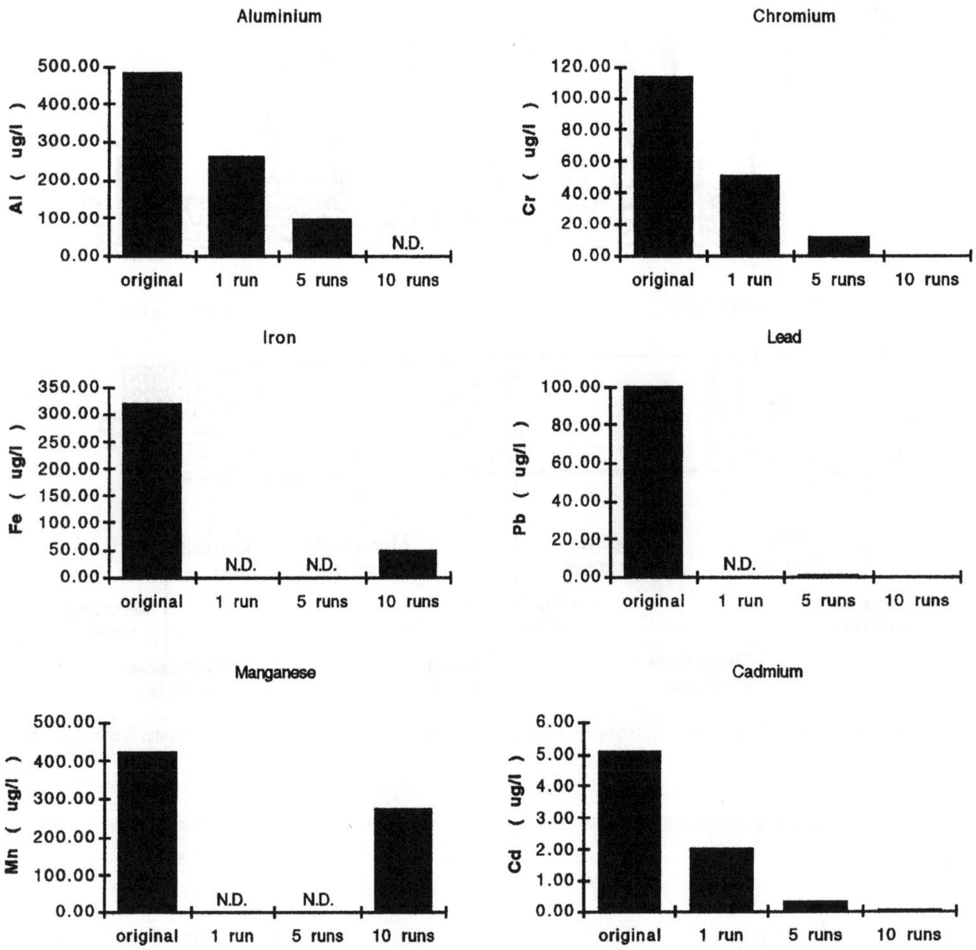

Figure 3. The removal of Fe, Mn, Al, Cd, Cr and Pb by percolation down 133 $mm^2 \times 250$ mm long column of ~ 6 mm laterite at 10 ml min^{-1} (N.D.: Not Determined).

removing phosphorus. The control unplanted bed was however more efficient at removing phosphorus compared with the gravel planted bed.

Figure 5 shows the performance of the beds at removing Fe, Mn and Al. These metals, which showed least removal in the column experiments (Figure 3), exhibit good removal in the CW beds. Over 81% removal of aluminium occurred in all beds and analysis of variance at 95 % confidence limits indicated no significant difference between treatments. Similarly for iron over 95% removal was reported in all beds though statistical analysis did indicate that Fe was more efficiently removed in the planted gravel bed as opposed to the control unplanted bed. The bed containing laterite did not show a significant difference compared to the gravel planted bed. With regard to manganese a 50% reduction was achieved in the planted laterite bed, 40% in the planted gravel bed with the poorest removal being in the control unplanted bed (34%). Statistical analysis showed no significant difference in performance between treatments and the presence of plants did not seem to improve performance.

Discussion

The laboratory studies provide evidence that laterite is an efficient binder of P and Cr, Cd and Pb. With

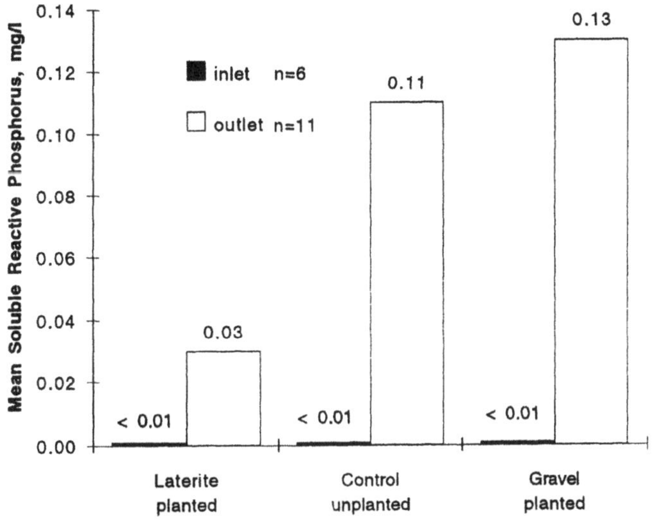

Figure 4a. Increase in Soluble Reactive Phosphorus on passage through University of Ulster, Freshwater Laboratory's pilot-scale experimental CW (14 day trial, 4 day retention time, flow rate 360 l d^{-1}.

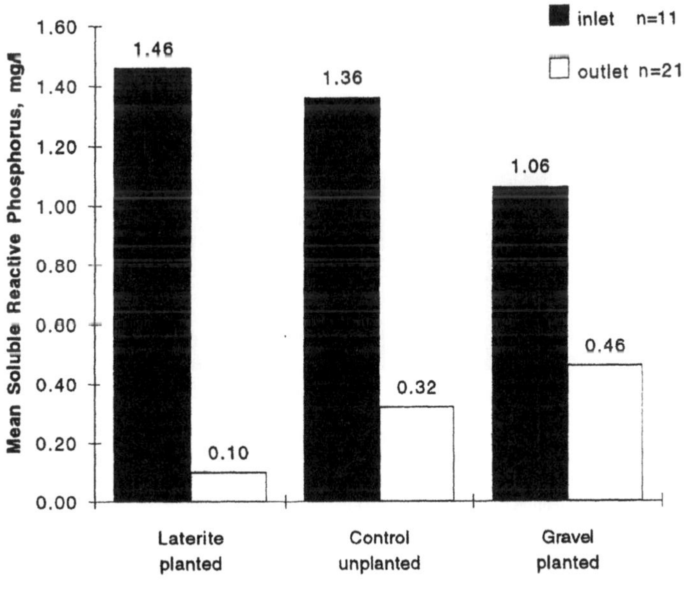

Figure 4b. Removal of Soluble Reactive Phosphorus on passage through University of Ulster, Freshwater Laboratory's pilot-scale experimental CW (28 day trial, 8 day retention time, flow rate 180 l d^{-1}).

regard to phosphorus the laterite bed has achieved a statistically significant improved removal (96%) compared to the other beds. This is well above the 30–40% Cooper (1990) quoted as the general experience with European systems and comparable with the 80–90% removal reported by Surface et al. (1993) during summer growth in reed beds treating leachate. There is some evidence that new beds are more efficient at removing phosphorus possibly due to the availability in the bed of binding sites which rapidly become exhaust-

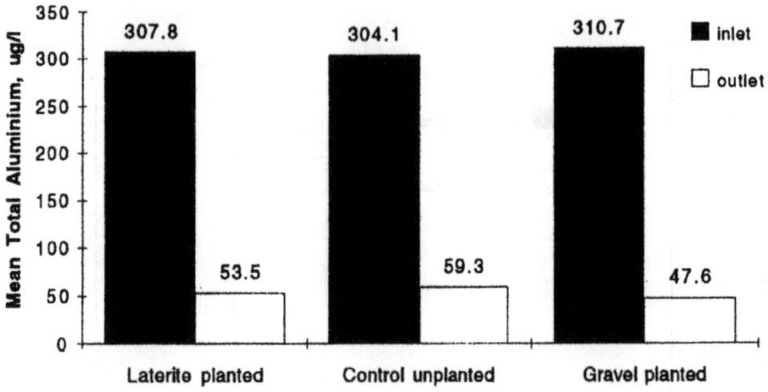

(a) Total Aluminium

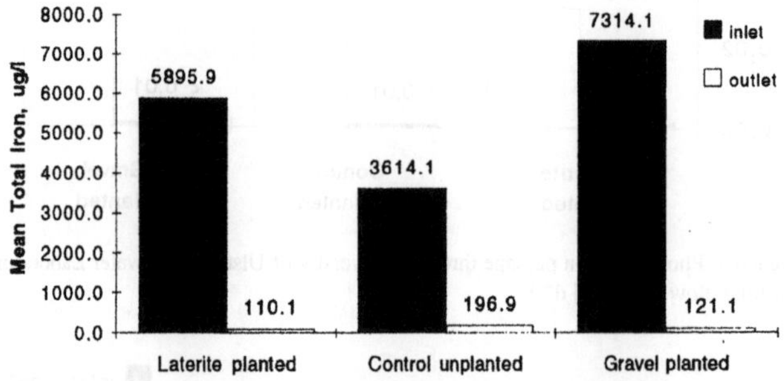

(b) Total Iron

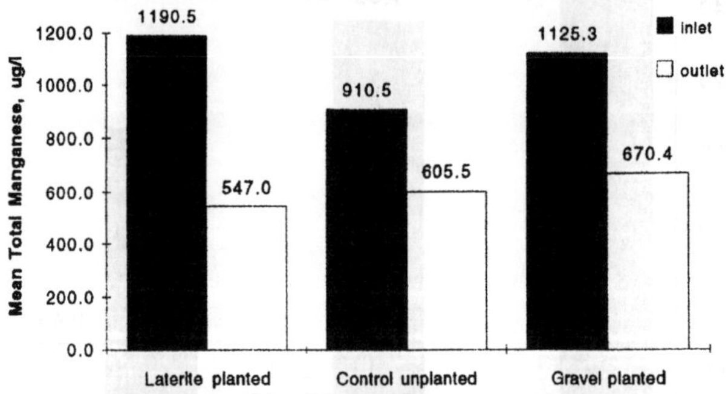

(c) Total Manganese

Figure 5. Performance of pilot-scale experimental beds at the University of Ulster, Freshater Laboratory. (a) Total Aluminium, (b) Total Iron, (c) Total Manganese. (14 day trial, 4 day retention time, flow rate 360 l d^{-1}, $n = 6$ inlet and outlet)

ed. This may partly explain the efficient removal of P in our unplanted gravel bed (Figure 4b). As stated our principle aim is to enhance and extend that P-binding capacity of the bed material with laterite.

Given that laterite is rich in Fe and Al it is not surprising that equilibria do not lie in extremis in favour of the bound form. The results from the CWs are in reasonable agreement with the laboratory studies though we must point out that in one sample during the CW trial Al actually increased marginally in the outflow. While good removal of Al is environmentally highly desirable even the slight increase in Al concentration found only once did not raise the level sufficiently to negate the use of laterite.

Statistical analysis on Fe and Mn removals reveals that the bed medium and not the plants have a significant role in heavy metal uptake during this trial. Throughout we are concerned with the difference between beds planted in gravel, with and without laterite addition. This is not to deny the role of plants in the whole system particularly in redox-mediated chemical and microbiological processes but to emphasise the possible enhancement of the role of the bed material in CW design. Other results on the removal of suspended solids, biochemical oxygen demand, chemical oxygen demand, ammonia, nitrate and other heavy metals will be reported elsewhere.

The beds are still comparatively young, and improvement in 'performance' over the first five years is often reported (Cooper et al., 1990). Further trials should determine whether heavy metal and phosphorus removal is sustainable. Already there is room for optimism that laterite-enhanced CWs will contribute to developing an effective 'alternative' local leachate treatment. Elsewhere CWs are already the preferred option for polishing small sewage effluents (see Green & Upton, 1994). Severn and Trent Water, one of Great Britain's largest water companies, expect to have over 100 such beds running in the near future.

Acknowledgments

This work was funded by the ENVIREG programme of the European Union administered by Environment Service, Department of the Environment (Northern Ireland) DoE(NI). We thank George Alexander, Clifford Henry and Eilish Clarke (DoENI) for their help and Mairead Diamond for help in preparing the manuscript.

References

Armstrong, W., J. Armstrong & P. M. Beckett, 1990. Measurement and modelling of oxygen release from roots of *Phragmites australis*. In P. F. Cooper & B. C. Findlater (eds), Constructed wetlands in water pollution control. Pergamon Press, Oxford: 41–51.

Brix, H., 1993. Macrophyte-mediated oxygen transfer in wetlands: transport mechanisms and rates. In G. A. Moshiri (ed.), Constructed wetlands for water quality improvement. Lewis, Boca Raton: 391–398.

Cooper, P. F., 1990. European design and operations guidelines for reed bedtreatment systems. EC/EWPCA Emergent Hydrophyte Treatment Systems Expert Contact Group. WRC Medmenham UK.

Cooper, P. F., J. A. Hobson & C. Findlater, 1990. The use of reed bed treatment systems in the U.K. Wat.Sci. & Tech. 22: 57–64.

Delgado, M., M. Bigeriego & E. Guardiola, 1993. Uptake of Zn, Cr and Cd by water hyacinths. Wat. Res. 27: 269–272.

Denny, P., 1980. Solute movement in submerged angiosperms. Biol. Rev. 55: 65–92.

Eyles, V. A., 1952. The composition and origin of the Antrim laterites and bauxites. Government of Northern Ireland, Memoirs of the Geological Survey. HMSO, Belfast, 90 pp.

Gray, A. V., 1985. Removal of phosphate at sewage treatment works and the implications on phosphate loadings into Lough Neagh. J. Inst. Wat. Eng. Sci. 39: 137–154.

Green, M. B. & J. Upton, 1993. Reed bed treatment for small communities: UK experience. In G. A. Moshiri (ed.), Constructed wetlands for water quality improvement. Lewis, Boca Raton: 517–52.

Green, M. B. & J. Upton, 1994. Constructed reed beds: a cost-effective way to polish wastewater effluents for small communities. Wat. Envir. Res. 66: 188–192.

Higgins, M. J., C. A. Rock, R. Bouchard & Wengreynek., 1993. Controlling agricultural runoff by use of constructed wetlands. In G. A. Moshiri (ed.), Constructed wetlands for water quality improvement. Lewis, Boca Raton: 359–367.

HMSO, 1978a. Manganese in raw and potable waters by spectrophotometry (using formaldoxime). In Methods for the examination of waters and associated materials. HMSO, London. ISBN 0 11 751328 8

HMSO, 1978b. Iron in raw and potable waters by spectrophotometry (using 2-4-6 tripyridyl-135-triazine). In Methods for the examination of waters and associated materials. HMSO, London. ISBN 0 11 751327

HMSO, 1980. Aluminium in raw and potable waters by spectrophotometry (using pyrocatechol violet). In Methods for the examination of waters and associated materials. HMSO, London. ISBN 0 11 751464 0

Howard-Williams, C., 1985. Cycling and retention of nitrogen and phosphorus in wetlands: a theoretical and applied perspective. Freshwat. Biol. 15: 391–431.

Lawson, G. J., 1985. Cultivating reeds (*Phragmites australis*) for root zone treatment of sewage. Contract report to the Water Research Centre. ITE Project 965 Cumbria, UK.

Lienard, A., C. Boutin & D. Esser, 1990. Domestic waste-water treatment with emergent hydrophyte beds in France. In P. F. Cooper & B. P. Findlater (eds), Constructed wetlands in water pollution control. Pergamon Press, Oxford, UK: 183–192.

Martin, C. D., G. A. Moshiri & C. Miller, 1991. Mitigation of landfill-generated leachate incorporating in-series constructed wetlands in a closed loop design. Abstract G.6. In Constructed wetlands for water quality improvement Symposium. Pensacola, Florida.

Moshiri, G. A., 1993 (ed.). Constructed wetlands for water quality improvement. Lewis, Boca Raton: 632 pp.

Murphy, J.& J. P. Riley, 1958. A single-solution method for the determination of phosphorus in sea water. J. Mar. Biol. Ass. UK, 37: 9–14.

Murphy, J. & J. P. Riley, 1962. A modified single-solution method for the determination of phosphorus in natural waters. Analytica Chimica Acta, 27: 31–36.

Ogden, M. H., 1993. The treatment of septage using natural systems. In G. A. Moshiri (ed.), Constructed wetlands for water quality improvement. Lewis, Boca Raton: 525–533.

Schierup, H. H. & V. J. Larsen, 1981. Macrophyte cycling of zinc, copper, lead and cadmium in the littoral zone of a polluted and a non-polluted lake: I. Availability, uptake and translocation of heavy metals in *Phragmites australis* (CAV) Trin. Aquat. Bot. 11: 197–210.

Surface, J. M., J. H. Peverly, T. S. Steenhius & W. E. Stanford, 1993. Effect of season, substrate composition, and plant growth on landfill leachate treatment in a constructed wetland. In G. A. Moshiri (ed.), Constructed wetlands for water quality improvement. Lewis, Boca Raton: 461–472.

Vymazal, J., 1993. Constructed wetlands for wastewater treatment in Czechoslovakia: state of the art. In G. A. Moshiri (ed.), Constructed wetlands for water quality improvement. Lewis, Boca Raton: 255–260.

Wood, R. B & C. F. McAtamney, 1994. The use of macrophytes in bioremediation. Biotechnol. Adv. 12: 653–662.

Hydrobiologia **340**: 333–338, 1996.
J. M. Caffrey, P. R. F. Barrett, K. J. Murphy & P. M. Wade (eds), Management and Ecology of Freshwater Plants.

Backwater habitats and their role in nature conservation on navigable waterways

Nigel J. Willby* & J. W. Eaton
Department of Environmental & Evolutionary Biology University of Liverpool, P.O. Box 147, Liverpool L69 3BX, England
**Address for correspondence: Division of Evolutionary & Environmental Biology, Institute for Biomedical & Life Sciences, University of Glasgow, Glasgow G12 8QQ, Scotland*

Key words: refuge design, boat, channel, management

Abstract

In Britain, lightly trafficked canals frequently contain diverse, productive macrophyte communities. These represent important habitats for macroinvertebrates and fish while having a high intrinsic value in nature conservation terms. As recreational boat traffic increases, fragile macrophytes are progressively eliminated and the biomass of the remaining species is greatly reduced, thereby adversely affecting weed-associated animals and ultimately simplifying the structure of the whole ecosystem. From the viewpoint of aesthetics, nature conservation and fisheries management, ecological enhancement of these traffic impacted ecosystems is desirable but options are limited by channel size and the intensity and type of disturbance. Backwater areas connected to the main channel but apparently remote from traffic influences ought however, to provide a minimally-disturbed refuge for macrophytes and dependent organisms. An extensive field survey was undertaken to test this hypothesis and evaluate the potential for exploiting backwater sites as 'off-line' nature reserves. Principal determinants of vegetation structure and species diversity are identified and discussed and are used to prescribe a set of ideal characteristics for prospective backwater nature reserves and to forecast likely management problems.

Introduction

The British canal system, a relic of the Industrial Revolution, currently comprises ca. 2800 km of relatively narrow (10–15 m wide) and shallow (1.2–1.5 m deep) artificial channels which connect with navigable rivers and are now exploited extensively for recreational boating and angling. Lightly used canals frequently support diverse and productive macrophyte communities due to the favourable continuous low intensity disturbance regime created by low density boat traffic (Murphy & Eaton, 1983), and have attained regional or national importance for nature conservation (Hanbury, 1986; Nature Conservancy Council, 1989). Many unnavigated canals have undergone hydroseral succession to reed swamp of little conservation interest, but others have retained diverse plant communities where an intermittent disturbance regime has been created by occasional dredging and weed cutting (e.g. Lousley, 1976). Some waterways support important populations of rare or uncommon macrophytes, such as *Luronium natans*, *Potamogeton compressus*, *P. friesii*, *P. trichoides* and *Callitriche truncata*, several of which have suffered serious population declines in Britain or Europe over the last few decades (Willby & Eaton, 1993; Stewart, Pearman & Preston, 1994). These unnavigated and lightly trafficked canals, a small fraction of the total system, are vulnerable to ecologically detrimental increases in traffic density unless protected through designation as Sites of Special Scientific Interest. They are also increasingly the subjects of development projects which include restoration or increase of boat traffic. Clearly conflict can arise in such cases between the needs of water plant conservation and those of recreational development.

One attractive potential solution to this conflict of interests, already implemented on the Montgomery Canal, is to exploit small backwater sites, such as lock sideponds and redundant winding holes, mooring basins, branches and timber ponds, as 'off-line' refuges for aquatic plants and associated invertebrates eliminated from the main channel by the increased disturbance. The rationale is that boat traffic density will be diluted within a backwater by the effective increase in cross-sectional area of the channel, with screening offering additional protection if necessary. Unfortunately it is impossible to assess critically the long-term effectiveness of this approach until a canal has been fully restored and reopened to traffic for some time. An investigation of existing backwaters on canals where navigation is long established can, however, be used to predict the value of refuges once subjected to renewed traffic and assist the design and choice of future sites and management requirements. This paper therefore aims to define (i) the basic differences between backwater and main channel vegetation; (ii) the main environmental determinants of vegetation structure and species richness in backwaters; (iii) habitat features which maximise the conservation interest in a backwater, and (iv) the overall scope for maintaining conservation interest in backwaters on heavily trafficked canals.

Methods

A preliminary cartographic analysis revealed that the canal system managed by the British Waterways Board contains 1000–1200 backwaters (defined as a >50% increase in channel width), ranging in shape from simple widenings to complex loading areas and in size from 0.02–5.00 ha, though the majority (80%) are <0.1 ha.

Aquatic vegetation was surveyed between July and September 1990 in 179 randomly-selected backwaters, which were widely distributed and representative of a range of traffic densities and channel dimensions. For comparative purposes, surveys included a 150 m section of mainline canal located a short distance upstream of each backwater. Submerged macrophytes were sampled by grapnelling (Murphy, Hanbury & Eaton, 1981). Samples were collected from 10 randomly determined points in main channel sections and at similar density (i.e. one throw ca. $15m^{-1}$) in backwaters. The abundances of emergent and floating-leaved vegetation were assessed at the same points by calculating areal cover of all species within a 2 m wide transect running perpendicular to the bank into open water, values then being converted to mean % cover for the whole site. In large, wide-mouthed backwaters (>0.2 ha), vegetation was mapped by measuring areas of individual stands. To ensure a comprehensive species list for each site, rare or highly patchily distributed species not recorded in transects were assigned a nominal cover value of 0.1%. All cover values were subsequently converted to estimated above-ground biomass ($gDWm^{-2}$) using either published figures for biomass recorded under comparable conditions, or, when these were unavailable, empirically derived data obtained by drying material clipped at the substrate surface. Data were collected on a number of environmental variables (Table 1).

Since the abundance of individual species and biomass groups followed highly skewed distributions which could not be satisfactorily normalised by transformation, pairwise differences between backwaters and adjacent mainline sites were assessed using a Wilcoxon signed ranks test. Comparisons of the abundance of individual species were confined to those present in at least 10 backwater and/or mainline sites (67 species). Their frequencies of occurrence were compared using a simple sign test on data recoded to presence/absence. The relationship between biomass and species richness of backwater vegetation and environmental variables was investigated using stepwise linear regression, from which a subset of the most influential environmental factors was obtained. Variables were admitted until none of those remaining were significantly related ($P = <0.05$) to the residual variation in the vegetation data.

Results

Plant biomass in backwater and main channel habitats is compared in Table 2. In most cases backwaters clearly supported considerably higher biomasses of emergent vegetation than the adjacent main channels, but biomasses of submerged and floating-leaved vegetation did not differ significantly between the two habitats. Since emergent vegetation was the overriding biomass component, the overall result was a significantly higher total standing crop in backwaters than in the main channel.

Comparing species compositions in backwater and main channel vegetation (Table 3) highlights the greater emphasis on emergent species in backwaters, with most of the difference in overall biomass being

Table 1. Measured environmental variables and stepwise regression functions for backwater vegetation

Variable	Code	Units/scale & notes
Water Quality		
Turbidity	NTU	nephelometric turbidity units } on sample collected in backwater at point furthest from main channel
Total Suspended Solids	TSS	mgl^{-1} } on sample collected in backwater at point furthest from main channel
Conductivity	COND	μScm^1
NWC Class	QUAL	1=1A, 2=1B, 3=2, 4=3; Department of the Environment & Welsh Office (1986)
Physical Attributes		
Surface Area	AREA	m^2
Perimeter	PERIM	m
Interface	WIDE	m = width of interface between main channel and backwater
Length	LONG	m = maximum distance from main channel
Cross-sectional area	XSEC	m^2 } depth measurements by plumb line at 1 m intervals
Maximum depth	ZMAX	m } depth measurements by plumb line at 1 m intervals
Mean depth	ZAVE	m } depth measurements by plumb line at 1 m intervals
Channel profile	BNK	1=box-shaped, 2=asymetric, 3=trapezoidal
Bank hardening	HARD	0=0%, 1=1–25%, 2=26–50%, 3=51–75%, 4=76–99%, 5=100%; as % of total perimeter with artificial hardened bank
Shading	SHAD	0=0%, 1=1–5%, 2=6–25%, 3=26–50%: visual assessment of water surface
Aspect	(eg)NE	principal compass bearing perpendicular to site apex
Disturbances		
Boat traffic		
- backwater	BOAT	0=unused, 1=occasional use, 2=used at least daily for boat manoeuvring
- main channel	MY	movements yr^{-1}; Murphy & Eaton (1983) using 1988–90 data
Vegetation reduction		
- by livestock	TROD	0=no grazing, 1= 1–5% bare ground, 2= 6–25%, 3=>25%
- by manage./herbivory	CUT	0=none, 1= cut in past 6 weeks or large populations of carp, water voles or crayfish, 2= cut in past 3 weeks or stocked with grass carp

Variable	Regression function
Biomass	
Total	y = 6.444 + (–0.526 BOAT) + (–1.563 BNK1) + (–0.240 MY) + (–0.326 ZAVE) + (0.85 XSEC) + (–0.729 CUT)
Emergent	y = 6.495 + (–2.555 BNK1) + (–0.478 ZAVE) + (–0.948 CUT) + (–0.326 SHAD) + (1.051 XSEC) + (–0.019 MY)
Submerged	y = 2.274 + (–0.404 TSS) + (0.158 ZMAX) + (–0.023 MY) + (0.457 BNK1)
Floating-leaved	y = 2.333 + (–0.599 NTU) + (–0.339 BOAT) + (0.338 TROD)
Species Richness	
Total	y = 2.644 + (–0.16 HARD) + (–0.133 TSS) + (0.274 LONG) + (0.263 TROD) + (–0.265 BOAT)
Emergent	y = 1.687 + (–0.144 HARD) + (0.291 LONG) + (–0.607 BNK1) + (0.289 TROD) + (–0.288 BOAT) + (–0.355 NE) + (–0.562 NW)
Subm + flg-lvd.	y = 2.156 + (–0.006 MY) + (0.388 XSEC) + (–0.061 HARD) + (–0.182 TSS) + (–0.217 QUAL) + (–0.15 SHAD) + (–0.215 BOAT)

Variables listed in functions in descending order of importance. All regressions significant at P=<0.001. N=179
Response variables and all continuously distributed environmental variables log-transformed ($\log_e[x+1]$), except MY, ZMAX and ZAVE which were square root transformed
Scores of ordinal variables unchanged. Dummy values defined for nominal variables (aspect and bank type) to avoid colinearity

Table 2. Comparison of vegetation biomasses in backwater and main channel sites

Biomass	Backwater		Main channel		P
(gDWm–2)	mean	S.E.	mean	S.E.	
Total	291.3	21.7	149.6	13.6	<0.001
Emergent	258.5	21.7	123.2	12.6	<0.001
Submerged	19.4	3.4	15.9	2.6	ns
Floating-leaved	13.4	2.9	10.7	1.8	ns

Significance of ranked pairwise difference tested by Wilcoxon signed ranks test. N=179.

attributable to the greater abundance of highly productive stand-forming monocotyledonous species such as *Glyceria maxima*, *Typha latifolia* and *Sparganium erectum*. Backwaters also supported a greater frequency or abundance of many herbaceous marshland species, reflecting the availability of shallow marginal habitats which in the main channel are often removed through bank erosion or engineering. Consequently total species richness was often relatively high in backwaters. Lemnids were significantly more abundant in backwaters, reflecting the virtually static water conditions compared to the main channel, while the greater abundances of *Callitriche stagnalis*, *Ceratophyllum demersum* and *Elodea nuttallii* implies that backwaters represent a refuge for aggressive but fragile elodeids, which are normally restricted by boat traffic in the main channel. By contrast *Acorus calamus*, *Alisma lanceolatum*, *Nuphar lutea*, *Sparganium emersum* and *Potamogeton pectinatus* were characteristically more common in the deeper and generally more turbid and turbulent environment of the main channel.

The results of stepwise regression analysis are summarised in Table 1. Estimated total standing crop showed strong negative correlation with the intensity of boat usage of backwaters, mainline traffic density and average depth and was usually lowest in steep sided sites. The requirements of emergent vegetation are closely reflected in this function. Submerged crop was strongly negatively correlated with total suspended solids (adj. $R^2=0.40$; $F=59.83$; $P=<0.0001$) and mainline traffic density, but was generally greatest in deep, steep sided sites, reflecting the increase in colonisable volume and the restriction on growth of marginal reed swamp. The function for floating-leaved plant biomass reflects the negative response of large, robust species with submerged foliage (*Potamogeton natans*, *Nuphar lutea* and *Sparganium emersum*) to high turbidity and physical damage from boat manoeuvring. The positive reaction to livestock activity relates to the proliferation of *Glyceria fluitans* along shallow margins exposed by cattle poaching. Total species richness was negatively correlated with bank hardness, total suspended solids and both channel and backwater boat movements but increased with the degree of remoteness from the main channel and livestock activity. With the exception of total suspended solids, emergent plant diversity follows a similar function, but is additionally reduced in steep sided sites due to the elimination of soft, shallow margins and in those with a northerly aspect. Species richness of submerged plus

Table 3. Summary of major floristic differences between backwater and main-channel sites based on ranking tests of pairwise differences

Frequency of occurrence and/or abundance significantly greater in: Backwater	Main channel
Emergent	
Angelica sylvestris	*Acorus calamus*
Cardamine pratensis	*Alisma lanceolatum*
Carex otrubae	
Epilobium hirsutum	
Epilobium palustre	
Galium palustre	
Glyceria maxima	
Impatiens capensis	
Iris pseudacorus	
Lycopus europaeus	
Mentha aquatica	
Myosotis scorpioides	
Ranunculus sceleratus	
Rorippa nasturtium-aquaticum	
Solanum dulcamara	
Sparganium erectum	
Stachys palustris	
Stellaria alsine	
Typha latifolia	
Veronica beccabunga	
Floating-leaved	
Lemna gibba	*Nuphar lutea*
Lemna minor	*Sparganium emersum*
Lemna trisulca	
Submerged	
Callitriche stagnalis	*Potamogeton pectinatus*
Ceratophyllum demersum	
Elodea nuttallii	

Significant pairwise differences ($P<0.05$) were detected by a sign test (frequency of occurrence) and Wilcoxon signed ranks test (abundance by biomass)

floating-leaved macrophytes increased with cross sectional area but was negatively correlated with bank hardness, suspended solids and both channel and backwater boat movements and also declined with increased shading and decreasing water quality. There was no evidence that backwater area *per se* was an important determinant of species richness, but the significance of the area-dependent factors, length of backwater and cross-sectional area, is analogous to a species-area effect.

Discussion

The predominance of emergent vegetation in navigation backwaters, as found here, appears to be a consequence of the production of shallow water habitats by sedimentation (Peck & Smart, 1986; Bhowmik & Adams, 1989). The mixed lemnid/elodeid flora present in backwaters which still retain open water is typical of eutrophic standing water habitats such as ponds and ditches (Hejny & Husak, 1978; Nature Conservancy Council, 1989). By contrast, species associated more strongly with the main channel (Table 3), together with others such as *Potamogeton perfoliatus* and *Sagittaria sagittifolia*, are characteristic of deep water habitats and the margins of sluggish lowland clay-based rivers (Haslam, 1978; Holmes & Newbold, 1984). Brierley, Harper & Barham (1989) reported comparable differences in vegetation between backwater channels and the main river of the navigable R. Nene, Northants.

Emergent vegetation provides valuable habitat and bank protection in canals, yet there is very rarely sufficient area available for colonisation for even the largest stands to attain the functional and conservation importance of the extensive areas of reed swamp found in other wetland habitats. Hence, in backwater refuges, the maintenance of diverse submerged or floating-leaved plant communities of high nature conservation value may be a more sensible priority. On the evidence of stepwise regressions, maximum submerged biomass will occur in sites with low suspended solid loads, large maximum depths and steep bank profiles, situated on canals carrying a low density of traffic. Maximum species richness, which is more likely to be found at an intermediate range of plant biomass (e.g. Al Mufti et al., 1977; Wheeler & Shaw, 1991), will generally be found in sites with low turbidity, low shading, soft banks, infrequent boat manoeuvring, a large cross sectional area and high water quality, located on lightly trafficked canals. Therefore if backwater reserves are to be sited on more heavily trafficked canals (>6000 my), it will be critical to ensure that values for variables other than traffic are optimal.

In all but the most isolated or long backwaters, main channel traffic promotes sediment accumulation in backwaters (Peck & Smart, 1986) through the deposition of solids previously suspended by boats. This, combined with detritus flushed out from the main channel, creates a very soft, organic-rich and highly anaerobic substrate rich in hydrogen sulphide and methane, inimical to the growth of rooted water plants (Weisner, 1991), but favourable to the development of extensive floating rafts of monodominant *Glyceria maxima* (Lambert, 1947), which are a common feature of canal backwaters. Frequent boat manoeuvring is extremely disruptive, since it concentrates the damaging effects of boat movements (Murphy, Willby & Eaton, 1995) in an often small area and maintains a high suspended solid load through remobilisation of the very soft bed. Benthivorous fish such as carp (*Cyprinus carpio*) and bream (*Abramis brama*), which at high densities are known to have destructive effects on submerged vegetation (Wright & Phillips, 1992), also frequently congregate in backwaters on navigable waterways. Redundancy or infrequent management of backwaters allows tree canopy shading and leaf litter inputs to increase, greatly accelerating shallowing and adding to oxygen demand. Given these facts, backwater refuges must either be physically isolated from the main channel or of a suitable size or shape to incorporate areas unaffected by traffic. Management to maintain a favourable rooting substrate and control reed swamp encroachment and tree shading is also imperative.

There is, at least in theory, scope for engineering and managing backwaters to develop their water plant conservation potential. Turbidity, siltation rates, bed disturbance and wave action may be reduced, either by screening the mouth of the site with an earth bund, rock gabion, steel piling or reed fringe reinforced with geotextile matting, or by installing a silt screen or trap to promote settlement of suspended clay particles. However, the capital cost of recreating favourable conditions on a scale large enough to support viable plant populations and the subsequent need for interventionist control of dense beds of *Elodea* spp., *Ceratophyllum demersum* and of *Glyceria maxima* reed swamp, may make such options unattractive, especially as mechanical dredging or weed cutting may prove technically difficult and expensive in small inaccessible sites.

Consequently all except the largest backwaters appear to have rather limited potential for botanical conservation on British navigable canals. A more practical alternative may be to restore currently reed-choked branches and basins as off-channel long-term mooring sites for private boats. Since such craft move infrequently, they could create the desired low intensity, non-interventionist disturbance regime analogous to that which exists on lightly trafficked waterways and be compatible with both the nature conservation and recreational development potential of the waterway.

Conclusions

1. Most smaller British canal backwaters are too unstable and prone to siltation and reed swamp encroachment to have useful potential for water plant conservation.

2. Larger backwaters, sufficiently isolated from the main channel to create conditions of clear, high quality, well illuminated water of a wide range of depths, a firm substrate for rooting and low disturbance by boats and cyprinid fishes, have potential for conservation of diverse floating-leaved and submerged plant communities.

3. On heavily-trafficked canals, physical control of sediment movement into backwaters is essential if rapid shallowing and habitat loss is to be avoided.

4. Canal branches and basins restored or constructed for use as private boat moorings may combine nature conservation and recreational development functions, where design dimensions create and sustain appropriate plant habitats.

Acknowledgments

We are grateful to the British Waterways Board for providing financial support to NJW and particularly to Roger Hanbury and Jonathan Briggs, Environmental & Scientific Services, Gloucester, for useful discussions. The views expressed in this paper are, however, the sole responsibility of the authors.

References

Al Mufti, M. M., C. L. Sydes, S. B. Furness, J. P. Grime & S. R. Band, 1977. A quantitative analysis of shoot phenology and dominance in herbaceous vegetation. J. Ecol. 65: 759–791.

Bhowmik, N. G. & J. R. Adams, 1989. Successional changes in habitat caused by sedimentation in navigation pools. Hydrobiol. 176/177: 17–27.

Brierley, S. J., D. M. Harper & P. J. Barham, 1989. Factors affecting the distribution of aquatic plants in a navigable lowland river; the River Nene, England. Reg. Rivers: Res. & Mgmt 4: 263–274.

Department of the Environment & Welsh Office (1986) River Quality in England & Wales, 1985. London, HMSO.

Hanbury, R. G., 1986. Conservation on canals: a review of the present status and management of British navigable canals with particular reference to aquatic plants. Proc. EWRS/AAB 7th Symposium on Aquatic Weeds. Loughborough.

Haslam, S. 0M., 1978. River Plants. Cambridge University Press. 396 pp.

Hejny, S. & S. Husak, 1978. Higher plant communities. In D. Dykyjova & J. Kvet (eds), Pond Littoral Ecosystems: Structure & Functioning, Berlin: Springer-Verlag: 23–62.

Holmes, N. T. & C. Newbold, 1984. River Plant Communities-Reflectors of water and substrate chemistry. Nature Conservancy Council, Peterborough, England.

Lambert, J. M., 1947. Biological flora of the British Isles. *Glyceria maxima* (Hartm.) Holmb. J. Ecol. 34: 310–344.

Lousley, J. E., 1976. Flora of Surrey. David & Charles.

Murphy, K. J., R. G. Hanbury & J. W. Eaton, 1981. The ecological effects of 2-methylthio triazine herbicides used for aquatic weed control in navigable canals. I. Effects on aquatic flora and water chemistry. Arch. Hydrobiol. 91: 294–331.

Murphy, K. J. & J. W. Eaton, 1983. Effects of pleasure boat traffic on macrophyte growth in canals. J. Appl. Ecol. 20: 713–729.

Murphy, K. J., N. J. Willby & J. W. Eaton, 1995. Ecological impacts and management of boat traffic on navigable inland waterways. In D. M. Harper & A. J. D. Ferguson (eds), The Ecological Basis for River Management, John Wiley: 427–442.

Nature Conservancy Council, 1989. Guidelines for selection of Biological SSSIs. NCC. Peterborough, England. 288 pp.

Peck, J. H. & M. M. Smart, 1986. An assessment of the aquatic and wetland vegetation of the Upper Mississippi River. Hydrobiol. 136: 57–76.

Stewart, A., D. A. Pearman & C. D. Preston, 1994. Scarce Plants in Britain. JNCC, Peterborough, England, 515 pp.

Weisner, S. E. B., 1991. Within-lake patterns in depth penetration of emergent vegetation. Freshwat. Biol. 26: 133–142.

Wheeler, B. D. & S. C. Shaw, 1991. Above-ground crop mass and species richness of the principal types of herbaceous rich-fen vegetation of lowland England & Wales. J. Ecol. 79: 285–302.

Willby, N. J. & J. W. Eaton, 1993. The Distribution, Ecology and Conservation of *Luronium natans* (L.) Raf. in Britain. J.Aquat. Plant Mgmt 31: 70–76.

Wright, R. M. & V. E. Phillips, 1992. Changes in the aquatic vegetation of two gravel pit lakes after reducing the fish population density. Aquat. Bot. 43: 43–49.

Hydrobiologia **340:** 339–343, 1996.
J. M. Caffrey, P. R. F. Barrett, K. J. Murphy & P. M. Wade (eds), Management and Ecology of Freshwater Plants.

Experimental revegetation of the regulated lake Ontojärvi in northern Finland

S. Hellsten[1], J. Riihimäki[1,2], E. Alasaarela[2] & R. Keränen[3]
[1] *VTT, Communities and Infrastructure, Water Engineering and Ecotechnology, P.O. Box 19042, FIN-90571 Oulu, Finland*
[2] *Water and Environmental District of Oulu, P.O. Box 124, FIN-90101 Oulu, Finland*
[3] *Institute of Ylä-Savo, FIN-74300 Sonkajärvi, Finland*

Key words: regulated lakes, macrophytes, revegetation, erosion, restoration

Abstract

Water level regulation causes large-scale ecological changes in the littoral areas of lakes in northern Finland. If the summertime water level is raised, intensive erosion processes begin, causing a sudden decline in shore vegetation. The need for shore protection is obvious in areas of high recreational value. At lake Ontojärvi, planting experiments with littoral helophytes and bushes were carried out during the years 1990–92. All the experiments were carried out in the eroded sandy areas, which were partly protected by mechanical barriers. Several plant species were planted on the shore which had been treated with different peat mixtures, etc. The frequencies of the different species were followed monthly. After the first summer, the average survival rates were about 45 % due to the drying of seedlings. A rapid decrease in the survival rates took place during the high water level period in 1991 at which point only 20% of the planted individuals were alive. The best results were obtained for the helophyte *Carex rostrata* Stokes, of which 30% had survived erosion. Tall willows (*Salix phylicifolia* L.) were also erosion-resistant with a survival rate of 80%.

Introduction

In Northern Europe there are numerous lakes in which the water levels are regulated. About 10% (33 522 km^2) of the total area of Finland is covered by lakes and over one third of this area (11 900 km^2) is regulated, mainly for hydroelectric purposes. Regulation is usually achieved by raising the water level during the summer (0.5–3.5 m) and lowering the winter minimum by 2–7 m. During the first few years of regulation, the littoral zone is subject to considerable erosion, depending on the rate of water level uplift (e.g., Alasaarela et al., 1989). At the same time, the changes in vegetation are obvious, including a decrease of the former macrophytic vegetation as reported in several Scandinavian lakes (e.g., Rørslett, 1985). If the erosion is continous the shoreline remains without vegetation and also without the erosion sheltering properties of littoral vegetation. Eroded shore areas without changes in the water level fluctuation or mechanical barriers, which diminish the effects of waves and currents, are a hostile environment for young seedling of plants. The aim of this study is to find possibilities for promoting the succession of vegetation by planting seedlings of typical shore plants and applying several soil treatment methods to the regulated Lake Ontojärvi.

Materials and methods

Study area

Lake Ontojärvi (102 km^2) is situated in the Oulujoki watercourse (Figure 1). Ontojärvi has been regulated for hydroelectric purposes since 1951 with an amplitude of 4.4 m (Figure 2). At the beginning of regulation, the mean summer water level was raised by one meter.

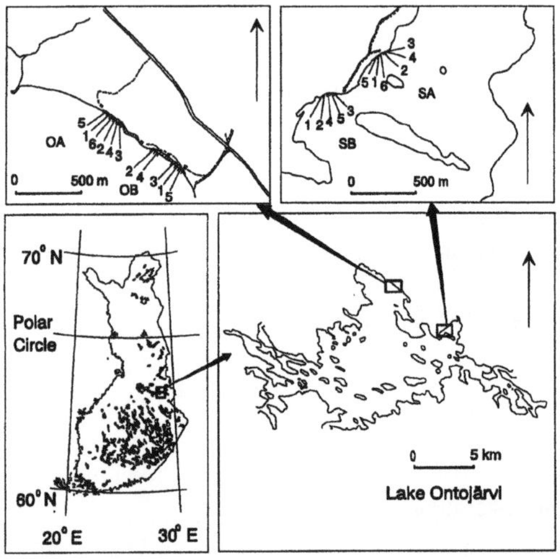

Figure 1. Location of Lake Ontojärvi in northern Finland. Study areas are enlarged in boxes, refer to the text for details.

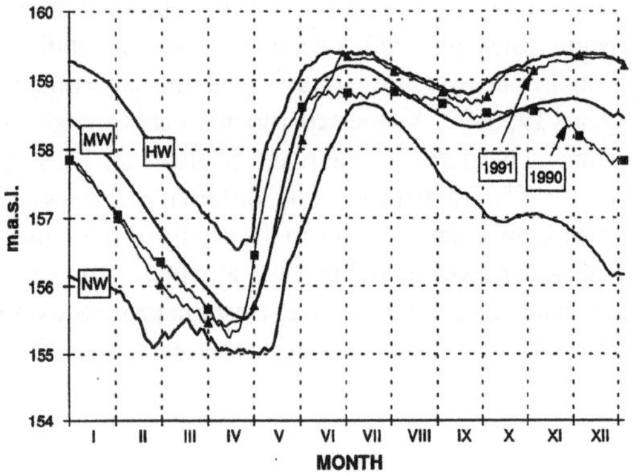

Figure 2. Water level fluctuation in lake Ontojärvi. MW, HW and NW lines calculated from daily values observed during the years 1964–91. Daily values during main years of the research project presented by filled squares (1990) and filled triangles (1991).

The water level uplift initiated an erosion of shoreline, which caused instability in the sandy shore areas.

Shore vegetation planting experiments

Open (O) and sheltered (S) experimental areas were established in the sandy eroded shore areas of Lake Ontojärvi (Figure 1). OA refers to an open (fetch = 3.9 km) shore area without any mechanical protection, while OB (fetch = 2.7 km) stands for an area protected by groins made of stones. A sheltered shore SA (fetch = 0.8 km) was without protection, while SB (fetch = 1 km) was protected by a chain of floating timbers (four parallel timbers stabilised by floating barrels). All the research areas were divided into six subareas (6 × 10 m), in which the soil was manipulated by the following methods (Figure 1). The first subarea (1) was a non-treated control area. In the second subarea (2), fertilized peat (VAPO C1) was mixed in to the soil (2.1 kg m^{-2}) to improve its fertility and moisture content. The rapidly growing oat (*Avena sativa* L.) was used as a protective plant against erosion in the third subarea (3). Peat treatment and oat were used together in subarea four (4). Subarea five (5) consisted of bunches of willows (mainly *Salix phylicifolia* L.) bound together into a carpet (Begemann & Schiechtl, 1986). An erosion carpet (Rejtex©) manufactured of easily degradable recycled fibre was used in the sixth (6) subarea.

All the aquatic macrophytes planted in the subareas were erosion-resistant species collected from areas situated near the experimental plots (cf. Lester et al., 1986; Comes & McCreary, 1986). *Carex rostrata* Stokes was planted as a piece of turf in two rows (50 cm between rows, planting depth 5–10 cm). *Juncus filiformis* L. was planted as a bunch of 3–10 individuals in four rows (25 cm between rows, planting depth 5–7 cm). Pieces of roots of *Phragmites australis* (Cav.) Trin. ex Steudel were planted in two rows (50 cm between rows, planting depth 10–15 cm). *Calamagrostis* sp. (mainly *Calamagrostis purpurea* (Trin.) Trin) was used as turfs of 1–5 individuals in four rows (25 cm between rows, planting depth 5–10 cm). Some typical shore bushes were also used as experimental plants. *Salix phylicifolia* was planted in two rows (50 cm between rows, planting depth 20–30 cm). Small seedlings of *Alnus incana* (L.) Moench were planted near the foot of shore cliff in one row (distance 50 cm, planting depth 10–20 cm). All the species were planted according the natural zonation observed in other parts of the lake. The planting order from the shoreline outwards was *Alnus-Salix- Calamagrostis-Carex/Juncus-Phragmites*. *Carex* was only used in sheltered (S) and *Juncus* in open (O) areas. In the sixth subarea (willow carpet), only tall (1.5–3 m) cuttings of *Salix phylicifolia* were planted near the foot of the littoral shelf.

The experimental areas were prepared during May 1990 and the planting took place during June 1990. The areas were analysed twice per month in 1990, monthly in 1991 and once in 1992. All the living (green) plants were included in the survival rate estimation. The difficulties in distinguishing living plants from dead ones during spring and autumn caused a fluctuation in the survival rate.

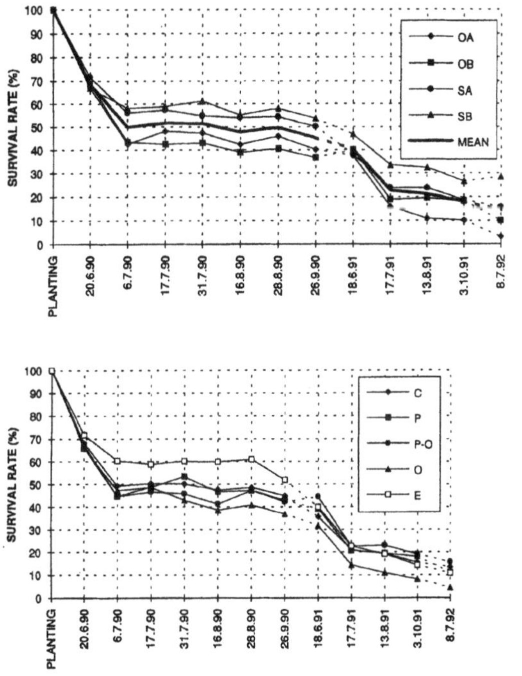

Figure 3. A. Mean survival rates of species planted in different sub-areas and total mean. Refer to the text for details. B. Mean survival rates of species planted at research sites with different treatments. C = control areas, P = peat treatment, O = oat as shelter plant, P-O = both treatments together, E = erosion carpet.

Results and discussion

The mean survival percentages in the different research areas are presented in Figure 3a. Quite soon after the planting, almost half of the seedlings became desiccated. The water level was almost 50 cm below the normal, and it reached the foot of the shore cliff and the level of planted vegetation only in the sheltered areas (SA, SB) (Figures 1, 2). Survival rate was 15% higher in that area (Figure 3a). The effect of ice-erosion was easy to recognize in the sheltered areas (SA, SB), where mortality was 10–15% during the first winter (Figure 3a). The high water level and wave erosion during the second summer of the research caused sudden changes in the survival rate which dropped in all areas (Figures 2, 3a). Only in the sheltered area protected by floating timbers (SB) did the survival rate stay above 20%.

The differences between the different treatments were surprisingly small (Figure 3b). Only the erosion carpet provided a better survival rate during the first summer, but the second-year erosion caused a rapid decrease in the survival rate (Figure 3b). Oats as a protective plant provided the lowest survival rates for planted helophytes, because ducks grazed on young oat seedlings and also plucked at the other plants.

The mean survival rates of the different species at the end of each year of the research programme provided quite a confused overall view (Figure 4). *Carex rostrata* was the only aquatic macrophyte which survived fairly well, but *Calamagrostis* also showed notable persistence against erosion. *Phragmites* and *Juncus* disappeared quite rapidly, even though the latter is quite common on open shores. Small willow seedlings also survived well, but the survival rate of *Alnus* was low although it also grows also on the eroded shores of regulated lakes (Figure 4). As a point of special detail tall willow cuttings (*Salix phylicifolia*) survived fairly well showing a survival rate of 85% at the end of 1991.

The zonation of littoral vegetation is usually formed by water level fluctuation (e.g., Spence, 1982; Wilson & Keddy, 1985). In this study, the planting levels were exactly the same as elsewhere in the littoral zone of lake Ontojärvi (S. Hellsten, unpublished data). In regulated lakes, the living area of macrophytes is usually narrower and moves downwards due to wave erosion (e.g., Rørslett, 1984). The effect of down-dwelling ice during the wintertime also causes scouring and freezing of the bottom sediment, as has been described for several other lakes (Rørslett, 1984, 1987; Erixon, 1981). In the uppermost part of the littoral there is an obvious possibility of desiccation due to regulation (e.g., Rørslett, 1984). The ecological environment, especially on sandy shores, is therefore extremely harsh for all living plants. On open shores, the foot of the littoral shelf is situated at a higher level (159.40 m.a.s.l) compared to sheltered ones (159.00 m.a.s.l.), and the possibility of drying out causes a higher mortality (Figure 3a). Sheltered shores with a lower height of the littoral shelf foot provide better moisture conditions, but they are also more easily eroded in the case of a high water level. Although species were used which are quite resistant against erosion, we did not get good results. In fact, most of the helophytes are typical C-strategists (e.g. *Phragmites, Juncus*) which are good competitors, but slow in extending their living areas (see Grime, 1977; Murphy et al., 1990). All the species with a good resistance against erosion are small R-strategists (e.g. *Ranunculus reptans, Eleocharis acicularis*) which are difficult to use as experimental plants.

The substrate quality or, more precisely, the grain size and the amount of organic matter, is one of the

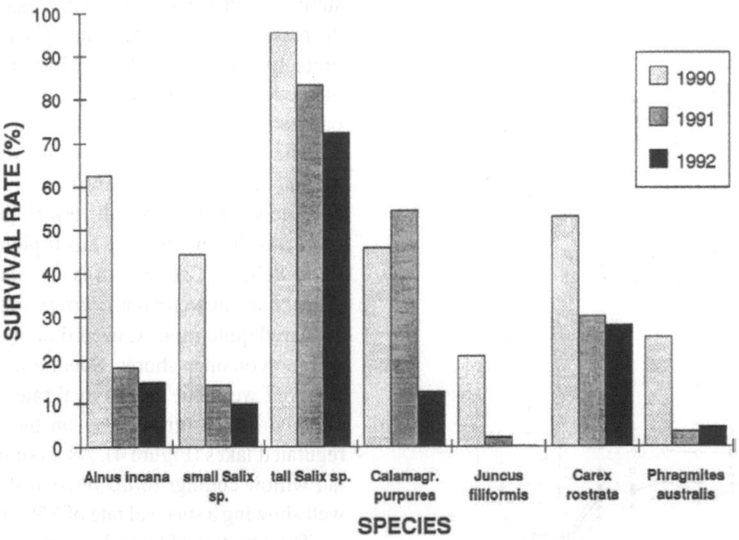

Figure 4. Mean survival rates of different species at the end of different research years.

most important factors affecting aquatic macrophytes (e.g. Barko & Smart, 1983). If the erosion is at low level, the content of organic matter in the sediments is usually high. Low erosion can be affected by erosion barriers, as described in Allen et al. (1984). Another way to improve sediment quality is to add fertilizers (Fowler & Maddox, 1974; Broome et al., 1988) or organic matter, as in our study. Our results showed clearly that these methods were not beneficical in lake Ontojärvi, where intensive wave erosion flushed the shoreline regularly. Bache & MacAskill (1984) found that the use of protective plants gave a good shelter for young seedlings whereas, in our case, duck grazing reduced the value of oats as a protective plant. The use of an erosion carpet was successful in experiments carried out by Allen et al. (1984) but in lake Ontojärvi, wave erosion broke the carpet or it was covered by a thick sand layer. Tall seedlings of willow yielded one of the most promising results in our experiments. Willows are easily collected and resistant against flooding. The willow carperts did dry or erode during the first year of our study. The carpets should have been covered with wet sand for better results. Comes & McCreary (1986) and Bache & MacAskill (1984) also found similar carpets to be convenient for shore protection. Willow carpets are also commonly used for stabilizing river banks (Bagemann & Schiechtl, 1986). In our experiments we did not shape the littoral shelf in any way, as was recommended by Allen & Klimas (1986). A milder slope should give better protection against erosion, as has been seen in lake Oulujärvi studies.

Acknowledgments

This project is part of a larger project called 'Development of the Regulation of the Oulujoki Watercourse' launched by the Water and Environment District of Kainuu. The project was financed by the National Board of Water and Environment, the IVO hydropower company, the University of Oulu and VTT. Dr Björn Rørslett (NIVA, Norway) kindly made his literature review available to our project. We would also like to thank anonymous referees and Mrs Sirkka-Liisa Leinonen, M.Sc., for correcting the English.

References

Alasaarela, E., S. Hellsten & P. Tikkanen, 1989. Ecological aspects of lake regulation in northern Finland. In H. Laikari (ed.), River Basin Management – 5, Pergamon Press, Oxford.

Allen, H. H. & C. V. Klimas, 1986. Reservoir shoreline revegetation guidelines. Technical report E-86-13, US Army engineer waterways experiment station. Vicksburg, Miss. 87 pp.

Allen, H. H., J. W. Webb & S. D. Shirley, 1984. Wetlands development in moderate wave-energy climates. Proceedings of dredging '84, Waterway, Port, Coastal and Ocean division, American Society of Civil Engineers, November 14–16, 1984, Clearwater, Fla.: 943–955.

Bache, D. H. & I. A. MacAskill, 1984. Vegetation in civil and landscape engineering. Granada, London: 317 pp.

Barko, J. W. & R. M. Smart, 1983. Effects of organic matter additions to sediment on the growth of aquatic plants. J. Ecol. 71: 161–175.

Bagemann, W. & H. M. Schiechtl, 1986. Ingenieur biologie: Handbuch zum naturnachen Wasser- und Erdbau. Bauverlag GMBH, Wiesbaden und Berlin: 216 pp.

Broome, S. W., E. D. Seneca & W. W. Woodhouse, 1988. Tidal salt marsh restoration. Aquatic Botany 32: 1–22.

Comes, R. D. & T. McCreary, 1986. Approaches to revegetate shorelines at Lake Wallula on the Columbia River, Washington-Oregon. Technical report E-86-2, US Army Engineer Waterways Experiment Station, Vicksburg, Miss.

Erixon, G., 1981. Aquatic macrophytes and their environment in the Vindelälven river, northern Sweden. Wahlenbergia 7: 61–71.

Fowler, D. K. & J. B. Maddox, 1974. Habitat improvement along reservoir inundation zones by barge hydroseeding. J. Soil Water Conserv. 22: 263–265.

Grime J. P., 1977. Evidence for the existence of three primary strategies in plants and its relevance to ecological and evolutionary theory. Am. Nat. 111: 1169–1194.

Lester, J. E., C. V. Klimas, H. H. Allen & S. G. Shetron, 1986. Shoreline revegetation studies at lake Texoma on the Red River, Texas-Oklahoma. Technical report E-86-1, US Army Eng. Waterways Experiment Station, Vicksburg, Miss.

Murphy, K. J., B. Rørslett & I. Springuel, 1990. Strategy analysis of submerged lake macrophyte communities: an international example. Aquatic Botany 36: 303–323.

Rørslett, B., 1984. Enviromental factors and aquatic macrophyte response in regulated lakes – a statistical approach. Aquatic Botany 19: 199–220.

Rørslett, B., 1985. Regulation impact on submerged macrophytes in the oligotrophic lakes of Setesdal, South Norway. Verh. int. Ver. Limnol. 22: 2927–2936.

Rørslett, B., 1987. Niche statistic of submerged macrophytes in Tyrifjord, a large oligotrophic Norwegian lake. Arch. Hydrobiol. 111: 283–308.

Spence, D. N. H., 1982. The zonation of plants in freshwater lakes. Adv. ecol. Res. 12: 37–125.

Wilson, S. D. & P. A. Keddy, 1985. Plant zonation on a shoreline gradient: Physiological responce curves of component species. J. Ecol. 73: 851–860.

Hydrobiologia **340**: 345–348, 1996.
J. M. Caffrey, P. R. F. Barrett, K. J. Murphy & P. M. Wade (eds), Management and Ecology of Freshwater Plants.

Enhancing river vegetation: conservation, development and restoration

S. M. Haslam
Department of Plant Species, University of Cambridge, United Kingdom

Key words: river, restoration, conservation, development, sense of place

Abstract

The welcome increase in projects increasing river vegetation prompts theoretical consideration. To restore is to bring back what was there before – but at what period? 1860 to 1940 is suggested, as 'traditional'. Traditional rivers differ greatly between river types and between rivers: in water, structural and biotic characters, and Sense of Place. Aiming at these, holistically, is aiming at restoration. Aiming at part, or trying to convert rivers to a Standard Recommended river is, at best, enhancement. The latter development lessens the unique variety and special features of rivers that are the heritage of each country.

Introduction

It is axiomatic that no restoration can ever be perfect; it is impossible to replicate the biogeochemical and climatological sequence of events over geological time that led to the creation and placement of even one particle of soil, much less to exactly reproduce an entire ecosystem. Therefore, all restorations are exercises in approximation, and in the reconstruction of naturalistic rather than natural assemblages of plants and animals with their physical environments (cited in Maurizi & Poillon, 1992. 'Natural', though, is appropriate to North America, not to Britain, see below).

This paper provides some preliminary thoughts on restoration and enhancement. Altering rivers from straight, uniform, trapezoid channels to habitats fit for abundant plants and animals is an entirely welcome development, but any changes made to rivers involve theoretical considerations (as well as those of operating heavy machinery etc.).

Discussion of terms

(Shorter Oxford English Dictionary *used where relevant*)

Enhancement is the raising or increasing in value, importance or attractiveness. Therefore any change increasing appropriate biota is enhancement. Not all such changes are restoration, however. Restoration is, technically, restoring to a former state, or the representation of the original form of a ruined building, extinct animal, etc.: the representation of the original form of a ruined river, perhaps. Ecologically, 'restoration is returning the system to a close approximation of the pre-disturbance ecosystem that is persistent and self-sustaining' (though dynamic in its composition and functioning) (Maurizi & Poillon, 1992). This American definition must be altered, for Britain. In much of North America pre-disturbance ecosystems existed until quite recently, and there are still innumerable sites there termed pristine (and which are indeed pristine, in the English sense, compared to European rivers). British rivers have been altered by man since pre-Roman days, the influence being great over most of England by Domesday Book (1086). To restore is to return to a former state, so to what former state? That of pre-disturbance is impossible, since no one knows even what it was. That before the first main drainage (*c.* 1840 to 1860) would be unsuitable for public health, as fevers, digestive and liver diseases, rheumatism, etc. were all too prevalent, especially in slow lowland river areas (Haslam, 1991). That after the second main drainage (starting *c.* 1945) is unsuitable as the rivers degraded so rapidly thereafter. That leaves the period *c.* 1860 to 1940, in which is included the classic descriptions of Butcher (1927, 1933). 'Traditional' could perhaps be used for the rivers of that period, and restoration towards that state would be eminently satisfactory. It follows that while Americans can reasonably hope to restore (*sensu stricto*) a fair proportion of their damaged rivers, the British can only have this as an aim. Rehabilitation is, in English, synonymous with restoration, but in American use it refers more to enhancement or improvement, not to restoration. Conservation is the preservation from destructive influences, and the official charge and care for, e.g. rivers. Conservation is both preservation and care, a dynamic process. It is the maintenance of a river habitat in its present state (or aiming at this), whether that state is deplorable or excellent. Conservation is, therefore, appropriate for the better river habitats, restoration, for the worse ones, while the awful, though best restored, are benefited by any enhancement.

Finally, development is the working out, unfolding or new form of that which is already there. It is now used both for what is there in the field, and for what is there in the developer's mind (e.g. building development). Too many restoration schemes are only development, some even reaching what Anthony Trollope described as 'Doomed to that kind of destruction which is called restoration'.

Restoration difficulties

Ascertaining the earlier state of a given river, or even of a river type is at best difficult, at worst impossible. The primary ingredient of a river is water; water which differs in quantity and quality. No lowland, and few highland British rivers have the water regime they had half a century ago. In faster-flowing streams this cannot be restored during the present demands for abstraction and land drainage. In slow streams, especially clay ones, the earlier ponding weirs could be re-inserted, but not high enough because of the level required for drainage. Water loss from rivers has been great.

In quality, before the 1950s, much of Britain had near-clean streams. Some had severe pollution (e.g. Merseyside, S. Wales coal streams, Birmingham conurbation), but most had only local town and village effluents. Although, since then, the worst areas have been partly cleaned, the typical rural stream is now polluted: from busy and large roads, from Sewage Treatment Works effluents now with much household chemicals and, often, small Industrial Estates, from farm agrichemicals and (unintended) silage, slurry and sewage, from run-off from recently built areas. Traditional water quality is that derived from run-off influenced by the underlying rock, soil and traditional land uses. Land use changes (to more intensive arable, more roads, peat drainage, etc.) have altered this quality, quite apart from the more direct influences of pollution. STW water (chemically very different to natural river water) enters the river and may form a large proportion of this river water.

As with loss of quantity, loss of quality is almost impossible to restore, at the present time. Therefore this the primary attribute of the river can be enhanced (as by weirs for depth in clay stream, or by buffer strips to clean out many agrichemicals), but cannot, in practise, be restored.

The solid features of a river are usually easier to change than the water: the position and patterns, the banks and bed. Enhancement, even restoration, is possible. Replacing a stream on the base of a valley floor, inserting bends, altering bank shape and substrate type are all feasible. What, though, were these features like in a given river in traditional times? What, even, were they like in a given river type? Discovering these is not easy. The river features are determined by the rock type, water force, and upstream-downstream position of the site, as influenced by centuries of land use and management (and mis-use). The general patterns of good 1970s examples are shown for Britain in Haslam & Wolseley (1981) and for more of Europe in Haslam (1987). For proper data, landscape paintings and old photographs can be sought (but may be partial and have sites chosen for artistic rather than ecological reasons. A river worn bare by many picturesque boats may not be typical of all of that river type). Old descriptions (e.g. Butcher, 1927, 1933) and large-scale maps can be useful, as can studying present ecology, in and outside Britain, in relation to degree of impact. If, though, the river is improved without reference to the earlier state, it is enhanced, not restored.

The third component is the living one, the plants and animals. With less water and that differing in chemical composition, and little evidence on traditional structural or species composition, restoration is unlikely. Consider, for instance, the earlier recorded catches of immense numbers of fish and fowl! Enhancement is certainly possible, restoration to the best state of the 1970s, may be possible. It is most important that enhancement is within the proper vegetation for each given stream type and size. Neither other species, nor non-local strains should be imported (the former lead to non-traditional communities, and latter contaminate

Figure 1. Bland designer-river (Haslam, 1990)

Figure 2. Lowland limestone stream with intensive (though not extreme) impact (France), (Haslam, 1987).

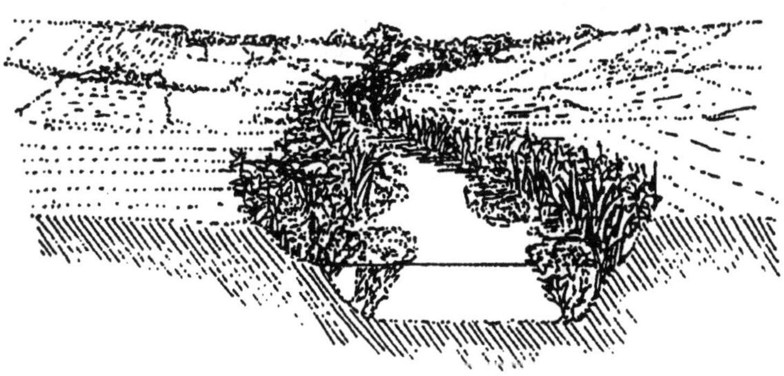

Figure 3. Lowland limestone stream in a more traditional condition (France), (Haslam, 1987).

the local gene pool). The best method, if it is possible, is to allow natural colonisation from a good upstream reach and any remaining propagules in the site. The next, is to transplant species from other equivalent parts (as close as possible, in location as well as in rock, water force and size); except that polluted reaches had better receive transplants from other polluted ones, or the species may not survive (e.g. Kohler et al., 1974). Except in urban areas, species should not be chosen for any other reasons, neither for ornamentation, since it is always possible to have pleasing patterns from the 'proper' vegetation, nor for supposed 'cleaning' value, since *Phragmites australis* is a traditional river species in only a few habitats, and various other species are as effective (e.g. Hammer, 1989, Reddy & Smith, 1987).

River development

Each river is unique, each group of similar rivers is unique. Human impact tends towards uniformity: clean river types differ much, grossly polluted ones are all without macrophytes, undisturbed rivers differ much, uniform straight trapezoid channels are remarkably alike. This is also the danger of a basic 'Restoration Scheme'. which may result in a bland, dull 'recommended' lowland river, with bends, trees and an emergent fringe, though without diversity, buffer strip, or Sense of Place (Figure 1). Sense of Place includes the subtle features which centuries of slightly different management produce (Figures 2, 3). German rivers differ from Danish, Schleswig Holstein is more Danish than German, reflecting history. Crossing the west Franco-Belgian border, in the same catchment, streams are clearly different. This is intrinsically important, important in conservation, and easier to conserve than to re-develop or restore.

Maurizi & Poillon (1992) described a 'restored' Oregon river, with again abundant fish. However, the pollution was partly retained, though diluted by augmented summer flow, and only partly stopped at source. The water regime was altered by this augmentation and by many dams, so also altering temperature, migration, etc. Game fish (anadromous) were back, and fostered, but other fish were sparse. The conclusion was that improving one parameter (pollution), in the absence of a systematic effect to recreate a fluvial system's diverse and abundant plant and animal communities, is not restoration.

Similarly, in Britain, producing an approximation to the native traditional river type in water, structure, input and therefore biota is restoration. Changing rivers in other ways is development. If the result is in part towards restoration, it is also enhancement.

Conservation is for rivers in a satisfactory state. Most rivers subject to the pressures of large human populations need care to be sustainable, whether this is preventing pollution, disturbance, water loss, etc., or improving structure, organic matter, substrate etc.: processes which, in a natural state, were self-sustaining. Do No Harm is the first principle. Until more is known, development should be restricted to highly degraded rivers, where further damage is very unlikely. Schemes should be monitored after one, two, five and ten years, and only when there is a good accessible data base of the long-term results of schemes should satisfactory habitats be changed.

References

Butcher, R. W., 1927. A preliminary account of the vegetation of the river Itchen. J. Ecol. 15: 55–65.

Butcher, R. W., 1933. Studies on the ecology of rivers. 1. On the distribution of macrophytic vegetation in the rivers of Britain. J. Ecol. 21: 58–91.

Hammer, D. A. (ed.), 1989. Constructed wetlands for waste water treatment. Lewis Publishers, Chelsea, Michigan.

Haslam, S. M., 1987. River Plants of Western Europe. University Press, Cambridge.

Haslam, S. M., 1990. River Pollution: an ecological perspective. Belhaven Press, London.

Haslam, S. M., 1991. The Historic River. Cobden of Cambridge Press, Cambridge.

Haslam, S. M. & P. A. Wolseley, 1981. River vegetation: its identification, assessment and management. University Press, Cambridge.

Kohler, A., R. Brinkmeier & H. Vollrath, 1974. Verbreitung und indikator wert der submersum makrophyten in den fliessgewässern dem Friedberger Au. Ber. bayer. bot. Geo. 45: 5–36.

Maurizi, S. & F. Poillon (eds), 1992. Restoration of aquatic ecosystems: science, technology and public policy. National Research Council, National Academy Press, Washington, D.C.

Reddy, K. B. & W. H. Smith (eds), 1987. Aquatic plants for water treatment and resource recovery. Magnolia Publishers, Orlando, Florida.

Hydrobiologia **340:** 349–354, 1996.

J. M. Caffrey, P. R. F. Barrett, K. J. Murphy & P. M. Wade (eds), Management and Ecology of Freshwater Plants.

Bankside stabilisation through reed transplantation in a newly constructed Irish canal habitat

J. M. Caffrey & T. Beglin
Central Fisheries Board, Mobhi Boreen, Glasnevin, Dublin 9; Office of Public Works, East Drainage Maintenance, Trim, Co. Meath, Ireland

Key words: canals, transplantation, erosion, reeds, stabilisation

Abstract

In January 1989 a major breach in an embanked section of the Grand Canal occurred. As a result, a 2.5-km long section required complete reconstruction. This work was completed in approximatley 12 months, at a cost of IR1.5 million. The canal was rewatered in March 1990. The banksides, above normal water level, were dressed with a layer of moss peat and seeded with a mixture of grasses. The grass roots failed to bind the peat to the sub-layer of Puddle Clay and significant erosion resulted in the season of treatment. In order to halt the harmful erosion and to expedite the natural reed colonisation process, roots and rhizomes from established monocotyledonous plant colonies, external to the canal, were acquired. This paper presents the findings from transplantation trials using *Schoenoplectus lacustris*, *Glyceria maxima* and *Phragmites australis* from river and lake habitats and comments on the efficiency and cost-effectiveness of this operation. The value of reeds in amenity watercourses is discussed.

Introduction

In January 1989 a major breach along an embanked section of the Grand Canal, near Edenderry (Figure 1), was reported. A 400-m long section of the north bank was seriously damaged and most of the water from this 30-km long level discharged to the adjacent bog. Repair and reconstruction works commenced immediately and, although navigation in the level was suspended for 1989, most of the channel was rewatered within days of the breach event. 28 000 tonnes of sod peat were used to build-up the bed and banks of the damaged sections of canal and a 2.5-km long section was designated for extensive reconstruction work.

An impermeable membrane (2000 gauge H.D.P. Pondliner) was used to line the newly constructed canal bed and this was further protected by a 30-cm deep layer of Puddle Clay, imported from Geashill in County Offaly. This clay is impermeable, tenacious and relatively easy to work with. It is regarded as being among the most suitable materials available for retaining water in canal and other aquatic habitats (N. Somers pers. comm.). Reconstruction work was completed within 12 months, at a total cost of IR$1.5 million, and the repaired section was rewatered and opened for navigation in March 1990.

The banksides, above normal water level, were subsequently dressed with a 7.5-cm deep layer of moss peat, which was seeded with a mixture of grasses. This aimed to enhance the aesthetic appeal of the canal and had no function in stabilising the banksides. By the end of the 1990 summer season, significant erosion of this peat and the newly laid Puddle Clay that occupied the wave-wash zone along both banks was evident. While most of the grass seeds had germinated and grown, the roots were not sufficiently deep or strong to bind the peat to the sub-layer of Puddle Clay. By the following March it was apparent that some form of bankside stabilisation was required if more serious erosion was to be avoided.

It was decided that the natural bankside consolidation process could most efficiently and cost-effectively be expedited by 'seeding' the banks with roots and rhizomes from established monocotyledonous plant colonies. The species selected for transplantation were *Glyceria maxima* (Hartman) Holmb., *Schoenoplectus*

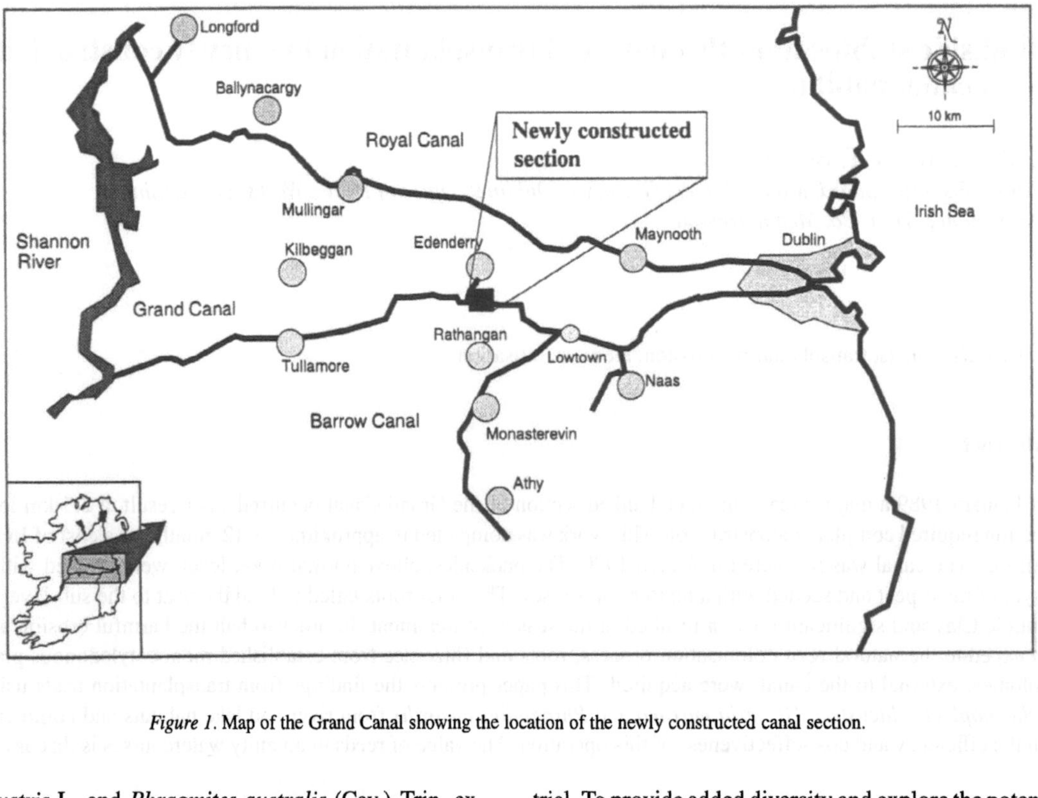

Figure 1. Map of the Grand Canal showing the location of the newly constructed canal section.

lacustris L. and *Phragmites australis* (Cav.) Trin. ex Steudel. All three species have intricate and consolidating root masses (Bonham, 1980; Haslam, 1987), are indigenous to the Grand Canal (Caffrey & Monahan, 1992) and all were readily available in adjacent aquatic habitats.

This paper describes the methods used to acquire and plant the monocotyledonous species involved, presents results from the transplantation trials and comments on the efficiency and cost-effectiveness of these operations.

Materials and methods

A 2.5-km long section of the Grand Canal, including the newly constructed breach section, was used for the transplantation trials. Because both banks were planted, this represented a 5-km long trial site.

Plant sources

Based on results produced by Bonham (1980) it was decided to use *Phragmites* and *Schoenoplectus* in this trial. To provide added diversity and explore the potential of another helophyte for bankside stabilisation, it was decided to use *Glyceria maxima*, a fast-growing plant that is ubiquitous along the canals (Caffrey & Monahan, 1992).

Material for transplanting was collected at a variety of locations. On the River Boyne and the River Deel (located within 30 km of the trial site) dense, encroaching stands of *Glyceria* and *Schoenoplectus*, respectively, presented impediments to water flow. These channels had been designated by the Drainage Division of the Office of Public Works (O.P.W.) for maintenance in 1991 and, therefore, dredging equipment was available on-site. Rhizomes were carefully loaded into trucks and transported to the breach site. Roots and rhizomes for *Phragmites* were collected from White Bay in Lough Ennell (located 35 km north west of Edenderry) where O.P.W. were removing stands of the species to facilitate the safe mooring of angling boats. Likewise, the dredged plants were transported by lorry to the canal near Edenderry.

During the transplantation operation 30 truck loads of rhizomatous material was transported from the excavation sites to the canal bankside. A tracked hydraulic excavator operated at each site while a crew of three canal staff planted the rhizomes. Two trucks were employed transporting the rhizomes in order to maximise the potential of the staff and equipment. A full-time ganger was appointed to ensure the smooth running of the operation.

Preliminary trials

Preliminary transplanting trials were conducted in May 1991 and aimed to determine planting protocols for the species. Dredged *Glyceria* rhizomes and cut *Phragmites* plants were used during these trials. In respect of the latter, stems measuring approximately 0.7-m long were hand cut and the cut end was inserted 10 cm into the substratum, just below normal water level, using methods described by Lewis & Williams (1984). A total of 100 cut plants were planted in this manner and their progress monitored over the subsequent weeks. Prior to planting the *Glyceria* rhizomes, a shallow (0.2 m) trench, 100-m long and 0.5-m wide, was excavated at water level. Clumps of rooted plant material, each measuring *circa* 3-m long, were laid by hand in the trench and firmed into position with imported top soil. Large-scale transplanting operations were conducted between late September and early October 1991 when the plants had set seed and were dying-back.

Main trial

In late September and early October, prior to the arrival on site of the dredged plant material, planting bays or trenches were prepared on both sides of the 2.5-km long section. A detailed examination of in-channel locations for naturally colonising stands of *Glyceria*, *Schoenoplectus* and *Phragmites* in adjacent canal habitats revealed that each had different preferences with respect to levels of immersion in water. *Schoenoplectus* almost invariably grew in the water and rarely colonised the moist or temporarily exposed banksides. For this reason, planting bays for this species were located below normal water level. *Glyceria* prefered to grow at or marginally above normal water level, while *Phragmites* grew best on drier ground, from where it could extend latterally into the water. Based on these observations planting bays for *Glyceria* were located at water level while those for *Phragmites* were positioned approximately 25 cm above water level. Because water levels in the channel fluctuated widely, it was difficult to accurately gauge where normal water level was located.

Planting patterns

Because the three plant species used in the trial rarely grow in mixed vegetation assemblage in nature (Caffrey, 1990), each was allocated alternate 100-m long sections. Within each section approximately 25 planting bays, measuring 3m in length and separated from each other by 1 m, were prepared. Dredged roots and rhizomes for each species were placed into each bay and firmed into place with imported top soil. Care was taken to minimise damage caused to the plants' rooting structures during dredging or planting.

Monitoring

The trial site was surveyed on a number of occasions between May and September 1992. On the latter occasion the percentage survival among transplanted species was determined. One hundred percent survival was recorded where actively growing plants were observed in each of the 25 planting bays in each 100-m long sections. Shoot density, for each species, was recorded in 1992, 1993 and 1994. Counts from 30 quadrats (0.5 m^2 each) were made for each of the three species on each sampling occasion.

Results

Preliminary trials

Practically all of the *Glyceria* that was planted in May 1991 survived and many of the plants flowered and set seed in that summer. Cattle and sheep grazed some of the vegetation, although this exhibited significant regrowth within the same season, after exclusion of the livestock. The roots of this vegetation firmly consolidated the underlying soil, preventing further erosion from wave action.

By June 1991 practically no growth among the hand cut and planted *Phragmites* plants was recorded and *circa* 20% of the stems were missing. The remaining plants were green and appeared healthy. No roots were yet evident on the five plants removed for examination. By July none of the *Phragmites* plants remained. According to the local lock keeper, the site had been vandalised by local children.

Main trial

In April 1992 the first stems and leaves from transplanted *Glyceria* rhizomes were observed. *Schoenoplectus* and *Phragmites* were later to emerge, these first showing in early June and late June, respectively. Both *Glyceria* and *Schoenoplectus* subsequently showed vigorous growth and large numbers of plants from both species flowered and set seed in the 1992 season. Both species also expanded laterally and, at a few locations, adjacent clumps of *Glyceria* merged to give continuous bands. In no case did *Schoenoplectus* clumps coalesce, although significant lateral expansion into the unplanted zone between bays was recorded. In addition, quite a number of *Schoenoplectus* plants were observed growing in the channel, beyond the planting zone. It is probable that these represented rhizomes that had broken from the parental root masses and secured a foothold in the soft substrate. The fact that no such stands of *Glyceria* were recorded probably reflects the ability of *Schoenoplectus* to better withstand complete submersion.

Phragmites plants were slow to emerge and did not flower or set seed in 1992. Where vegetation did emerge from a clump, only one or two stems were recorded (Table 1). This contrasts with *Glyceria* and *Schoenoplectus* where mean numbers of stems recorded in September were 63 and 42 m^{-2}, respectively.

In 1993 and 1994 the area of canal colonised by *Glyceria* and *Schoenoplectus* expanded significantly and continuous vegetation bands, rather than discrete vegetated bays, were evident. In places, *Schoenoplectus* encroached deeply into the channel and presented an obstruction for navigation. The density of both plant species also increased significantly and, by September 1994, dense swards of bankside vegetation were present (Table 1). Only isolated *Phragmites* plants were recorded during 1994.

The overall percentage survival among *Glyceria* and *Schoenoplectus* plants was high, while that recorded for *Phragmites* was low (Table 1). Within the *Glyceria* and *Schoenoplectus* stands the plants were firmly rooted and the perennating rhizome system was well developed. This served to firmly bind the substratum and offered a high level of protection from erosion to the banksides. This was not the case with *Phragmites* and the banks at *Phragmites*-transplanted sites were subjected to erosive forces from wind-induced and boat-generated wave action.

Discussion

Bankside reeds serve a number of important functions in freshwater habitats. Reedbeds increase the productivity of a watercourse. This may be achieved directly through the growth and decay of the vegetation or indirectly by providing habitats for periphyton, macroinvertebrates, fish and wildlife (Armour, Duff & Elmore, 1991). They also provide important spawning substrates for fish and areas in which fry and adult fish can shelter (Caffrey, 1993). From a bankside stabilisation viewpoint, reedbeds absorb and dissipate wave-wash energy (Pearce & Eaton, 1983; Lewis & Williams, 1984), whether this is generated by boats or wind action. The rhizomes and roots of reed species perennate the soil and bind it firmly in place. In so doing they prevent erosion or substrate slippage. Along recreational fisheries reed beds serve as a backdrop against which anglers are less easily seen by fish. Furthermore, they provide protection from wind and a degree of isolation or seclusion for the angler. Bankside reedbeds also have the capacity to extract nutrients and pollutants from water that passes through them (Otte, 1991).

Work conducted by Bonham (1980) on the River Thames and Norfolk Broads revealed that *Phragmites australis*, *Schoenoplectus lacustris*, *Typha angustifolia* and *Acorus calamus* were very effective in absorbing and dissipating the energy from wind- or boat-induced waves. Both *Phragmites* and *Schoenoplectus* grow abundantly along the Grand Canal. *Typha* is less common and no *Acorus* is present. *Glyceria maxima* is ubiquitous in Irish canal habitats. Westlake (1981) has shown that, at any time during the growing season, terminal buds on the rhizomes of *Glyceria* plants can grow into aerial shoots, from which new rhizomes will develop. When these shoots are well separated and free from intra- or inter-specific competition, as in their 'new' canal habitat, each can produce between one and three rhizomes, typically 10 to 30 cm long, and ending in new aerial shoots. This process, under favourable conditions, may continue exponentially and up to 30 new shoots may be produced from one rhizome in three months (Westlake *op. cit.*). In productive habitats, free from competition, rhizomes of *Phragmites australis* may extend up to 2 m each year and produce up to 10 aerial shoots from each rhizome in their first year (Haslam, 1972).

Rhizomes and roots from these rhizomatous species were transplanted rather than attempting to grow them from seed. This reflects the poor success achieved using the latter technique and the relatively high suc-

Table 1. Percentage survival (1992) and density (no. shoots/leaves m^{-2}, N = 30) of emergent monocotyledonous plant species transplanted along exposed bankside sections of the Grand Canal in 1992, 1993 and 1994. Standard errors are given in parenthesis.

Plant species	% survival	No. shoots		
	1992	1992	1993	1994
Glyceria maxima	89.3 (16)	63 (14)	149 (22)	251 (29)
Schoenoplectus lacustris	81.2 (11)	42 (17)	106 (34)	189 (48)
Phragmites australis	12.8 (5)	1 (1)	0.5 (5)	0.6 (2)

cess that accompanies transplantation (Brooks, 1981; Newbold et al., 1989). The plants were transplanted at the end of the growing season, although Brooks (*op. cit.*) suggests that the best time to transplant is during or at the end of the plant's dormancy period.

As part of this extended field trial, drainage maintenance programmes operated by the East Region Drainage Maintenance Divsion were planned to coincide with the schedule for reed transplantation. This obviated the necessity to specifically divert heavy machinery to rhizome excavation operations, at considerable cost, and meant that plant species which presented in-channel problems at one location could be usefully employed at another. In circumstances such as these, reed transplantation for bankside stabilisation and habitat creation is far more cost-effective, and probably more efficient, than reseeding or employing geotextiles. In the present trial healthy swards of *Glyceria* and *Schoenoplectus* had grown within one year of rhizome placement and within two years, continuous bands of helophyte vegetation protected the newly constructed banks and provided a diverse wildlife habitat.

The poor survival among *Phragmites* transplants was unexpected. This may reflect the fact that this plant grows best on the drier slopes of banks (Newbold et al., 1989) and is relatively intolerant of water movement (Haslam, 1972). During planting operations rhizomes from *Phragmites* were located roughly 25 cm above water level. Water levels, however, fluctuated widely during the winter and early spring months, often submerging the rhizomes beneath 0.25 m of water. Work with this species by Haslam (1972) and Newbold et al. (1989) has shown that the cut or exposed portions of rhizomes are prone to fungal attack under wet conditions and that whole cuttings may be killed as a consequence. This probably contributed to the poor result achieved.

Acknowledgments

The authors would like to express their gratitude to the Office of Public Works, East Drainage Maintenance and Inland Waterways Division for their co-operation during the course of this project. We would specially like to acknowledge the field assistance of Catherine Monahan and Marie Fallon. We would also like to thank Dr P. Fitzmaurice for his helpful editorial comments.

References

Armour, C. L., D. A. Duff & W. Elmore, 1991. The effect of livestock grazing on riparian and stream ecosystems. Fisheries 16: 7–11.

Bonham, A. J., 1980. Bank protection using emergent plants against boat wash in rivers and canals. Report No. IT 206, Hydraulics Research Station, Wallingford, 12 pp.

Brooks, A., 1981. Waterways and wetlands. A practical conservation handbook. British Trust for Conservation Volunteers, Berkshire, 186 pp.

Caffrey, J. M., 1990. The classification, ecology and dynamics of aquatic plant communities in some Irish rivers. Ph.D. Thesis, University College, Dublin, 254 pp.

Caffrey, J. M., 1993. Aquatic plant management in relation to Irish recreational fisheries development. J. aquat. Plant Mgmt, 31: 162–168.

Caffrey, J. M. & C. Monahan, 1992. Aquatic plant management in Irish canals. Annual Report, 1991. Office of Public Works commissioned report, Central Fisheries Board, Dublin, 42 pp.

Haslam, S. M., 1972. *Phragmites communis* Trin. J. Ecol. 60: 582–610.

Haslam, S. M., 1987. River plants of Western Europe. Cambridge University Press, Cambridge, 512 pp.

Lewis, G. & G. Williams, 1984. Rivers and wildlife handbook. Royal Society for the Protection of Birds, Kettering, 295 pp.

Newbold, C., J. Honnor & K. Buckley, 1989. Nature conservation and the management of drainage channels. Nature Conservancy Council, Peterborough, 108 pp.

Otte, M. L., 1991. Contamination of coastal wetlands with heavy metals: factors affecting uptake of heavy metals by salt marsh plants. In J. Rozema & J. A. C. Verklesj (eds), Ecological Responses to Environmental Stresses. Kluwer Academic Publishers, Dordrecht, The Netherlands: 126–133.

Pearce, H. G. & J. W. Eaton, 1983. Effects of recreational boating on freshwater ecosystems – an annotated bibligoraphy. In Waterway Ecology and the Design of Recreational Craft, Inland Waterways Amenity Advisory Council: 13–68.

Westlake, D. F., 1981. The development and structure of aquatic weed populations. Proceedings Aquatic Weeds and their Control, 33–47.

MIX
Papier aus verantwortungsvollen Quellen
Paper from responsible sources
FSC® C105338

If you have any concerns about our products,
you can contact us on
ProductSafety@springernature.com

In case Publisher is established outside the EU,
the EU authorized representative is:
Springer Nature Customer Service Center GmbH
Europaplatz 3, 69115 Heidelberg, Germany

Printed by Libri Plureos GmbH
in Hamburg, Germany